Praise for *The Small-Scale Poultry Flock*

"As usual, Harvey never ceases to amaze with his in-depth knowledge of poultry keeping. I love that, as knowledgeable as he is, he still constantly searches for new 'tweaks' to improve his farm efforts and the quality of life for his birds. He writes from that direct experience and it certainly shows in this wonderful go-to book for any aspiring poultry keeper. So many people claim to be chicken husbandry experts, but Harvey, he's the real deal."

—JEANNETTE BERANGER, senior program manager, The Livestock Conservancy

"Amid the new and urgent focus on holistic, inputs-free food production married to ecologically robust land management, this expanded edition of Harvey Ussery's classic guide to all things chicken should have a place on every small farm or homestead bookshelf. Dedicated to the conviction that what is best for the animal and the land will always be best for the farmer, Ussery points the novice 'flockster' toward best practice, while he inspires the long-time poultry keeper with a deeper commitment to ecological cooperation."

—SHAWN AND BETH DOUGHERTY, authors of *The Independent Farmstead*

"Harvey Ussery has spent a lifetime developing and showcasing a truly viable poultry model that is ultimately carbon-sequestering, hygienic, neighbor-friendly, and food-secure. . . . This book is about a call to heritage, to the wisest of wise traditions in food security and relationships. Harvey brings the latest tools and practices within the grasp of any aspiring flockster. It is this functional spirit that makes this book a classic in the small-scale poultry rearing genre."

—JOEL SALATIN, Polyface, Inc.

"*The Small-Scale Poultry Flock* is the perfect book for all chicken keepers, from the folks just starting out to the most experienced. It covers everything from the history of the species and coop designs to feeding, breeding, keeping predators at bay, processing meat, and much more. Harvey's writing style is easy going and feels like an old friend giving you sound advice. This is a comprehensive resource to keep on your bookshelf for many years to come."

—JESSI BLOOM, author of *Free-Range Chicken Gardens*; coauthor of *Practical Permaculture*

"It is refreshing and inspiring to read a book on chickens that has original content and creative techniques. This can only come from personal insights and hands-on experience. Harvey Ussery eloquently describes how to keep and employ a family flock of chickens.

"Chickens have migrated with humans for over 3,000 years all over the planet. It's time to bring them back home. Let them work in our communities and production gardens where they can feed not only themselves and humans, but also the soul and soil of Nature, enabling an abundant ecology. Food security with a consistent supply chain can be as close as your own backyard. Just read and use the practices described in *The Small-Scale Poultry Flock*. You will not be disappointed. High fives to Harvey!"

—PAT FOREMAN, author of *City Chicks;* founder, The Gossamer Foundation

"Written by a self-described 'old hick with chickenshit on his boots,' *The Small-Scale Poultry Flock* is a welcoming and decisive guide to the poultry-keeping experience. Harvey Ussery's natural approach is that of a partnership with his flock, in what he terms 'an integrated food independence enterprise.' The author shares straightforward, encouraging information written from the viewpoint of someone who desires to share the knowledge that has come out of three decades of hard-won experience. Anyone considering a natural approach to producing eggs and meat will cherish this must-have reference, enjoyable to sit down and read cover-to-cover, but also perfect for answers on the go."

—*FOREWORD REVIEW*

"Ussery provides an encyclopedia of chicken and other fowl care, encompassing everything from anatomy and species selection to feeding, breeding, and selling in the local market. . . . Altogether, there's no better introductory reference on the joy of home-raising chickens."

—*Booklist*

"This colorful and informative volume on small-scale poultry farming provides a comprehensive reference for homesteaders and urban farmers, covering the details of raising chickens for eggs and meat. Most useful for intermediate poultry keepers, the volume offers professional advice in flock planning, housing, feed, health, processing, and developing small commercial opportunities, and provides detailed practical information, including step-by-step photographs of important processes and procedures."

—*Book News Review*

"Here's the ultimate book for those who want to know everything there is to know about raising poultry. And every detail is backed up by the author's own (often entertaining) experiences. I could not find—in this encyclopedic array of chicken knowhow—one detail that I could quibble with."

—Gene Logsdon, author of *Letter to a Young Farmer* and *Holy Shit*

"This book is packed with practical advice on raising poultry by someone who has not only done it all but has learned from his broad experience and knows how to communicate that wisdom clearly and in a lively, readable style. Harvey Ussery has written one of the most comprehensive guides out there, but what places it above the rest of the crowd is that he shows you how to work with nature rather than against it in ways that will minimize work while ensuring the health and happiness of the flock. Whether you're a beginner or an old-time poultry farmer, you shouldn't go any further without this excellent manual."

—Toby Hemenway, author of *Gaia's Garden* and *The Permaculture City*

"*The Small-Scale Poultry Flock* is about establishing a free-range poultry flock fully integrated into a healthy homestead ecosystem. Based upon the author's decades of hands-on experience with many breeds and species, it covers all the basics about raising poultry, and fills some important gaps not usually covered well enough elsewhere, including chicken behavior, poultry breeding, raising chicks with broody hens, managing free-ranging, dealing with predators, using electric net fencing, feeding poultry with home-grown feeds, and integrating the poultry with soil mineral balance, gardens, lawns and pastures, orchards, worm bins, and soldier fly (larvae) production. If you want to raise chickens and can afford just one book, I recommend this one."

—Carol Deppe, author of *The Resilient Gardener*

"Ussery's outstanding book is certain to withstand the test of time both for its encyclopedic and practical information, and for its acknowledgment that the future of our culture and our food security is in the hands of the small farmer and backyard producer. If you are starting out with your first flock, this is your book. And when you've been keeping poultry for 30+ years, this will still be your best book."

—Shannon Hayes, author of *The Grassfed Gourmet Cookbook* and *Radical Homemakers*

"Anyone interested in practical, experienced, insightful information about how to select, breed, care for, manage, feed, protect, process, eat, or market small-scale poultry flocks for their own eating pleasure or selling to others—and have FUN—should read this book."

—Frederick Kirschenmann, distinguished fellow, Leopold Center for Sustainable Agriculture; author of *Cultivating an Ecological Conscience*

"There is a revolution going on, and it is the popular return of keeping poultry to provide food for our home tables. Ussery's *The Small-Scale Poultry Flock* helps lead the way by integrating the small flock with its natural environment, the homestead, or small farm. Nowhere else will you find such valuable information on putting poultry to work in the garden, producing much of their feed, and producing healthful food for ourselves."

—Don Schrider, author of the American Livestock Breeds Conservancy's *Chicken Assessment for Improving Productivity* and *Storey's Guide to Raising Turkeys*

The Small-Scale Poultry Flock

The Small-Scale Poultry Flock

An All-Natural Approach to Raising and Breeding Chickens and Other Fowl for Home and Market Growers

REVISED EDITION

Harvey Ussery

Chelsea Green Publishing
White River Junction, Vermont
London, UK

Project Manager: Alexander Bullett
Project Editor: Fern Marshall Bradley
Copy Editor: Deborah Heimann
Proofreader: Angela Boyle
Indexer: Shana Milkie
Designer: Melissa Jacobson

Printed in the United States of America.
First printing October 2022.
10 9 8 7 6 5 4 3 2 1 22 23 24 25 26

Our Commitment to Green Publishing
Chelsea Green sees publishing as a tool for cultural change and ecological stewardship. We strive to align our book manufacturing practices with our editorial mission and to reduce the impact of our business enterprise in the environment. We print our books and catalogs on chlorine-free recycled paper, using vegetable-based inks whenever possible. This book may cost slightly more because it was printed on paper that contains recycled fiber, and we hope you'll agree that it's worth it. *The Small-Scale Poultry Flock,* Revised Edition was printed on paper supplied by Versa Press that is made of recycled materials and other controlled sources.

ISBN 978-1-64502-101-8 (paperback) | ISBN 978-1-64502-102-5 (ebook)

Library of Congress Cataloging-in-Publication Data is available upon request.

Chelsea Green Publishing
85 North Main Street, Suite 120
White River Junction, Vermont USA

Somerset House
London, UK

www.chelseagreen.com

Dedicated to
Lottie Mae Croom Ussery (1919–2012)
and Robert Marshall Ussery, Sr (1921–2020)
to Heather
who started our first flock
and to
Ellen
love of my life
sine qua non

. . . how we eat determines,
to a considerable extent,
how the world is used.

—Wendell Berry

Contents

Introduction

In the ten years since *The Small-Scale Poultry Flock* was first published, I have continued—as in the previous thirty years—to experiment with new ways to raise and manage poultry. For a major part of the decade I bred my own chickens almost exclusively—and with more rigor than ever before. Much of the new material in this revised edition focuses on small-scale breeding. I hope many of my readers will be inspired to breed their own stock. It is accepted wisdom that gardeners who save their own seeds are rewarded over the years with varieties ever more adapted to their specific climate and soil conditions—and to their approach to garden management and nuanced preferences for what they like to eat. In the same way, *breeding your own* can result not only in *improvement* generally (healthier, more vigorous, and more productive stock) but in better *adaptation* of your flock to your specific conditions and goals.

I have expanded the discussion of using natural mothers to raise the hatchlings produced by your breeders. There is no joy like it, especially if you have children to share the miracle.

I'm getting older (happens every day!), and it will not be long before my wife Ellen and I move from our beloved homestead of four decades—so in the past couple of years I have simplified my poultry keeping and greatly reduced the size of our flock. But my strategies for keeping poultry at a smaller scale may be helpful to flocksters unable to keep poultry at the scale I once did, with flocks of dozens and big electric net perimeter fences. I especially hope my discussions of a greatly improved static chicken run and of "The Ultimate Mobile Shelter" will be useful.

Keeping poultry is more likely to be successful when our birds live in conditions that more closely emulate the ecology in which they evolved. Yes, "the ecology" often comes across as a highfalutin term for something big and impossibly complex "out there." Just remember that "the ecology" begins in our own backyard.

PART ONE

Getting Started

CHAPTER 1

The Abundant Ecology

Although this is a book about keeping poultry, let me begin with a story from my garden. In the spring of 2010 Ellen and I added a small pool to our garden. A few years previously we had planted a black gum tree at the back of the garden, and as it grew it became our preferred place to sit and rest and enjoy our green companions. What better spot to add a bit of water to enhance both the garden's peace and beauty and its biological diversity?

We dug a hole for the pool, 6 feet by 8 feet (2 by 2.5 m) across and 2 feet (60 cm) at its deepest; lined it with purchased pond liner; and filled it with water from our well. I planted two pots with small rhizomes of water lilies and a few sprigs of elodea, a water weed, as an oxygenator, then sank the pots to the bottom of the pool. Ten days later I introduced seven goldfish.

I had waved aside all the pond supplier's blandishments for adding filters, chemicals for killing algae, pumps for circulating the water—and I never fed a *flake* of fish food. Yet the pool thrived. All these years later, there is an active, healthy population of goldfish, and the increase in water lily biomass has been literally thousands-fold. (Every couple of years I lift the water lily pots and cut the rhizomes back severely. Last time I did, the cuttings overflowed a 5-gallon bucket.) The pool has never supported wriggling populations of mosquito larvae, nor has its surface become a scummy algal bloom.

Species multiplied in and above the garden, using the pool as habitat and resource: blacksnake, frog, toad, swallows and bats hunting in the air overhead—all helping to balance insect populations (and in the case of the blacksnakes, gulping the occasional vole as well). How marvelous—we just put in the pool, and these new neighbors showed up, eager to help.

Where did the fertility for all that growth come from? Why did the pool thrive rather than fester? Well, *magic*, of course—that is, the closest to magic we are likely to find. The vision of that magic, right in my backyard,

Figure 1.1. One of the new neighbors who showed up to share the beauty and the bounty of the pool.

Figure 1.2. At home in the abundant ecology. Photo courtesy of Mary Perrine.

was profoundly informative. I saw the surrounding web of life, endlessly complex, inexhaustibly dynamic, more clearly than ever before. And I recognized that I—along with every living creature—was at its center, with the web spreading out to the horizon. Excess algae? Mosquito larvae? They were eaten by the goldfish and, in their season, the tadpoles. Fertility? Courtesy of dust and pollen flung in by the wind, fish poop (which started as algae, which started as sunlight), and droppings of bird and bat and insect.

What I saw in that epiphany I call *the abundant ecology*. Our vision of our work on the homestead or small farm tends to be post-Fall, that is, following Eve and Adam's loss of Eden: *Ours* is the work of "making it happen." We see what *we* must do to improve the soil and make it more fertile—believe that the burden of dealing with crop-damaging insects is *ours* to bear—and all too often are suckered into the premise that solutions in garden and flock must be *purchased*. But my garden pool is a real-world case where fertility and renewal generate *spontaneously*, where species come into finely honed, self-correcting *balances*, in which nobody gets a free ride but in which, despite "red in tooth and claw" *competition*, the whole seems determined as much by *cooperation*. Why not make *that*—the tendency at the heart of nature toward spontaneous abundance—the starting point for all projects in the homestead? Including the keeping of a flock of poultry.

Abundant Diversity

For a vision of how the abundant ecology works spontaneously at geographic scale, consider the American bison—before the arrival of European settlers a key-

Figure 1.3. Instant ecology: Just add water and stir! Photo courtesy of Mary Perrine.

stone species on the prairie, one of the most complex ecologies on Earth. The bison ranged the grasslands in herds of up to a million individuals. A *million*? Grazing by that many big hungry animals must have been devastating, right? In fact, the periodic intense grazing by the herd was vital to the prairie ecology in many ways—fertilization by millions of gallons of urine and millions of pounds of manure being only the more obvious. The tramping of seeds of prairie species by the hooves of the grazers ensured firm contact with the soil, supporting germination and renewal of the sward. More importantly, think of the response to heavy grazing by the prairie grasses themselves, in places as tall as a mounted rider. In a matter of days the herd ate the grasses, which before grazing had a root mass several times that of their top growth, leaving the plants severely unbalanced. They compensated by shedding massive quantities of rootlets. As the shed roots decomposed, they created openings in the soil through which oxygen (as essential to soil organisms as to us) could move and rainwater could rapidly penetrate, rather than eroding the soil. What remained was a permanent addition to *humus*, complex carbon compounds that resist further decomposition and, while no longer feeding plants directly, increase water retention, assist the chemical bonding of nutrients with plant roots, and improve soil texture.

Imagine this kind of mass grazing event repeated several times per year, as the herd returned to the grassland, year after year. The result over millennia was topsoil of a depth measured in feet rather than inches, ever higher in organic matter content and thus resistance to drought, of a fertility rarely matched anywhere on Earth.

Soil improvement at colossal scale was only part of the bison's contribution to the prairie. They preferred

Figure 1.4. Keystone soil builder on the prairie. Photo by Jack Dykinga.

grasses, sedges, and rushes and tended to avoid the broad-leaved flowering plants, the forbs. Such preferential grazing favored diversification of forbs (sunflower, milkweed, lupine, sage, clover, daisy, yarrow, goldenrod, thistle, echinacea) and follow-on emergence of an enormous array of associated insect and bird species. And note, diversification paid off even at the unseen level, with well-balanced microbial predator–prey relationships virtually eliminating disease from the landscape.

Prairie dogs proliferated on the close-grazed prairie, where they more easily spotted approaching predators. The digging of their millions of burrows mixed mineral-laden subsoil into the topsoil, and when abandoned, burrows provided habitat for countless animal species, nesting owls among them. The abundance of prairie dogs supported many predator species such as ferret and coyote. Wolf packs culled the bison herd of old and sick individuals, while grizzly bears and other opportunists also fed on the kills.

How odd, all this growth, renewal, health, and abundance, emerging *spontaneously* without benefit of plow or fence or dollops of NPK salts or blasts of pesticides. And how amazing that the bison—though unquestionably *keystone* in an extraordinarily rich ecology—were not following some master plan they'd agreed on: They were just *looking for something to eat.* Looking for mates, looking to make babies.

But truly, we should think of the iconic bison, however majestic, as bit players in the prairie drama—it was *microbes*, and other organisms working unseen at the surface of the soil and below, that created the miracles of ever-increasing fertility and ecological health on the prairie. A teaspoon of that prairie soil might have contained a billion bacteria, hundreds of yards of fungal filaments, thousands of protozoa, and hundreds of nematodes (who often get a bad rap in gardening circles, though the vast majority play beneficial roles in soil ecology).

The good work of microbes began above ground. Whatever fell to the soil surface—whether poops or shed leaves or winter-killed top growth of perennials or dead animals—was immediately colonized by hosts of organisms. What the "first response" organisms could not break down or digest, they passed in turn to other members of the soil community, in an ongoing feast of decay, of the great return to earth. Which is to say: There was no *waste* from the system—every creature's "trash" was another's "treasure."

The Abundant Ecology in the Homestead

Like the bison, I am *looking for something good to eat* in my backyard. The work of garden and orchard and flock is not a hobby for me, but an essential part of nourishing my *life*. For me, the miracle of the prairie's spontaneous abundance, all its players in harmonious balance, is the starting point for the work of gathering in good things to eat, for the striving in each new day after vitality and health. And the way of the buffalo informs my highest aspiration: to leave this bit of ground I nurture more fertile—the web of life above and below the soil more vital, more bountiful—than when I came to it.

Unlike the bison, I do not *graze* my way to greater soil fertility. But I emulate the bison's transformative role when I take cuttings off the pasture—bringing on the shedding of roots under the sod and increase in soil carbon—for use as nutrient mulches in garden and orchard. And I *grow fertility* right in place by *cover cropping*—the dense covers of plants put more into the soil than they take out. Some cover crops (legumes) specialize in adding nitrogen to the soil, synthesizing it out of the air itself—talk about spontaneous abundance!—but they *all* add soil carbon as their roots decay. No, I don't work at the scale of millennia, but as the seasons roll I know that every round of cover cropping counts—I *see* the cumulative loosening and deepening and darkening of the soil.

I am determined that, as on the prairie, there will be *no waste* on this homestead—*all* organic residues are cycled through compost heaps and nutrient mulches, completing their return to earth—with trillions of soil microbes along for the ride. And with *zero* runoff of toxins to the wider ecology.

Remember the almost complete absence of disease or decimation by insect populations out of control on the prairie, and keep that aspect of the abundant ecology in mind if ever you are tempted to spray toxic chemicals onto plants you are planning to *eat*. My experience here has been that, the less my garden resembles a monoculture, the less likely disease organisms will proliferate. As for insects? I try to ensure that, as on the prairie, there are abundant flowering plants of all sorts, in all parts of the season—to *boost* insect populations. That's right, the key response to crop-damaging insects is not *Kill those insects!* but *How can we encourage more insects?* With increasing diversity of insect populations (and of players such as blacksnake and bird and toad and frog that showed up at our pool), balances between predator and prey species are far more likely.

I'm not spinning fairy tales here: Reliance on the working of the abundant ecology—in lieu of purchased, artificial solutions—*really works*. In the first couple of years gardening on our homestead I purchased fertilizer (pelletized dried chicken manure, mind you, never once odd-smelling chemical granules from a bag) and sprayed Bt for cabbage-family insects (never once toxic chemicals). In the almost four decades since, though, I have purchased *nothing whatever* for either soil fertility or insect control. I have grown my own fertility using cover crops in all four seasons of the year, turned trash to treasure using mulches and composts, captured my chickens' manure as soil nutrient rather than water pollution, and supported insect diversity by beautifying our garden—that is, by growing lots and lots of flowering plants.

During that time the soil has become ever more friable and less subject to drought, with an organic matter content that has tested as high as 8.35 percent, which is well above the recommended minimum for gardening.[1] We have crop damage from insects who get out ahead from time to time, but that happens less every year—and even when it occasionally does, our garden *always* gives us as much as we can eat. (Admittedly, what we bring to the table doesn't have that piquant tang from pesticides.)

The Flock in the Abundant Ecology

Throughout this book, I propose that best husbandry of your flock is to be found at the heart of the abundant ecology. That's where *Gallus gallus*, wild ancestor to domesticated chickens, evolved—where it fed, where it helped create the complex balances in its environment, where its nearly perfect health was honed, where the incessant predation it suffered challenged it to reproduce an abundance of progeny (early in life and in a hurry) as a survival strategy.

I propose that the key to health is not "fixing" birds with medications but rather providing the flock with a lifestyle that relies on their robust health. That giving your flock access to the kinds of natural foods that sustained *G. gallus* not only saves on feed costs but provides nutrition unmatched by anything from a bag. And that breeding *your own* replacement stock from the best of the best of your birds is both easy and *fun*.

The purpose of a home or small-farm flock is greater than providing us with eggs and chicken dinners. Remember how profoundly the bison contributed to soil fertility, and consider how we too can engage our flocks in the great work of soil nurture and ecological health. With wise management, their natural behaviors to *find something good to eat* can be harnessed for the real-world work of the homestead: tilling of weeds or of established sod to prepare new ground for planting, tilling in cover crops, making compost. Let's welcome them as partners in the great work of soil nurture, and then reach for ever more harmonious balance in our little corner of the ecology.

Figure 1.5. Keystone soil builders on the homestead. Photo (*left*) courtesy of Bonnie Long. Photos (*right*) by istockphoto.com.

CHAPTER 2

The *Integrated* Small-Scale Flock

The poultry husbandry I describe in this book is based on allowing the flock, *within constraints imposed by domestication*, to live as closely as possible to the way the natural chicken would live on its own in the wild. And on the observation that, if we do so, the birds in return can help us improve the homestead or farm and make it more productive. One reason those ideas sound so good to me now is because that's not exactly how Ellen and I started.

How We Got Started

We met in a Zen monastery (bet you weren't expecting that one!). True love blossomed when we discovered a shared passion for—compost. After we married and lived a couple of years far too close to Washington, DC, we followed the call of the compost westward and ended up in a two-hundred-year-old house on 3 acres in a crossroads rural village in northern Virginia—we called our new home "Boxwood" after the prominent border of mature boxwood out front. Though as said compost was foremost in our minds, my daughter, Heather, living with us at the time, was thinking *chickens*. We approved her project and, not long after our move (spring of 1984), we ordered twenty-five New Hampshire chicks through the mail. Later that summer we ordered a mixed batch of bantams (miniature breeds) as well. Even at that early date, we were enamored with the idea of breeding our own chickens naturally, and had read that bantams make great mothers. (See chapter 4 for more about bantams.)

We brooded the chicks in a cardboard appliance carton and at four weeks moved them to a small shed. Though we lost most of that first group to a weasel and had to start over, it was not long before we had a lively flock in a wire run off the coop. After supper we would take dessert out on the grass by the run, enjoying the show. Our neighbors probably thought us peculiar—but we found the chooks a lot better 'n television. And not so long after that we had the thrill of every beginning flockster: our first egg.

But I was troubled by our project. Within two weeks of being allowed onto the grass in their wire-enclosed run, our scampering young chooks had eaten every last blade of grass. They didn't look quite so pretty against a patch of bare dirt festering with chicken poops—nor did they seem happy with the change. Each time I threw in a cricket, or grubs from the garden, or lettuce and cabbage trimmings, their obvious hunger for these foods made the dry stuff from the feed bag look ever more lifeless in comparison. Memories of my grandmother's flock, which she allowed to range over a 50-acre farm in the North Carolina piedmont, made our chicken run seem more and more a prison—or more accurately, a concentration camp.

Then Ellen told me she'd heard about a farmer in the Shenandoah Valley named Joel Salatin who was raising his chickens on pasture—growing meat birds in movable pens, following beef cattle with laying chickens that ate fly maggots out of the cow pies. Knowing that somebody was managing to "do chickens" somewhat as Granny had once done excited me—and crazy stuff like chickens eating out of cow poops appealed to the kid in me. I had to give this guy a call! I did so, clueless how much I was imposing on the time of a busy farmer and prolific writer, but Joel was both gracious and encouraging. He urged me to explore options for the Great Liberation—the exodus of the flock out of the dead-zone chicken run. He was pleased with my obvious enthusiasm for giving my birds a more natural life on grass. "There aren't many of us out there," he said, "but we know there's a better way."

I built a 10- by 12-foot (3- by 3.5-m) Polyface-style mobile pen (*Polyface* is the name of Joel's farm) and began raising batches of broilers on our acre or so of pasture (*broiler* is another word for a young meat chicken). Dinner guests remarked in astonishment, "Man, chicken was never like this!" Before long I discovered electric net fencing, use of which allowed me to do what Granny did with her free-ranging flock, but within the limits imposed by having close neighbors on either side.

Poultry husbandry ever since has been, for me, about giving the flock the most natural, happiest lives possible, and discovering all they can give as partners in the work of the homestead. Some key ideas have guided our progress.

Imitating Nature

The great mistake of modern agriculture is assuming we can *control* the processes of growth and yield, and that problems or needs that arise can be dealt with using one-for-one solutions—most likely needing to be purchased. Disease prevention in flocks? *Antibiotics.* Soil fertility? *Supercharger chemical salts.* Crop-damaging insects? *Blast those guys with pesticides.* Runoff pollution? *Not my department.*

We have lost the intuitive understanding that natural systems are highly diverse and interconnected, one species supporting others and benefiting from them in return; and that therefore the practice of agriculture must see "the whole problem of health in soil, plant, animal, and man as one great subject."[1] The key to success with the homestead flock is a return to that vision. The setting for the flock is not our backyard but the abundant ecology—another way of saying that, if *Gallus gallus* evolved to a state of near-perfect health ranging the landscape and eating live, self-foraged foods, the key to success with our flocks is a life as open and unconfined as our situation allows, with access to as wide an array of natural feeds as possible.

Imitation of nature means, most of all, fitting flock husbandry into the cycles of reproduction and growth, death, decay, and renewal—in which nothing is squandered; no spillover from one sphere of activity poisons the whole; and the potential of the underlying foundation, the soil, to support ever more robust, abundant life *increases* with each round of growth and death.

Ah, but let's qualify this *imitation of nature.* Yes, the subtitle of this book promises "An all-natural approach to raising chickens . . ." I don't think that is misleading—certainly in contrast to models of husbandry involving a high degree of confinement and use of manufactured feeds exclusively, the one offered here is far more "natural." But it would be naive to strive for a "natural" husbandry that ignores the reality that, while the lifestyle of *Gallus gallus* was indeed "all-natural," the birds in our backyard are not in fact *Gallus gallus*, they are *Gallus gallus domesticus*. That difference has critical implications for their proper care.

We can only speculate about precisely how a partnership emerged between *Gallus gallus* and *Homo sapiens* (initiated perhaps more by the former than the latter). Since that alliance formed, "*all* natural" has never been an option, and chicken husbandry has ranged from the almost totally laissez-faire (as in my grandmother's flock) to the entirely manipulated (as seen in industrial poultry production, which results in the birds that are truly monstrous). Flocksters must find their own best "fit" on that spectrum, guided neither by dreamy notions of "at home in Eden" nor arrogant illusions of control: "I'm the Big Boss and I'm in charge here." In every phase of husbandry, the sweet spot on the spectrum will be defined for each flockster by their own circumstances, goals, resources, and management style.

In reflecting on your flock's best fit on that spectrum, consider five biological realities chickens share with all other animals, and ways in which their "natural lifestyle" necessarily changes in the context of domestication:

They eat: *The very best foods* for chickens are the foods they would seek out by preference *if allowed to range freely*—no manufactured feed, however well produced, can match a free-range diet. The best-fed flock would therefore be one that ranges freely and eats wild foods exclusively. Though my grandmother's free-ranging flock came close to that Edenic ideal, most of us—myself included—must necessarily rely to some extent on purchased *manufactured* feeds (that are inevitably, to some degree, *compromised*). Wise feeding requires best-fit choices among flawed options.

They poop: The manure from *Gallus gallus* created no festering problems because it was part of the fertility cycle in nature. Wherever manure was dropped in the landscape, a host of microorganisms leaped on it; and in the process of feeding on it neutralized it as a vector for disease and broke it down into basic components for return to the earth. To the extent that our flocks are free to roam outside during the day—unquestionably the best option—their poops as well will become resource rather than problem. But our chickens do as much of their pooping on the roost at night as they do during the day. Intervention on our part is required to keep the manure from accumulating into an unwholesome mess. Fortunately, as we will see, the best solution is the one that most closely mimics *natural* recycling processes.

They seek shelter: *Gallus gallus* had no housing, but she did seek shelter at night, typically in trees. Thus the "all natural" behavior of your birds would be to find a roost in trees when the sun goes down. You do *not* want your chickens roosting in trees.

Somebody wants to kill and eat them: Most animals are prey to at least one species—often many—looking for dinner. In nature, death is more likely from predation than from old age. Your flock will come to grief if you are blind to the hazards of a fully "natural" life—be prepared to intervene as much as required to foil predators (for example, by making sure your chooks *do not roost in trees*). Later in this book, I recommend one high-tech (and scarcely "natural") solution to prevent predation.

They reproduce: *Gallus gallus* had behaviors that controlled who mated whom, in ways that favored both genetic diversity and the breeding of the most fit. That does not mean that free-for-all mating is the smart choice if you want to breed your own replacement stock. Should you choose to *breed your own*, know that to ensure a combination of *both* genetic diversity *and* selection of the best of the best as breeders, you must be prepared to intervene, to control the process, more assertively than in any other area of husbandry. Fortunately, adaptive strategies that *Gallus* evolved in the face of unrelenting predation—*reproducing early in life, with lots of progeny, in a hurry*—make it easy to breed chickens. I hope many of my readers will take advantage of that opportunity, finding inspiration and guidance from the significant new emphasis in this revised edition on the art and practice of breeding your own birds.

A Meditation on Tool Use

In chapter 1, I said that my highest aspiration is to leave the bit of ground I'm responsible for *better* when I am gone than it was at my arrival. But, hmm, all that cover cropping and composting and planting for habitat diversity—sounds like a *lot* of work to me. Time to tool up, get some whirling steel on-site to work more efficiently and speed things up, no? What better example of a mechanical partner than the garden tiller. My dear wife Ellen gave me one decades ago—a small wheelless model with a two-cycle gas engine. I found it would prepare a fine seedbed—in soil already so "mellow" it could be prepared for seeding with a garden rake to

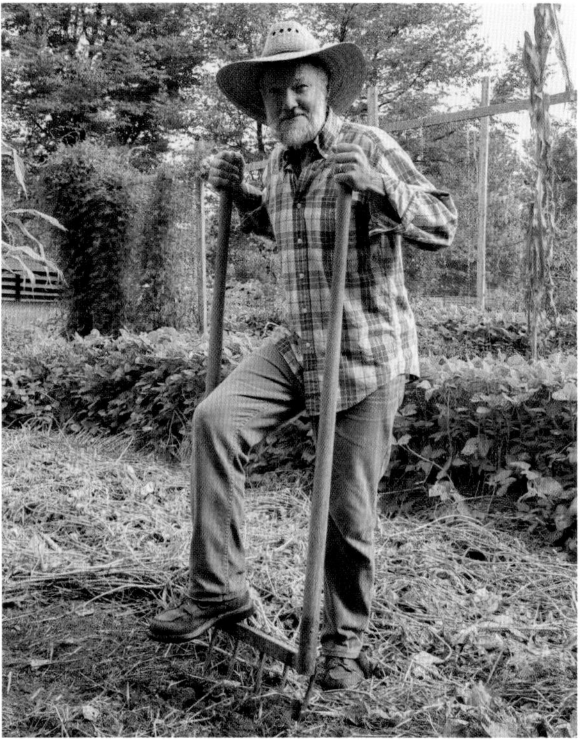

Figure 2.1. The broadfork loosens the soil without breaking down its structure. Ah, and no jarring vibrations, no noise.

begin with. In compacted clay soil, heavily grown up in weeds—or in a cover crop ready to be tilled in—the tiller was a nightmare: As the tines smacked against the tight clay of new ground, it bucked wildly. When I tried to turn under a mature cover crop such as 4-foot-tall (120 cm) rye or sprawling vetch, the tines would become completely bound with tough stringy stems and roots within minutes, leading to lengthy time-outs to clear them, usually after removing them from the axle entirely to do so. Any mechanical failure was beyond my handyman skills and required hauling the tiller for repair, waiting for completion of the work, hauling it back—to say nothing of shelling out the bucks for the job. Wait a minute, I thought mechanizing was supposed to *save* me time! As for making garden work *easier*: I found that wrestling a shuddering, bucking machine emitting an ear-splitting whine added a high level of stress to body and nerves—it *added to* the amount of *work* required for the job. What a joy it was to exchange the tiller for a few well-chosen hand tools, more than adequate to the tasks encountered in home gardens: scythe and sickle, hoe and spade, broadfork and garden cart. Using such tools becomes a sort of dance—to the music that had been drowned out by the tiller's deafening whine: the songs of birds and the lowing of cattle on the next-door farm.

Ah, but even with best-loved tools there remains a *lot* of work to produce more of our food on the homestead. That's why I propose that best husbandry for the backyard flock—putting them right at the center of the abundant ecology, doing exactly what they would most want to do in the wild—is also the smart way to enlist some highly efficient help with the work of the homestead.

"Efficiency"

Modern agriculture is said to be more efficient, as proved for example by the immense quantities of dressed broilers and eggs produced in its factories. But it's a peculiar notion of efficiency that touts production of 1 calorie of food energy for 10 calories of energy input—in contrast with a model in which much of the energy to produce eggs and dressed poultry comes directly from the nearest star. And a system that produces $1.29-per-pound chicken while polluting waterways is hardly a model for efficiently providing for human needs.

Biological efficiency offers itself anytime we're inclined to make use of it. When a single enterprise can provide free fertilization of the garden, control of slugs, tillage of weeds, management of "waste," with a dozen eggs thrown in as a bonus—what could be more efficient than that! The key is to avoid sticking the flock in some isolated corner all to themselves—and instead to make them *happy* by giving them good work to do.

Closing the Circle

This book is about patterns of poultry husbandry that are more independent of outside purchases from far away, and more dependent on home efforts and local resources. It sees the homestead or small farm as an energy system, and wise usage as seeking out all available energy resources on the home place, while avoiding energy leaks out of the system, thus creating a more closed and resilient whole. It assumes that local resources and home-based solutions will become apparent as we stop looking to purchased inputs to fill every need.

Poultry are arguably the starter livestock par excellence—easier and cheaper to set up on a small scale, with perhaps more to offer as working partners than any other livestock option. They reproduce quickly and prolifically, adding yet another dimension to independence from outside inputs. If my place can't host a herd of bison, I'm going to make my poultry flock its keystone species instead. In practical terms, that means the ultimate efficiency is to find every opportunity to fit the flock into larger patterns of diversity, individual and ecological health, resource utilization, and recycling of "wastes." Pulling that off is a lot easier in the green seasons of the year than during the winter, so let me tell you about two approaches I've used in caring for the winter flock, to illustrate the abundant opportunities for closing circles on the homestead.

I *hate* confining my birds, but for years I kept the flock inside my main poultry house in winter because I saw no alternative—if released to the pasture as they are in the green season, the constant scratching of the chickens would destroy the dormant sod. The nightmarish result is all too easy to imagine: a bare stretch of frozen dirt, dotted with chicken poops just waiting for the next rain to run for every natural water system between here and the sea. I had early on discovered the key to best, most natural, manure management—deep litter over an earth floor—but I assumed that solution worked only *inside* the walls of the henhouse. I saw no way to bring it *outside*.

One of our several garden areas was behind the greenhouse. A constant in my gardening practice has been the planting of more soil-building cover crops every year. One autumn many years ago I had grown a fine lush mix of cover crops over the entire greenhouse garden—crucifers for their "cleansing" effect on the soil, peas to set nitrogen, small grains for their addition of root biomass. I knew the cover would be a

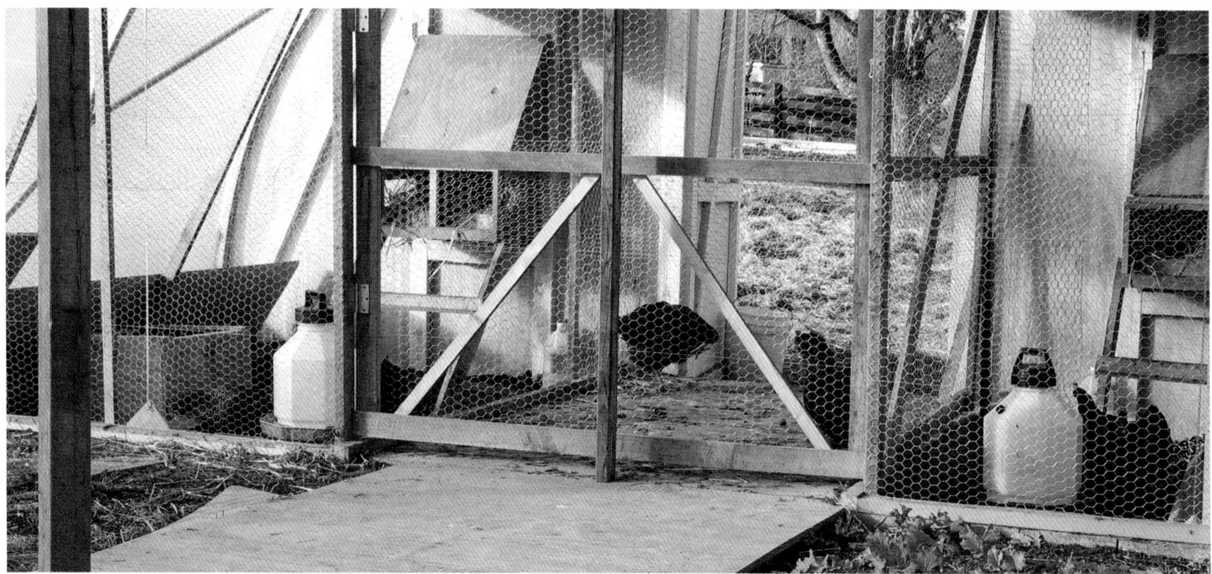

Figure 2.2. Two pens, each 8 feet by 8 feet, give access to an outdoor feeding/exercise yard. Each pen has its own door—the flock can be separated into two groups if needed—and either door swings into blocking position to keep the birds out of the greenhouse crop beds. The plywood platforms cover vermicomposting bins.

great boost to the garden's soil when I sent in my tiller chickens come spring.

But why wait until spring? Suddenly I saw how to get an early start on the flock's tillage work—*and* to bring deep-litter practice outside.

I built two 8- by 8-foot pens to house our mixed flock—chickens, ducks, and geese—in the far end of our 20- by 48-foot greenhouse. The project was easily done: The greenhouse itself provided shelter from the brutalities of winter; all that was required in addition was some chicken wire on light wooden framing to keep the flock out of the greenhouse's growing beds. I furnished the pens with the usual henhouse accessories: dustboxes for dust-bathing, nestboxes mounted above floor level on the end wall of the greenhouse, roosts, and hanging feeders. I left an earth floor in the pens and covered it 8- to 12-inches (20 to 30 cm) deep with oak leaves. I found it was okay to leave the waterers inside the pens if I had chickens only in them. When ducks and geese were part of the greenhouse flock, it was better to position the waterers outside to avoid wet litter—they're messy with their water. On especially cold nights, I moved the waterers to the basement to prevent freezing.

On the far side of the pens, the greenhouse's end door opened directly onto the garden. I propped that door open each morning, giving the birds access to the outside during the day.

I installed a door on each pen, so I could manage two separate subflocks if needed. Each door was designed to swing into blocking position, on the near side of the pens, to keep the birds out of growing beds filled with lettuces, chicories, and cold-hardy cooking greens.

I surrounded the garden with electric net fencing to protect the birds and confine them to their work.

Figure 2.3. All the comforts of home: Each pen is furnished with laying nests, dustbox, waterer (sometimes kept outside), hanging feeder, and deep litter. Not shown are the two-by-four roosts that are installed each evening (fitted onto the dustbox covers and the framing of the pens).

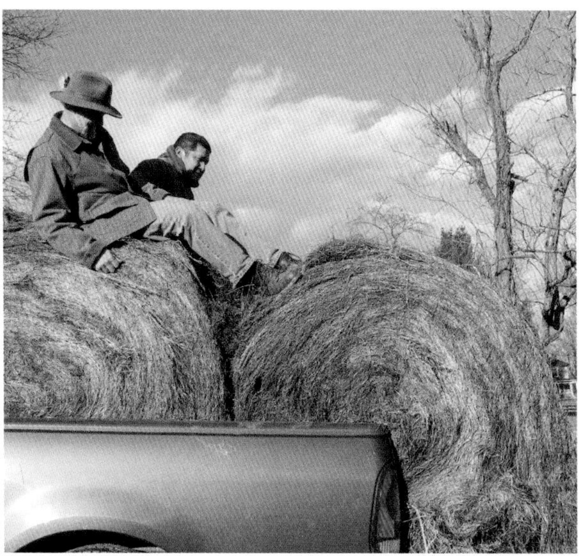

Figure 2.4. It was easy (and fun) to kick the big round bales off the pickup and roll them out . . .

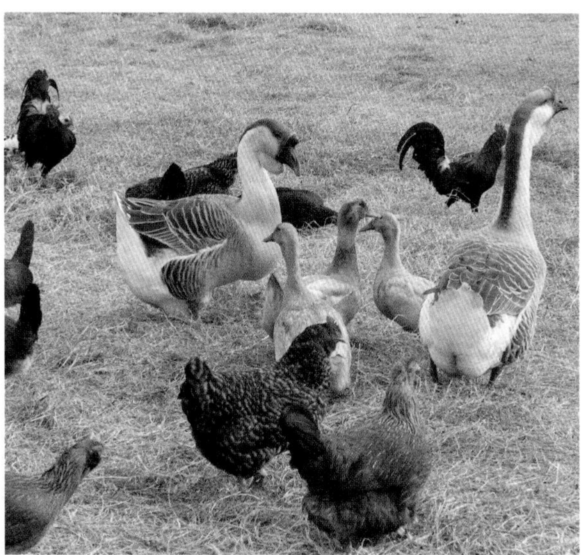

Figure 2.5. . . . for a deep, absorbent duff over the entire winter yard.

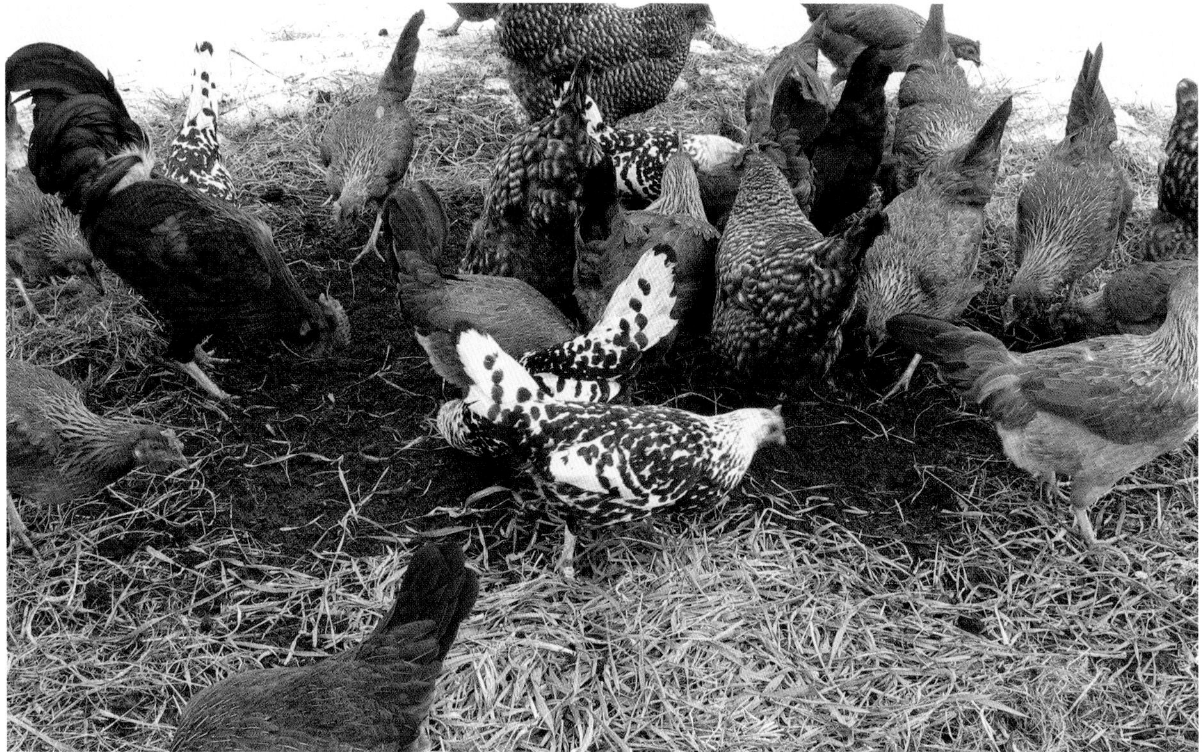

Figure 2.6. Vermicompost and green forage from inside the greenhouse kept the birds well fed, contented, and busy preparing their winter yard for garden crops come spring.

When I first put them in the greenhouse pens, they feasted on the green forage while tilling in the cover crops I had grown. And to be sure, the last shred of green disappeared, leaving that intolerable patch of bare unprotected dirt. Ah, but I was ready for the next phase—turning the space in effect into a giant compost heap, using several loads of round-bale hay from a nearby farm, spoiled for use as feed but just the thing for a deep-litter yard.

I would have preferred using my scythe to cut the necessary mulching material off my own pasture, but at that time in the year there were no more cuttings to be had. Any number of other organic debris types—autumn leaves, spent crop plants, feed mill residues like corncobs and husks—would have served as well.

The deep organic duff absorbed the poops, preventing their loss as runoff pollution while retaining their mineral fertility and boosting the diversity and health of the garden's soil food web. The soil never froze under the heavy mulch, so earthworms and slugs remained active, furnishing live foods for the chickens. From time to time I also added bedding from vermicomposting bins inside the greenhouse—the birds ate the worms while working the worm compost into the mulch—and fed them "salads" from the greenhouse, either green forage cut from beds planted to cover crops or trays of greened sprouts. In late winter the seeds in the round hay bales (they're the reason you should *never* use hay as a garden mulch, right?) began sprouting, and the birds cleaned up the sprouts—more free food.

Using this strategy, my flock stayed healthy and contented all winter long. Poultry are cold-hardy, so they enjoyed the sunshine, fresh air, and exercise in their winter yard in all but the nastiest weather. When the weather turned hostile, they retreated inside—as they did when night fell, and I would shut the greenhouse doors. My layers maintained both quality and quantity of egg production at levels never matched in previous winters, and none of the cocks suffered from frozen combs.

I had no way of precisely measuring the temperature difference inside the greenhouse, but theoretically the body heat of the flock—maybe 400 pounds (180 kg) or so of live bird at times—moderated the overnight low temperatures. Further, the carbon dioxide in the birds' exhalations could well have boosted the growth of the winter crops, which metabolize CO_2. (If the idea of factoring in chicken breath seems far-fetched to you, note that gardeners in the Netherlands, known for expert greenhouse growing, spend good Dutch money to buy bottled CO_2 to pump into their greenhouses and boost crop growth. Why they don't just get some chickens is beyond me.)

By spring the flock had assisted the breakdown of the hay into something between a finished compost and a mulch, the soil underneath becoming increasingly friable as the cover crop roots decayed, and with not a weed in sight. I moved the flock onto the pasture, laid out the garden beds, and without further ado the new growing season was off and running.

How better than in that mulched winter yard could I have enlisted my flock to address so many needs at once? How more completely could my concerns—for soil fertility; feeding, contentment, and health of the flock; and prevention of spillover pollution into the wider environment—have become "one great subject"?

You don't have a greenhouse, you say? Then know that I've used the same strategy for my flock in the main henhouse, the "Chicken Hilton." Again, I gave the birds access to a winter foraging yard in front of the coop—heavily mulched with autumn leaves, spent crop plants, and other decomposables—and protected by electric net fencing. Not only did the flock find natural foods beneath the duff throughout the winter, but working the mulched yard was great for their mental health—in lieu of the unrelieved boredom of confinement inside for months. In the spring I planted potatoes in the newly tilled and fertilized soil, and followed the potato harvest with a mixed crop of sunflowers and sorghum—more feed to cut and throw to the chickens (and in the case of the sunflowers, support for pollinators). How could you improve on easy, all-in-one husbandry that provides so much free, high-quality feed?

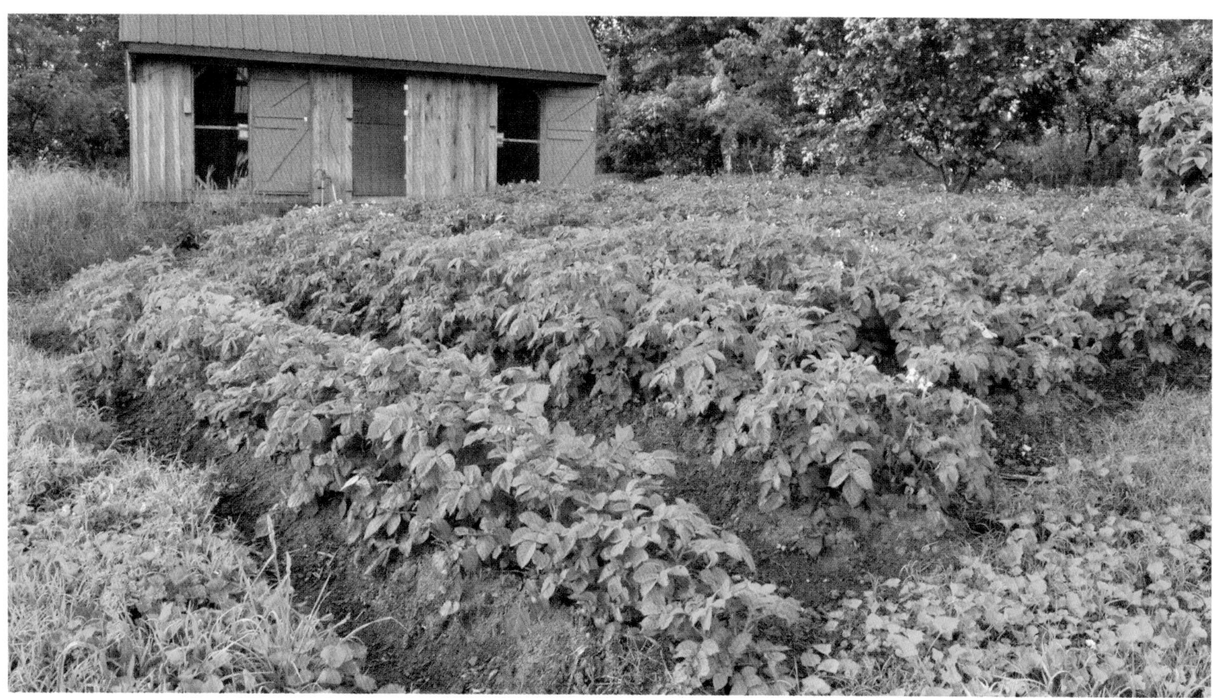

Figure 2.7. Best-ever potato crop in the chooks' well-tilled and -fertilized winter yard.

Figure 2.8. After the potato harvest, there was still plenty of time in the season for "freebie" crops of sunflower and sorghum to feed the flock.

CHAPTER 3

Your Basic Bird

This chapter is an introduction to the domesticated chicken. It is not a zoological treatise: Each topic here about derivation, anatomy, and behaviors of *Gallus gallus domesticus* ties in, later in the book, to practical strategies to take full advantage of our chickens' natural inclinations and talents.

There is disagreement about the ancestry of the domesticated chicken. It is descended largely from *G. gallus*, the wild Red Junglefowl of Southeast Asia (see figure 3.1); though it is possible that our backyard version was the result of a cross between *G. gallus* and *G. sonneratii*, the Grey Junglefowl. Domestication may

Figure 3.1. The Adam and Eve of chickens: Red Junglefowl cock and hen. Photo (*left*) istockphoto.com; photo (*right*) courtesy of Dr. L. K. Yap.

have occurred as early as ten thousand years ago in modern Thailand or nearby regions of Southeast Asia.

For our purposes the precise time and place of origin, and details of ancestry, are less important than a couple of other points of interest. Chickens are birds of the order Galliformes, along with wild relatives such as partridge, grouse, and prairie hen. Interestingly, except for domestic waterfowl and pigeons, other domesticated fowl are all Galliformes as well—turkeys, guineas, pheasants, quail, and peafowl.

Galliformes are primarily ground-dwelling species. Though they all can fly, they do so for only short distances—to escape predators, get over barriers, or roost for the night in trees. The most significant thing for us about their ground-dwelling nature is the fact that food—a diversity of both plant and animal species—is something to be found primarily *on the ground*. If we give our chickens—and turkeys and guineas—the opportunity, they will do what self-respecting Galliformes are used to doing: hustle up a lot of their grub on their own. As well, some of these species—chickens prominent among them—*scratch* for food such as worms and grubs beneath the surface of the ground. We will see how, as smart flocksters, we can "capture that technology" for our own purposes.

Though domesticated chickens and Red Junglefowl are so closely related that they easily hybridize along the edges between their native habitat and human settlements, a couple of key genetic mutations have occurred since *G. gallus domesticus* split off from its *G. gallus* ancestor.

Like all wild birds, the Red Junglefowl hen laid eggs only during the breeding season—egg laying was exclusively about reproducing the species. Selective breeding following domestication seems to have altered a hormone receptor that relates reproduction to change in day length. From that point egg laying in chickens was no longer tied to a specific breeding season but continued through much or all of the year. Other mutations seem to have increased capacity for appetite, growth, and body size. The result of these genetic changes was a fowl capable of greater production of eggs and flesh than required for mere survival of the species.

Nomenclature

A *hen* is a mature female chicken, one year or older; a younger female is a *pullet*. A male up to a year old is a *cockerel*; an older one is a *cock*. The distinctions are more about size than sexual maturity: Cockerels begin mounting the females in the flock as early as three months of age, even earlier for particularly precocious breeds; and onset of lay for pullets averages twenty-two weeks or so—whereas hens and cocks reach their adult weights at about one year.

A *chick* is usually understood to mean a young chicken, still in the down stage (not yet feathered). But then what is a young chicken from the just-feathered stage to adulthood? If we specify gender, we can use *pullet* or *cockerel*. Technical terms in the butcher's trade might be *poussin* (from the French word for "chick"), a tender young chicken no older than twenty-eight days at slaughter, or *spring chicken*, a young table chicken a little less than 2 pounds. But I really do not know a good, generic term for a chicken at this stage of life. In this book I am most likely to use the awkward term *young growing birds*.

I sometimes use *chook* to mean "chicken." The term originated in Australia and New Zealand, but is being used increasingly among North American flocksters as well, with an implied sense of affection.

And what is a *flockster*? That is my own word, to define someone who is enthusiastic about keeping chickens or other poultry as enjoyable, respected, and productive partners.

Finally, what is a *rooster*? A prudish euphemism. The cock has been for a long time, in many cultures and languages, a symbol of resurrection, of the sun, and of the male sexual member—doubtless because he stands up proudly at dawn to greet the sun as it ushers in the renewal of the day. Though from its origins in old Anglo-Saxon until the King James translation of the Bible (it was the "cock" that crowed three times, not the "rooster"), the word for the male chicken and for the penis were one and the same—that usage was no longer tolerable in the priggish Victorian era. By the nineteenth century in the United States and Australia, the male chicken was typically a "rooster"—an especially silly euphemism, if one had to be found, since female chickens "roost" the same as males. Note that the euphemizing compulsion became even more extreme—the culinary terms *leg*, *breast*, and *thigh* were renamed *drumstick*, *white meat*, and *dark meat*. The leg was even at times referred to as the *second wing*! Ah well, at least we still describe someone who is brashly self-assertive as "cocky."[1]

The Working Model Chicken

Good management requires an understanding of the chicken's basic working systems. In fundamental ways they are similar to our own—the digestive processes for converting food to energy and tissue, for example, or reproduction based on specialized male and female cells, each carrying half of the DNA code. We need to pay closest attention to the points at which we differ—a digestive system that does not include chewing, for example, and reproduction based on that biological marvel, the egg.

Dimorphism

Notice in figure 3.1 that the Red Junglefowl is strongly *dimorphic*—that is, the cock and the hen are distinct in many ways. The cock is larger; and his comb and wattles are much larger than those of the hen. He is more extravagantly colored, and bedecked with large showy feathers around the neck, over the saddle, and arched out from the tail—as befits his courtship displays. The hen has tighter feathering, in subdued colors—

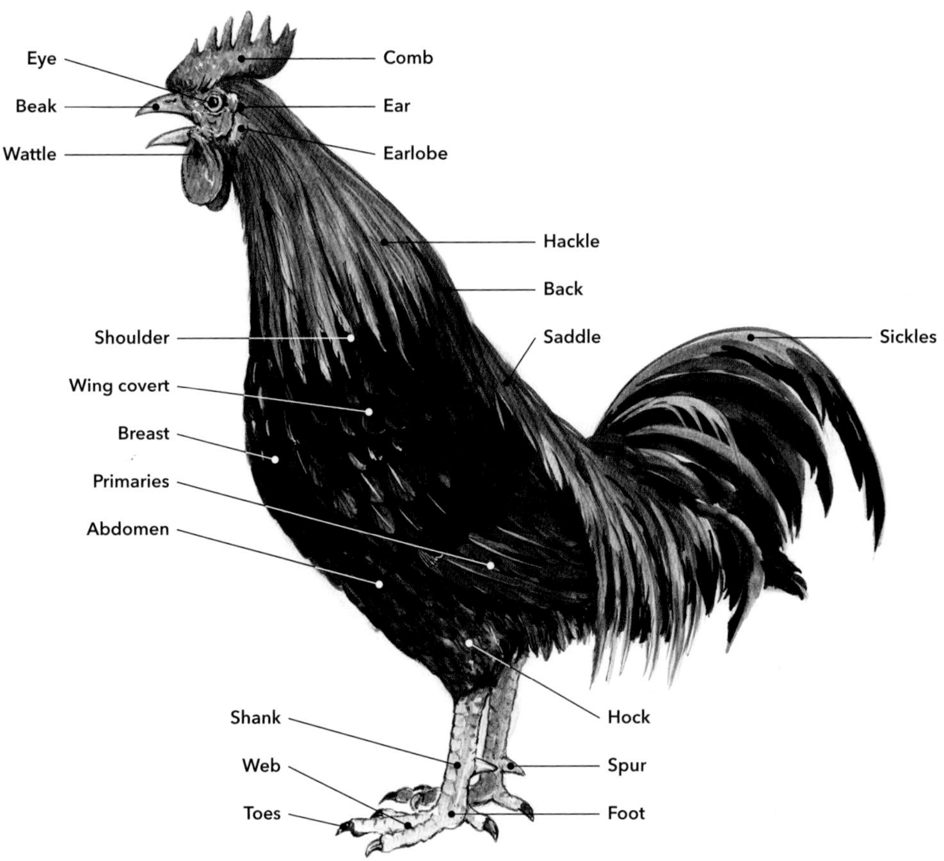

Figure 3.2. The chicken's basic parts list. Illustration by Elayne Sears.

camouflage that reduces her vulnerability to predators while incubating her eggs and rearing her young.

This dimorphism is characteristic of domesticated chickens as well. In most particolored breeds, the male has the more conspicuous coloring. Even in solid-colored varieties, in which cock and hen share the same basic color, the cock is likely to exhibit a deeper tint, with more gloss. The cock's hackle, saddle, and tail feathers are more prominent, with a higher sheen. Hackle, saddle, and covert feathers are typically pointed and narrower in the cock—equivalents on the hen are broader, with round ends.

Anatomy

The illustrations in this section show the key features of the external anatomy, including the spur and the larger comb and wattles of the cock, and the internal organs, both digestive organs (the same in both genders) and the reproductive organs of the hen.

COMB AND WATTLES

As noted previously, the larger comb in the male serves to attract females. Physiologically, the chicken's comb helps radiate heat out of its body as blood circulates through this exposed tissue—such dissipation is second only to panting in helping the bird adjust to hot weather. Thus, breeds with large single combs might be preferred in hotter climes. Since large combs are more subject to frostbite in winter, however, far northern flocksters might prefer breeds with minimalist combs.

The comb can be an indicator of health—changes in color or texture could signal a problem. Combs

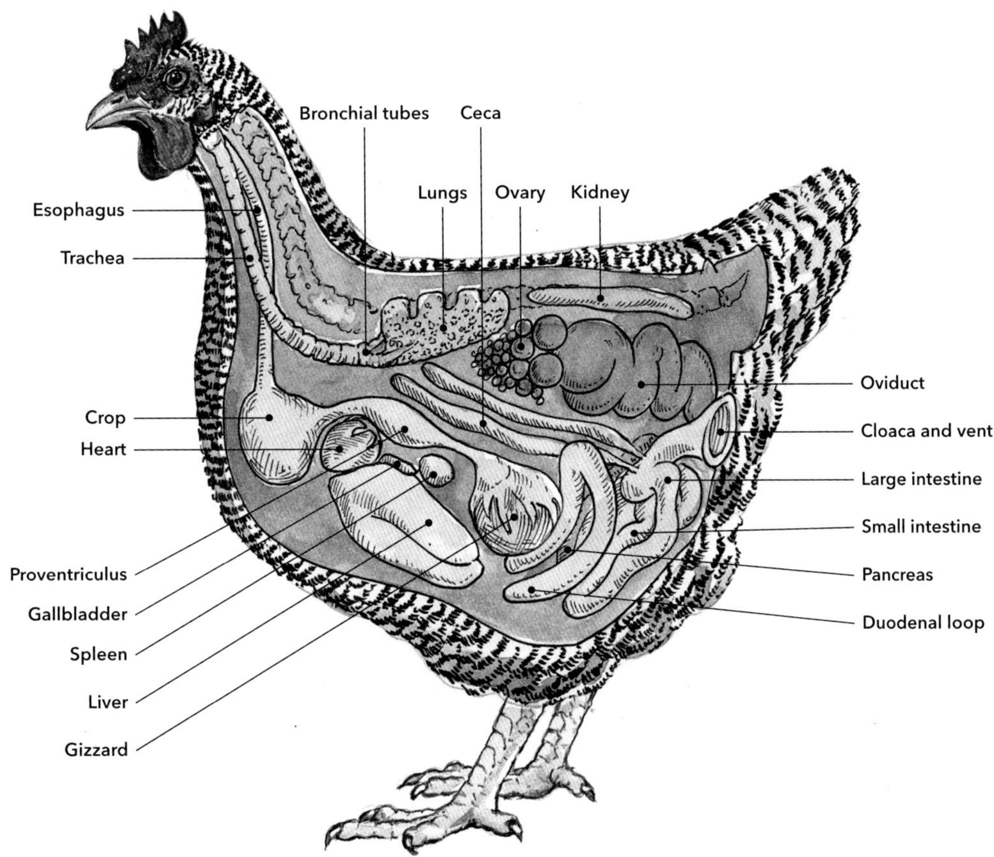

Figure 3.3. Digestive system and, in the hen, reproductive organs. Illustration by Elayne Sears.

of pullets turn a brighter red when they are about to start laying.

Selective breeding has resulted in a diversity of comb styles: single, rose, pea, cushion, strawberry, buttercup, and V-comb, as illustrated in figure 4.1.

SHANKS AND FEET

The color of the shank varies by breed. Shanks of ancestral Red and Grey Junglefowl are "clean"—lack feathers—but a mutation emerged among certain Asian breeds that resulted in feathering of the shank as well. When feather-legged breeds arrived in Europe in the 1800s, breeders enhanced the trait.

Growing from the back of the shank of the cock, above the back toe, is the horny spur: shorter and blunter in a farm breed cock, longer and quite sharp in a game cock such as the Old English Game—and capable of killing a rival. Hens of more ancient breeds—in my experience, Old English Games and Icelandics—occasionally grow spurs as well, much shorter than the cock's but needle-sharp.

Most breeds of chickens have four toes, but a minority, such as Dorkings, Faverolles, Houdans, and Silkies, have five.

FEATHERS

One of the distinguishing features of birds as a class among other vertebrates is feathers, or plumage. Feathers constitute the most complex of all *integumentary systems* among vertebrates—the covering of the body, including the skin and its associated appendages, such as hair or feathers or scales. Feathers help protect, waterproof, and insulate the bird—and serve also as sensory apparatus. Feather color and pattern can either enhance the bird visually—as for mating displays—or help it blend into its background to escape the notice of predators.

It is essential that the feathers remain in good condition to fulfill their functions, so chickens spend a lot of time *preening* their feathers: Their tails, which support the large tail feathers, also contain the uropygial gland, or preening gland. The bird expresses preening oil from the gland and uses it to dress the feathers, cleaning and conditioning them in the process and waterproofing them with the oil—partially in the case of chickens, almost completely in the case of waterfowl.

Even given the best of care, however, feathers become worn and broken and lose function. A periodic *molt* is necessary—the shedding of *all* the feathers and growing new ones. Be prepared for this annual ritual—the birds can look appalling missing half their old feathers and with new ones coming in like porcupine quills. A chicken does not molt in its first year, since its feathers are still so new. Thereafter, however, it molts once a year, in fall or early winter. Timing of the molt varies from year to year, and individual hens follow their own schedules rather than molting simultaneously.

Feathers are almost pure protein, so replacing them all requires a *lot* of resources. Despite the mutation discussed previously that led to egg laying by domesticated hens year-round, it is not surprising that they lay fewer eggs, and may cease laying entirely, during the molt. If your chickens are well fed and free of stress, don't worry about them when they begin shedding feathers in the fall.

Perhaps the most important thing to appreciate about feathers is just how tough they make our birds. They protect so well from the weather that many keepers of turkeys allow them to roost outside even through the harshest parts of winter. Chickens are not quite that hardy, but their feathers insulate them so well in winter that we're better advised to provide their housing with plenty of ventilation rather than heat it. On the other hand, in summer the feathers inhibit escape of heat from their bodies, so shade is essential to prevent overheating.

DIGESTIVE SYSTEM

Figure 3.3 shows the main features of the chicken's digestive system. Food is picked up into the mouth by the *beak*. The bird has no teeth—hence anything in short supply is said to be "as scarce as hen's teeth"—and does not chew its food. The mouth adds saliva, containing digestive enzymes, as it passes the food on into the *esophagus*, or *gullet*, which deposits it into a flexible pouch-like enlargement of the esophagus called the *crop*. An important function of the crop is storage: If the chicken finds a treasure trove of food, more than can be accommodated in the digestive organs at the moment, she can take advantage of the windfall by stuffing quite a bit of excess into her crop—like a chipmunk stuffs its cheek pouches. If you pick up one of your hens after she's been feeding, you can feel the distended crop, slightly to her right at the base of the neck—the size of a golf ball or larger. In addition to storing the food for later digestion, the crop bathes it with fluids, which soften it and ready it for digestion. Hard foods such as seeds might remain in the crop twelve hours or so.

When the digestive system is ready, contents of the crop are drawn into the first of two parts of the chicken's "stomach," the *proventriculus*, which mixes them with hydrochloric acid and various enzymes that start the process of digestion. But the food still hasn't been chewed, right? That's the next part of the process: The mix of softened food and digestive juices enters the *gizzard*, or *ventriculus*, the organ in avian species that takes care of both the grinding and mashing of food that humans accomplish with teeth and tongue and the early stages of digestion that take place in the stomach. The gizzard is made of two sets of powerful muscles surrounding a pouch with a thick, tough lining that holds the food. One opening into the gizzard from the proventriculus brings in the food; another passes it on into the small intestine for continued digestion and absorption of nutrients.

The grinding of the food in the gizzard depends on the presence of *grit*—bits of stone the bird swallows—inside the pouch. Birds eating a natural diet with a lot of whole seeds and fibrous green plants need plenty of grit—more than confined birds eating commercial feeds do. The pieces of grit are retained in the gizzard until completely worn away, so the chicken needs to continually renew its supply. If foraging outside, she will likely pick up all the grit she needs on her own. However, it is easy to supplement with purchased granite grit, which is cheap, by way of insurance.

The soupy puree from the gizzard enters the *small intestine*, which adds digestive enzymes from the *pancreas* for protein digestion and *bile* from the *gallbladder*, synthesized in the *liver* and essential for the digestion of fats and absorption of fat-soluble vitamins A, D, E, and K. By the time this liquefied food has passed the length of the small intestine, most of its nutrients have been absorbed through the intestinal walls and into the bloodstream.

Reading the Poops

How does a mother judge the state of her children's internal health? One way is to see what their bowel movements look like. In the same way, you can judge whether your chooks' digestion is efficient and healthy with a form of divination I call "reading the poops."

A dropping from a chicken who is eating a diversity of natural foods is gray—or greenish if the bird has been eating a lot of fresh plants—and firmly shaped, with a coating of white uric acid crystals from the kidneys. In the henhouse you would notice something else about it: It does not present a foul stench to the nose at standing level.

A poop from a bird who has been eating commercial feeds, with limited access to fresh greenery and other natural foods, is loose and splattered, black or darker in color, with a foul odor even at standing height.

There is one caveat to the above reading of the poops: The ceca empty their contents directly to the outside two or three times daily. Cecal droppings are looser, mustard to dark brown in color, and smell worse than normal droppings. So do not be concerned if you see such droppings here and there, so long as most of the birds are producing poop that looks like the healthy one described above.

The residual fecal material fills the two *ceca* (singular *cecum*), where some of its water is reabsorbed, and where fermentation breaks down remaining fibrous material, a process that produces several fatty acids and eight B vitamins, some of which are absorbed.

In the *large intestine*, or *colon*, more water is absorbed, concentrating the fecal matter, which then moves into the *cloaca*, a stretchy pouch that serves as the final staging area for the poops. Note that the eliminatory functions that are separated into bladder and bowel in our case are combined into one in the cloaca. In other words, chickens do not urinate. Their kidneys do extract nitrogenous metabolic waste as ours do, but they excrete it as uric acid, in the form of a coating of white crystals over the surface of the feces, which are expelled through the *anus* by contraction of the cloaca.

REPRODUCTIVE SYSTEM

Birds suffer high rates of predation, a threat they survive through an extremely effective adaptation: the egg. Egg laying and the hatching of young in sizable groups—young who mature at an astonishingly rapid rate—allow for reproductive rates far beyond those of mammals, ensuring that enough offspring live long enough to reproduce despite the onslaught of predators. This high reproductive rate is one reason that raising chickens is so easy and so practical.

For us, there is a significant benefit to our chickens' reproduction via eggs. An embryonic mammal receives nutrients for growth on an as-needed basis via the womb. That is not an option for the chicken embryo, who must make its entire growth—from fertilized speck on the yolk to fully formed chick ready to hatch—entirely out of nutrients stored in the egg. The egg *must* be a powerhouse of nutrition—a major reason it is rightly treasured as a food.

Reproduction on the cock's side centers on two *testes*—held deep in the body cavity, as opposed to externally as in male mammals—which produce sperm in massive numbers. Two ducts called *vas deferens* carry sperm from the testicles to two elongated bumps on the back wall of the cloaca called *papillae*. When the cock copulates with the hen, he ejaculates sperm from the papillae into the hen's cloaca, from which they are moved by muscular action up into the *oviduct*.

Reproduction on the hen's side (see figure 3.3) begins in the ovary. Its thousands of reproductive cells develop sequentially into a cluster of tiny *yolks* or *ova*, the most advanced of which grow rapidly when yolk material is added. *Ovulation* is the release of the most mature yolk into the second part of the hen's reproductive system, the *oviduct*, a convoluted tube about 25 inches (63 cm) long. If the hen has been mated by a cock, his sperm will fertilize the ovum in a brief stay in the first part of the oviduct, the *infundibulum*. The yolk next moves into the longest part of the oviduct, the *magnum*, where over the next three hours the *albumen* or egg white is added. The inner and outer *shell membranes* are added in the next section, a constricted part of the oviduct called the *isthmus*. In the next section, the *shell gland* or *uterus*, the shell is laid on, largely as a deposit of *calcium carbonate*. The hen mobilizes almost half the calcium to make the shell out of her bones, so a large intake of calcium in her diet is essential to support both shell strength and bone density. After about twenty hours, the finished egg moves into the last segment of the oviduct, the *vagina*, a muscular sheath whose contractions force the egg out when it is time to lay. Before expulsion, the vagina coats the egg with the *bloom* or *cuticle*, which both eases passage and upon drying helps protect the porous eggshell from invasion by bacteria.

When you imagine the laying of that egg, your reaction might well be one of consternation: *Oh yuck—you mean the egg is laid through the poop-hole? You'd think nature would have come up with a better design than that!* Actually, she did. The contraction of the vagina to expel the egg coincides with a prolapse of the uterus that completely occludes the cloaca—there is no contact with the interior surfaces of the cloaca as the egg is expelled directly to the outside.

Of interest is the presence of *sperm host glands* at the juncture between the vagina and the shell gland. These glands serve as storage sites from which sperm are moved by the oviduct up to the infundibulum.

Storage of sperm in the glands enables continuing fertilization of ova long after the hen's last mating with a cock—unlike mammalian sperm, the cock's sperm can survive at the hen's body temperature for two weeks or even longer. The function of the sperm host glands isn't just a fun fact—it will have practical implications later, in chapter 26.

Note that the laying cycle is not likely to be keyed exactly to a twenty-four-hour day, even though it is strongly influenced by photoperiod (the daily cycle of light and dark periods). Since it takes twenty-five or twenty-six hours for a new egg to form, the hen—who starts the cycle in the morning—will lay her egg later each day. She will not lay much later than three o'clock in the afternoon, however, and when the laying of the egg occurs this late in the day, she ceases production for a day or so to reset the clock and begin laying once again in the morning.

The Inner Chicken

An understanding of natural chicken behaviors will help you better care for your birds. The more time you spend observing your birds, the more you will understand the nuances of their social organization, what makes them tense, what makes them content—even the language they speak. You will be a more skillful flockster—and will have more fun.

Flocking

Most Galliformes are *flocking* species: A major survival strategy is living together in closely bonded groups. The flock has a hierarchical structure, with a place in the pecking order for each individual. Cooperation and mutual protection are key parts of flocking behavior. A hen who finds a trove of good things to eat will not only start gobbling them herself but will call the rest of the flock to come and get in on the feast. The first to spy a predator calls an alarm, triggering a rush by everyone for cover.

Understanding flocking will prepare you for behaviors that might otherwise be unsettling. Hens who seem to be fighting may well simply be having a "discussion" about relative position in the flock's hierarchy, not a cause for concern. To be sure, introducing new members from elsewhere sparks more intense discussions with potential for injuries, so requires careful monitoring and occasional intervention (more on this in chapter 20).

An especially distressing flock behavior is the instinct to zero in on a bird who is sick or injured—at the extreme, a literal pecking her to death. While gruesome, in a natural setting this behavior is a survival strategy—elimination of the weak sister could make the flock less vulnerable to predators. As a flockster, do everything you can to prevent stress in the flock, which can bring on elimination behaviors. Isolation of a temporarily disadvantaged member can help give it the opportunity to recover and then rejoin the flock.

Language

Communication within the flock involves a complex set of unique vocalizations, which in the broadest sense we could consider a language. Studies of vocalizations of Red Junglefowl have found them to be largely the same as in domestic chickens, with at least two dozen distinct calls, which seem to combine into more complex signals coded to environmental and behavioral context.[2]

Communication begins even before hatch: The pipping of a chick in the egg alerts the mother hen that it is beginning to break out of its shell. Among adults, there are unique vocalizations to mean "food here"; to express contentment, disturbance, fear, or frustration; or to issue threats. Alarm calls can signal whether the predator is on the ground or in the sky. Crowing is an especially complex vocalization, varying with the position of a cock in the dominance–submission hierarchy, or whether he is trying to attract the attention of a hen. A cock in your flock may engage in crowing duels with one in a neighbor's flock hundreds of yards away.

Pay close attention to their vocalizations and you will learn to recognize the chicken talk used for specific situations and needs. A frantic note from a brooder chick communicates distress to the flockster, instantly distinguishable from its usual clear peep. A mother hen on pasture has a call to let her chicks know she has found food, another to signal "follow me," and another to help a chick find its way back to her if it has wandered away. You will learn the call of a cock when "tidbitting" his hens, described in the next section—you may even notice that the intensity of his food calls varies with the palatability of the tidbit he is offering (a nutrient-dense cricket being signaled more excitedly than a berry, for example). Hens foraging contentedly together engage in a quiet conversational clucking. And some hens use a sharp, distinctive cry to signal to the world their extraordinary accomplishment of having just laid an egg.

Sexual Behaviors

Red Junglefowl cocks have several mating behaviors we may see in our own flocks. Their brassy *cock-a-doodle-doo* call is issued both as invitation to hens as prospective mates and as warning to rival cocks in the area. Should another cock not be deterred, a serious fight may ensue to establish access to the hens. Given their long, sharp spurs as weapons, the fight can result in the death of the loser. Recognize that aggression in pursuit of domi-

Sex in the Barnyard

Temple Grandin, well known for her work with domesticated animals, has remarked on the behavior of cocks called "dancing": The cock performs a strutting display—circling the hen, wings spread stiffly toward the ground and quivering—which persuades her to squat and erect the shoulders of her wings, welcoming his advances. When the hen's cooperation is encouraged by dancing, mating is not violent and the hen is not injured, even when the cock's spurs are long and sharp. Grandin believes that, having ignored this dancing behavior when breeding our chickens, we have bred modern cocks who have "forgotten" how to dance, with the result that mating is carried on with more violence, sometimes injuring the hen.[3]

I have always watched for dancing behavior, and favored for use as breeders cocks who danced for the ladies.

An elderly neighbor who kept chickens for many years offered further light on these mysteries. Sitting by his chicken pen, he would observe the flock's behavior four or five hours at a time. I bet he knew more about chicken mating behavior than any expert in any ag college in the country.

"So if you got two roosters," he told me, "the top guy is gonna have his pick of the hens, and he'll have his own special group that are his. He'll look out for 'em, and he'll tread 'em. The other rooster can tread the other hens, but the top guy will keep him away from the hens he's picked. Well, sir, after about four hours there'll be a switcheroo—suddenly the top guy will be treading a different group of hens!"

If my friend was accurate in his description of natural mating behavior, think of the implications: The dominant male gets his pick of the hens—that is, priority when it comes to passing on his genes. But the flock has the instinctual wisdom to know that the subordinate cock also has his role to play in ensuring genetic diversity—in keeping some wild cards in the hand—and affords him the opportunity as well to pass his genes on to the future flock.

nance in mating success is normal behavior. In a settled flock with plenty of space, cocks will typically work out dominance–submission places in the hierarchy, with little subsequent conflict. Occasionally the competition gets deadly serious, however. (See chapter 20.)

Once he's taken care of rivals, the cock courts the female by "dancing" in a circle around her, his stretched wings stiff and pointed to the ground. If the hen is impressed, she allows herself to be mounted for copulation. Look for this behavior on the part of cocks in your flock as well.

The junglefowl cock—along with other galliform males—also engages in a behavior called "tidbitting": If he finds a choice bit of food, such as an insect, he bobs his head, makes a throaty, burbling call, and picks up and drops the tidbit repeatedly, to attract the hen to come receive the treat. He may even pick up the tidbit and offer it to the hen with his beak. Though tidbitting is doubtless a bid for romance, I like to think as well that the cock has the instinctual wisdom to know that the hen needs the protein boost for nutrient-rich eggs that produce sound offspring. In any case, tidbitting is about mating and species survival, and is an endearing behavior you will enjoy seeing in your flock.

As for the hen, both among junglefowl varieties and among domestic chickens, she takes exclusive responsibility for making a hidden nest, assembling a clutch of eggs, incubating them, and nurturing her young. The instinct to brood may be so powerful that the broody hen can even be used to hatch eggs of other fowl species and nurture the young, or to adopt a clutch of chicks not her own. (More information on working with mother hens is in chapter 27.)

Are There Dangerous Behaviors?

As you work around your flock, are there any behaviors of concern, any precautions you need to take? I address elsewhere the special cases of hens who may give a fierce protective peck if disturbed on the nest, and of the care to be taken to avoid triggering aggression on the part of cocks. But as a general matter, you should keep in mind that chickens are naturally curious, and—lacking hands with clever fingers—their typical way of checking out something of interest is to *peck* it. This behavior is rarely a problem—in winter, a hen may peck at the snow clinging to a boot, for example. But let me pass on three experiences as cautionary tales.

I was once squatting close to the ground, attending to some chore, when a hen, curious about the glint in my eye, *pecked it.* Because I couldn't see the peck coming in from outside my peripheral vision, I had no chance to blink, and the peck was a direct hit. The pain was stunning, but even worse was the terror that she had done serious injury. Fortunately that turned out not to be the case, but I've since been cautious any time my face was down at chicken level.

I sometimes wear an earbob, and one day last year while I was hanging the waterer inside a mobile shelter, a hen wondered, "Hmm, what *is* that thing?" and *pow!*—pecked it right out of my earlobe. (And wouldn't you know, it was the one with the fire opal—my favorite!)

Then there was the time I was out collecting eggs and suddenly couldn't wait for a little "urgent release." So I figured why not, a little liquid nitrogen added to the deep litter has to be a *good* thing, no? I unlimbered, and just as I was breathing the sigh of blessed relief—a hen jumped up and pecked me, ah, where I would have least wanted to be pecked. So, you guy flocksters out there: You have been warned!

Other Thoughts on Behavior

Remember what was said about the role of predation in shaping gallinaceous fowl. Chickens are extremely sensitive to anything they read as a predator in the environment. Avoid crowding them or making sudden moves within the confines of the henhouse. Avoid carrying large flat objects such as an empty feed bag or large scrap of plywood—they see it as a flying predator and freak out. Because of their instinct to be out of reach of the predator while asleep, chickens will feel more secure with roosts on which to spend the night.

Tidbitting behavior could be your key to making friends with your birds. Especially in the case of a skittish cock inclined to be too assertive in defense of his ladies, you can offer special treats—crushed hard-boiled egg or grubs from garden beds, for example—to encourage more friendly relations. It really works.

Keying on natural behaviors is a lot smarter than trying to figure out everything on your own. The first year we had chickens, my daughter, Heather—who usually took care of them at the time—was away for the weekend, and I was doing the chores. Each evening when I went out to shut up the coop, I found that some chicks, just days out of the shell, had somehow escaped the fence and, with the coming of evening, had hidden under the coop, cheeping miserably. And each evening I crawled under the coop to rescue them, belly-down in the dried litter that sifted down through cracks in its floor. After Heather returned, I asked her, "Say, have you had any problem with those chicks getting out of the fence and being stranded when the chickens go in for the night?"

"Oh sure," she replied matter-of-factly. "I just get the mama hen and hold her by the edge of the coop outside. She clucks and the chicks come to her, and I pick them up and put them inside."

Never assume you've seen it all when it comes to unique behaviors in your flock. We had a bantam Old English Game cock named Charlie Brown who couldn't mate the hens because his feet were badly deformed. However, he would dance most handsomely for the ladies and would get quite excited when a more fortunate cock mounted a hen, standing close to the action and eagerly clucking advice. Goldilocks, a Golden Sebright hen, exhibited an inexplicable but endearing behavior during her long life here (about seven years). For about three weeks each spring, every time I entered the henhouse she would swoop up and land on my shoulder—and stay contentedly perched there as long as it took for the chore that had brought me in.

A friend passed on my favorite chicken behavior story: His flock ranged freely during the day and returned to roost in the henhouse in the evening. One windy night while watching television with his family, he heard a persistent tapping at the front door. Opening the door revealed one of his hens looking up at him expectantly. Following her out to the henhouse, he found its door blown shut, with the flock clustered outside. Think about it: In a crisis, the flock had analyzed the nature of the problem, formed a committee to decide on the best course of action, and commissioned an emissary to carry an appeal for assistance.

I hope this chapter has given you some information about chickens that will guide your management practices. But remember, most of what you need to know you will learn by spending time with your flock and observing their behaviors.

CHAPTER 4

Planning Your Flock

There is no ideal flock in the abstract—your ideal flock is the one that best achieves your purposes, within the limitations of your own situation. Thinking through your goals and management choices is likely to make for a smoother start with fewer surprises. Here are some questions to address.

"Is It Necessary to Have a Rooster?"

One of the most common questions I get from folks contemplating starting a small flock is: "Do I have to keep a rooster with my hens for them to lay eggs?" The answer is simple biology: Just as a woman ovulates on a regular cycle whether or not she has a mate, a hen ovulates (makes an egg) in *preparation* for reproduction, whether or not there is a cock around to mate her and fertilize her eggs.

Thus it is not necessary to have a cock in your flock of laying hens. If, for example, you live cheek by jowl with neighbors who might object to the crowing of a cock, you can keep a much quieter flock of hens only. Though eggs will not be fertile absent the attentions of a male, the hens will lay just as many eggs. And they will form their own hierarchical social structure without a cock.

Our own preference is to include one or more cocks in our flock. As with most avian species, the boys are the more spectacular of the genders, and we enjoy their flash of color and long graceful sickle feathers. We also like the way the cock completes the natural social structure of the flock, how solicitous he is of the welfare of his ladies, and even his clarion call.

If you plan to have a small flock, say up to a couple of dozen hens, including only a single cock should be sufficient if you just want him there to complete the flock. If you want to be sure all the eggs are fertile, one cock can service from eight to twelve hens. If you include more than one cock in your flock, see chapter 20 for cautions about dealing with aggression.

Flock Size

Proper size for your flock depends first on your production goals and your level of experience. To estimate egg production, as a rough average you can anticipate two eggs per day for every three layers in your flock. Numbers needed to produce dressed poultry should be pretty straightforward if you raise batches of meat birds to put in the freezer—how many do you want to eat, and on what schedule?—but the question gets more complicated if your table fowl will be produced by necessary culling of an ever-changing, mixed-ages flock. In any case, if you are a complete beginner, it's better to start with a smaller flock and work up as you gain experience.

An estimate of flock size might begin with the number of eggs your family eats on average. Imagine a family of four who are content with an average of an egg a day per person. Given the rough three-to-two ratio above, six good layers would fill the need for eggs for such a family. If they want to eat on average two eggs a day per person—and chances are good they will, once they start eating the best eggs anywhere—then they would want a flock of a dozen hens.

That's a good start on calculating flock size, but remember that production will drop a good deal in winter, and decide whether you want to increase flock size to compensate (and preserve some of the summer surplus with one of the methods in chapter 29). Remember also that some breeds will not meet the three-to-two production level. You might well choose a less productive breed such as Dorking or Brahma because you love their look or temperament, or one that lays eggs with pastel or deep brown shells such as Marans or Ameraucana (a breed that is often incorrectly referred to as Araucanas).[1] But in that case, you should add more hens to the flock. And be assured that you will want to share the world's best eggs with friends and relatives—I would factor another dozen eggs a week into your calculation for that purpose alone.

Now, what about meat birds? As noted, if you want to raise batches of meat hybrids, the math is obvious: How many do you want to put on your family's table, and on what schedule? But let's consider the question assuming that your table fowl will come exclusively from culling your first group of started chicks, and that you want to raise two dozen layers. A good plan might then be to start with fifty chicks "straight run"—that is, in the natural ratio of male and female, which for chickens as for humans is roughly half and half. (In chapter 6, I explain why I always order chicks straight run, and never as sexed pullets only.) Starting with the fifty chicks, you'll have about two dozen cockerels to cull for the table—almost enough to have chicken from your own backyard every other week. If you want to eat more chicken but still end up with two dozen started pullets, increase the number of chicks to seventy-five, or a hundred. But in this case, order the first fifty straight run, and the remaining ones as all-cockerels. (There should never be a problem with ordering sexed cockerels only. Again, see the discussion about ordering chicks from hatcheries in chapter 6.) The more chickens you want to cull as table birds, the more sense it would make to start with the initial fifty, then start subsequent batches of all-cockerel chicks later in the season.

Some questions about flock size are less obvious. If you plan to pasture your flock, consider that the pasture will not support as many birds in the drier part of summer when the grass does not grow as fast. If you want to supply as much of your flock's feed as possible out of home resources, recognize that the limits on the amounts of foods you can make available may set limits on flock size—you can come closer to being self-sufficient for feed with a smaller flock than with a larger one.

During our nearly four decades at Boxwood, we raised our flocks on a little more than 3 acres, about an acre of which is in pasture over which I usually ranged our flocks using electric net fencing. Flock size fluctuated, but we typically had a layer flock of about two dozen, raised two or three times that number as table birds each growing season, and in many seasons grew between a dozen and two dozen waterfowl—sometimes ranged separately on areas of lawn—for slaughter each fall. With these numbers, we were able to rotate our flocks over available pasture (or lawn) about three times in the growing season.

Profile of Your Flock

Check your local zoning regulations regarding keeping of poultry where you live. Far too many localities forbid the keeping of poultry. Many more place certain restrictions, especially regarding number of birds, whether keeping cocks is allowed, and setback distances between flock housing and property lines.

If local ordinances are not to your liking, you may be able to change them. When you look for allies, remember this phrase: *Out of the mouths of babes . . .* Ellen and I participated in an effort in our county to liberalize its zoning ordinances about keeping livestock. The debate before the board of supervisors seemed a close thing—until two young brothers, desperate to keep their pet goats, made their appeal. After the younger boy, eleven years old, told the supervisors earnestly: "My goat doesn't smell bad, and she's never bitten anybody"—it was scarcely a surprise when they liberalized the county's codes regarding family livestock.

Another approach where restrictive ordinances are concerned is to keep your flock "under the radar." I've heard from many, many flocksters who quietly keep a small laying flock despite the fact their locality technically prohibits doing so. In such cases it is especially important to encourage a benevolent attitude toward your project among your neighbors—if they don't complain, it is rare that the powers that be will send in the posse.

There are times when the regulations can work to your benefit. If a neighbor unreasonably objects to your keeping a flock under any conditions, you may find that local ordinances guarantee your doing so as a right, as my daughter, Heather, discovered. (See the sidebar "Heather's 'City Cousins' Flock.")

Whatever the regulations, relations with neighbors are among the most important keys to happy poultry husbandry; do everything you can to make sure your flock does not create problems for them. You will not make friends if your birds are free-ranging your neighbor's garden and prize rose beds. I've already mentioned omitting a crowing cock from the flock, but you may need to forgo geese and guineas as well—they love to vocalize. Notice I didn't say, "And make sure your poultry operation doesn't smell"—of *course* it's not going to smell, because you're going to practice the sensible manure management you'll learn about in chapter 8.

Sharing with neighbors an occasional dozen of the best eggs they've ever tasted will go a long way toward making sure they're on your side where the flock is concerned. Even more effective is initiating their children into the joys of poultry husbandry—feeding the chickens, gathering the eggs, or watching chicks grow in the brooder or with a mother hen.

Heather's "City Cousins" Flock

Don't simply assume that you're prohibited from keeping poultry where you live—check the ordinances. When my daughter, Heather—our resident flockster during the several years she lived with Ellen and me—moved with her mother to Greenville, North Carolina, she missed her chickens. So, on her own initiative, this thirteen-year-old marched herself down to city hall and asked at the help desk whether regulations allowed for keeping poultry inside city limits. The clerk, having never encountered the question before, had no clue. But she looked up the ordinance, reporting—to her own surprise—that Heather would be permitted to keep up to four domestic fowl, no specifications regarding species or gender.

Despite the latitude about species and gender, I advised Heather that—given the close presence of neighbors on either side of her postage-stamp lot—a low profile was best: no turkeys, no cocks. My father and I made a cage—something like a rabbit hutch, wire on light wooden framing, mounted at chest height on four legs. We placed it in an open-front garage, where it received sunlight but was protected from wind and rain.

We populated the cage with four bantam hens I had started for her. In addition to feeding them purchased layer feed, Heather made sure they got fresh greens every day as well, whether trimmings from the kitchen or grass cut from the lawn.

Those petite hens kept Heather and her mother supplied with all the eggs they needed in the remaining years before Heather left for college.

Choice of Species

Though *backyard flock* likely makes us think first of chickens, there are other species of domestic poultry that might fit your interests, limitations, needs, and goals. Where their management differs from that of chickens, I make specific references to the care of turkeys, guineas, ducks, and geese—and discuss those species in more detail in chapter 24.

Pheasants, peafowl, or pigeons might suit you as well, but I have no experience with them and cannot advise about their care.

It may be useful to know the average life spans of common domestic fowl:

Chicken: 7–8 years
Goose: Up to 80, with a possible record of 104, but an average of 20–22
Duck: 10 or less
Muscovy: 8–10
Turkey: 10
Guinea: 10–15

Standard or Bantam?

Most flocksters interested in a productive homestead or farm flock choose the standard or large breeds—they produce larger eggs and more meat. But the bantam or miniature breeds might be the best fit in some situations, especially if space is limited. One of my correspondents found that, while standard-sized chickens would trash her garden, her diminutive bantams were effective bug eaters while being "quite gentle" on the growing crops.

Bantams probably originated on the isle of Java. Brought to the West, they were used to develop the first standardized bantam breeds there: Nankins, Rosecombs, and a little later Sebrights—known as "true bantams" because they have no large-breed equivalents. Fanciers, enamored with the idea of miniature chickens, began crossing the bantam gene into standard breeds. Crossing the hybrid offspring back to the standard fowl, and selecting always for small size, yielded miniature versions of the standard breeds, one-fifth to one-fourth standard size but with other traits of the standard breeds intact.

Bantams lay small eggs (though larger in proportion to body size than those of standard breeds), but they can supply all the eggs the family needs. A flock of bantams, like standard breeds, requires ongoing culling. Although small, they make tasty additions to meals.

Factors Influencing Breed Choice

A *breed* is a related group of chickens all close to the same size, conformation, and carriage, and with shared comb style and (with a few exceptions) feather color and patterning. Originally, unique breeds tended to emerge in regions where farmers were selecting for the same production traits in the same climate and farming systems—and swapping the results of their breeding efforts. These days there is a bewildering array of breeds and varieties to choose from, though many variations are only skin deep, having emerged from exhibition breeding for feather style and color and comb shape, rather than production traits.

Breed choice has a great deal to do with goals for your flock—what you want to get out of the project. I assume that the reader of this book is looking for a breed that is naturally sturdy and robust, productive, and economical to keep. Such qualities are to my mind more easily found among the traditional breeds rather than the latest "superhybrids" (such as Cornish Cross among meat hybrids; and among layer hybrids, crosses onto White Leghorn, and various sex-link hybrids with such names as Red Star and Black Star). The following are some of the traits to consider as you contemplate breed choice for a homestead or farm flock.

Broodiness

Broodiness refers to the set of instinctive behaviors for hatching a clutch of eggs and nurturing chicks. Most modern breeds have "forgotten" how to brood, while hens of some of the traditional farm breeds and most of the historic breeds have a greater inclination to be mothers. Broodiness is a boon if you want to hatch your own stock using natural mothers, but an annoyance if you don't, or if you want to maximize egg production—a hen lays no eggs while incubating and raising her young. If you want both good egg production and the opportunity to hatch new stock with natural mothers, keeping a subflock of broody hens (such as Old English Game or Malay or, among bantam breeds, Nankin or Silkie) is a good compromise. See chapter 27 for more information on working with mother hens.

Comb and Feather Style

Among the many variations in chicken breeds, a couple that stand out are differences in comb structure and plumage—not only of color and pattern, but of texture and of feather placement on the body. Such variations may be today largely aesthetic points on which exhibition breeders compete, but they can have practical implications the homesteader might consider. For example, the most common comb, the *single* comb (like a vertical serrated fleshy blade attached to the top of the head), is more subject to frostbite in severe winters—thus, far northern flocksters might choose breeds with more compact *rose* combs or *pea* combs, or even the minimalist *cushion* comb of the Chantecler, bred for cold Canadian winters. On the other hand, large single combs help dissipate heat better in hot weather, so they might be a better choice for more southerly flocksters.

Placement of feathers can affect self-reliance. Feathers on the legs, when bred to extravagance, can inhibit efficient scratching, reducing intake of self-foraged foods.

Comb Styles

These drawings illustrate most of the comb styles you will see among the many breeds of chickens. Most of the breeds mentioned in this book are categorized by comb style in the following list. Some breeds have been bred with more than one comb style—for example, you may see Leghorns, Anconas, Minorcas, and Dorkings with either single comb or spiked rose comb.

Single comb: Australorp, Cochin, Delaware, Faverolle, Java, Jersey Giant, New Hampshire, Old English Game, Orpington, Plymouth Rock, Rhode Island Red, Sussex
Rose comb: Wyandotte
Spiked rose comb: Dominique, Hamburg
Pea comb: Ameraucana, Araucana, Asil, Brahma, Buckeye, Cornish, Shamo
Cushion comb: Chantecler
Buttercup comb: Sicilian Buttercup
V-comb: Crèvecoeur, Houdan, La Flèche
Strawberry comb: Malay

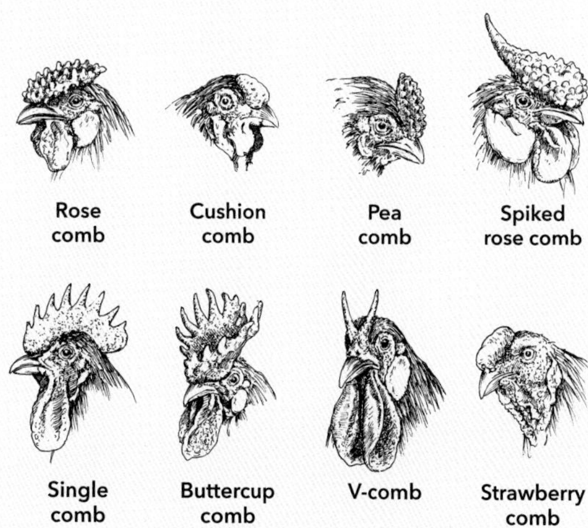

Figure 4.1. Comb styles. Illustration by Elayne Sears.

Some feather-legged breeds such as Brahmas, however, retain their ability to forage—indeed, the Faverolle, a utilitarian feather-legged breed developed in France, will lay and grow much better if given free access to the outdoors but will languish and lay poorly in confinement. The extravagant headdress on crested breeds such as the Polish, Crèvecoeur, and Houdan limits vision, making them more vulnerable to predation.

Even feather color can be a practical issue. All-white plumage stands out more than any other color pattern to the eyes of predators, so all-white breeds would be a poor choice if you want to free-range your birds. The barred gray pattern is said to be the best camouflage, followed by the black-breasted red pattern. (See "Breeds We Have Raised," on page 39.)

Eggshell Color

Most hens lay either white eggs or brown eggs. That the latter are nutritionally superior to the former is one of those facts that "everybody knows" but which is in fact not a fact. The misunderstanding probably arises from the *fact* that supermarket eggs (most of which are produced by white-egg layers) are notably inferior to farm eggs (more often produced using brown-egg breeds). However, there is no *inherent* difference: If layers of white and brown eggs are managed and fed the same, there is no difference nutritionally between their eggs.

Figure 4.2. Eggs of Black Copper Marans and Pale Sky Easter Eggers.

If you have a preference as to eggshell color, you can make your choice free of nutritional guilt.

If you get carried away with a preference for brown eggs, you might want to try one of the breeds known for extremely dark, almost chocolate-brown eggs—more commonly the Marans and Welsummer, more rarely the Barnevelder and Penedesenca. I raised Cuckoo Marans for years. They never laid chocolate-brown eggs, but many were extravagantly speckled—a nice look as well. (See figure 29.1.)

Incidentally, I said *most hens* lay white or brown eggs. Ameraucanas, and a few other breeds with genes from Araucanas, lay eggs that are tinted pastel—greens, blues, occasionally even pink. The colors are never as emphatic as the pictures in the catalogs touting "Easter Egg Chickens!" would have you believe. Still, the novelty of the rainbow effect appeals to some—especially children.

As I write, I have two each of Black Copper Marans hens, layers of the darkest of all brown eggs, and of Pale Sky Easter Eggers, one of several hybrids bred specifically for pastel-tinted eggs (see the eggs in figure 4.2).

Some breeds that lay eggs with greater color appeal tend not to lay as well as more traditional farm breeds—the visual special effects may come at some cost to production. But that is one of the joys of keeping your own chickens—*you* get to decide what is just the right mix of eye appeal and daily payoff of eggs.

Temperament

Some breeds (the lighter-bodied layer breeds as a group, such as Leghorns and Hamburgs) are more flighty or excessively nervous; others, more docile and laid back (such as the Cochin, Buff Orpington, and Dorking). Cocks of some breeds, such as Rhode Island Red, are said to be especially aggressive. You might therefore consider reputed temperament before settling on a breed. Be aware, however, that there is probably no more "your mileage may vary" trait than temperament; your management style and treatment of your birds have far more to do with temperament than innate disposition. If you are calm and respectful around them, they will likely be so in return.

Winter Production

Egg laying declines dramatically in winter—indeed, some breeds cease production altogether. Our Cuckoo Marans hens, for example, laid only an egg per week or so in winter, while the Old English Games quit laying entirely. (Remember that it is not natural for chickens to lay eggs in winter at all, so we flocksters should be grateful for what we do get.)

Some breeds hold production better in winter than others—though as in most things chicken, the source of your *strain* (based on traits selected for by a specific breeder) may have more influence on rate of lay in winter than *breed*. But for planning purposes, refer online to "the Henderson chart," an attempt to summarize numerous traits of dozens of chicken breeds—from standard breed sizes to temperament to expected annual egg production, with notations for breeds that lay better in winter.[2] Breeds we have included in our flocks specifically to keep egg production up in winter include Partridge Chantecler, Delaware, New Hampshire, Rhode Island Red, and Wyandotte. Reliable winter laying was a major factor in my choice of Icelandics, as related in the next chapter.

Choosing Your Breed(s)

There is no "right" or "better" breed. While the fit of a given breed with your goals and needs will largely determine your choice, intangible factors will influence you as well—how much you like the look of a given breed, or its temperament. Your project will be more successful if you truly enjoy it, so choose the breed that speaks to you.[3] If you respond to a beautiful, more mellow breed, such as Brahma, in preference to a breed that is twice as productive of eggs, such as the flightier Leghorn, there is nothing illogical about your choice.

Note that, in the age of the computer, many hatcheries let you order chicks individually—it would be easy to put together an order for a kaleidoscopic array of breeds. And remember, if you're not pleased with the breed you start with, it will be easy to switch to another when it's time to rotate stock.

A couple of breed organizations are especially useful for putting you in touch with breeders of poultry stock. The Society for the Preservation of Poultry Antiquities (SPPA) is dedicated to perpetuating and improving rare breeds of poultry. Their well-organized *Breeders Directory* helps members locate stock from fellow breeders.

The Livestock Conservancy (TLC; formerly American Livestock Breeds Conservancy) is dedicated to the proposition that traditional breeds will be conserved not through heroic efforts to keep them alive essentially as zoo specimens or genetic libraries, but through growing them for their *economic* (production) traits. Only when homesteaders and small farmers appreciate the mealtime virtues of traditional and historic breeds will they seriously commit to their conservation. Check TLC's list of breeds (of all farm livestock, not just poultry) in danger of extinction: You can help with the effort to conserve vanishing breeds, either by breeding heritage breeds yourself or by buying your stock from breeders who do.

One of the most important questions is whether your emphasis will be on egg production only or putting dressed poultry on the table—or both. But remember that you'll want to integrate your flock with the total work of small-scale or homestead food production—what this book is all about—so I recommend one of the sturdier, more self-reliant breeds typical of small farms in earlier eras.

In the following overview of breed types, I will ignore the fancy types—of interest mostly as ornament or in the show ring—and focus on breeds likely to be of more use in small-scale food production projects. If you are interested in learning more about traditional breeds, there are some great resources online, including the website for Glenn Drowns's Sand Hill Preservation Center.[4]

Laying Breeds

Most of the traditional breeds developed for high egg production are small, putting more resource into laying than growing large frames and muscle mass. They

Some Useful Traditional Breeds for a Small-Scale Flock

There are hundreds of breeds of chickens from which to choose. This list highlights some that are useful as sturdy, productive breeds on homesteads and small farms, categorized by type. Inevitably, categorizing certain breeds—Ameraucana and Australorp, for example—as either layer or dual purpose is somewhat arbitrary. This list is by no means definitive—I have omitted the more ornamental breeds, such as Buttercup—and you may well discover breeds that strike your fancy not listed here. But checking out these breeds should be a good beginning toward choosing a breed or breeds with which to start a small-scale home or market flock.

Note that I have included the category "game" breeds—that is, breeds that have been used for cockfighting. I include them not to encourage the "sport" of fighting cocks—indeed, I strongly oppose it—but because the game breeds have made important contributions to the development of modern breeds and should be conserved; because some breeders prize them for their beauty and spirit; and because the hens are almost certain to be superb mothers.

Layer Breeds

- Ameraucana
- Ancona
- Barnevelder
- Hamburg
- Leghorn (White Leghorn most common, many variants)
- Minorca
- Welsumer or Welsummer

Meat Breeds

- Brahma
- Cornish or Indian Game
- Naked Neck or Transylvanian Naked Neck or Turken

Dual-Purpose Breeds

- Australorp
- Buckeye
- Chantecler
- Dominique
- Dorking
- Faverolle
- Houdan
- Jersey Giant
- Marans (Cuckoo Marans most common)
- New Hampshire
- Orpington
- Plymouth Rock (including White, Barred, and Partridge varieties)
- Rhode Island Red
- Sussex
- Wyandotte (several color varieties)

Game and Other Broody Breeds

- Aseel or Asil
- Cochin
- Malay
- Modern Game
- Nankin (bantam)
- Old English Game
- Shamo
- Silkie (bantam)

lay white eggs; come into lay early; tend to be flighty around people; and almost never go broody. Ancona, Hamburg, and Minorca are among the easier to find of the traditional laying breeds. The queen of this group is the Leghorn. Innumerable white strains of this breed have been developed for use in the egg industry, but traditional strains—active and hardy in addition to being prolific layers—are readily available. Leghorns also come in many colors other than white—Light Brown Leghorns, both male and female, are among the most beautiful chickens that ever graced a homestead—and with either single or rose combs.

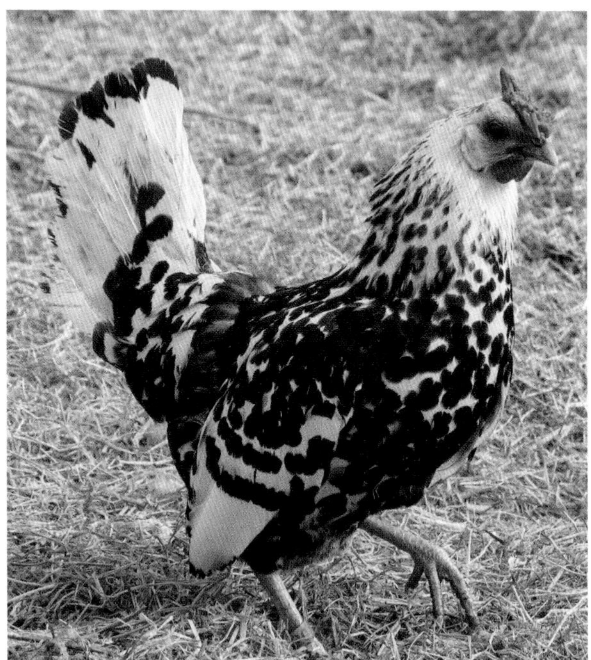

Figure 4.3. This Silver Spangled Hamburg hen is an example of the smaller-bodied traditional egg breeds.

When selecting a layer breed, remember that the production figures you see are widely variable breed averages, suitable as a rough guide only. It is said that an Australorp hen holds the world laying record—364 eggs in 365 days! But the average flock of Australorps will not lay as many eggs as the average Leghorn flock—not by a long shot.

Meat Breeds

A few traditional breeds were developed primarily as table fowl: Jersey Giants (White and Black) in America; Cornish in England; and Brahma, Cochin, and Shamo in Asia. These days, though, such breeds have been supplanted by fast-growing hybrids, foremost among them the Cornish Cross, currently the foundation of the broiler industry, grown worldwide at an estimated sixty-six billion or more per year. A fundamental question is whether *meat chicken* at your house will mean fast-growing broiler strains, grown in batches to fill the freezer, or the product of routine culling of a dual-purpose flock, discussed next.

Dual-Purpose Breeds

The most typical farm chickens in the past were breeds now called *dual purpose*. While such breeds do not lay as well as the egg-specialist breeds, nor grow as fast as meat hybrids, they serve as the best compromise of the two: reliable layers that grow fast enough to a generous size to make good table fowl. They tend to be more tranquil than the flightier egg-specialist breeds—a bit easier on the nerves. Most breeds in this group lay brown eggs; in some, occasional hens may retain the broody trait and make good mothers. These breeds are likely to be your best bet for a small-scale flock, and there are many to choose from: Buckeye, Delaware, Dominique, New Hampshire, Plymouth Rock, Rhode Island Red, Wyandotte, and more from America; Chantecler from Canada; Faverolle, Houdan, Orpington, and Sussex from Europe; Australorp from Australia. Many of the traditional dual-purpose breeds are seriously in need of conservation breeding.

Breeds We Have Raised

The following are some of the breeds we have worked with during close to four decades to make up our typical mixed flock of adult layers, young growing replacement birds that are continually culled as table fowl, and a working subflock of broody hens:

New Hampshire: Our first flock of chickens were New Hampshires, developed out of Rhode Island Reds by Andrew Christie in the 1920s. Christie bred for a superior meat bird, ensuring their hardiness by keeping his breeders in pasture shelters through his harsh New England winters. His success is attested by the fact that the New Hampshire was at one time the dominant broiler in the poultry industry, and that numerous broiler crosses were made with the New Hampshire as half of the mix. We took care to locate the Newcomer strain, developed by Clarence Newcomer in the 1940s for better egg production. (I'm not sure it's even possible to find the original Christie strain.) Our New Hampshires were vigorous and hardy, laid plenty of large brown eggs, and were fine table birds. Some of the hens went broody and proved to be excellent mothers.

If You Want to Grow a Meat-Bird Hybrid . . .

The Cornish Cross—developed out of crosses between traditional Cornish, massive and broad-breasted but slow growing, and White Plymouth Rock for faster growth—has become the foundation of the commercial broiler industry and is often the meat bird of choice in local pastured broiler markets as well. Even many homestead flocksters choose this hybrid for its astoundingly fast rate of growth—to as much as 5 pounds (2.3 kg) dressed weight in eight weeks, some strains reaching slaughter size in as little as forty-four days. But that seemingly miraculous growth comes at a cost: Development of muscle tissue outstrips all other systems, and the Cornish Cross suffers leg ailments, low vitality, and heart problems. It is as well exceptionally lazy as a forager, not to say stupid, taking only limited advantage of self-feeding opportunities on pasture. My buddy Mike Focazio told me he saw one of his Cornish Cross broilers tentatively peck up an earthworm—and then spit it out!

In recent years hybrid meat strains have been developed that are hardier and more resilient than the Cornish Cross, and much better suited to taking advantage of natural forages if allowed to range. They require a longer grow-out—an additional two to four weeks—but are for that very reason superior in flavor to the Cornish Cross. Some such strains come out of the *Label Rouge* system of range-produced poultry in France, while others are hybrid crosses made in the United States—all are to my mind a better choice if you want to raise a meat hybrid for your freezer. While some have fanciful names such as Freedom Ranger, Red Ranger, Royal Broiler, and Silver Cross, the important things

Figure 4.4. Cornish Cross chicks.

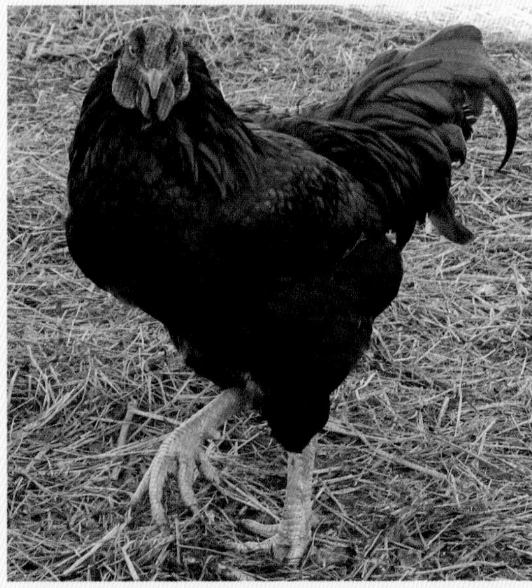

Figure 4.5. Dark Cornish.

Figure 4.6. Freedom Ranger broilers, typical of hybrids bred for the French *Label Rouge* system of certified free-range broiler production

to key on when selecting a broiler strain are the more moderate rate of growth and the adaptation to ranging.

I raised Cornish Cross for several years to produce plump roasters for the table. But in early summer one year, when I had a large batch on our pasture, we had a sudden, unseasonable heat spike. When I arrived on the pasture, I found many of my Cornish Cross, right at slaughter size, prostrate with heat exhaustion—I lost twenty-two of them all told. They had sat on their butts in the shade of their shelter and *died* rather than walking 10 feet (3 m) for a drink of water! I turned 180 degrees to another section of pasture on which I was ranging a group of New Hampshires, the same age as the Cornish Cross to the day: They were scooting about like water bugs, running across their entire enclosure when they wanted a drink of water. It was hard to imagine them the same *species* as their sadly compromised cousins nearby. From that time, I never raised another batch of Cornish Cross.

Rhode Island Reds: Both males and females of this breed are a rich, lustrous red, darker than the New Hampshire. They are among the best layers of all brown-egg breeds and an excellent dual-purpose breed. This breed has a reputation for being more aggressive, especially the cocks. Both New Hampshire and Rhode Island Reds lay well in winter.

Barred Plymouth Rock: A quintessential rock-solid American farm breed, they have sharply defined parallel bars of dark gray and lighter color (not quite white). These are among the best of dual-purpose breeds.

Cuckoo Marans: The Marans—developed as a dual-purpose breed about a hundred years ago in the town of Marans, on the western coast of France—is known for extremely dark brown eggs (see figures 4.2 and 29.1). The Cuckoo color variant has much the same look as the Barred Plymouth Rock, though its barred pattern is not as sharply etched. We found them an excellent table fowl, though their winter egg production was poor. Some of our Marans hens were excellent mothers.

Welsummer: Another dark-egg breed is the Welsummer, bred in the village of Welsum in the Netherlands in the early years of the twentieth century. In our experience they were not as large nor as vigorous as the Marans. The color pattern is black-breasted red—the one most typically seen in "cock at dawn" illustrations and photographs. (The male has a flaming red hackle and saddle feathers, with a black breast and tail. See figure 4.8 as an example of this pattern in both cock and hen.)

Delaware: This was one of the crosses of the New Hampshire (with Barred Plymouth Rock) when it was a mainstay of the broiler industry. It is a handsome bird, mostly white, marked with black barring in the hackle and tail feathers (see figure 4.9). Though the Delaware was bred to be a meaty chicken and a good layer of large eggs, it is hard these days to find stock that lives up to its full potential. (For more about Delawares, including a photo of cock and hen, see "Bad News, Good News" on page 235 and figure 25.1.)

Figure 4.7. Trio of Cuckoo Marans. Note the auto-sexing colors: The cock is noticeably lighter than the hens.

Partridge Chantecler: This was bred in Canada in the early twentieth century as an exceptionally cold-hardy dual-purpose fowl (due in part to its minimalist comb and wattles, less subject to frostbite). Though not champion layers, we sometimes kept a few in the flock because they hold their egg production well in winter. Most of our Chantecler hens went broody and were competent mothers.

Light Brahma and Buff Cochin: Both these Asian breeds are large and make good table fowl, though they are slow to mature. Neither is a champion layer, though both tend to go broody and make excellent mothers. I prefer clean-legged chickens (both these breeds are fully feather-footed), but if you're looking for a breed that is both beautiful and among the most docile of all chickens, either could be a fine choice for you.

Buff Orpington: A heavy general-purpose fowl, this is another extremely mellow breed for those seeking docile backyard companions. Though average in egg production (of larger-than-average eggs), they hold their production well in winter. Orpington hens incline toward broodiness more than those of most breeds and are excellent mothers.

Silver Spangled Hamburg: An ancient breed originating in Holland, this is one of only a couple of "egg specialist" breeds we have raised. Though too small to be considered a meat breed, they are visually striking (see figure 4.3) and prolific layers (though the eggs are small). It is a breed that was in earlier times known as "thrifty," both because they are economical eaters and because they forage well.

Dark Brown Leghorns and Light Brown Leghorns: For several years I kept a few individuals of these non-industrial Leghorns. If you are looking for handsome birds that will reliably keep egg production up, either would be a good choice.

Australorp: These are striking coal-black birds with bright red single combs that were developed out of Orpington in Australia in the early twentieth century (see figure 4.9). Though a bit small for a dual-purpose

Figure 4.8. Old English Games. Note how similar their black-breasted red pattern is to that of the Red Junglefowl in figure 3.1. Photographs courtesy of Bonnie Long.

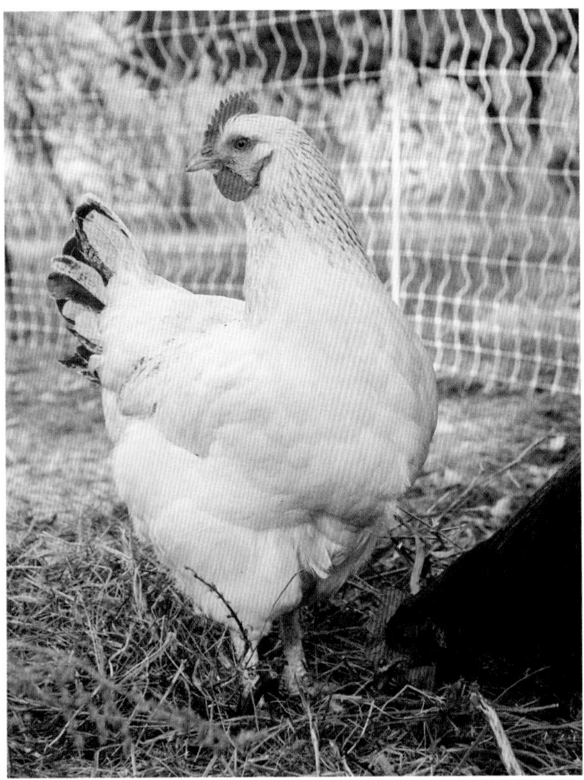

Figure 4.9. Delaware and Black Australorp hens. Photo courtesy of Bonnie Long.

breed, this breed's egg production is excellent. We have a friend whose Australorp hen successfully hatched a clutch of eggs she laid in a bucket of rusty nails!

Wyandottes: This is another quintessentially American dual-purpose farm breed, which originated in New York State in the 1870s. We've kept several of the color variations available, especially the Silver Laced. Consider Wyandottes if a docile breed with a rose comb (less subject to frostbite) and good winter egg production appeals to you.

Dorking: If you want a bit of history in your backyard, consider keeping Dorkings, brought to England by the Romans in Julius Caesar's invasion, more than two thousand years ago. Dorkings are short-legged, and males have quite large single combs. (There are rose-combed varieties as well.) They are scarcely champion layers—though friends of ours had a Dorking hen who at six years laid more eggs than anyone else in the flock. They were valued in England in earlier times as an excellent meat breed. Dorkings are among a few breeds distinguished by having five toes.

Old English Game (OEG): If I knew that starting tomorrow I could purchase neither feeds nor chicks from outside sources and could choose only one breed of chicken for my flock, I would take the Old English Game (see figure 4.8). Like Dorkings, the OEG has a long lineage, going back a thousand years. Though much used during that time as "game" chickens for cockfighting, they have as well been valued as a utilitarian farm breed. While hopelessly unproductive by modern standards for rate of growth and egg production, they can feed themselves if given enough ground on which to range. Though small, they are surprisingly plump under their feathers, and rich in flavor—at one time in England they were considered the standard against which all other meat fowl were judged. The hens are among the best chicken mothers on the planet and for many years were the backbone of my working broody subflock. Despite their aggressiveness toward one another, I found the cocks friendly toward both my visitors and me. Old English Games have contributed their vigor, hardiness, and longevity to innumerable modern breeds.

Hmmm, On Second Thought . . .

At the time of the first publication of this book, I made the previous "if you were on a desert island" statement—trying to fit in *all* the traits that would be essential in a breed. Many would consider my choice an odd one—the Old English Game. But the thought experiment included the supposition that I would have *no* access to replacement chicks from elsewhere, nor to artificial incubators, and that my flock would have to feed themselves *entirely* by foraging. In that light, the Old English looked to be the best candidate, despite its shortcomings (chief among them the complete absence of winter laying).

But I subsequently discovered a "does it all" breed that came closest to being my hypothetical entirely self-sufficient flock. I'll tell you about them in the next chapter.

CHAPTER 5

My Seven-Year Love Affair with Icelandics

Choosing a breed eventually comes down to: Which breed is most likely to fulfill your goals for keeping a flock of chickens? If your main interest is eggs, you want a breed that will reliably produce plenty of eggs, year-round. If it is Sunday dinner chicken, you want a breed that will grow quickly to good slaughter size. If ideas mentioned in chapter 2 about putting the flock to work appeal to you, seek out an active, go-getter breed. If you want to breed your own replacement stock, you need a breed whose hens reliably go broody and make good mothers. Even if your choice is driven by aesthetics—you want a breed that will look good in the backyard—that is a perfectly sound basis for deciding on a breed. But suppose my preference is for a breed that *does it all.* Eggs, baked and braised chicken, hearty broth, partners to help get the garden work done, hens who hatch their own chicks, and eye-catching color and style to boot—*why can't I have it all?*

Figure 5.1. A hardy flock of Icelandics exhibits a kaleidoscopic array of plumage styles. Photo courtesy of Lisa Richards.

When I've imagined the ideal flock that "does it all," I've thought first of my grandmother's flock. Ranging her fields and pastures, they fed themselves almost entirely from foraged natural foods. (The occasional handful of grain she threw her flock served not as the mainstay of their diet, but just a means of keeping them fixed on the coop as the place to come home at night.) Granny never bought in batches of chicks through the mail. Occasionally one of her hens would disappear, then show up three weeks later with a clutch of chicks in tow. Think about it: Cost of replacement stock, *zero.* Cost of scratch grains, *a few dollars a year.* Granny

Figure 5.2. Icelandics, the "do it all" flock.

Figure 5.3. An Icelandic flock is highly variable for feather color and pattern and for comb styles.

kept that self-feeding, self-replicating flock going for decades—every egg, every pot of chicken 'n' dumplings, on her family's table essentially *for free*.

I said in the previous chapter that, if I had to select *one* breed that comes closest to doing everything I want a flock of chickens to do, I would choose Old English Game. My experience with them convinced me they would be nearly self-sufficient in terms of feed if given enough foraging space. And I knew I could count on the hens as devoted and protective mothers. I had to accept, of course, that focusing on those virtues meant overlooking a couple of big shortcomings of the OEG. Egg production declines among *all* breeds during the dark days of winter. But OEG hens quit laying *entirely*—for a full three months or even more. And yes, the hens go broody—at 100 percent. Though that was a great thing for *Gallus gallus* in the wild, it creates problems in a domesticated flock. If you only need 20 percent of your hens on "broody duty" hatching and rearing chicks, believe me, fighting broodiness in the other 80 percent of an all-OEG flock would be a major pain. That's why I always thought of my OEG hens as a working *subflock*, and only suggested adopting OEGs as my sole breed as a thought experiment.

Then I learned about Icelandics.

The Story of Icelandic Chickens

As with Old English Games, the history of Icelandics goes back more than a thousand years: When the Norse settled Iceland in the tenth century, they brought along their farmstead chickens—even today known in Iceland as *Íslenska landnámshænan*, "Icelandic chicken of the settlers." In the following centuries those chickens were selected to sustain themselves almost exclusively by foraging their own natural feeds on homesteads and small farms—raising grain to feed poultry was not an option. In the absence of artificial incubators, hens were selected for their mothering skills. In short, they were selected as chickens who could "do it all," including continuing to lay during the long, dark winter, when the egg supply was especially precious. As with all livestock that feed themselves largely by foraging, the resulting strain was on the small side (mature cocks weigh 4½ to 5¼ pounds [2 to 2.4 kg]; hens, 3 to 3½ pounds [1.4 to 1.6 kg]), but the culls of excess males and of old females were a vital part of the family's fare.

Icelandics do not have a standard "look"—for comb style, shank color, feather color and pattern. Properly speaking, they are not a *breed* but a *landrace*, a strain selected in a geographic region with a characteristic climate and ecological conditions, for a particular suite of utilitarian traits (though with little thought to standardizing the "look"). A major bonus to keeping Icelandics is thus that visually they are strikingly variable, with individuals displaying both single and rose combs; every color and pattern in chickendom; and variable shank colors. Some Icelandics, both male and female, sport an eye-catching crest (see figure 5.7), while the heads of others are smoothly feathered. If I'm going to have only one kind of chicken in my backyard, why settle for a single look—why not a visual kaleidoscope! (See figures 5.2 and 5.3.)

For a millennium, the only chickens in Iceland were of this robust landrace. But industrial breeds such as Leghorns were imported in the 1930s to boost production in larger-scale operations. Inevitably the imports were crossed with the natives, with their extinction as a unique regional strain a growing threat. Conservationists in the 1970s sought out farms with true native stock and founded a national organization dedicated to maintaining their purity. Eventually, traditional Icelandic stock was imported into other countries, including four importations, early in this century, into the United States.

Starting My Icelandic Flock

Finding Icelandic stock was difficult, since neither hatcheries nor breeders in TLC and the SPPA offered them. After many fruitless searches came to dead ends, in early 2013 I found my way to an online group dedicated to supporting the true Icelandic strain in the United States. Two experienced breeders (one of whom had made two of the four importations from Iceland) furnished my starter stock of seven breeders: four pullets and three cockerels, four to six months old, ready to start breeding almost immediately. By the time proven broodies in my existing flock settled in the nest, I had plenty of Icelandic hatching eggs, fertilized and ready to incubate. Two of my Icie pullets subsequently went broody as well and proved to be excellent mothers. (See chapter 27.)

After that first hatching season, I phased out the last of the hens from my previous mixed heritage-breed flock and for the following six years kept Icelandics exclusively.

Homestead Partners

I found that Icelandics came closer to my ideal do-it-all flock than any of the dozens of breeds I had raised in the past—certainly they were among the hardest working, doing real-world work that would otherwise have cost me significant time and effort. Instead of laboriously turning my compost heaps myself, I unleashed my Icies onto them and let them do the turning. (In their native land they are also called *haughænsni* or "pile chickens" because of their preference for such debris heaps.) In our bit of woodlot, which includes young nut trees hungry for fertility, I didn't even have to haul away the finished compost: I spread under the trees all the compostables that accumulate on any working homestead—autumn leaves, spent crop plants, weeds, even felled "weed trees" such as ailanthus (worthless as firewood)—plus truckloads of woodchips from power line clearing operations, and I watched with satisfaction as the flock shredded the material in their pursuit of grubs and other detritivores. When they were on pasture, I typically rotated them to prevent damage to the sod, of course (as I describe in chapter 11); but, had I needed to till up new ground for gardening, I would first have laid out such "debris fields" (like the one in figure 5.6) on that ground and let my flock kick off the process by killing the existing sod while tilling in a goodly application of finished compost.

And speaking of pasturing the flock: I used electric net fencing (see chapter 11) to keep them inside large ranging areas. Though I have read complaints from some Icie owners that this breed cannot be contained inside electric net, that was not my experience. So long as my birds had enough range to "keep them busy"—with lots of available natural forage—they had little inclination to fly over the fence. Clipping a wing (see the sidebar "Clipping Wings" on page 103) sufficed to keep the occasional rogue flier safe inside the perimeter.

I found consistently that my feed costs dropped by half when I released the flock from winter quarters onto the ranging areas I've described.

Egg production never matched that of egg-laying champs such as Leghorns, Minorcas, and Rhode Island

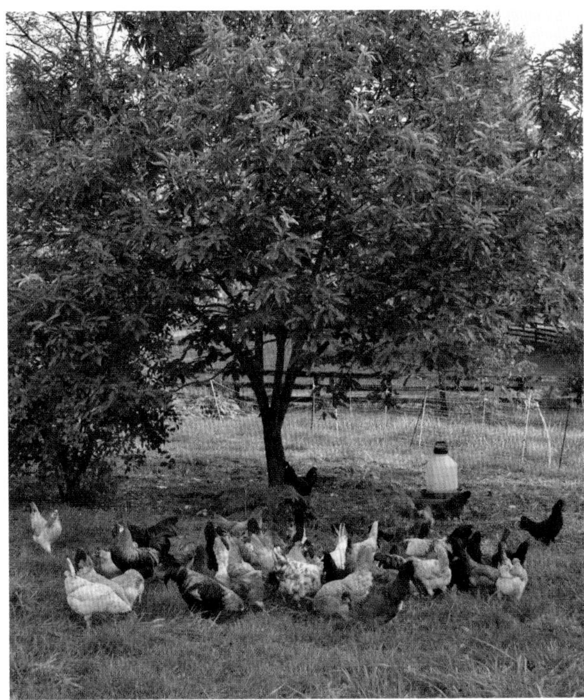

Figure 5.4. Flock of Icelandics support a young chestnut tree: They love working debris piles (such as the one near the trunk), turning them into compost to feed the tree as they feed themselves on crickets, grubs, and the like. As well, they break the life cycle of the chestnut weevil, by finding and eating the larvae as they burrow into the ground in the fall, or the adult weevils as they emerge in the spring to mate and lay eggs.

Figure 5.5. Right out of the nest, Icie chicks learn from Mama how to find nutrient-dense feeds in decaying debris heaps.

Reds, but was impressive for such a self-sufficient flock. Especially gratifying was their winter egg production. It decreased of course in October and November, at the time of the molt, but even then they gave us all the eggs we needed. Interestingly, their rate of lay seemed to rise and fall according to the condition of the ground in their winter range—higher when the soil was unfrozen and they could scratch up worms and grubs, and dropping when it was frozen. I learned to plan ahead: Laying down a debris field in the fall deep enough to prevent freezing helped keep winter production up.

Icelandic eggs are white to cream and small, though surprisingly large for such small hens. (Egg weight from my older hens averaged 1.75 ounce [50 g], same as for commercial eggs graded Medium.)

My Icies were healthy, hardy, and robust, not only during winter as I expected, but as well in our hot, humid mid-Atlantic summers. Losses to illness or the sort of unexplained death I call JCOS (Just Crapped Out Syndrome) were rare.

Breeding My Icelandic Flock

Breeding my own replacement stock was not new to me, but in my Icelandics conservation project I intended to breed with a lot more rigor, especially regarding maintenance of genetic diversity within the constraints of relatively small sets of breeders. I settled on a breeding system called clan mating as the best fit for me. From the beginning I separated my seven starter breeders into three "clans" or "families"—just a pair each in two of the three, and a trio in the third. (For more on clan mating and its specific application in breeding my Icelandics, see chapters 25 and 26.)

Figure 5.6. Icelandic flock working a "debris field" of wood chips.

In that first hatching season my hens hatched seventy-two healthy chicks; in subsequent years up to eighty-five. No, I wasn't growing freezer-filling batches of "meat birds," but I think of *any* bird needing to be culled as *meat chicken*—and, as I'll emphasize in chapter 25, conscientious breeding requires a *lot* of culling. Understand that Icelandic carcasses are small—for example, seven-month-old cockerels dressed out at an average 2¼ pounds (1 kg)—but the flesh is fine-grained and unusually flavorful, even in comparison with most other breeds we had previously raised on range.

Given the number of males in the flock (typically six—two each in the three clans) I was especially pleased to find that Icelandic cocks are not overly aggressive. They spar to establish dominance—completely normal behavior—but are not as inclined to "take it to the limit" as Old English Game cocks.

Figure 5.7. This broody Icelandic hen sports a crest of feathers atop the head, yet another plumage variation of this striking landrace.

If Icelandics Sound Good to You . . .

If you're interested in a fascinating breed conservation project—or just a colorful flock that lights up the backyard—consider Icelandics. But remember they are active birds, and be prepared to give them a lot of space in their coop and plenty of natural range. A tightly confined Icie flock would be miserable, and you would not enjoy keeping them at all.

No commercial hatchery I know of offers Icies. TLC's online directory lists some dedicated conservationist breeders. The Facebook group "Icelandic Chickens (*landnámshænan*) Official Preservation Organization" has a list of breeders willing to share hatching eggs or chicks. Glenn Drowns includes Icelandics in his astounding collection of heritage breeds. His description of Icies parallels my own experience: "They are superb foragers with a most pleasant temperament. This is perhaps the ideal breed for someone who wants a diversity of color, but only wants one breed. They are not a terribly large chicken but are extremely feed efficient." See the Resources section for more information and website listings for these organizations.

Be extremely cautious whatever your source if you want pure Icies: In this variable landrace, crossing in from non-Icelandic stock, accidental or otherwise, might well not be obvious in the progeny. Quiz potential sources in detail about the source of their stock and their breeding practices—you're looking for someone who is *fanatic* about the purity of this unique landrace.

Yes, Icies are broadly variable, but it is useful to remember a few absolutes as indications of outcrossing: Though shank color differs, true Icelandics will never have feathering of the leg. Face patches range from white to light lemon yellow but will never be solid red. Some birds in an Icie flock will have a crest of feathers atop the head (though never as extravagant as in Polish or Crèvecoeur), but true Icies will never show a "beard" or "muffs" (elongated feathers at the sides of the face and between beak and top of the breast, as in Ameraucana or Faverolle) nor ear tufts (as in true Araucana).

The most significant difference between Icelandic and OEG hens involved brooding and mothering skills. In my experience, among the latter incubation performance at 100 percent can simply be assumed, and the hens are without exception devoted and fiercely protective mothers. As noted, some of my Icie hens did not go broody, and this was a good thing. About the right number continued laying with nary a thought of motherhood—but enough did go broody to ensure all the hatching required in my breeding project. There was a wide and unexpected variability in performance, however, both for incubation and for nurturing the chicks. The very best Icie broodies matched OEG hens in performance, while some were only "good enough"—and some earned a trip to the stewpot. (See chapter 27 for more about selecting for improving the broody trait.)

CHAPTER 6

Starting Your Flock

Which came first—the chicken or the egg? Most beginning flocksters prefer to start with the chicken—either just-hatched chicks, started birds just out of the brooder up to onset of lay, or adult birds—though a few go-getters might prefer to start by hatching eggs in an incubator. I have never used an artificial incubator, but my friend Don Schrider shares his experience in appendix H.

Sources of Stock

There are many sources of stock for your flock, each with its own advantages and disadvantages. Whichever you choose, try to find stock that is true to its origins, with emphasis on the breed's economic (production) qualities. For example, almost any hatchery will offer "New Hampshire Reds," but most of that stock should more properly be labeled "Production Reds," for they are seldom true to Andrew Christie's rugged original, nor to the Newcomer strain (as described in chapter 4). When I bought my first New Hampshire chicks, I followed a tip from a friend: The hatchery he suggested offered run-of-the-mill "New Hampshires" in its catalog, but one of the hatchery's employees was still maintaining a breeding flock of Newcomer strain and I was able to special-order.

Local Sources

Check the classified ads in local newspapers, check bulletin boards at your post office or farm co-op, and ask friends and relatives about local flock owners near you with good stock they're willing to sell. County fairs often have poultry on display—talk with the exhibitors. Find out from leaders of the local 4-H club who has poultry projects. Vendors at farmer's markets who sell eggs might sell live birds as well or might be in a network of local small producers who do.

All such purchases via direct contact have the advantage that you can see the birds themselves, and ask their owners about their flock goals and management. Inspect the birds by hand to be sure they are well fleshed, and check under the feathers for lice and mites (tiny creepie-crawlies), especially on the skin around the vent. Even if you are not (yet) an expert, it will be obvious whether the birds have the bright eye of health, or are dull-eyed and listless.

Most farm co-ops offer poultry stock, typically just-hatched chicks in the spring and ready-to-lay pullets later in the season. These are likely to be truly mass-market birds, sometimes of mediocre quality, and you can be sure the staff knows nothing about their breeding. Selection is usually limited, more typically highly hybridized layers (though sometimes a couple of the more common traditional breeds as well).

My First Batch of Chicks

Long before Ellen and I married and moved to Boxwood, I lived for a while out in the sticks by myself, mostly off the land. One day on a rare town trip I heard an advertisement on my radio for "twenty-five free chicks" at the local farm co-op. When I stopped to check it out the offer really was: Buy a bag of chick feed and get the chicks free. When I said to the clerk that seemed like quite a deal, he laughed and said, "They're excess layer cockerels—won't be much meat on their bones!" But the kid in me recognized a fun project—I paid for the feed and grabbed my box of free chicks.

I had zero experience brooding chicks, but it turned out to be easy enough: I just kept them in a closet, with a lightbulb for warmth. When they were feathered, I set them up in a spacious part of a barn on the place, over a deep mulch of waste hay. They were fine companionship for a lonely hermit.

As they grew, I found the clerk had been right—they weren't especially meaty. But all the same I began putting them on the table, a welcome alternative to the groundhogs from the surrounding fields and bluegills from a nearby pond that had been gracing my table. (Hmm, I even remember eating a possum who proved a bit too slow.)

Not long afterward an acquaintance paid me an unexpected visit. We took a walk through lovely hay fields, shimmering in the breeze the way they do when just ready to cut. I asked her if she'd like to stay for supper. I guess I'd been off to myself way too long: When she accepted, I nonchalantly picked up my hatchet, walked out to the barn with my friend, and proceeded to make dinner. From scratch.

When I saw her next, she admitted diffidently she had never accepted a dinner invitation that turned out quite like mine.

Breeders

I referred in the previous chapter to two organizations dedicated to conservation of traditional breeds, the SPPA and TLC. Membership in either will put you in touch with breeders of most breeds you are likely to be interested in.

Be especially careful buying stock from someone breeding for competitive exhibition. To be sure, some breeders who show their poultry are dedicated to preservation of their production traits and would be fine sources of stock. For example, I got my starter stock for Old English Games from a fellow member of SPPA dedicated to conservation of Old English as a utility farm fowl (rather than as a fighting breed)—selecting for a more plump, rounded body style (not the leaner body preferred for the fighting pit); for enhanced egg production; and for retention of the broody instinct, for which this breed is noted.

For many breeders, however, the birds they show are essentially works of art sketched in DNA, *not* productive participants in the home economy. I have met show breeders who have no compunction about mixing in "a little of this and a little of that" to fine-tune the plumage or conformation of their birds, heedless of any loss of production capabilities or possibility they are creating a genetic cesspool. It shouldn't be a surprise when such breeders produce the proverbial grand champion show hen who produces only a couple of dozen eggs per year.

Hatcheries

It is possible to order just-hatched chicks through the mail from one of many commercial hatcheries around the country. I like to order from smaller, regional, family-operated hatcheries—especially those committed to conservation breeding—which often give more personalized service. On the other hand, the huge mega-hatcheries typically feature larger selections of breeds and more choices of shipping dates, and are more likely to allow requests for small numbers—even individual chicks—of different breeds in the same order. Hatcheries I have ordered from tailor their offerings toward the productive home flock (rather than exhibition), and most furnish decent stock and attentive service. When I get consistently poor results with chicks from a particular hatchery, or indifferent or clueless service, I avoid it like the plague thereafter. If you know experienced flocksters, heed their advice about hatcheries that are reliable, or not.

In the past, the required minimum number of chicks in a hatchery order was typically twenty-five (for maintenance of body heat in transit). These days the use of warming packs in the shipping boxes may be an option, enabling the shipment of small orders of half a dozen or so.

STRAIGHT RUN OR SEXED?

If you order chicks from a hatchery, you will frequently have the option of choosing "sexed" or "straight run." Sexed chicks are separated by professional sexers trained to detect minute, breed-specific differences between newly hatched cockerels and pullets. You can order all pullets if you want layers only in your flock—or all cockerels, which are cheaper and reach larger butchering size, if you want to raise a batch for the freezer. As mentioned previously, a straight run (not sexed) batch should be the natural hatch rate—roughly 50 percent each males and females—though flocksters who receive batches heavily weighted toward cockerels swear somebody is stacking the deck.

If you do order straight run chicks, remember that they will *have* to be culled sometime before maturity—a flock with large numbers of excess males is a *bad* idea. Don't be naive about this fact of life—include it in your plans for the flock before it's a last-minute crisis. Don't think it will be easy to give away your excess cockerels—"free-to-good-home roosters" are as much in demand as ants at a picnic. And unless you know someone who does custom butchering, don't assume that "of course" you'll be able to find that service for a fee when the time comes.

Though the option of placing all-pullets orders is convenient, flocksters inclined to consider the question more deeply may conclude that the choice of straight run or sexed chicks is more complicated than it first appears. Since orders overall are heavily weighted toward pullet chicks, it becomes impossible for hatcheries to sell the unwanted cockerel chicks—however many "cockerel specials" they offer. It is simply a fact of life in the business, therefore, that excess cockerels are killed, by the hundreds of thousands—by conveying them alive into a high-speed grinder; with "controlled atmosphere killing" (using carbon dioxide, argon, or nitrogen); sometimes by simply dumping them into a barrel and leaving them to suffocate. The reader may well choose otherwise, but my choice—since learning that my pullets-only orders *necessitate* the treatment of living creatures like disposable garbage—has been to make straight run orders exclusively.

DEBEAKING

Many hatcheries offer the option of debeaking—chopping off half the upper beak of the newly hatched chick. Debeaking is routine in the poultry industry as a "necessary" alteration to prevent cannibalism: relentless pecking at one another, to the point of death, both in the brooder and among mature layers. But such behavior emerges *only* in situations of extreme stress—which is to say, debeaking is based on an expectation that the management system will be a profoundly stressful one. The reader who agrees that our duty, to the contrary, is to give our birds a stress-free life can be assured that a flock managed as recommended in this book will *never* require debeaking. Let's recognize debeaking for what it is: mutilation, plain and simple—mutilation that may well result not only in chronic pain the rest of the bird's life, but in decreased ability to forage, and an inability (in the absence of a fine point on the beak) to properly preen the feathers and rid the skin of external parasites.

You may read—in literature targeted not at the industry but at you, the backyard, small-scale flockster—why you might want to debeak and how to do it. My question to anyone who gives that advice: *What is it about your management that causes crazed behavior among your birds?*

VACCINATION

Many hatcheries offer vaccination of chicks as an option. If so, it will likely be for Marek's disease; occasionally a hatchery will also offer vaccination for coccidiosis. The additional charge is minimal, and some flocksters choose to have their chicks vaccinated at the hatchery. Vaccines for other diseases—laryngotracheitis, Newcastle disease, bronchitis, avian encephalomyelitis, fowl pox, infectious bursal disease—may be purchased and administered by the flockster.

Marek's is a viral disease that kills more chickens than any other disease. Coccidiosis is caused by any of nine species of protozoans and can cause weakening of growing birds, slow growth, and death. That sounds scary, and the temptation is to choose vaccination by way of insurance. However, I have never had chicks vaccinated—for any disease, ever. Consider: Both the Marek's virus and cocci protozoans are virtually universal—that is, to be found anywhere chickens are raised. Since my flocks have never had a problem with either, it is reasonable to assume that neither is the *cause* of the associated diseases (in the same way we can say that viral organisms do not *cause* the common cold, since they are universally present in the environment). Healthy chicks gradually exposed to cocci develop an immunity to them (just as healthy children develop immunities when challenged by exposure to normally present pathogens). The Marek's virus is called out of dormancy by stress. In contrast to the enormous, highly confined (which is to say, highly stressed) flocks of the poultry industry, small home flocks—receiving normal exposure to universally ambient pathogens and managed to minimize stress—do not require vaccinations to thrive. Since there is as well a small chance of harmful effects from vaccines, I avoid them.

Be forewarned that you may feel pressured by recommendations on the hatchery order form to vaccinate. For example, Murray McMurray advises routine vaccination of all chicks: "Vaccinating your birds for Marek's is another appropriate step in strong poultry management. Don't take any chances. Let us vaccinate your chicks prior to shipment of your order. Don't forget to choose vaccination as you go through the check out process." It helps to resist that pressure if you know that an equally large-scale hatchery, Ideal, takes a more nuanced approach: "IDEAL recommends that you vaccinate your flock only if you are mixing young and mature poultry, have ever been diagnosed with Marek's Disease, or have ever had any chickens vaccinated for Marek's Disease live on your property. If you do not meet any of the criteria listed previously we do not recommend introducing Marek's Disease to your property. The vaccine used is a live virus vaccine."[1]

Enrollment of the hatchery you order from in the National Poultry Improvement Plan (NPIP) might be of more importance to you than vaccination. Breeder flocks in the program are certified to be free of several poultry diseases. Note, however, that smaller breeders may choose not to get involved in maintaining the paper trail required for NPIP certification, while still taking care to ensure their stock is disease free.

Starting Chicks in a Brooder

Many flocksters, even those with no experience whatever, start their flocks with just-hatched chicks in a homemade brooder—a nursery for baby chickens. Doing so is not at all difficult, so long as you remember they are dependent on you for their every need: *You are Mama.* If you frequently monitor your babies, and heed what their behavior tells you, all will be well.

The Chicks Are in the Mail

People are often surprised that live chicks can be sent through the mail. The key to the mystery is nature's provision for the fact that the earliest chicks in a natural hatch may be out of the shell hours or even a whole day before the last ones hatch: Just before hatch, the chick absorbs the last of the yolk material and thus has in reserve sufficient internal resources to remain comfortably on hold without feed and water while waiting for its slower siblings to hatch—easily two full days, even up to three. It is during this on-hold period that chicks can be shipped through the mail.

Do note that the chicks must not be fed or watered before shipping—they remain in their suspended state only during the no-intake period. Once they begin feeding and drinking, their metabolism shifts to a more active phase, and they must have feed and water regularly.

Note the ship date for your order on the calendar so you will be home to receive it. Shipping usually takes two days, but orders sometimes arrive the day following shipment. Advise your postmaster or letter carrier that you are expecting the shipment. You might want to have the postmaster call so you can pick up the chicks at the post office yourself, rather than waiting for your carrier's delivery. Open the box to check the chicks in the presence of the postal employee—most hatcheries insure their orders but require substantiation of a claim by postal personnel. Sometimes one or two chicks die in transit—and many hatcheries include one or more extra chicks in case of such losses. I have had losses beyond this level only a couple of times in dozens of shipments over the years.

Setting Up the Brooder

The down of newly hatched chicks does not insulate as well as the feathers of adult chickens, so they are vulnerable to rapid loss of body heat. An artificial brooder is *any* arrangement that substitutes for a mother hen's maintenance of warmth for the chicks and protection from wetting and from harsh drafts. You can purchase a brooder, like the one shown in figure 6.2, complete with heat source, feeding troughs, wire floor, and clean-out tray. Or you can save your money and house your babies in a brooder as simple as a cardboard appliance carton—we brooded our first batch of chicks in the carton from a new refrigerator. You might temporarily partition off a corner of a toolshed or stall in the barn, such as those in figures 6.3 and 6.4. In our grandmothers' time many a batch of chicks was started in a box behind the woodstove.

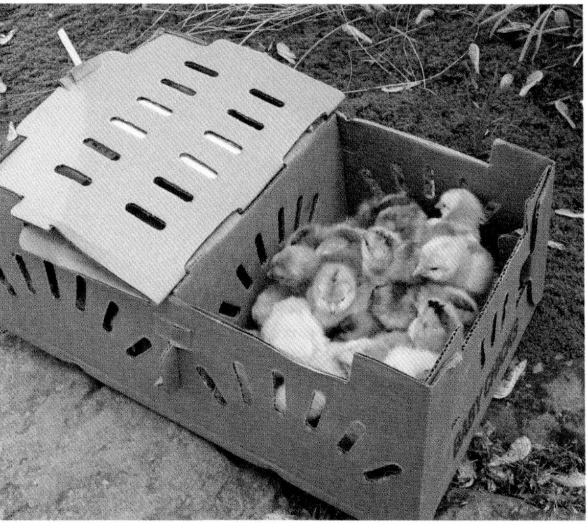

Figure 6.1. Just-hatched chicks can be sent through the mail before they begin to eat and drink.

Some flocksters who are concerned about piling up (chicks crowding into a corner so tightly that they suffocate) avoid corners by setting up the brooder as a long strip of cardboard about 2 feet (60 cm) wide, fastened into a circle. One of my correspondents uses 14-inch metal roof flashing set as a circle of any needed size, with the ends fastened together with metal siding screws; it can be rolled up and stored when not in use. I have never used this option, since I've never had a problem with chicks piling up, which usually occurs as a result of some stress such as being frightened. Small batches of chicks should be free of such stresses in a well-managed brooder. The brooder I usually use for colder parts of the season is four pieces of scrap particleboard, topped with a large piece of scrap cardboard, screwed together using corner cleats when in service, and disassembled and hung flat on the poultry house wall when not. In warmer parts of the season, I simply shut off the smallest section of my poultry house with a wire partition and a wire-on-frame door.

Figure 6.2. This purchased brooder for small batches of chicks includes its own heat source, feeders that reduce waste, and clean-out tray.

Figure 6.3. This brooder is a corner of a horse stall, set off with hay bales, topped with a mesh gate. Not pictured is a hanging 250-watt bulb for heat.

Figure 6.4. This brooder is an end of a toolshed, temporarily partitioned off with plywood and furnished with an adjustable hanging waterer, waste-inhibiting feeder, grit, rheostat-controlled heating element, and wood-shavings litter.

Figure 6.5. Homemade plywood hover in a brooder, with short legs that raise it a few inches above floor level, shown here with top open to reveal interior heat lamps. Feed and water are outside the hover, and chicks self-regulate for temperature.

The brooder must have a heat source. Most brooders for small home flocks rely on 250-watt heat lamps, available at the local farm supply, *securely affixed at least 18 inches (46 cm) from any flammable surface*. Some folks prefer a red lamp (thought by some to reduce the chance of injurious pecking at each other known as "cannibalism"); others use a clear one. Again, if the chicks are not stressed, I've never seen much difference between the two. Another option is a hanging warmer with an electric heating element on a rheostat control, available from larger hatcheries and poultry supply houses. The brooder should be well ventilated but should permit no direct drafts onto the chicks. Since warm air rises, constant airflow can be assured by having small openings at floor level and a vent at the top of the brooder.

A more elaborate setup, especially for larger groups of chicks, features a hover like the one in figure 6.5: an insulated structure of metal or plywood, suspended a few inches above floor level, that contains the heat source. The feed and water are offered outside the hover.

Chicks self-regulate body temperature by venturing out to eat and drink, then retreating to the greater warmth under the hover as needed. Such a setup can encourage faster feathering, since the chicks spend more time in the cooler temperatures outside the hover.

Predators must be excluded, of course. Most of the more obvious ones—fox, raccoon, possum—should already be excluded by the housing in which you set up your brooder. A couple of predators of chicks are especially difficult to exclude from the housing and so require special vigilance. Rats are no threat to an adult chicken but will eat helpless chicks, sometimes pulling them down their burrow holes. Snakes have a taste for chicks as well, so make sure they have no place to hide. (More on neutralizing these threats in chapters 7 and 21.) Don't forget your pets as potential predators: Make sure the family dog or cat can neither get at the chicks nor frighten them by trying.

Introducing the Chicks to the Brooder

Make sure the brooder is completely ready by the time your chicks arrive: They will have had a stressful trip, so you want to get them into the more compatible conditions of your brooder without delay. Turn on the heat source a couple of hours in advance of expected delivery, so the temperature is nice and cozy when you put in the chicks.

The floor of the brooder should have a layer, several inches deep, of absorbent, high-carbon litter. I prefer coarse, kiln-dried pine shavings. Straw is acceptable but should be replaced as it becomes damp, to prevent growth of molds. I don't like fine shavings or sawdust, which tend to pack down—I prefer the higher oxygen level in a coarse medium, which helps inhibit anaerobic pathogens. Shredded paper and cardboard work well if they are available, but the floor should *not* be a slick surface such as sheets of newspaper or cardboard, which can cause the chicks to slip and injure their joints—even more likely for ducklings and goslings. Not only is a thick litter in the brooder the best provision for management of droppings, but it gives the chicks something to scratch in and entertain themselves with, preventing the stress of boredom.

The first priority for the chicks is giving them a drink of water. Don't wait for them to find the waterer on their own: Remove them individually from the shipping box, gently dip their beaks in the water for a few seconds until you see them swallowing, then release them onto the brooder floor. Be sure the water is lukewarm, not cold.

Some flocksters offer electrolytes—available from hatcheries and poultry supply houses—in the first drink. My "country boy" version of this restorative boost is the addition of some honey (a quarter to a half cup in half a gallon of water), a couple of tablespoons of raw apple cider vinegar, and garlic—a couple of fresh cloves squeezed with a garlic press. I keep this solution on offer the entire first day, then switch to plain water.

Managing the Brooder

Remember that you are a second-best substitute for a mother hen. Be as vigilant to their needs as she would be—*frequent monitoring of brooder chicks is the key to success*. For example, the standard recommendation is to keep brooder temperature at 95°F (35°C) the first week, then decrease by 5°F (2.8°C) each week. But I have never found it necessary to use a thermometer in the brooder. Simply observe the chicks' behavior and apply the Goldilocks principle: If they huddle together under the heat source, as in figure 6.6, it is too cool; if they are hanging out around the edges, away from the

Figure 6.6. These chicks are huddling under the heating element to keep warm–time to turn up the dial.

heat source, it is too warm. Adjust the temperature, raising or lowering the heat lamp (or turning on or off a second lamp) or dialing a different setting on the electric heater. Chicks scooting about over the litter like water bugs on a pond indicates that the temperature is "just right." (Of course, like all babies, chicks need plenty of sleep, so don't be perturbed to see immobile chicks beak-down in the litter.)

Watering

One thing to monitor especially closely is watering. Do *not* use an open waterer—that is, any waterer into which a chick can clamber and get wet or drown. Even in moderate temperatures, a soaked chick can rapidly chill and die. The best waterers for small brooder batches are the vacuum-seal type, featuring a reservoir that is filled, then screwed onto a base and inverted. The raised edge around the perimeter of the molded plastic base creates a trough that is easy for the chicks to drink from, but narrow enough they can't climb into it and get wet. A hole in the base allows water to run out of the reservoir until the hole is covered by the rising water level—the resulting vacuum prevents further flow until the hole is again partially opened by the dropping of the water level as the chicks drink. Three designs are shown in figures 6.2, 6.4, and 6.7.

Figure 6.7. To minimize wetting of brooder litter by waterfowl hatchlings, I use a 3-gallon waterer on ½-inch wire mesh over a catch basin. Either I dig the basin into the earth floor of my henhouse or set up cinder-block steps to give ducklings and goslings easy access. Photo courtesy of Bonnie Long.

The litter should never be soaking wet, a condition conducive to disease organisms. If water splashed by the chicks does wet the litter, either remove it or spread it so it will dry.

Typically there is little problem with wetting of the litter when brooding chicks, guinea keets, or turkey poults. Waterfowl hatchlings, however, are exuberantly messy with their water, so wet litter is inevitable when brooding them. Frequent removal of wet litter is necessary. A labor- and litter-saving alternative is to set the waterer on a wire-on-wood frame over a catch basin, like that in figure 6.7—most of the splashed water will end up in the basin, emptied as needed. There will still be some wetting of the litter, so monitoring its condition frequently is necessary.

Like chicks, ducklings and goslings in the downy phase can quickly chill and die if they get soaked. Later on you will want to provide water for them to bathe in, but in the brooder use a waterer that does not allow them to wade.

Some flocksters like to keep a light on brooder chicks even at night, to encourage more feeding and faster growth. I figure chickens naturally fit the same circadian rhythms I do, so I turn off the brooder light at night and let the chicks sleep. Note, however, that I use a heating element to keep them warm. If you use a heat lamp to warm the brooder, don't be concerned about the full-time presence of light.

Sterility?

The distinction between *sanitation*—essential for preventing disease and distress—and *sterility* is critical. The waterer, for example, should be cleaned frequently—but boiling it or soaking it in bleach is unnecessary. The most important distinction between the two relates to proper *litter management*. I've already mentioned the need to make sure the litter doesn't get wet from the waterer. But it will become damp as well as it absorbs the moist droppings of the chicks—who produce a considerable amount of them. Your nose will be your best clue that

"Mature" Deep Litter in the Brooder

If my comments on management of brooder litter are not scientific enough for you, here are quotations summarizing research on deep litter conducted at the Ohio Agricultural Experiment Station in the 1940s.[2]

> *Sanitation in brooder houses has been largely restricted to the everlasting use of the scoop shovel, fork, broom, and spray pump. What's new is the discovery of how to let nature's sanitary processes do a better job using built-up litter . . .*
>
> *The prevention or control of coccidiosis by starting day-old chicks on old built-up litter could have been prophesied years ago. It has long been recognized that chicks exposed to small dosages of coccidia at an early age developed a resistance which gave protection against heavier dosages to which they are often exposed from 4 to 12 weeks of age. Built-up litter has thus proved the most practical and effective means by which this resistance can be established.*
>
> *A second reason why built-up litter could have been expected to limit coccidiosis is the fact that nearly all, if not all, living organisms including bacteria, protozoa, etc., have their parasites. Old built-up litter would seem to offer a favorable medium and conditions for the functioning of the parasites and enemies of coccidia and perhaps other diseases, too . . .*
>
> *The first experimental evidence with reference to the use of built-up litter as a sanitary procedure was secured by the Ohio Station in 1946 when it was first used in the brooder house. During the three years previous when the floor litter was removed and renewed at frequent intervals, the average mortality of 10 broods, or a total of 18,000 chicks, was 19 percent. During the succeeding three years with the use of built-up litter, the average mortality of 11 broods, or a total of 10,000 chicks, was 7 percent. Seldom did a brood escape an attack of coccidiosis before the use of built-up litter. Afterward there was no noticeable trouble from coccidiosis in 11 consecutive broods started and raised on the same old built-up floor litter. Old built-up litter is floor litter which has been used by two or more previous broods of chicks.*

the litter has become overloaded—when you get that first whiff of ammonia, top off the litter with a nice thick layer of new shavings or whatever other clean, absorbent litter you are using. (And after that, add the fresh high-carbon material a little *before* the "whiff" point: Ammonia begins to challenge respiratory functions at levels too low for us to detect by its distinctive smell.)

If you are doing successive broods of chicks in a brooder, do not be misled by advice you will likely see to "clean out and sterilize" between batches—advice that comes out of an obsession we've developed about the threat of "germs," and the delusion that we can defeat them with heroic feats of cleanliness and the use of sterilizing chemicals. So long as the litter is maintained in good condition using the management described previously, *it is in fact good practice to leave it in place to "ripen" between batches*—that is, to become more alive with beneficial microbes.

You might want to skip forward to chapter 8, "Manure Management in the Poultry House," for a fuller discussion of the benefits of the deep-litter system. In my practice, *manure management with deep litter begins in the brooder*. I brood chicks on deep litter that the adult flock has been working "forever," topped off with a 2-inch (5 cm) layer of kiln-dried pine shavings. Beneficial microbes are already at work in the established litter and proliferate into the chick litter on top. Are there pathogens present as well? Probably. There are almost certainly the cocci that cause coccidiosis—as said, almost universally present wherever chickens are raised. The exposure to modest numbers of cocci in the litter challenges the chicks' immune systems, and they develop natural resistance to them.

My own experience is borne out by many of my poultry-keeping correspondents who raise successive batches of broilers to serve local markets. They report that subsequent broods of chicks do *better* than the first batch of the season, as the litter ripens. Indeed, unanimously positive reports from those who make a practice of leaving the litter in place incline me more than ever to see the usual "sterilize between batches" as mere superstition.

Cannibalism

In the literature you may see scary references to *cannibalism*—chicks in the brooder may constantly peck at one another's feathers and toes. Once raw wounds develop, everybody starts zeroing in on them—and things get ugly, fast. It is often recommended to use an infrared lamp to prevent this gruesome abuse. The most extreme prevention is debeaking, discussed previously. But cannibalism emerges only among chicks under enormous stress, as in massive industrial brooder operations. Assuming that the chicks' basic needs are met—proper temperature and ventilation, easy access to the waterers and feeders, and sufficient protein in their feed—the only inducement to cannibalism would be overcrowding or boredom. Give them plenty of room to run around in, and a litter that they can scratch and have fun in, and you are unlikely ever to have a problem with cannibalism.

Feeding Brooder Chicks

Chicks are notorious wasters of feed, so choose (or make) a feeder with holes for the chicks' heads or a lipped edge to inhibit scratching in the feeder or billing out of preferred bits, which scatters the finer portions of the feed into the litter. Some flocksters prefer a hanging feeder, which can be raised as the chicks grow.

Though I know of some producers for local broiler markets who practice otherwise, I think the best advice for the beginner is to feed brooder chicks free-choice—that is, keep feed in the feeder at all times. Be aware, though, that the finer portions of the feed sift to the bottom—they should not be left to go stale. From time to time, pour these "fines" into a shallow container, mix with a bit of milk, whey, broth, or water, and let the chicks clean it up. (And make sure they do so quickly—if it sits around too long, it will become moldy.)

As for feed formulations, see the discussions in chapter 16, if you plan to buy your feeds, and chapter 17, if you would like to make your own. It is worth remembering that modern superhybrids—meat strains ready to slaughter at seven or eight weeks, or layers who begin laying at sixteen or seventeen weeks—have higher protein requirements than traditional farm breeds.

Like adult chickens, chicks need grit in their gizzards to grind their feeds—the only difference being grit size (say, the size of radish seeds). Choose the "chick size" of granite grit if you buy it; or find a deposit of coarse sand on your place. Sprinkle the feed lightly with grit each time you feed. After the first couple of days, furnish the grit in a separate container, free-choice. They will know how much of it to eat.

Pasting Up

Remember what was said in chapter 3 about "reading the poops" as a guide to your birds' well-being? Here in the brooder is your first chance to practice this form of divination. A condition called *pasting up* or *pasty butt* is sometimes seen in chicks in the brooder: The expelled fecal matter is not a neat little dropping that gets incorporated into the litter, but a viscous mess that sticks to the down around the vent, becoming larger as the chick continues to poop (see figure 6.8). As it dries, it tends to occlude the vent and if left unattended can even cause the death of the chick—simply because it cannot poop.

If you find that your chicks are pasting up, take remedial action immediately. First aid consists of holding the chick gently in one hand while you carefully pull the accumulated feces from around the vent. It may help to soften the deposit first with warm water. You may pull out some of the down it's stuck to, but that is no problem. Indeed, you might want to pull out a bit more, to clear enough down to reduce the chance expelled poop will stick. Be careful, though—if you simply yank the down out, you may tear the skin. Feeding a little raw cornmeal or fine oatmeal can help clear up pasty butt. But the most important thing you should do is heed what pasty butt is telling you: "Something is *wrong*!" Then make changes to put things *right*—rather than repetitively treat the symptom.

A good starting point is to note that I have *never* had a single case of pasting up in a chick on pasture with a mother hen—this condition is *not* an inevitable part of a chick's early growth. The hen makes sure her chicks stay warm enough, gathering them under her breast and wings for a warming session whenever needed, and helps them find high-quality natural feeds such as green plants, wild seeds, insects, slugs, and earthworms from day one. Imitate her good work.

Being chronically too chilled can cause chicks to paste up, so make sure the brooder is warm enough. But in my experience the major cause of pasty butt is mediocre feeds. If you do not have an alternative to purchased feeds for your chicks, then follow the mother hen's lead and provide them in addition the widest range of natural feeds you can—figure 6.9 shows some of the possibilities. In my experience the beneficial effects of such feeds offset the negative effects of poor feed, and pasting up is rarely a problem.

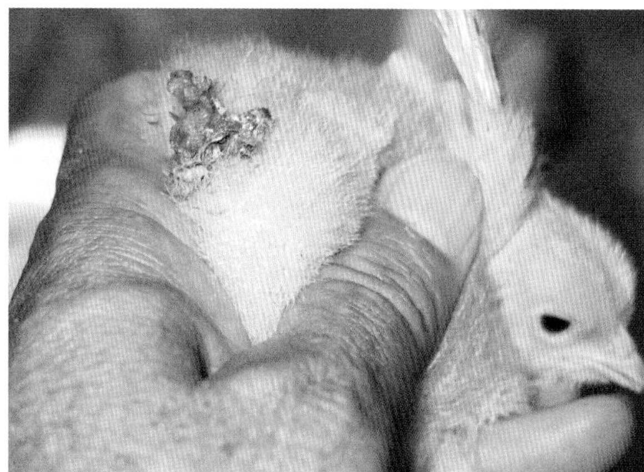

Figure 6.8. Pasting up requires immediate first aid—but also reconsideration of feeding and management practices.

Figure 6.9. I like to offer live feeds from day one. Here, three-week-old Freedom Ranger chicks enjoy dandelions, garden thinnings, and grass clumps—all with soil attached—as well as crushed hard-boiled egg and live soldier grubs. Hmm, is there a lesson to be learned? They're ignoring the feed in the feeder, prepared with such effort and expense by yours truly!

Mixed Brooder Batches?

The conventional wisdom is that all chicks in the brooder should be of the same age—and certainly the same species—lest older brooder mates, or those of faster-growing species, bully the younger and smaller members. Like so many cases where small home flocks are concerned, there is some room for bending the rules. I have occasionally brooded different ages and species together in one brooder without significant problems. Indeed, one year I raised in a single brooder a total of one hundred hatchlings—made up of two separate batches, hatched one week apart, each batch consisting of ducklings, goslings, *and* chicks. Though the older goslings got a bit bossy toward the end of the brooder phase, no harm was done, and nobody seemed seriously stressed. (I never combined mixed batches more than a week apart in age, and wouldn't expect that I could much exceed that interval without problems.)

Graduation from the Brooder

How long do chicks need to remain in the brooder? The answer depends on the point in the season, and how you manage them afterward. You can think of three weeks as an approximate minimum, and they should not need longer than five, even in the early part of the season when night temperatures are still cold. If you want to make the extra effort, when the weather is good you can give the chicks day outings in a sunny part of the pasture or lawn, inside a circle of 1-inch poultry wire, and return them to the security and warmth of the brooder at night.

The chicks should be completely past the downy stage when moved from the brooder—that is, completely covered with feathers. Be aware, however, that at this stage, even though visibly they are fully feathered, feathering is still rather sparse and not as insulating as it will become later. As well, they are inexperienced and do not understand how deadly even a light rain can be, so attentive monitoring at this stage is especially important. We once had a new batch of layer pullets on the pasture just out of the brooder. They were happy to be in the big wide world and left their open shelter to explore in the early morning, getting soaked in a drizzling rain. When I went out to feed, I found comatose chicks scattered about the pasture. Grabbing them up and heaping them in my shirttail, I rushed them into the poultry house and under the brooder lamps. Most recovered after they had warmed up, though some died from the exposure. Even running about in dewy grass can be too wetting, and too chilling. I subsequently kept the shelter door closed until I arrived on the pasture to supervise.

Where you move the chicks when they're ready to leave the brooder is up to you. Many seasoned flocksters advise that you should keep them in a group separate from the older established flock. My own experience is that, despite some initial hazing from the older birds, feathered chicks out of the brooder have no great problem integrating with the larger flock. For details on integrating young growing birds with adults, see chapter 20, especially regarding key feeding issues should you choose this option.

Do You Need to Identify Your Flock?

Decide as you get started whether you need to permanently identify birds in your flock. Some tracking of the flock is necessary for effective management as it ages and changes composition through the years. Most keepers of the homestead flock, however, will not need elaborate records and may never need to identify individual birds. If you are not too particular about breeds, tracking your flock might be as simple as switching breeds when you replace your layer flock. ("Let's see now, the Wyandottes and Barred Rocks are the old girls, and the New Hampshires and Australorps are the young 'uns.")

For more elaborate record keeping requiring detailed identification, see "Identification of Breeders" on page 242.

This chapter has focused on starting young chicks, a period in which they are vulnerable and require a lot of careful attention. Once they are well started, they are robust, and their care gets easier. Requirements for housing and water are easily met, and the most natural means for managing their manure also happens to be the simplest. They will be happiest of all if given plenty of room to scoot around in.

PART TWO

Basic Care

CHAPTER 7

Housing

Gallus gallus—the Red Junglefowl, ancestor of domestic chickens—did not have housing; and I've heard from intrepid flocksters who follow suit, allowing their chickens to live outdoors, fending for themselves and roosting in trees at night. That is the cheapest housing option, but it certainly opens the field to the neighboring predators. Equally problematic, chickens tend to revert to the feral state if not kept in close association with their keepers. A tree-roosting flock would be too wild for my taste. It is *housing* as much as any other factor that keeps our birds *G. gallus* ***domesticus***.

Chickens' notions about housing conditions parallel our own. In summer we want shelter from the broiling sun. In winter we want to be out of the wind. We can bundle up against cold temperatures per se, but a sharp wind in the winter cuts to the bone. Our chickens, bundled up in a layer of feathers, among the best of all natural insulation, are ready for the cold—but windchill is as much a challenge to staying warm for them as for us. And think how miserable you feel when getting soaked in a windy rain, even in summer. Despite their feathers, getting wet while exposed to wind and low temperatures can be lethal.

But it is good to remember tree roosting when thinking about housing. The fact that some flocks roost in trees—in some cases by preference—reminds us that our birds are tough critters. It's important to banish the notion of pampering them.

One of the most frequent questions I encounter about chickens is, "How do you heat their house in the winter?" In my mid-Atlantic climate (Zone 7a, where winter temperatures drop into the teens), I add *no* heat in the winter housing. Nor have I installed artificial insulation nor doubled the walls of the henhouse.

Flocksters considerably to the north of me should need no artificial heat or insulation either (though I can't advise keepers of fowl in Alaska or northern Canada). Indeed, *adding either would probably be more detrimental than beneficial*: A warm, airtight house will be a damp house—both the manure and the chickens' exhalations accumulate moisture in the coop—and dampness encourages both disease pathogens and molds.

Henhouse Design

This chapter does not provide a blueprint for building a coop but instead helps you think through the options. There are as many design possibilities as there are flocks, and you can find some great ideas under "Books" in the Resources section. Materials left over from other projects, begging to be used, influence design—as do management style, flock size, climate, the nature of local predation, and more.

There are no limits when it comes to housing. Flocksters of my acquaintance have made chicken coops

Figure 7.1. This year-round portable housing made from heavy framing, PVC pipe, poultry wire, and heavy canvas tarp houses up to fifty hens and guineas who range freely inside an area protected by electric net fence. Photo courtesy of Jon Wilson.

Figure 7.2. Plastic covered hoophouses are widely used for poultry. Caroline Cooper of British Columbia assembles this one to house her flock in winter, then takes it apart for storage the rest of the year. Made from 2 × 6 lumber and 20-foot (6-m) PVC pipes. Buckwheat hull bedding is 18 inches (45 cm) deep. Photo courtesy of Caroline Cooper.

from every imaginable converted outbuilding; discarded recreational vehicles, cars, and school buses; and pickup truck caps. They have mounted them on poles and bermed them into the earth to keep them cooler in the summer and warmer in the winter. Hoophouses have become a popular choice for lightweight structures that are easily assembled, most often for winter use, and just as easily disassembled for storage. (Figures 7.1 and 7.2 show a couple of examples.) Native materials and primitive techniques—adobe, wattle and daub, cordwood masonry—may fit your needs and will be cheaper than building exclusively from purchased dimensioned lumber.

If you provide for the fundamentals—shelter from wind, rain, and predators; generous space per bird; plenty of ventilation and sunlight—all else is secondary. Let your imagination run wild. Indulge your whimsy. Have fun.

How much space to allow per bird depends on how the house will function. If it is simply sleeping quarters for a flock that spends most of its time outdoors, a small space can accommodate many more birds. If the flock is confined to the building (for example, in winter), be generous with space. I recommend a minimum of 3 square feet (.3 square m) per adult chicken, ideally up to 5 square feet (.5 square m) or more.

Make flexibility a key part of housing design. Unless you *know* your flock is never going to expand, allow extra space from the beginning—or design for ease of adding on to the coop. I like plenty of interior partitions and doors, made of poultry wire on light wood framing. Partitions make possible separation of groups within the flock—younger birds or breeders from layers, waterfowl or guineas from chickens, brooding hens from everyone else—or turning one section into a brooder for chicks. When not needed, the doors between partitions can be removed from their hinges and hung on a wall.

Remember your own convenience. If the coop is small, hinged access to the nestboxes from the outside beats crawling into a cramped interior to collect eggs. Cleaning out an interior too low to stand up in comfortably is a strain. Electricity in the coop is convenient if you want to make the investment. I don't want run-

Did He Say *Ventilation*?

Don Schrider is a champion breeder of Light Brown Leghorns and Dark Brown Leghorns; he has written for such poultry publications as *Backyard Poultry* and headed up TLC's groundbreaking Buckeye breeding project (as described in "Bad News, Good News" on page 235). Here he shares his thoughts on what I mean by *plenty of ventilation*:

> *I house about three dozen show-quality Light Brown and Dark Brown Leghorn chickens. I make my pens from frame-and-mesh, bolt-together steel panels used to make dog kennels. I use tarps (secured with plastic cable ties) and half a roof (corrugated flexible roofing panels—brand name Ondura—attached with pipe clamps) to block the wind and keep rain and snow off the birds. The roosts are set up high, about 4 feet (1.2 m), so that the birds use their wings—if I had Orpingtons or another heavy breed I would have lower roosts so the birds could climb up. Feed and nestboxes are under the roof and protected by the tarps. Bedding is straw, and it should be thought of as a compost area. I feed whole corn and toss it into the bedding in the afternoon daily so the birds turn the bedding, get exercise, and get a warming snack before bedtime. You can't overestimate the value of clean, fresh air and exercise for the health*

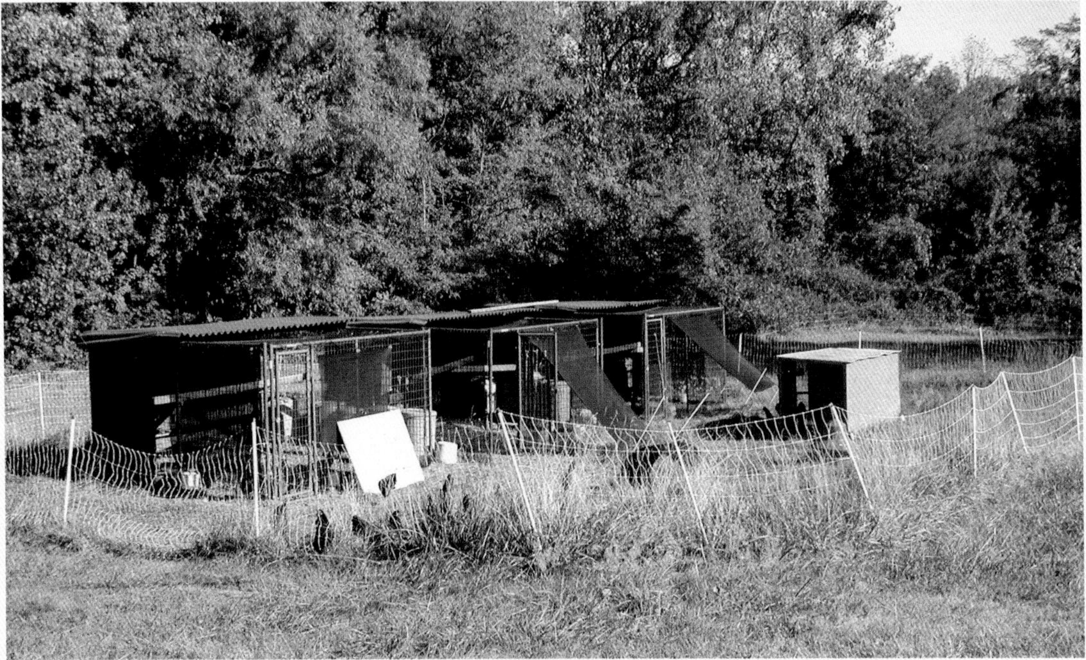

Figure 7.3. Don's open-air pens in summer. Note the three shade cloths that provide additional protection from the sun, and the electric net fencing to deter predators. Photo courtesy of Don Schrider.

Figure 7.4. When the snow flies . . . Photo courtesy of Don Schrider.

of chickens and to keep them warm by causing their bodies to work.

Frostbite is minimal because no moisture can build up as it does in a closed pen. Combine a lack of moisture in the winter air with exercise, a late-day heat-generating meal, and a circulatory system at full function—and you have birds that are less likely to frostbite.

This system is superior to closed houses at least as far north as Pennsylvania.

—DON SCHRIDER

Figure 7.5. . . .the pocket created by the half roof and a wraparound tarp protects the birds from snow, rain, and harsh wind. (One end of the tarp is visible at the left rear of the shelter.) Photo courtesy of Don Schrider.

Figure 7.6. Sheltered interior space includes roosts, nestboxes, feeder, and straw litter. Photo courtesy of Don Schrider

ning water inside the poultry housing, but I do like having a hydrant just outside the door.

While beginners tend to overestimate their flock's need for artificial heat in winter, they often underestimate their need for *shelter from the hot sun in summer*. Siting the coop in the shade of trees is a good idea. If their house is the birds' only access to shade, maximizing ventilation (discussed next) is even more critically important.

If feed is stored in the coop, make sure it is absolutely rodent-proof. Metal trash cans provide secure feed storage. Stout plywood may do the job as well, though rodents sometimes chew into wooden containers. Reinforcement with metal roof flashing or quarter-inch hardware cloth should solve the problem.

Ventilation and Sunlight

We don't want to live in "dark, damp, and stuffy"—we want to bring the sun and fresh air inside our house. Ventilation and sunlight in the poultry house are just as important.

A complete exchange of air in the coop four times a day should be the absolute minimum—to keep the interior from getting too damp (conducive to growth of molds and pathogens, and to respiratory ailments) and to get rid of gases such as ammonia and carbon dioxide while maximizing oxygen. Keeping in mind the necessity to shelter the birds from the direct blast of the wind and from rain, it is pretty much the case that, even in winter, the coop *cannot have too much ventilation*.

Since so many flocksters are not prepared to understand just how literally to take that last statement, I share with you my friend Don Schrider's description of the minimalist housing he uses for his champion birds—year-round—not far from where I live in northern Virginia, Zone 7a (see "Did He Say *Ventilation*?" on page 69).

Remember the flock's need for sunlight as well. Bring plenty of sunlight into the interior using wire mesh over all exterior openings. It's easy to install solid doors that open outward, and additional inner doors of wire mesh on light frames that can keep the flock inside while maximizing sunlight and airflow (as shown in figure 7.7). Whether or not windows have hinged or sliding sash, permanently install wire mesh in the openings. When the sash is open, ventilation increases, but predators are excluded.

Poultry Housing at Boxwood

Our original chicken house was on our place when we moved in close to four decades ago—8 feet by 18 feet (2.5 by 5.5 m), with a wooden floor. Over the years we changed the window layout and replaced the siding and metal roofing (see figure 7.7). After we built a larger henhouse (see figure 7.8), we altered the original coop for use as the "Breeders Annex," adding interior partitions for isolating breeders and broody hens in the breeding season. (More on breeding and working with mother hens in part 6.) The two windows are glazed with double-walled polycarbonate left over from construction of our greenhouse, and remain open almost all the time. Half-inch hardware cloth is permanently installed in the window frames to exclude predators. There is a single doorway, with two hinged doors: The solid outer door swings outward and can be latched open; the interior door (wire mesh on wood frame) swings inward and can either be latched open (allowing the flock access to the outside) or closed (to keep them inside while maximizing sunlight and ventilation into the interior). Placement of door and windows furnishes maximum cross-ventilation, while the corrugations in the metal roofing allow flow of air up through the living space and to the outside.

Figure 7.7. Our original chicken house, later modified to serve as the Breeder Annex. I do not know its original purpose, nor why its builder chose to elevate it on low piers of mortared fieldstone.

Figure 7.8. Our main poultry house, the Chicken Hilton. For maximum ventilation, the solid outer doors stay latched open full time, even in winter, except when there is driving rain or snow. If the flock must be shut in, inner mesh doors can be closed to confine the birds and exclude predators.

The Chicken Hilton

This is a sketch of the layout of our main poultry house, a pole barn construction with plenty of headroom, 24 feet long and 13 feet wide. It has an earth floor, covered deeply with oak leaf litter, with a barrier of 24-inch roof flashing dug in 18 inches (45 cm) around the perimeter to block attacks by digging predators. For additional ventilation a ridge vent runs the entire length of the roof's peak.

Interior partitions (A, B, C, D, and E): The interior was initially divided into three separate sections (A, B, and C, each 8 by 13 feet [2.5 by 4 m]) using chicken wire on 2 × 4 framing. Partition framing includes 40-inch-wide (1 m) doorways, into which I can either install partition doors—wire on light framing—or remove the hinges and hang them on a wall.

I subsequently installed additional wire-on-frame partitions to subdivide sections A and B, for additional sections D and E. A and B are now 8 by 8 feet (2.5 by 2.5 m), and D and E are 5 by 8 feet (1.5 by 2.5 m). Removable partition doors into D and E are 36 inches (91 cm) wide.

Doors: We placed three doorways in front, 32 inches (81 cm) wide by 72 inches (1.8 m) high, one in the center of each interior section. I made for each doorway a solid outer door that swings

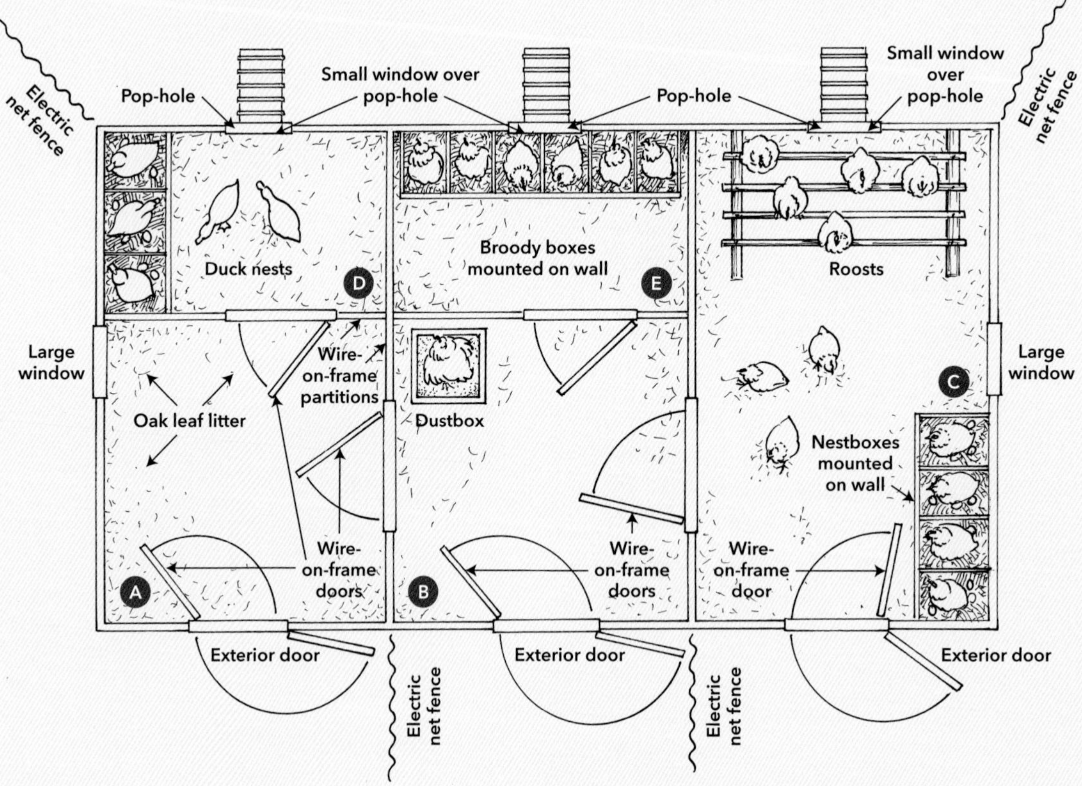

Figure 7.9. Main poultry house layout. Illustration by Elayne Sears.

outward and can be latched open. The outer doors are almost always open, even in winter, except when there is driving rain or snow, or when temperatures in the teens coincide with heavy winds. The inner doors, ½-inch hardware cloth on wooden framing, swing inward to give the flock access to the outdoors; or they can be latched closed to keep the birds inside (and keep predators out) while allowing maximum ventilation.

Pop-holes: At the rear of each section is a pop-hole—a sliding chicken-scale door (24 inches [60 cm] wide and 18 inches [45 cm] tall) that can be opened to give the flock access to the outdoors through the rear.

Electric net placement: Note that I can anchor the ends of electric net fencing at any point along the outside of the house, allowing separate flocks access to separate pasture enclosures. For example, I can install one section of fencing such that on one end it encloses the door of section A, surrounds a 3,000-square-foot (280-square-m) pasture enclosure, and terminates on the rear left corner of the building. With strategically placed additional runs, I can provide access for separate subflocks into a maximum of three ranging areas outside—thus maintaining their separation whether inside or out. *That* is flexibility, discussed in more detail on the next page.

Windows: A 24- by 42-inch (60- by 101-cm) window is set into each end wall. Smaller windows, 20 by 24 inches (50 by 60 cm), are set in the middle of each section in the rear, over the pop-holes. All window openings have ½-inch hardware cloth permanently installed. Together with the wire mesh doors in front, the side and rear windows provide maximum ventilation in every direction.

In the winter I block all five windows with plywood, creating pockets of still air in roosting areas. At the same time, there is plenty of airflow through the mesh doors in front and out the ridge vent up top. Like Don Schrider, I have almost no problems with frostbitten combs since I started keeping the henhouse more open in winter—in contrast with earlier years when I kept the coop closed as tightly as I could at night.

Roosts: Figure 7.9 shows a ladder-style set of roosts filling the rear of one of the sections. I subsequently changed my provision of roosts, as discussed in "Accessories" on page 75.

Nests: Nestboxes for the hens are mounted on a wall, well above floor level, with a landing perch in front that the hens use when entering the nests. All the nests were designed to be fitted with doors when I need to use them as trapnests (special nests with doors that close after the hen enters, trapping her so a record can be made of her production, discussed in detail in chapters 25 and 26). When there is no need to nest-trap, I put the doors into storage.

Ducks have a nest at floor level for laying eggs at times other than their breeding season. In late winter, however, as the breeding season approaches, I keep the ducks and geese outdoors full time, with a pasture shelter and sturdy, spacious nest units.

Broody boxes: I mounted a set of six broody boxes on a wall at a comfortable working height. These special nestboxes (described in more detail in chapter 27) provide privacy for hens incubating eggs, but require no loss of floor space for the rest of the flock.

Dustboxes: The deep litter is usually dry and fine enough for the chickens to use for dust-bathing. But just in case, I provide a 24- by 24- by 16-inch (60- by 60- by 40-cm) dustbox, which can be placed in one section or another as needed. (See appendix B for step-by-step construction.)

Flexibility: You are unlikely to make a duplicate of my Chicken Hilton. But study its layout carefully, and strive for the same flexibility, the same broad range of options, that underlies the design. Here are some examples of management flexibility with this design.

At the beginning of winter, I might have both partition doors in place between sections A and B, and between B and C. Ducks and geese spend the night in section A and during the day have access through that section's front door to graze a patch of cold-hardy rye and wheat enclosed by an electric net fence. A 140-gallon tank in the fenced area is for their bathing and watering outside. Meanwhile I have a group of excess cockerels awaiting slaughter in section B. Their front door gives them access to a heavily mulched winter feeding yard in a separate fenced enclosure. (I set up a temporary roost for their use.) The layer flock is in section C. Since the other two front doors are inside fenced areas, I need easy access to the door into C for my own use. However, during the day I open the pop-hole out the back of C, giving the layers access to yet a third fenced area out back, another heavily mulched winter yard.

Late winter comes, and I start keeping the ducks and geese outside—that's a simple matter of shutting the front door to A and parking a shelter for their use in their grazing area. I slaughter the last of the excess cockerels and remove the partition door between B and C to give the layer flock more space.

I receive a big batch of chicks right at the end of winter and set up a brooder temporarily in section A. (It disassembles and stores flat on the wall the rest of the year.) As my mother hens go broody, I move them into the broody boxes in section E and set them on eggs I've collected from my breeders for hatching.

A coop much smaller than my Chicken Hilton may serve your needs. If you design for flexibility, however, even a small structure will allow for changes in management and for meeting different needs as you go through the seasons.

We designed the new poultry house ourselves for a larger flock and for maximum flexibility, especially for breeding our own stock and for keeping waterfowl in addition to chickens. See the sidebar "The Chicken Hilton" on page 72 for its layout and discussion of the flexibility it makes possible.

The standard response to our main poultry house, with a profile higher than any sane flockster would choose, is: "That looks like a horse barn!" But the large amount of headroom increases air circulation—and offers intriguing possibilities for multispecies housing.

My main regret about this building is that I opted for pole barn construction, based on white oak and red oak poles, which I treated myself using sodium tetraborate, brand name *Bora-Care*. (I will not use lumber pressure-treated with heavy metals.) The deep litter in the house is highly bioactive and may, after a few more decades, defeat my attempts to preserve the posts with the boron. I wish I had spent the additional money up front for a concrete-block foundation—solid against both predators and rot.

Multispecies Housing

It is possible to accommodate more than one livestock species in the same housing. I referred previously to the large amount of headroom in our main poultry house. I've often fantasized about imitating the dovecotes or pigeon lofts of earlier times: turning that overhead space into a pigeon loft, releasing the pigeons to fly free during the day and forage their own food at no cost to me, while providing them a protected space to roost at night and to rear their squabs to be harvested for the table. The chickens, at floor level, would provide manure management by scratching the pigeons' droppings into the litter.

Joel Salatin's Raken House (housing *ra*-bbits and chic-*kens*) is an excellent example of the benefits of multispecies housing. The breeding rabbits live in wire cages suspended at waist height, over a large flock of layer hens, over 12 inches (30 cm) of wood chips as a deep litter. The chickens turn the droppings and urine of the rabbits into the litter, speeding its conversion to compost for the farm's fields. Offal from butchering the rabbits is thrown to the chickens as a protein feed.

Some flocksters set up a stall in the barn for their chooks' quarters and give them free rein to clean up spilled feed, denying it as a resource for rodents, and to scratch through the manure of horses and ruminants, ridding them of parasites and their eggs.

Our poultry flocks have often been a mix of fowl species, which from time to time offers challenges. If not carefully managed, waterfowl may soak the litter. (See chapters 8 and 9 for management strategies.) In late winter, when the testosterone starts rising in preparation for the breeding season, cock guineas may harass the chicken cocks relentlessly—I have gone so far as to hobble them (with loops of baling twine) to give their chicken brethren some peace! The important question of the disease called blackhead is discussed in the section on turkeys in "One Big Happy Family," on page 197—be sure to read it if you are considering a mix of chickens and turkeys. But for the most part, keeping a mix of fowl species in the same housing has not been a problem for us.

Accessories

Just as we furnish our own houses to round out their comfort, the chickens' quarters will need a few "furnishings." The best-furnished chicken house is the one that provides both maximum comfort and security to the birds, and maximum convenience and efficiency for us.

Roosts

Chickens have a strong instinct to roost: to get up off the ground, out of reach of predators, for their night's sleep. Roosts (perches) should be provided for all chickens—and guineas and turkeys—except for fast-growing broiler hybrids that are too heavy and clumsy to utilize them. While chickens denied a roost will settle down for the night on the litter in their pen, they will be more content and feel more secure with a raised place to perch at night.

There is nothing complicated about providing roosts. The cheapest option is to cut some sturdy saplings around 2 inches (5 cm) in diameter—leave them rough—and attach diagonally in the corners of walls in the sleeping area. Plastic or metal pipes are not good choices—they are too smooth for grasping securely, and metal gets too cold in winter. A simple wide "ladder" of 2 × 4s—or 2 × 2s for smaller numbers of birds—is easy to make. Round off the edges of the roosts so the chickens can grasp them more easily and comfortably. Lean the ladder-roost against the wall without attaching, so it is easy to swing out of the way for adding or removing litter.

Allow 8 to 10 inches (20 to 25 cm) of roost space per adult bird, 18 inches (45 cm) between a roost and a wall, and 12 inches (30 cm) apart both vertically and horizontally if the roosts are in the ladder configuration.

Note that, after I started keeping Icelandics, I stopped using ladder-style roosts. The birds insisted on roosting as high as possible and ended up crowded along a joist, 8 feet (2.5 m) above the nighttime quarters. I installed enough additional joists to give the entire flock room on their preferred higher roosts; and cut the former wide "ladders" and reconfigured them as true ladders, single 2 × 4s with stubby cross pieces, to give the birds access up to the joists. That was a great solution for the birds, who were much more content, though it required the wearing of an old hat and overshirt anytime I was inside the henhouse after the birds had gone to roost—to guard against the occasional hazard of getting "bombed."

Nestboxes

A hen's instinct is to hide her eggs from predators. Nest design should satisfy this instinct—the nest should convince her that she has found a hidden, secure place in which to lay. The top and sides should provide a sense of isolation, and keep the interior fairly dark. Line the nest well with straw or wood shavings, which not only keeps the eggs clean but cushions them to prevent breakage. Fasten a retaining strip of some sort, about 4 inches tall (10 cm), along the bottom edge of the front of the nest—otherwise hens scratch out the nesting material and you'll have to renew it frequently.

Provide plenty of nests in relation to the number of layers—if hens urgently needing to lay crowd a nest, there is increased chance that the jostling will crack an egg, an invitation to egg eating. One nest for each four to six hens should be sufficient. Make nests 12 to 14 inches (30 to 35 cm) in each dimension. Even when provided with plenty of nests, though, hens will sometimes perversely crowd together into one of them to lay.

I have always provided individual nests as described previously. Some flocksters prefer a communal nestbox—a single, larger interior space that can be accessed by more than one hen at a time. In either case a top over the nests, slanted at a 45-degree angle down from the wall, is a good idea.

You can buy manufactured nests, but I don't recommend them—both because of their expense and the flimsy construction of those I've tried, in comparison to home versions you can build or rig yourself. See appendix A for instructions for homemade nestboxes. While those instructions are specifically for making trapnests (for testing layer performance and for selecting hens as breeders), you can omit the doors and their tracking strips to make ordinary nestboxes.

Flocksters I've spoken to have used 5-gallon buckets, cast-off plastic milk crates, wooden boxes, or a few straw bales to define a private nook in a dark corner.

I make my nests of plywood, mounting them on a wall to save floor space. Setting the nests above floor level is another strategy to prevent egg eating. In my experience this unwelcome behavior may start with the cock of the flock, who gives an exploratory peck to an egg he finds in a nest at floor level. If a crack in the shell results, other flock members peck curiously as well. It's not long before the birds discover there's something good to eat inside, and egg eating becomes a nasty habit that spreads in the flock. Mounting nests on the wall helps prevent that first exploratory peck. Landing perches affixed in front of wall-mounted nest entrances make it easier for hens to enter the nests.

I don't like solid bottoms in nests. The lining material such as straw eventually gets dirty and disintegrates, and a solid bottom makes cleaning out more of a chore. Instead, I make the bottom of the nest of quarter-inch hardware cloth. As it disintegrates, the lining sifts down through the wire mesh, making the nest somewhat self-cleaning as I renew the straw on top from time to time.

Chickens sometimes roost in the nests, or on top of them; and—since they do much of their pooping at night—the nests get soiled, or the tops become caked with a layer of dried droppings. Younger birds are more likely to roost in the nest, since the older hens—higher in the social hierarchy—out-compete them for preferred real estate on the roosts. Providing more than adequate roost space can help. A slanted top over the nest will prevent chickens' roosting there. If a slanted top has not been built on the nest, it can be added, in the form of scrap plywood or even a piece of stiff cardboard, tacked into place over the nest.

It may be necessary to block access to the nest itself at night, again with a piece of plywood or cardboard. In my experience if I block access each night for a week or so, the birds get used to sleeping on the roosts, and I can discontinue blocking the nests with no further incursions into them at night.

Collecting eggs as frequently as practicable is a good idea. With frequent egg collection, the eggs will stay cleaner in wet weather, with less time for hens to track mud onto eggs already in the nest; and there will be fewer eggs in the nest at a given time, with less chance that jostling by hens will crack one as they come and go.

If a hen is on the nest, she may make a fuss when you try to check under her for eggs. Generally she will make no serious objection to an exploratory hand underneath—even a peck at that hand will rarely be more than a token. Occasionally a hen will indeed put serious intent into the peck—a bit painful but bearable—though I have lost blood to *guinea* hens on the nest.

Dustboxes

A major chicken ritual is dust-bathing: Chooks find a dusty spot somewhere and thrash about in it, fluffing the dust up under the feathers and onto the skin, in the process ridding themselves of external parasites. Outside, chickens hollow out a depression in dry ground and dust-bathe in it. Inside, they use the driest parts of the litter for bathing. But in case they have no other access to dust-bathing—during a rainy spell, or when the litter is not fine enough—it's a good idea to furnish them a dustbox.

You can bang together a dustbox out of scrap materials on hand. Do remember that the chickens will scatter the dusting material out of the box if it's too shallow—a deeper box will help retain the contents, as will adding a lip around the edges. My dustboxes are simple edge-nailed plywood boxes, 24 inches (60 cm) square on the bottom, 16 inches (40 cm) deep, with 2-inch (5 cm) plywood strips around the top edges as a lip. Instructions for building them are in appendix B.

Fill with 4 inches (10 cm) or so of any dusty, nontoxic material, renewing it as needed. Sphagnum peat moss is excellent, though it must be purchased and there are questions about its sustainability as a resource. Coir (the granular residue of long-fiber extraction from coconut husks) mixed with dried and sifted clay soil works well. Wood ash is an option, though I do not add it at more than 2 parts out of 6 or 7. The most finely granulated litter from the henhouse is excellent. A few handfuls of garden lime, diatomaceous earth, or pure sulfur powder make the mix even more effective against exoparasites. *Wear a good dust mask when handling dusting materials.*

In lieu of building a dustbox, my friend Don Schrider covers a bit of ground outside with a piece of scrap plastic culvert, which keeps it dry and dusty even in rainy weather (see figure 7.11).

Figure 7.10. In case my chickens don't have sufficient access to dust-bathing elsewhere, I provide a dustbox as a backup.

Figure 7.11. Don Schrider ensures constant access to dust-bathing outdoors with this section of scrap plastic culvert, which keeps the clay soil dry even when the weather is not. Photo courtesy of Don Schrider.

Dropping Boards?

Poultry do half, or more, of their pooping at night. So even if we release our flocks to the outside during the day, we must deal with manure management in the coops. A common recommendation from traditional poultry husbandry guides is the use of *dropping boards* under the roosts: The droppings fall on the boards and are scraped off by the flockster and moved to the compost heap. An alternative approach is the use of a *dropping pit*—a framed pit underneath the roosts, topped with wire mesh. The droppings fall through the mesh, which prevents the birds' getting into them. Periodically the accumulated droppings are removed and composted.

While sources I respect continue to advocate these poop practices, my own reaction to dropping boards and dropping pits is sheer incredulity. Scraping the boards, or mucking out the droppings from a pit, would be unpleasant and weary work. The basic premise of both designs, as far as I can tell, is that the flock's droppings are vile, nasty, and threatening; the birds should be kept separate from them at all costs. Fortunately, there is a far better approach to manure management in the poultry house: *deep litter over an earth floor*. That subject is so important that I discuss it separately in the following chapter on manure management in the poultry house.

Preview that chapter on another important question as well before finalizing your henhouse design. If you have an existing building to convert to poultry housing, avoid the effort and expense of building new. But if you are going to build from scratch, my strong recommendation is to *leave an earth floor in the coop*. Not only will you save the expense of framing and installing a floor, but you will be ready to create the conditions for best manure management.

Rodents

With either a floor or a perimeter barrier in place, and wire mesh secured in window frames and on inner doors latched shut at night, you should have few worries about predators entering the coop. An exception is rodents, who are a lot more difficult to exclude than, say, raccoons. Mice are no threat as predators, though they can be a serious threat to your feed supply. Rats can be even more voracious eaters of your expensive feed but are also a threat to chicks, whom they drag down their burrow holes.

My best advice is to focus on *preventing* a rodent infestation of the poultry house, not carrying on a war of attrition after they move in. Two key practices in the poultry house should help keep rodents at a minimum. First, I try to give my rodent friends *nowhere to hide* inside the poultry house. I try to keep all storage space such as shelves as open as possible, so a mouse or rat nest stands out. I make sure nothing is lying around at floor level other than the deep litter—constantly turned over by the chooks, leaving it unusable for tenancy by rodents. If the birds do occasionally scratch up a mouse nest, the mouse doesn't have a chance—one of the hens snatches it up, then plays a desperate game of keep-away with her sisters while trying frantically to gobble it down. As for baby mice—"pinkies"—they're *caviar.*

The second strategy for rodent prevention is: *Don't feed them.* For more on that topic, see chapter 15.

Access to the Outdoors

The shelter you provide your flock is a house, not a prison. Give your birds access to the sunshine, fresh air, exercise, and natural foods to be had outside. Best of all—for maximum health, contentment, and foraging opportunities—is getting the flock out on *natural range*. This is such a desirable option that I discuss it in three separate chapters later in the book. But ranging is not an option for many flocksters, so let's here consider best access to the outdoors for their flocks.

I cannot readily imagine any situation where full-time confinement of chickens inside the coop is absolutely necessary. Even in situations where space is extremely limited, it should be possible to construct a double-decker coop and run—living quarters above on stilts, wire-enclosed run in the footprint below. If space is available, of course, provision of a larger run is better.

The problem with a static run—whether large or small, and whatever the size of the flock—is that eventually the chickens, hungry for green, eat every last blade of grass. Not only is the result unsightly, but the accumulating manure virtually poisons the soil, serves as breeding ground for flies and vector for disease pathogens, and runs off in the next rain as pollution to groundwater and streams. The responsible flockster will come up with more wholesome alternatives, such as the three suggested next. For *any* of them you will need a perimeter, which could be versatile electric net fencing. For ideas about designing and installing a more conventional chicken wire fence, however, see the sidebar "And If You Must: The Better Static Run" on page 94.

The Composting Run

Once you've read chapter 8, imagine taking the concept of manure management using deep litter *outside*. That is, cover the entire enclosure with a deep litter, using whatever organic "wastes" come to hand—fallen leaves, spoiled hay, crop residues, flower bed prunings—turning the run in effect into a giant compost heap, as pictured in figure 7.12. The organic debris will absorb the poops, retaining them for soil fertility applications, while preventing their running off as pollutants—a win–win solution in every way. And remember how well it works as a winter exercise/feeding yard.

Chicken Pie

Make the coop the center of a foraging-ground "pie," with "slices" defined by electric net or poultry wire fencing, either temporary or permanent. Release the chooks into one slice at a time, rotating through the entire area around the coop. If the flock is not too large in relation to the total available space, slices grazed earlier will recover completely before the birds are released onto them later in the rotation. Almost inevitably, the area immediately around the entrances to the coop become worn free of cover, so lay down a thick organic mulch in the bare spots as described.

Dueling Gardens

If you both keep chickens and grow a garden, you have the basics for a solution I call *dueling gardens*. Make the run and the garden each the same size; enclose each with wire or electric net fencing; and set the coop in between them, with separate doors opening onto the two spaces. In the first season, release the birds onto one of the spaces—set up as a composting run as described—and garden in the other. Alternate the use of the two plots in the following season. The soil in the new garden side will be pre-enriched with compost. Since it has been continuously under a heavy mulch for a year, you may not even have to till—just lay out the garden beds, and plant.

I've seen different recommendations about the waiting period to observe between application of raw manure on soil—deposition by chickens of their droppings in a garden space is in fact such an application—and harvest of vegetable crops, ranging from 60 to 120 days. Anyone serving produce markets should note that the standards of the USDA's National Organic Program specify a waiting time of 90 days for crops like corn that are not in contact with soil and 120 days for crops such as lettuces and carrots that do have soil contact. For growing in my own garden: I doubt there is a threat of pathogenic contamination from a well-managed homestead flock on soil abundant in oxygen and alive with beneficial microbes, and I have never observed a measured waiting time. After removing chickens from garden beds, I have simply planted and gotten on with the season.

Figure 7.12. If a static chicken run is your only option, be sure to cover its entire area with a mixed organic mulch. Practice responsible manure management while providing the flock endless entertainment. Photo courtesy of Bonnie Long.

CHAPTER 8

Manure Management in the Poultry House: The Joys of Deep Litter

If you are around any livestock operation, regardless of species, and you smell manure—you are smelling mismanagement.

—Joel Salatin

Repugnance for what comes out the far end of an animal is not merely cultural conditioning—our senses are warning us of potential danger: Feces can be a vector for disease. Joel's quote implicitly advises us to *trust* that repugnance: If it smells bad, it could be dangerous. But it also implies that *there are ways to manage manure so it doesn't stink*, giving us our most important hint that its threat has been neutralized. Properly handled manure, in other words, is not a danger.

Many readers of this book have already experienced the transformation of things yucky into something not only inoffensive but a valuable resource: the alchemy of the compost heap, which starts with manures and rotting vegetation and ends with compost, smelling as sweet as good earth, ready to fertilize the garden. *The compost heap is our model for making the same transformation in the henhouse.*

You assemble a compost heap from nitrogenous materials such as manures and spent crop plants, mixed with carbonaceous ones such as leaves and straw. Coarse materials will eventually compost, but if you make the effort to shred them more finely, the composting process speeds up considerably. Inconceivable numbers of microbes multiply in the pile, using the nitrogen in the manures and fresh green matter as a source of energy to break down the tough, fibrous, high-carbon materials into simpler components. The ideal balance of carbon to nitrogen in the mix is 25:1 to 30:1. Too much nitrogen is signaled by the smell of ammonia, meaning that some of the nitrogen—a potential source of soil fertility—is being lost to the atmosphere. (Ammonia is a gas of nitrogen and hydrogen, NH_3.) Moisture in the heap is essential to the microbes driving decomposition, though it must not be soaking wet—a condition that would inhibit decomposers while favoring pathogens.

Oxygen is also essential for the decomposers, so you turn the heap over completely at least twice during decomposition, maybe more. Heat is a by-product of the composting process—a well-made compost heap becomes amazingly hot. The result of this devoted effort is compost, one of the best possible fertility amendments the gardener can find.

It is possible to make the chicken coop in effect a slow-burn compost heap if you *leave the earth itself as the floor, and keep it covered deeply with high-carbon organic litter.* The sorts of decompositional microbes at work in the compost heap—and in the soil food web—migrate out of an earth floor into the deep litter; the slight wicking of moisture out of the earth helps them proliferate and thrive. (If you have an existing building with a wood or concrete floor to use for poultry housing, avoid the effort and expense of building new. You can still use deep litter to keep the henhouse sweet, with a couple of tweaks discussed in the next section regarding *use of straw as litter* and the need for *additional decomposition of a constructed-floor litter* in a compost heap before use in the garden.)

Oh, and all that laborious shredding and turning of the compost to assist its breakdown? Just leave that to the chooks.

Materials for Deep Litter

The poops laid down by the birds are rich in nitrogen, so naturally—as in the compost heap—we want plenty of carbonaceous material in the litter to balance it. In contrast to the ideal C:N ratio for a compost heap, however, *the higher the carbon content of the deep litter, the better.* That is, the more carbon in the mix, the more nitrogen-rich manure the litter can absorb before the C:N ratio shifts out of balance, resulting in production of ammonia.

The high-carbon material chosen for the deep litter depends on what is cheapest and most readily available to you. It should ideally be somewhat coarse, so the scratching of the chickens fluffs it up and incorporates plenty of oxygen, assisting its breakdown by microbes and discouraging growth of pathogens. I prefer oak leaves, but that's mostly because a close neighbor, who has half a dozen mature white oaks on her place, prefers to get rid of the accumulating leaves in the fall. She even hauls them over and dumps them in a big pile at my place. I say "God bless 'er!"

Kiln-dried wood shavings are excellent, with their extremely high carbon-to-nitrogen ratio (500:1), but are an additional expense if you must purchase them. For example, I recently bought compressed shavings at $6 per bale (expands to 8 cubic feet [.2 cubic m]) to use as brooder bedding. Buying enough to deep-bed the entire henhouse would be expensive indeed. Wood chips might serve—they too are extremely high in carbon and last a long time before they need to be replaced. Joel Salatin uses them as the litter in his Raken House—he cleans out only once a year, when even this coarse woody material has been reduced to compost by the microbes and the constant working of the chickens. Sawdust is satisfactory, though it doesn't fluff up as much as other materials. Whether using sawdust, wood shavings, or wood chips, be sure to use either kiln-dried or aged material—"green" woody materials may support the growth of molds, whose spores could be bad for your birds' respiratory systems and yours.

Note that old hay and certain crop residues such as soybean vines are *not* appropriate as litter materials—with a significant nitrogen content of their own, they do not effectively balance the nitrogen in the poultry droppings and quickly heat up.

What about straw? Many flocksters avoid the use of straw because, especially in the presence of the slight dampness of an earth floor, it can support the growth of *Aspergillus* molds, whose spores can cause serious respiratory problems. I have corresponded with flocksters, however, who report they use straw over an earth floor without problems. I have never used a litter of exclusively straw over an earth floor, though I have used it in a mix with a much higher proportion of oak leaves—with no mold problems. Note that there is no problem using straw as the litter over a wooden floor—the drier conditions in such a litter prevent growth of *Aspergillus*.

Nearby processing of agricultural crops may furnish other litter materials. Milling of corn, cane, buckwheat,

Figure 8.1. Deep litter: happy chickens, happy flocksters.

or peanuts, for example, may generate corncobs, chopped corn or cane stalks, or hulls that are available cheaply enough to be used as deep litter.

Alchemy

Over many years showing visitors through my poultry house, I have noticed that if my visitor has previously been in a chicken house, at some point she will sniff the air with a puzzled look and ask, "Why doesn't it *stink* in here?" When that happens, I know I'm on the right track with manure management.

But the transformation of "nasty" to "nice" is just part of the magic. Remember the comparison of the deep litter to an active compost heap—the process in deep litter is driven by the same happy gang of microbes. And among the metabolites of the microbes—by-products of their life processes—are vitamins K and B_{12}, in addition to other immune-enhancing compounds. The chickens ingest these beneficial substances as they find interesting things to eat in the litter. Don't ask me what they're eating, but chickens on a mature deep litter do little other than scratch and peck. This is alchemy indeed: *What started as repugnant and a potential vector for disease has been transformed into a substrate for health.*

Should you think I'm spinning fairy tales, know that scientific experiments have borne out the benefits of a bioactive deep litter. I urge you to read the full research bulletin from the Ohio Agricultural Experiment Station mentioned in chapter 6, but to summarize a point that is especially relevant here: One experiment compared two groups of growing pullets, both on old built-up deep litter but one group receiving a complete ration, the other fed a severely deficient diet. *Mortality and weight gain in the two groups were virtually identical.* In another experiment comparing pullets fed a severely deficient diet, groups on old, thoroughly bioactive litter suffered far lower mortality (7 percent as opposed to 23 percent) and achieved much higher weight gain (at 12 weeks, 2.34 compared to 1.64 pounds [1 to .7 kg]) than those on fresh litter. Both these and other experiments demonstrated: "Obviously, the old built-up litter adequately supplemented the incomplete ration."[1]

Deep-Litter Management

Plan for the use of deep litter when designing housing for your flock—deeper litter absorbs more manure and supports more microbes, so allow plenty of space for it. Aim for a depth of 12 inches (30 cm) if possible. Happily, in winter you can factor in as well the role of that thick layer of organic duff in insulating the coop from the frozen ground outside—and the heat generated in an active deep litter. The temperature is nothing like that of a well-constructed compost heap; but the warmth rising out of the pack moderates air temperature in the winter house. Caroline Cooper of British Columbia, Canada, sees temperatures of −13°F (−25°C) for two weeks at a stretch in a typical winter but finds that the bedding in her hoophouse, 12 to 18 inches (30 to 45 cm) deep, is warm to the touch a few inches below the surface.

The great thing about deep litter is that the birds do most of the work. But there are a few things requiring input and monitoring on your part as well.

Stocking Density

Joel Salatin makes this observation about stocking density on a deep litter: If you allow 5 square feet (.5 square m) per adult chicken, the birds' constant scratching will incorporate into the litter all the manure laid down, even in high-poop areas such as those under the roosts. At 4 square feet (.4 square m), there will be some capping of manure under the roosts—formation of a crusty layer impervious to the hens' scratching. At 3 square feet (.3 square m), there will be extensive capping. If there is capping of the manure in your coop, turn it over with a spading fork from time to time, and the chickens will break it up from the cap's underside.[2]

Let It Mellow

You will see advice that the coop should periodically be thoroughly cleaned out. But as the Ohio Agricultural Experiment Station experiments demonstrated, it is not fresh new litter that supports the health of the flock but "old built-up"—that is, highly biologically active—litter. Thus this important implication: *Never clean out the litter completely*. Once beneficial levels of microbial activity are established, don't get rid of them by a de rigueur "thorough clean-out." Over time, the buildup of the litter—or the need for compost for the garden—requires removing part of the litter. Leave plenty in place, however, to retain the benefits of the already active microbes and to "inoculate" the fresh material you add.

The Whiff Test

The caveat to the "mellow" rule against cleaning out too much of the litter is that inevitably the addition of nitrogen by the incoming poops will overwhelm the carbon in the mix—resulting in the generation of ammonia. Be alert to that first characteristic whiff: It is telling you that an imbalance must be corrected—both because nitrogen for soil fertility is being lost to the atmosphere, and because ammonia damages the chickens' delicate respiratory tissues. Reestablishing the necessary balance is simply a matter of generously topping off with your high-carbon litter material of choice.

Do note that ammonia's deleterious effects begin *below* the concentration our nose can detect (25 to 30 parts per million). With experience, you will learn to "read" the developing condition of the litter, so you can add fresh carbonaceous material *before* it starts generating ammonia.

Avoid Wet Litter

If you provide water inside, *avoid wet litter*. A soaked litter is anaerobic—deprived of oxygen—and more likely to support growth of pathogens. Wet litter also generates ammonia far more readily than drier litter.

Remember from chapter 7 that generous airflow through the coop prevents litter dampness. Wet litter is more likely around the waterer, so check conditions there often; scatter any wet litter out over the total litter surface, where the chickens' scratching will help dry it. As mentioned in earlier chapters, waterfowl are especially likely to wet the litter. (See chapter 9 for ways to cope with waterfowl and watering.)

Remember as well, however, that the busy critters in the litter need water for their work—monitor the litter to ensure that it is not powder-dry. Caroline Cooper reports that the winter air in British Columbia is extremely dry, so from time to time she carefully adds water to the litter to keep it active. If I have a waterer inside the chicken house, I frequently empty the small amount of water in its trough directly into the litter when rinsing it out.

Using the Compost

The deep-litter approach to manure enlists the flock in the great work of soil fertility. Over time—up to a year—the litter will be reduced by the action of chicken and microbe to a finished compost. When it's time to "harvest" the compost, fork aside the top, coarser material that "floats" over the finely granulated, slightly moist material below. *That's* the good stuff. Sniff a handful: Like any fine compost, it will smell of earth with not the slightest hint of raw manure. In my experience litter at this stage of decomposition is ready to use directly in the garden—it will not burn plants, will not inhibit seed germination, and visibly boosts the growth of crops.

In my experience, litter from a coop with a wooden floor is too raw to apply directly in the garden. Such litter should be further broken down in a conventional compost heap before use in the garden.

Disadvantages of Deep Litter

In close to four decades of relying on deep litter for best manure management, I have encountered only two potential disadvantages. Although the slight wicking of moisture from the earth into the litter is a benefit, we once had a summer of record-breaking rains, resulting in increased moisture in the soil under the deep litter. The litter was far from sopping wet but was considerably damper than usual—damp enough to encourage the growth of molds. Our birds suffered from several eye infections that season, and we lost an entire batch of nineteen guinea keets. Once I recognized the problem, I helped decrease the moisture content of the litter by adding a lot of shredded, thoroughly dry leaves and kiln-dried shavings.

The other potential disadvantage of deep litter over an earth floor—assuming the henhouse is not on a block perimeter foundation—is the absence of a wood or concrete floor as a barrier against digging predators such as foxes, coyotes, and dogs. As mentioned in an earlier chapter, my solution was to dig a barrier about 18 inches (45 cm) into the earth—metal roof flashing, but ½-inch hardware cloth would probably be a better choice—around the entire perimeter of the poultry house. That's a lot of digging (oh, my aching back!), but it prevents a lot of digging (by four-legged neighbors intent on dinner in your chicken house).

A Win-Win Solution

I cannot overemphasize the importance of deep litter in the henhouse for the most natural and therefore the most rational manure management. A deep-litter house is more pleasant for both owner and fowl, with the chooks doing most of the work. Microbial action in the litter turns what potentially causes disease into a substrate for health—indeed, ripe litter has demonstrable feeding benefits. And it boosts mental health as well—from the entertainment of happily scratching an endlessly interesting deep litter, in lieu of the stress of boredom. A deep organic duff insulates the floor of the winter poultry house, while the warmth of its decomposition moderates the chill. Finally, this magic process captures the fertility in the poops for soil building, always priority one on the home place. What better illustration of the coalescing of a diversity of natural processes into one dynamic whole—the abundant ecology—which is what this book is all about.

CHAPTER 9

Watering

Think of water as the most essential nutrient for your flock, and make sure there is never a lapse in their access to fresh, clean water. Layers drink more than nonlayers—eggs are 65 percent water—and serious water deprivation will greatly reduce production, in the worst case permanently. Fast-growing meat hybrids drink more water than their slower-growing traditional breed cousins.

I do not agree with the advice often seen to "sterilize" waterers with bleach or other chemical cleaners periodically—just give them a quick scour with a scrub brush, old toothbrush, or handful of coarse dried grass. And do clean frequently—the birds will drink more water if it is clean. Keep in mind what *you* want to see in a glass of water—as much as you can, make sure that's what the birds see in their waterer as well.

If you water inside the coop, remember to avoid wet litter. Avoid open waterers—the chooks kick debris into them as they scratch the litter. I recommend vacuum-seal waterers for chicks in a brooder. There are larger versions available for older birds as well, in either molded polyethylene or galvanized steel, from 2 gallons up to 14 gallons. All feature a reservoir fitted onto a base with a trough that allows easy access for drinking but not for splashing. Litter gets scratched into these waterers as well, but it is easily rinsed out of the trough, in lieu of dumping out the whole waterer. Setting the waterer on some sort of platform—about the height of the chickens' backs—helps keep the water free of litter.

When placing one of the large vacuum-seal waterers outside, find a patch of level ground to set it on—if it is tilted, the rising water level spills over the raised edge of the trough before it covers the hole in the base, so a vacuum never forms in the reservoir and all the water runs out. Be sure to *listen* to it before walking away. The larger waterers of this type feature a cap of heavy molded plastic with a handle, which seals the reservoir at its top by screwing down tightly onto a rubber O-ring. A bit of unnoticed trash on the O-ring will allow passage of air—whose faint whine or whistle alerts you that a vacuum cannot form inside the waterer, and that you must remove the cap to clear away the speck of trash, and then screw it back on before leaving the waterer.

Automating the Water Supply

Fill-and-tote waterers are adequate for most small flocks. The larger the flock, however, and the farther out on the pasture, the more inclined you may be to opt for automated watering of some sort. There are several options.

Various automatic watering systems are available for larger poultry houses or exhibition breeders with many individual cages to serve—nipples that flow only when the birds are billing them, small self-fill cups or bowls

Figure 9.1. Simple automated waterer based on a float-operated shutoff valve and rubber tub. Blocks provide access for smaller birds, while the insert prevents drowning if they fall in.

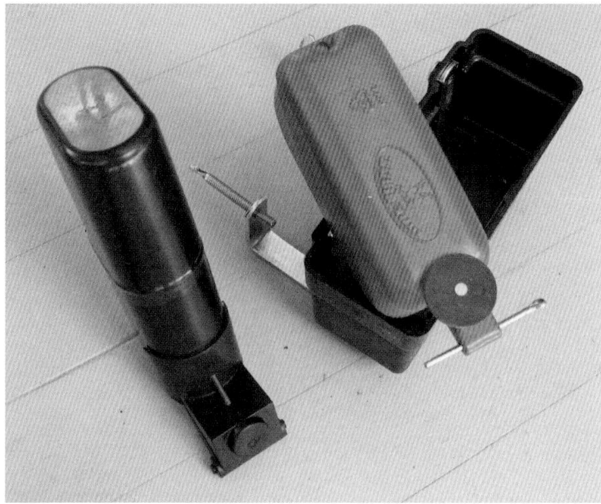

Figure 9.2. Before buying a float-operated valve to make an automated waterer, check how the float seats on the valve. Molded tabs (*left*) become unseated with the slightest jostling; a cotter pin (*right*) locks the float securely in place.

on individual supply lines—but such systems not only are overkill for most small flocks but may be impossible to set up efficiently on the range, which is where your birds deserve to be if possible.

A variety of automatic waterers for home flocks can be found online. Based on my experience, the one automatic waterer I strongly advise you avoid is a molded plastic trough, 18 or 36 inches (45 or 90 cm) long, with a flimsy float mechanism to regulate water supply. It is cheaply made and will not last.

The simplest version of automated watering is the one I use most for my birds on range: a molded rubber watering (or feeding) tub, 5 to 7 gallons, with an attached float-operated shutoff valve (like the one in a toilet tank) on the end of a supply hose that runs from an outside hydrant. Of course, the result is an open waterer—precisely what I warned against inside the henhouse. The difference here is placement on a grass sod: The water stays much cleaner when the waterer is on grass, and frequently moved. If left in one place, traffic around the waterer degrades the sod—the water always gets dirtier on bare ground.

The setup in figure 9.1 for automating watering requires no plumbing skills, nor is it nearly as finicky as some other options, which may require precise alignment, or clog with sediments or algae. Using Y-connectors readily available for garden hoses, you can serve any number of individual float waterers from a single hydrant. Here are some pointers if you want to give it a try:

The valve: The automatic watering valve I use is not made for watering poultry but to auto-fill a large-stock watering tank. However, it's easy to mount it onto a molded rubber tub, maybe adding a piece of scrap wood between the tub's top edge and the mounting bracket, providing a surface for the mounting screws to bite into. The valve could be rigged onto other vessels as well, so long as they are not too shallow—the float must have enough clearance from the bottom of the vessel to swing into its full-open position.

Float design: It is worthwhile ascertaining before buying a float-operated valve how the float attaches to the body of the valve (see figure 9.2). If the float is a single piece of molded plastic with tabs that snap into indents in the valve body, it is a piece of junk best avoided—the slightest nudge while cleaning or rinsing may unseat the float, leaving it incapable of shutting off the water. Reattaching the float may be difficult without unmounting the entire valve. A better design features a cotter pin inserted through matching holes in the end of the float and its seat on the valve, for a slip-free connection between them.

Hoses: I buy hoses from recreational vehicle suppliers that are manufactured for providing potable water. Many brands of garden hoses contain lead, which can leach into the water, particularly that initial flow, which has been sitting in the hose awhile. Hoses should be labeled specifically "Not for drinking" or "Safe for drinking" or equivalents—keep looking if the hose you're considering buying isn't labeled either way.

Making it drown-proof: This float-operated waterer works fine for adults, but young growing chickens may fall into it and drown. If you keep young birds with the adults, therefore, put something into the tank or waterer—a large rock, concrete blocks, or the cutoff bottom half of a plastic bucket, as shown in figure 9.1—onto which chickens can clamber if they fall in.

Watering in Summer

Chickens drink two to four times their normal amount in hot weather, and thus the ill effects of loss of access to water increase in summer. More frequent monitoring is essential to make sure they don't run out.

Chickens will drink more water if it is cooler, so place the waterer in shade if possible. If you use a hose to automate watering, remember how hot the water coming out of a dark-colored garden hose can be. That may be another reason for buying hose labeled for potable water: All such hoses I've seen are white, and in my experience the water in them stays cooler than in a dark hose.

Watering in summer is especially challenging if you are raising fast-growing meat hybrids, especially Cornish Cross, who in high heat may sit in the shade of their shelter and *die* rather than walk 10 feet (3 m) into the sunshine for a drink of water, as noted in chapter 4.

Watering in Winter

The "gotcha" of water deprivation can occur in winter as well, if you forget that the waterer can freeze, denying the birds their needed drink. There are water-heating options for preventing freezing—heat tape, heat lamps, birdbath warmers placed in the waterer, a heated base onto which the waterer is placed (if it's metal, not plastic). All such options require an electric socket nearby. The simpler option for the frugal homesteader is either to refill the waterer each day or to set it in the basement or elsewhere indoors overnight. In my area (Zone 7a), it is rare that daytime temperatures are low enough to cause freezing of the waterer. If that is a problem, two waterers can be used and switched in turn between the people house and the poultry house.

A freeze-proof yard hydrant makes winter watering more convenient (see the sidebar "A Frost-Free Yard Hydrant" on page 88.)

The *Water* in *Waterfowl*

The biggest differences between management of waterfowl and of chickens are watering issues. Chickens need water to drink; waterfowl ideally need water to splash in as well. Not only does the volume of water required increase dramatically when watering waterfowl, but if not carefully managed, the splash zone can become quite a mess. Watering needs for ducks and geese are the same so I discuss them as one topic.

Bathing

I strongly recommend that you provide an opportunity for your geese and ducks to bathe—not just to splash some water on themselves, but to swim and immerse themselves entirely in the water. Bathing is part of preening the feathers for waterfowl and serves the same function as dust-bathing for chickens: Geese and ducks use the water to drown external parasites under the feathers (as chickens use the dust to smother them), so bathing contributes to good health. But they also *love* it—bathing is good for their *mental* health. Enjoying

A Frost-Free Yard Hydrant

Many outside faucets must be shut off and drained for the winter to keep them from freezing. An exception is a freeze-proof yard hydrant, designed with a drain port or bleed valve at the base of the hydrant's vertical pipe to prevent freezing even when it is being used daily in the winter.

The supply line to that valve is installed well below frost line. When the hydrant's handle is up, linkage via a steel rod opens the supply valve and shuts the bleed valve. Returning the handle to the off position closes the supply valve and opens the bleed valve, allowing the water in the pipe to drain out into the unfrozen subsoil. Thus there is never standing water in the pipe that could freeze and expand, bursting the pipe.

There is one "gotcha" associated with this hydrant—ask me how I know! If you leave any sort of hose connector screwed onto the head of the hydrant, it may hold the column of water in the vertical pipe, and freezing will indeed burst it.

When installing the hydrant, make sure the bleed valve is surrounded by several inches of gravel above and below, with some scrap plastic sheeting over the top of the gravel. With this precaution, soil will never sift down around the opening of the bleed valve and stop it up, permitting water to remain standing in the pipe and freeze.

Figure 9.3. When this hydrant is shut off, water drains into the soil below frost line, leaving none standing in the pipe to freeze.

their exuberance in the water will boost your mental health as well.

If you plan to breed either geese or ducks, there is a further reason to give them the opportunity to bathe: For waterfowl, mating is easier in the water than on the ground. Indeed, among heavier breeds of both ducks and geese it may be impossible for the male to mount the female; and successful mating occurs only in water to buoy him up and help him perform.

If you have "wild water" on your place—a pond or stream—it may be possible to arrange access to it for your waterfowl. If you do not, it is easy to set up a "duck splash" for their use, like the one in figure 9.4. I use a couple of different stock watering tanks made of

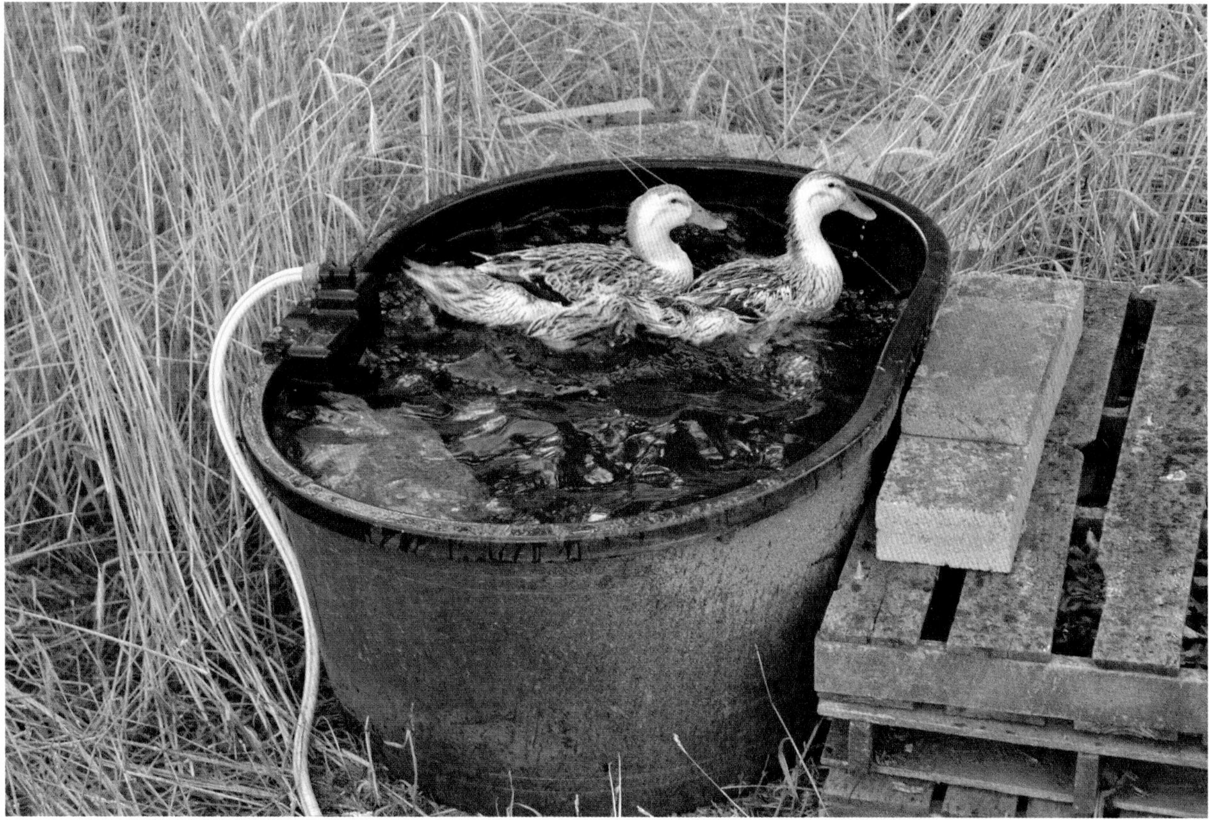

Figure 9.4. A stock tank on a supply hose with shutoff valve makes a great waterfowl bath.

molded rubber or fiberglass—one of 50-gallon capacity, the other 140—to provide bathing for ducks and geese. I use the same arrangement described previously: a supply hose to a float-operated shutoff valve attached to the tank—only the container is larger.

Cleaning: The bathing tank should be cleaned frequently—the ducks and geese will drink from it as well—especially in summer. Fortunately, clean-out is easy—I just open the tank's plug and let it drain, swipe the interior with a brush, rinse, screw in the plug, and walk away.

Access/egress: Unless you dig it into the ground, you will need to furnish steps up into the bathing tank for the waterfowl. I make mine by stacking wooden pallets and concrete blocks of various sizes. If you build one of wood, make the surface from slats or rough-sawn lumber, or score it well, so it doesn't get too slick with water, poop, or algae—waterfowl may injure their legs if they slip on such surfaces. Adults exit the tank without assistance, but ducklings and goslings in the down stage may die of drowning or exhaustion if they don't have something to climb onto. I place a stack of concrete blocks just below water level—once the young ones clamber onto it, they easily hop over the edge and onto the ground. In a mixed flock, provision of a submerged platform prevents drowning of young growing chickens as well.

Preventing drilling: Waterfowl, especially ducks, have a habit of drilling in wet soil—and there is bound to be plenty of that around the duck splash. If tearing up the sod in that area is a problem, set wood frames with attached ½-inch hardware cloth around the tank. A simpler option is to set the tank on a piece of hardware cloth large enough to cover the drilling zone. In this case, however, it is necessary to lift it free of the sod from time to time—if growth of grass through the wire is ignored, it will eventually be impossible to lift the wire.

The Rinsing Waterer

If it is not possible to provide a full bath for your waterfowl, *at a bare minimum you must provide water deep enough to submerge their entire heads.* Ducks and geese unable to rinse their eyes and nostrils frequently, especially when eating powdery commercial feeds, are more susceptible to eye infections and possibly even choking. The automated watering tub I use for my chickens on pasture (see figure 9.1) provides water deep enough for waterfowl to rinse their heads.

Waterfowl will splash water over themselves from such a waterer and vigorously work it with their bills, preening their feathers and inhibiting external parasites—though such water grooming is a poor second best to full bathing.

Winter Watering of Waterfowl

Providing water for waterfowl in the winter is more challenging than for chickens. Avoid watering them inside the poultry house if possible—they are incredibly messy with their water and will quickly soak the deep litter, which then is more subject to growth of molds and, because it is more anaerobic, of disease-causing pathogens.

If you must water inside, set a 5- to 7-gallon tub waterer on a wire frame over a large catch basin to reduce wetting of the litter (see figure 9.5). Again, the setup shown is a poor second best to real bathing—but a real bathing tank inside the poultry house would lead to soaking everything in sight. When I have used the setup shown in the past, it was only for a couple of geese and a trio of ducks—it would not be effective with many more than that. Empty the basin as needed

Figure 9.5. If you must water ducks or geese inside the poultry house in winter, set the tub over a catch basin.

using 5-gallon buckets carried outside. A better refinement if you are handy, of course, would be to plumb a drain line from the bottom of the catch basin to the outside.

Frequently disperse the wet litter around the catch basin onto the drier parts of the litter using a pitchfork. If you house your winter chicken flock with the waterfowl, they can help scatter it for drying, especially if encouraged with some scratch grains.

I strongly recommend keeping the watering of waterfowl outside, even in winter. Your experience may be different if you are far to the north of me, but I find that even in the dead of winter, maintaining the 140-gallon bathing tank for the waterfowl is not a great problem. That volume of water freezes slowly on even the coldest nights. The black color of the molded rubber absorbs solar energy as soon as the sun appears—it isn't long before the night's ice melts enough for me to break it up and toss it out in sheets.

In winter I don't fill the tank with the hose and float valve, which would freeze—using a bucket, I just top off as needed to replace water lost in the discarded ice. Clean-out in winter is infrequent, since algal growth slows so much in the low temperatures.

CHAPTER 10

Ranging Your Flock

Please allow your flock to range if that is possible on your homestead or farm. Ranging flocks forage the natural feeds most suited to their good health; thrive from fresh air, exercise, and sunshine; and have full scope to satisfy their curiosity and engage in instinctive social behaviors. No longer are they passive recipients of resources carried to them; instead, they are active participants in the ever-evolving abundant ecology around them, helping to drive renewal and fertility cycles and balance insect populations.

Ranging zones at Boxwood include stands of brush, a bit of woodlot, several lawns, and—most important—an acre of pasture.

Day Ranging

Some flocksters dispense with fences entirely and allow their flocks unrestricted ranging. That option maximizes the benefits of being out of confinement and works well where there is no serious threat from predators. Since many predators are active at night, day ranging typically involves letting the flock range where they will during the day, then religiously shutting them in their coop at night. (However far they range, they will return to the coop as the sun goes down.) Remember, however, that dogs can be daytime predators extraordinaire. (For more on predation challenges, see chapter 21.)

Another limitation on complete free-ranging is the presence of close neighbors: Our chickens will not make themselves, or us, any friends if they are eating the lettuces in a neighbor's garden, or scratching up a bed of prize roses.

Finally, the greater the freedom to roam, the greater the chance that hens will follow a natural inclination to hide their eggs, in lieu of using the nestboxes in the henhouse. Since they can be quite clever at doing so, there could be a considerable loss of egg production. (If you do find a nest hidden by a free-ranging hen, don't simply empty it of its eggs—the hen will conclude that a predator has found her nest and will find another nest site, even better hidden. If you leave some plastic or wooden eggs in the nest when removing the eggs, the hen will usually continue laying in it.)

A hen in a day-ranging situation may even decide to "hide a nest" (lay her eggs in a secret nest she has made)—and show up three weeks later with a clutch of chicks in tow, a surprise that may be more or less welcome.

Electric Net Fencing

The best way to maximize your flock's range, while keeping them where you want them and protecting them from anything on the ground with a nervous system, is *electric net fencing*. Because this type of fencing is such an ideal solution to the freedom/restriction conundrum for me, I discuss its use in detail in chapter 11.

Water and Shelter

Be careful not to let watering in range areas be "out of sight, out of mind"—ensure that your birds always have plenty of water. The farther a ranging flock is from the water hydrant, the more inclined I am to automate watering. The simplest version of automated watering is the float-operated one described in chapter 9—I set up such waterers as much as 200 feet (60 m) from the hydrant.

If your flock ranges full time on pasture, remember their need for a shelter from rain and for shade, and as a place to spend the night. Getting wet can be life threatening for chickens and guineas if it's windy and cold; and they have a deep instinct to be in a sheltered space at night.

Shade is essential for poultry on pasture. If the area where you have them penned does not have tree shade during all parts of the day, it is essential that a pasture shelter provide this retreat from hot summer sun. We learned this lesson one spring following a sharp temperature spike—when I went out to the pasture, I found my layer flock badly stressed, and several had begun attacking weaker members. One hen walked about in a daze, oblivious to the fact that other flock members were literally pecking her apart. She subsequently died, and we vowed never again to let our flock get that stressed by exposure to the sun.

Waterfowl, once they are well feathered, will ignore their shelter when it rains—they think the bath-from-heaven is terrific. Given the insulation of their heavy plumage, however, waterfowl have an even greater need than chickens for shade. They may or may not take refuge in the shelter at night. (Design of pasture shelters, a big topic, is discussed in chapter 12.)

Figure 10.1. A ranging flock is a happy flock.

Be especially careful if you have young growing birds on pasture—they have not learned the hazards of getting wet. Make sure young birds not long out of the brooder—who are still somewhat sparsely feathered—don't get lethally wet in a light rain or heavy dew.

Pasture Management

If the pasture improves rather than deteriorates as we range our flock on it, we know we're on the right track. Grazing by poultry—like grazing by ruminants—helps prevent the pasture's growing up in brambles and shrubs and following the progression to forest; it also stimulates vigorous new growth of the sward. Ranging chickens eat wild seeds, thus helping reduce the seed bank for weeds that might otherwise become too aggressive. The manure laid down by the flock fertilizes the sod—a benefit you'll never get from a power mower. We first saw a dramatic illustration of this fertility boost after we started raising our birds on pasture: I had fifty young broilers in a Polyface-style mobile pen, which I moved every day.[1] After a couple of weeks we had a nice spring rain. The band of rich green that emerged in the wake of the pen could have been laid out with a ruler.

It is desirable to have a diverse mix of plant species in the pasture sward—our goal should not be the monoculture desert seen in a typical suburban lawn. If you doubt the wisdom of encouraging as wide an array of ambient plant species as possible, answer me this: What do the following plants have in common—dandelion, lamb's-quarter, stinging nettle, burdock, and yellow dock? (No points for a dismissive, "They're all just *weeds*, for heaven's sake!") Answer: Each is at least 4 percent, and up to 12 percent, higher in protein than that quintessential high-protein fodder crop, alfalfa. Poultry relish them all. Further, each of these common weeds concentrates a different mix of minerals. If these plants are part of the pasture mix, our birds will more likely balance their mineral intake as they forage.

While the flock does much of the work of pasture management, there are a few contributions we should make to pasture improvement as well. It is best to rotate the flock over the pasture. Chickens especially wear at the pasture with their constant scratching and should be moved to a new plot before they damage the sod. As well, rotation helps ensure that every part of the sward gets its share of the fertility-enhancing deposition of poops and avoids a possible buildup of pathogens or parasites on the pasture.

Much less mowing is required on a pasture being grazed by poultry, but it's best to mow occasionally. Grass and other pasture plants are highest in protein and other nutrients when putting on rapid new growth after grazing or mowing. In the spring I like to let my pasture grasses grow as high as possible, then cut with a scythe, and rake up the long-stem grass for use in compost and mulches. A few days to a week later, I move the flock onto the rapidly regrowing grass.

Later in the season, when I would get less return for the labor of scything and raking, I occasionally use a power mower to prevent too heavy a set of weed seeds in the sward. Though nutritious "weeds," as said, are welcome in the mix, some get pushily aggressive if I don't head them off a bit with preventive mowing.

Grazing by the flock and occasional mowing by the flockster will do much to improve the pasture without additional seeding. However, you can certainly overseed the pasture to improve the mix if you like. After taking the flock off the pasture in the fall, I simply broadcast a mix of pasture perennials likely to do well in my area (timothy, orchard grass, Kentucky bluegrass, various clovers, alfalfa). Alternatively, I might broadcast my seed mix in the late winter and let the freeze–thaw cycle work the seeds into the soil. In either case, germination will not be as good as when drilling the seeds, so I use generous amounts of seed. Check with your extension agent about the best species for seeding pasture in your area, but I also encourage you to do your own research on species native to your area.

Many purchased pasture seed mixes are for introduced (rather than native) species. From time to time over the years I have seeded my pasture to grasses native to my region: big bluestem, little bluestem, broomsedge, Indian grass, switchgrass, side oats grama, eastern gamagrass, purpletop.

Figure 10.2. Lawns can be a resource, not just a chore.

"Pasturing" on the Lawn

Some flocksters pasture their flocks on their lawns. Indeed, one of my correspondents in the American Pastured Poultry Producers Association (APPPA) made her start in pastured poultry on the lawns of her ¾-acre house lot and for several years sold three hundred to four hundred broilers a year in local markets.

I have rotated many a waterfowl flock over four sections of lawn around our house during at least part of the growing season (see figure 10.2). That meant far less mowing; the birds fertilized the lawn; there were more trees for shade near the house than on the pasture; and the birds turned that lovely grass into terrific winter feasts and high-quality cooking fats.

And If You Must: The Better Static Run

I began this chapter with: "Please allow your flock to range if that is possible on your homestead or farm." For those flocksters whose circumstances do not permit widely ranging the flock, here I offer some thoughts on keeping your birds as safe and free of stress as possible, and as close as they will get to natural range inside a static fence. Figure 10.3 shows the chicken run I installed off a corner of the Chicken Hilton, quarters for my micro-flock of nine last winter.

- First and foremost, turn the run into a giant compost heap. That is, never be content with a bare dirt run, make sure it is constantly covered as deeply you can manage with the array of organic decomposables you would put into a compost heap. During their waking hours, your confined flock should be pooping in and shredding and finding critters to eat in such a "debris field"—the chooks find live detritivores in it even in winter. In the process, they become themselves drivers of the heap's decomposition, for a nice batch of compost whenever you need it.
- Use of rot-resistant posts would be best. I had to use construction-grade 2 × 4s, so I coated the

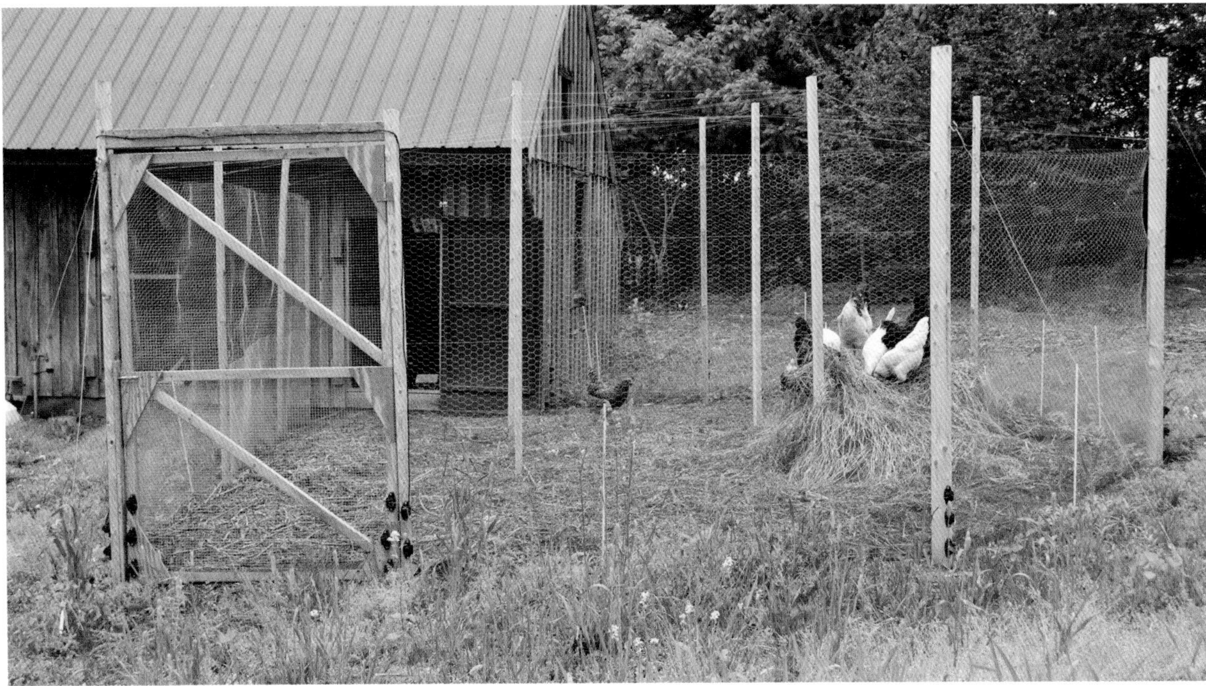

Figure 10.3. Super-secure composting run.

belowground ends with Bora-Care and the rest with a clear sealer.

- I attached guy wires on five posts where there was a sharp turn in the fence line. That may seem like overkill for such a short run of fencing, but we get wet, clinging snows here, which can put a lot of strain on inadequately supported chicken-wire fence.
- I made the door from 2 × 2s and plywood scrap for corner braces. (Use of triangular plywood corner bracing helps keep the frame from twisting.) It is clad with leftover ¼-inch hardware cloth.
- I dug the wire netting 8 to 12 inches (20 to 30 cm) into the ground to deter digging predators.
- For climbing predators, I ran electric wire at 3, 8, and 12 inches (8, 20, and 30 cm) around the perimeter of the fenced enclosure, using the same type of plastic insulators as shown in figure D.11 in appendix D. In the runs between posts, I used nonconductive 3-foot (90 cm) fiberglass rods to stand off the electric wires from the chicken-wire fencing. Notice that the position of the door made an unbroken run of the three charged wires around the entire perimeter impossible. To solve the problem, I ran insulated cable to carry the charge up and over the framing for the door. From the end of the insulated cable, I made two runs of charged wires, one across the low end of the door, one around the rest of the fence perimeter. To charge the fence wires, I used the same HotShock 5 energizer I used in the flock's "ultimate mobile shelter" (described in appendix D), plugged into an electric socket in the henhouse.
- As for aerial predators, I started by setting four additional posts in the interior of the run; then strung many runs of monofilament fishing line from the fascia boards of the henhouse to and among the tops of all twelve posts. (More about that strategy in "Aerial Predators" on page 202.) I had no losses, despite the frequent presence of a red-tail who made several hits on nearby flocks.

CHAPTER 11

Ranging Flocks Using Electric Net Fencing

Readers who have stuck with me so far will have noted that my poultry husbandry is as low-tech as I can manage. But there is one technological marvel that, in most of my decades as "the chicken man," has been fundamental: *electric net fencing*. Though electric fencing of any sort is not appropriate in most urban and suburban contexts, for me—on a couple of acres in a small rural village—it resolves the restriction/liberation conundrum, by giving my flocks plenty of range to forage while keeping them within needed limits. And it is incredibly effective at protecting them against predators on the ground. (It does not protect against aerial predation. For more about dealing with aerial predators, see chapter 21.)

Good electric net fencing is expensive, but I wouldn't cut corners on quality—money saved if you buy a cheap system vanishes if you lose chickens. If you buy the best equipment from a reliable company, the initial investment in a roll or two of netting plus a decent energizer and a few accessories will set you back several hundred bucks. However, that investment buys you one of the most useful of all tools for managing the homestead or farm flock, which with good care will last a long time. It buys you the ideal compromise between maximum health and well-being for your ranging flock and maximum protection from the heavy hitters in the neighborhood.

Since a good fence system is a considerable investment, thoroughly research the variables when planning your system. There are many options for netting and energizers, and potential gotchas for an inexperienced user. Before placing an order, be sure to talk with the technical support staff at your preferred supplier to make sure your choices are sound. If you are not met with courtesy and with precise technical information and clear answers to your questions, look elsewhere for a supplier for your system. (My preferred source is Premier.[1])

The larger the area on which you range flocks, the more prohibitive the cost of using electric net fence to protect them. Are there effective alternatives using single-strand electric fences? I have never experimented with this option myself (though I have used single-strand to protect chickens inside standalone shelters—that is, ones not inside an electric net perimeter). However, some members of APPPA use single-strand electric perimeters over a couple of acres or more to confine their market flocks. Some put up fences of only two strands, some with up to seven, usually with excellent deterrence of predators. It seems a fact of life for most, though, that single-strand fences allow some ranging of chickens outside the perimeter, where they are may fall prey to predators. It is for the individual flockster to decide whether such occasional losses are acceptable, given the savings involved and the more extensive range for the birds.

While my discussion of electric net will focus on its use on pasture, remember it can also be used to range flocks in woods or in brushy areas—though dealing with uneven ground or threading the needle among trees or shrubs may complicate setting up the fence. I have often used electric net on lawns, as seen in the background of figure 10.2.

Net Design

An electric net fence is a lightweight plastic strand mesh, with interwoven support posts, easily set up and easily moved. The vertical strands, for support only and carrying no charge, are securely attached at each intersection with the horizontals, which are twisted with several almost hair-thin stainless-steel wires that carry electric charge (except for the bottom one, of course, which lies directly on the ground when the fence is in place and thus must *not* carry a charge). The posts—fixed into the net at intervals of 8 to 12 feet (2.5 to 3.5 m) or so and made of plastic, to prevent grounding of the charge—end in steel spikes the user

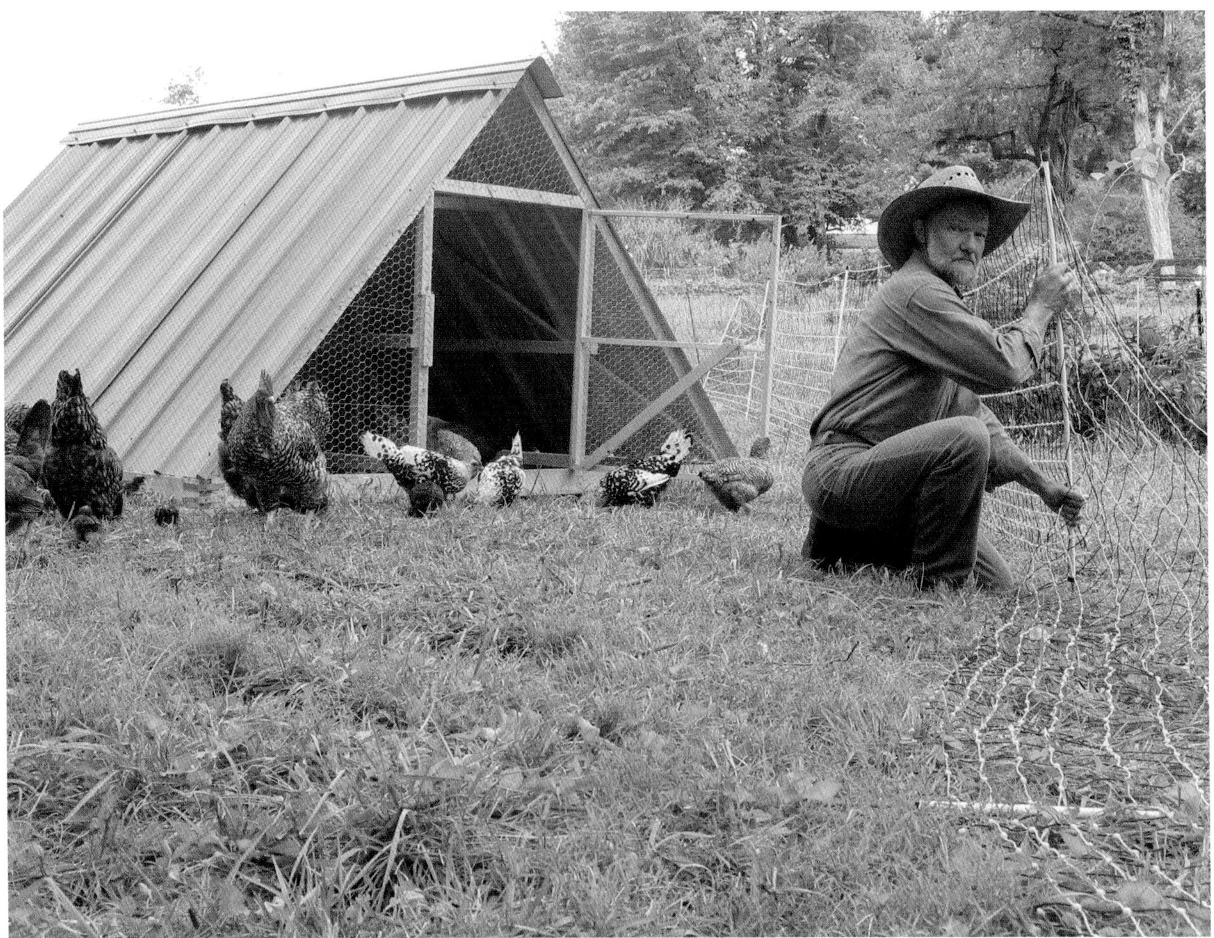

Figure 11.1. It is easy to set up a protective perimeter of lightweight electric net fencing.

pushes into the earth to erect the fence. Typically, the bottom horizontals are more closely spaced than those higher up, both to contain young growing birds and to make better contact with predators at nose level.

All the horizontals are twisted tightly together at each end of the net to share a common charge, ensuring that current remains available in the entire net, even if there is a break in an individual horizontal line. The braided strands on each end terminate up top in a metal clip, which can be connected to the companion clip on an adjacent net, to create multi-net perimeters of any size needed.

Electric mesh nets are available for other purposes as well, but those designed for poultry are available in heights of 40, 42, and 48 inches (1, 1.1, and 1.2 m). You might assume "the higher the better," but the 48-inch net is more difficult to handle when it's time to move the fence. I've always used 42-inch nets with excellent results. Of course chickens can fly that high—but they could fly over a 48-inch net as well. Once they get zapped a couple of times, however, they rarely do.

The standard length of poultry nets is 164 feet (50 m), though "half nets" of 82 feet (25 m) are available. Having a half net on hand can help make the end of a perimeter "come out right"—in lieu of doubling a full net back on itself when it is too long for the perimeter being laid out.

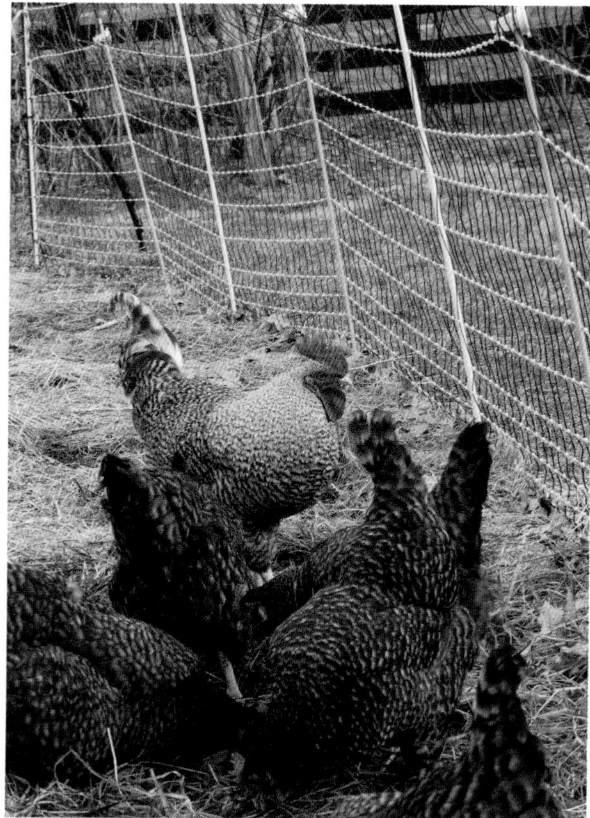

Figure 11.2. Except for the bottom line (*hidden in the litter in this photo*), all horizontal lines in the electric net carry electric charge. Support posts are an integral part of the net.

Setting Up the Fence

A tight fence makes better contact with a curious predator, or a chicken, than a sagging one and thus delivers a more effective deterrent shock. Some users prefer posts with double spikes: Not only do two spikes at the base of each post hold better in wet or loose soil, but the additional spike (welded onto the first in an upside-down *L* shape) offers a right-angle step for pushing on with a foot. I've always used the single-spike version, which is lighter and thus easier to move. In compacted or gravelly soil, I drive a hole for the spike with a piece of rebar and a small sledgehammer, then reinforce the fence as described later in this section.

Poultry netting will arrive properly folded and rolled. The end posts will be obvious—they're the ones with all the horizontals twisted together. Holding the entire bundle in one hand, start with one of the end posts and lay out the net flat on the ground, one panel at a time to the end. Now go back around the perimeter, poking the spiked posts in place to stand the fence upright. Mow the outside of the perimeter, as close to the fence as you can get. Now remove the fence and toss it on the ground, well inside the perimeter line you have mowed—and cut another swath with the mower, to the inside of the first. Stand the fence back up, in the middle of the double-mowed swath.

When mowing, set the mower blade low—the shorter the grass, the longer before it is necessary to

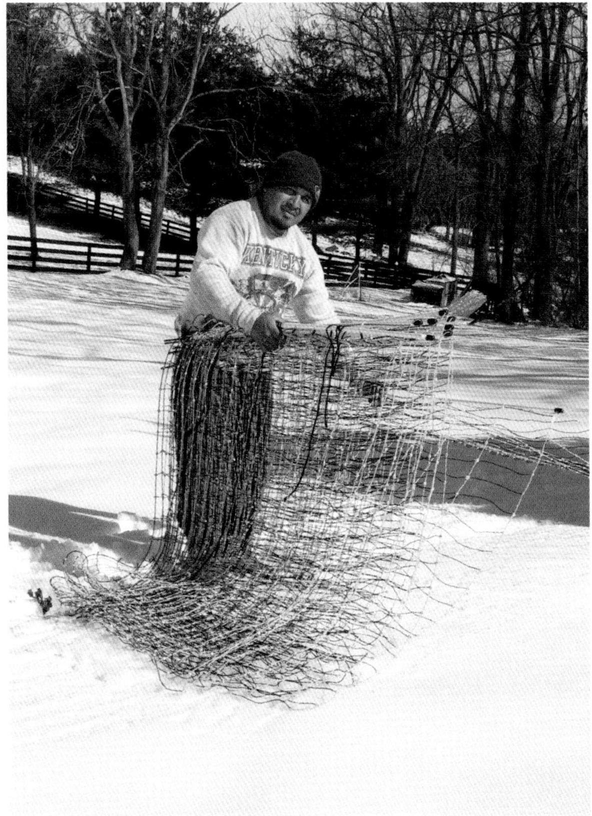

Figure 11.3. Holding the bundle of support posts in one hand, play out the net one panel at a time around the new perimeter.

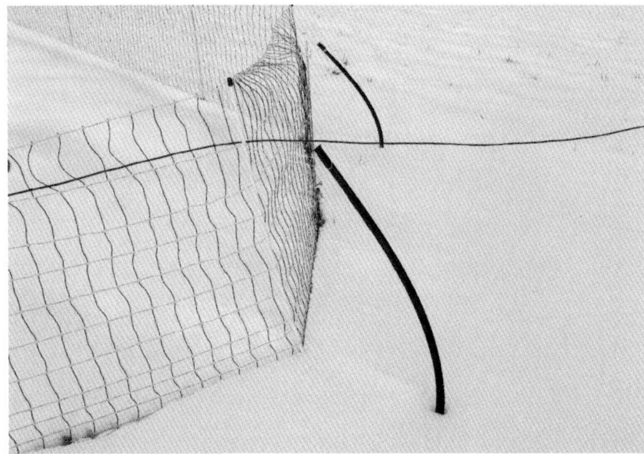

Figure 11.4. For a sag-free net, tie corner posts to support posts wherever there is a sharp turn in the fence line. Note as well the black insulated cable, which carries charge from a plug-in charger in the henhouse to this remote electric net perimeter.

mow the fence line again. If you prefer using a scythe and can mow a lawn with it, you can make the perimeter with your scythe.

Now tie every corner of the fence—not only right-angle corners, but anywhere it angles significantly—to a corner post, preferably with biodegradable twine. These more substantial corner posts, also with steel spikes, are available from your electric net supplier, but should be readily available in the electric fencing section of your farm co-op as well.

The resulting fence may or may not satisfy you, depending on how insistent you are on setting up a sag-free fence. I like a tight one, so in the middle of each panel—the section of mesh between posts—I place an additional ⅜-inch fiberglass rod, with a screw-on plastic insulator lifting the top strand, thus adding tension and removing sag from the fence. Be sure to use *coated* rods, available from electric fencing suppliers: Uncoated rods will weather over time, resulting in splinters that will stick into your hand. (In figure 11.2 the added mid-panel rods are the ones with the yellow insulators at the top.)

Plan to *rotate* your fenced areas over the available pasture to prevent excess wearing of the sod, break life cycles of pathogens and parasites, and ensure even application of the flock's manure as a fertility boost to the sward. Frequency of rotation depends chiefly on stocking density and the point in the growing season—the pasture will recover more quickly during lush spring conditions than during dry summer.

You can set up each perimeter in the rotation as an island unto itself, entirely separated from the poultry house—in which case you will need to use a mobile pasture shelter, discussed in chapter 12. Or you can anchor rotation plots onto the poultry house. Ideally, it will have multiple entranceways (if only pop-holes of chicken size). If the chooks always use the same entrance, the area around it turns into poop-dotted, unwholesome bare ground. If there is no alternative to continual use of the same entrance, cover the bare spot with deep mulch.

Remember when laying out the fence that you do *not* need to make a complete loop with the two ends of the fence attached to each other. Even if the far end terminates at the wall of a building, for example, contact with the fence at any point will cause a flow of electricity from the energizer through the body of the predator and into the ground.

It is rarely necessary to include a gate in the layout: At least in the case of the 42-inch (1.1-m) fence I use, I can easily swing a leg over to straddle the net, then follow with the other leg. Of course, you want to be *certain* the power is off before doing so!

Moving the Fence

Moving the fence is easy—*if* you do it properly. If you handle the fence carelessly when moving it, I promise you will weep with frustration. Most important, *avoid the temptation to roll up the net.* Begin by laying it out flat. Then, starting at one end, *fold* each panel in half as you draw the end post over to the next post. Repeat until all the panels are neatly folded, like pages of a book, with the posts in a bundle like the book's spine. Now pick up the gathered net and lay it out around the new perimeter.

A big challenge when moving the fence is keeping the flock from scattering to the four winds while you do so. Ducks and geese aren't much of a problem—once you have the new fence mostly set up, they are easy to herd into it. But herding chickens is like loading cats in a wheelbarrow—save yourself the aggravation of trying.

I have found it a good investment to have extra rolls of netting on hand, so I can set up my new perimeter—usually adjacent to the existing one—while the birds continue foraging in the old one. If they resist moving, I collapse the old fence—that is, pull it ever more closely around the flock—until the birds have no alternative but to enter the new paddock.

If you don't have additional nets, the best strategy is to leave the chickens shut up while you reconfigure the fence in the early morning. If they are in a mobile pasture shelter, move it into the new enclosure before releasing them. If the net is anchored on the poultry house, simply open the appropriate door or pop-hole onto the new enclosure.

Charging the Fence

When you inevitably get a taste of the fence, you will understand why it is so incredibly effective. The jolt you feel, however, will likely be when you're on your feet, with the soles of your boots giving you considerable insulation from the ground. If you hit it while down on one knee, more solidly grounded, the wizard in the fence will rattle your teeth. *That's* what the predator feels.

Keeping that hottest possible spark in the fence is thus the key to predator deterrence and is a matter of both equipment and management.

Energizer

There are many models of fence charger to choose from, depending on your needs. Remember, though, that electric netting requires a more powerful energizer than single-strand electric fencing. Err on the side of buying a charger with more voltage capacity than you need, rather than too little.

The major division is between chargers that plug into household current and those powered by a battery. I have routinely used both. My plug-in model is in the henhouse, which is wired for electricity. If the fence is anchored on the henhouse, I connect the charger's terminal to the end clip on the net. For freestanding fences, I pull current from the plug-in unit, up to 200 feet (60 m) out on the pasture, using insulated cable on the fencing rods with screw-on insulators, mentioned earlier, to make it more visible and prevent tripping. This is the arrangement I prefer, since the plug-in energizer will charge more nets and will take more weed load on the fence before losing voltage. I run the cable to a manual shutoff, so I don't have to walk all the way to the henhouse to cut power to a freestanding net. If there are multiple netted enclosures in the system, their cutout switches can be wired either *parallel* (all other nets in the system remain charged when an individual switch is open) or *in series* (nets farther down the line also lose power when a switch is open).

Battery units can be small or large, powered by batteries either disposable or rechargeable, from a couple

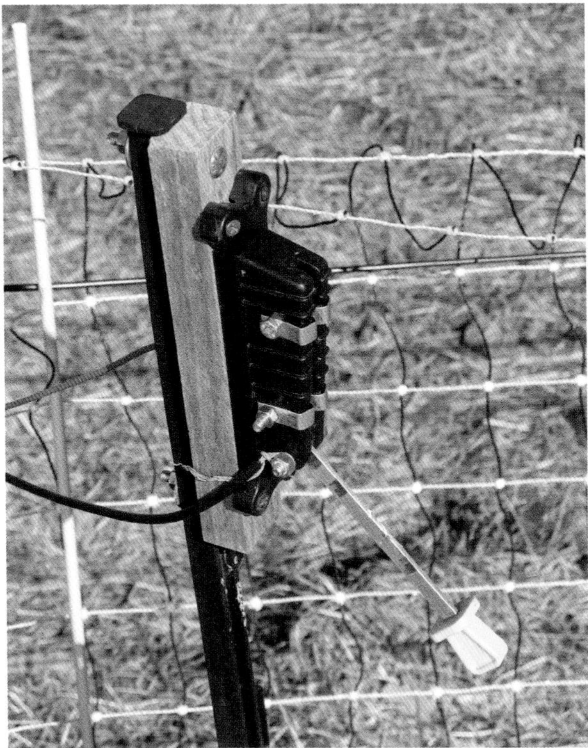

Figure 11.5. This is a manual cutout switch, used to kill power to a net temporarily—for example when entering the enclosure to feed. Here the lever is in the open position (no power to the fence).

Figure 11.6. Battery-powered energizer with solar panel. The rechargeable battery is inside the casing, at the bottom, and the controller is in the top part of the casing, under the panel. The warning sign is a good idea, required by law in some jurisdictions.

of D-size up to an automobile battery. Depending on its capacity, a battery charger may power multiple enclosures with many rolls of netting.

Some battery-powered energizers accept an add-on solar panel with a controller to maintain charge in the fence while trickle-charging the battery—at night or on heavily overcast days, the controller draws from the battery to charge the fence. For freestanding use in a site too remote to serve conveniently from household current, the solar-powered energizer is hard to beat. (The lead-acid, deep-cycle rechargeable batteries in these units can also be recharged on a plug-in trickle charger.)

Testing the Charge

Essential to making sure the fence will deter predators is knowing how much charge it is carrying, and that requires testing. Hint: The best way to test the net's voltage is *not* by grabbing it for a feel.

There is a country-boy version of testing voltage. Pull a long green stem of grass and lay its far end on the fence. Now slide the stem so the length between your fingers and the wire is progressively shorter. At some point you will feel a tingle of electric charge. Someone expert at reading that tingle will know approximately how strong is the spark in the fence. For greater precision, however, I recommend an electronic tester.

Please do *not* buy one of the dinky little testers with five feeble lights to signal level of charge, probably available at your local farm cooperative—it is impossible to tell whether the lights are on or off in full sunlight. Though it may cost more, buy a more sensitive tester with either much stronger indicator lights or an LED readout.

Nature of the Charge

You may be interested to know that the highest voltage I ever measured—on a single roll of net, completely free of weed growth, on the AC charger—was 9,700 volts. If your response is "*But 9,700 volts would kill you!*" understand that, while the *voltage* from this type of energizer is high, the *amperage* is correspondingly low—which is to say, the whiplash sting will wake you up but do you no real harm. (The fence does have its hazards, however. See "Hazzards of the Net" on page 106.)

Another technical matter: Electric net is designed for use with *low impedance energizers*. High impedance chargers, especially of the weed-cutter type, can seriously damage the fence, generating enough heat to melt the plastic strands, burn the wires in two, and—worst case—even start a fire in dry grass.

Robust Grounding

The strength of the jolt from the fence depends on how well grounded the energizer is. Don't rely for ground on a pipe driven haphazardly a foot or so into the soil. The ground for my plug-in unit consists of three 8-foot (2.5 m) steel rods driven full length into the earth in the dripline of the poultry house—even the night's dew dripping off the roof keeps the soil moist—joined with heavy-gauge wire, then connected to the ground terminal on the unit. Such a ground will never fail.

Ensuring a good ground for a battery-powered charger, especially when the ground is dry, is more challenging. Some battery units come with a mounting bracket with a metal stake that both holds the charger and serves to ground it. Do *not* rely on such a ground stake, which penetrates the soil only 8 inches (20 cm) or so. Use something like a galvanized 3-foot (90 cm), ½-inch (1.3 cm) grounding rod instead, driving it as deep as required for good ground, deeper in soil that is dry. (Not too deep, however—the ground rod may be hard to pull out when it's time to set up the fence elsewhere.) Pouring water around the ground rod—say, when rinsing a field waterer—will moisten the soil and improve grounding of the system.

Keep It Clean

While thorough grounding of the energizer is essential, accidental grounding in the fence will sap its charge. Check the fence line frequently to ensure that limbs have not fallen on the fence, forcing charged horizontals onto the soil. When installing the fence on uneven ground, make sure the lowest charged horizontal doesn't touch the earth. (Remember that the bottom horizontal, in contact with the ground, is a plastic strand only.)

Inevitably, as grass and weeds grow, contact with the bottom wire will reduce charge in the fence. When it drops to 3,000 volts, or 2,500 at the lowest—the recommended minimum needed to turn back predators—you should mow the fence line again. Simply stand the fence a couple of yards to the inside of the perimeter, to keep the birds from wandering while you mow. It often happens, however, that the point at which weed growth requires mowing will be about the time you want to rotate the flock to a new paddock anyway.

Challenges to Containing the Flock

Small hatchlings just out of the brooder may slip through the mesh of the fence. If both feet lose contact with the ground as they step onto a charged wire, they receive no shock. Newer designs of electric netting feature bottom horizontals closer together than they used to be, reducing the possibility of small birds slipping through with impunity. But some do—especially chicks—even with the newer netting. (Premier offers two versions of special chick netting, with mesh small enough to keep the smallest chick from slipping through. One version is nonelectrifiable—just a physical barrier—and the other can be electrified for around-the-clock predator protection.)

You might prefer to raise your hatchlings inside until they're too big to slip the net. Most of my young birds on pasture have been hatched by a broody hen or broody duck, and tend to stick close to Mama rather than seek adventure outside the fence. Yes, chicks wander outside on occasion, but I don't know that I ever had any losses as a result.

Clipping Wings

I would never clip the wing of a free-ranging chicken—making her more vulnerable to a predator. But if I am taking responsibility for protection with electric net, she *must* stay inside the fence. Thus I occasionally clip the wing of a rogue flier with a habit of leaving the enclosure.

Assuming you are right-handed, hold the bird and spread her wing with your left hand as in figure 11.7. Notice that the large flight feathers are in two distinct groups—the primaries in front and the secondaries behind. The secondaries are also called the *coverts* because they *cover* the rest of the wing when it is folded; thus, if we clip the primaries *only*, the clip is hidden and does not mar the appearance of the bird.

Simply shear off the feathers, about an inch from where they emerge from the follicle. If the feather is still new, on a younger bird or one who has just molted, there will still be blood supply into its shaft, and clipping will cause a bit of bleeding, though apparently not any pain. Otherwise, cutting off a feather is like trimming our own hair or fingernails.

Note especially that the clip is made on one side *only*. The idea is not to cripple the bird—a clipped bird will still be able to fly up onto her roost at night—but to imbalance her enough to discourage taking to the wing.

I make an occasional exception to clipping only primary feathers, and only on one side. Guineas can be *much* more determined fliers than chickens. If I have a normally clipped guinea who persists in flying out of the net, I first clip off *all* the flight feathers on the one side, both primaries and coverts. If she still flies out—amazingly, it happens—I then do what I call a radical clip, shearing *all* flight feathers on *both* wings. Unless she learns how to fly with her eyebrows, she absolutely will remain inside the net thereafter.

Figure 11.7. Clipping the wing of a "rogue flier."

As said earlier, chickens can fly as high as 42 inches (110 cm) or even 48 inches (120 cm) if they want to. Fortunately, once they've hit the fence a time or two with beak or comb, they almost never do. And it is possible to *train* the birds to the fence—that is, deliberately entice them to get a shock so they are shy of it thereafter, and less likely to fly over it. Just sprinkle a thin line of scratch grains right along the bottom of the fence—sooner or later everybody gets a taste of what's in the fence. Seem kind of mean to you? Yeah—but predators are way meaner.

Stopping Predators: Successes and Failures

The fur of many predators insulates against electric shock, and some could certainly jump over an electric net or walk right through it. How does the net deter furry predators like coyotes, large dogs, and bears? Fortunately, these guys lead with the supremely sensitive nose. Thus, if you set up and manage electric net properly, it gives an extraordinary level of protection.

In contrast with numerous flocksters I've known whose poultry husbandry should more accurately be termed "feeding the foxes," I can report that—with a single exception—we've had no losses to foxes in almost four decades. We often saw foxes trotting nonchalantly through the pasture, passing within yards of the flock on the other side of the net, but not even *looking* at the chickens—obviously they had already met the demon in my fence. A large dog once got tangled in my fence while trying to get at my flock. In his panic he broke one of the stout corner posts into *three* pieces before getting free—sans chicken dinner. A friend reports twice seeing her electric net fence turn back a black bear.

Electric net fence can be laid out in any configuration you need, but there is an important caveat: *Avoid long, narrow alleys*. I once lost a young goose to a cause unknown. Testing the fence proved it was plenty hot, there was no limb down on the perimeter—and geese, incredibly clumsy on the wing, should not have flown over it. A few days later I saw the apparent culprits—a couple of dogs using the hunting skills of the pack to attack the geese. They would not touch the fence—obviously they had learned to respect it—but using the pack mentality that lurks below the "good poochie" in all dogs, one of them would rush the geese inside a long narrow corridor between the coop and the main pasture enclosure, intent on forcing the panicked geese, normally reluctant fliers, over the fence—and into the waiting jaws of the other dog. Following that sad lesson, I always configured fences as large interior spaces—frightened birds would naturally retreat to the center, in lieu of panic flight over the net.

We once lost three hens inside a netted paddock on successive nights, their necks bearing the telltale marks of attack by a blood-feeding weasel. It could only have been a least weasel—incredibly small but admirably equipped to kill—who slipped under the bottom wire over extremely dry soil without getting a shock. That was when I installed the more robust grounding system previously described.

Do remember that keeping the fence line clean means more than just mowing periodically to cut back weed growth. I once had a dozen layers tilling new ground in preparation for putting in a blueberry bed. The ground was on a slight incline—just enough to cause uprooted grass and other debris from the hens' work to sift down in an accumulating little ridge covering the foot of the net. It nagged at me daily as it grew (*Better clear that line, boy!*) but you know how it can go on the homestead—always on the run. And lo and behold, one morning I arrived to find four splashes of feathers out over the pasture. I grabbed the fence to feel exactly what Mr. Raccoon had felt when he was looking for his dinner—*nothing*.

Caring for Electric Net Fencing

The service life of high-quality electric netting is said by the manufacturer to be ten years. My experience over the decades is that it can be considerably longer than that—*with proper care*. Never leave nets not in service on the pasture, exposed to weather and ultraviolet radiation—store them properly when not in use. Using

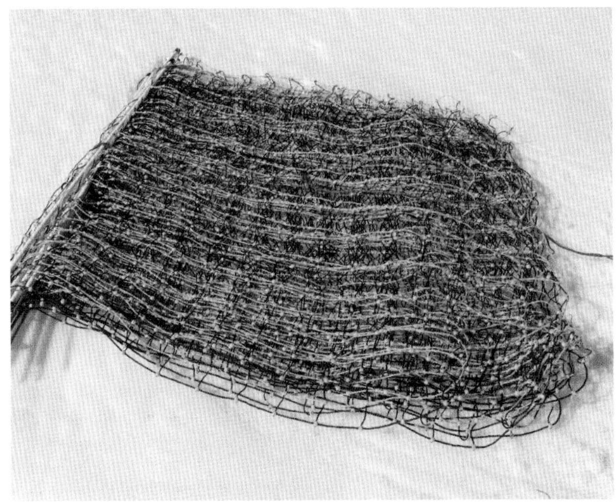

Figure 11.8. Tying a "lead string" before rolling the net for storage will reduce frustration next time you start laying it out.

Figure 11.9. One way to store nets to block wee rodents looking to make nests.

a string trimmer (weed eater) to clear the line would be lunacy. On a good day I can use my scythe to cut grass close to the fence. If I'm tired, though, not even the sternest reminders to "Be careful of the fence!" will prevent an eventual cutting of a couple of strands.

Netting is surprisingly strong and flexible, and can take a good deal of stretching without damage. There are limits, as we discovered one year after two record-breaking snowstorms that came back-to-back, causing several breaks in the mesh. On the other hand, we've had ice storms that deposited up to a quarter inch of ice on the strands, causing the fence to sag badly, but with no damage. Further accumulation would probably cause some line breakage, of course. Fortunately, breaks in individual strands are easy to repair. You can buy repair kits from your fence supplier, featuring small brass ferrules crimped around spliced-in lengths of fence strand to bridge the gap. It's cheaper, however, to cut snippets from a coil of single-strand (the braided plastic-and-wire type, which I keep on hand for other electric fencing applications), and simply knot them onto the broken ends on either side of a break.

When preparing the nets for storage, I recommend using what I call a *lead string* to keep the next deployment of the net tangle-free. Start as for moving the fence, with the net *folded* into a neat cluster of posts on one side, and the panels laid out flat beside them. Tie a piece of discarded baling twine around the cluster, a little longer than the length of the folded panels (see figure 11.8).

Yes, I said earlier, in the section on moving the net, you should never *roll* the net, you should *fold* it instead. Here's the exception: The final step in preparing the net for storage is to *roll* the *folded* net, starting with the posts, to make a neat bundle with the posts at the core and the tail of the lead string protruding. Tie the bundle near each end, and it's ready for storage. Next time you deploy the net, *hold on to the lead string as you roll out the bundled net.*

It's difficult to describe the problem that the lead line prevents, and you know what?—I'd bet you're going to ignore my advice until the first time you encounter that problem. Again, using the analogy of a folded net as a book, we can start either with the first page or the last page (the end post at either end of the net) when laying out the net. If we open into the *middle* of the book, however, it can be quite difficult to find our way out to the first or last page (one of the end posts). Once you've spent a quarter hour of frustration getting the net opened out properly, you'll understand the need for the lead line.

As for storage, by far the best advice I can give is: *Do not store nets on the ground, floor, or even on a shelf.* That is, do not store them *anywhere* accessible to your rodent friends—who will happily chew your nets into very expensive nests. The best option may be to hang them from the rafters of a storage shed (see figure 11.9).

Hazards of the Net

Potential dangers from electric net fencing may be structural or electrical. Attentive, careful handling of netting prevents snagging and tangling by the user, but animals can become entangled accidentally. For example, I once had to cut a fox out of a net (fortunately not powered on at the time) in which it had gotten hopelessly tangled.

With older versions of electric net, young waterfowl would occasionally get tangled after putting an exploratory head through the mesh. Because their anatomy is more front-loaded than that of chickens, it is more difficult for them to reverse out of the net. A few fatalities resulted when I didn't arrive on the scene in time to free them. I never had a duckling or gosling get tangled in the newer style of nets with more closely spaced bottom horizontals.

In the early years there were occasional deaths of wild animals on the fence: One or two box turtles, once even a possum, became jammed under the lowest charged wire, unable to escape the pulsing current, which burned holes in shell or hide. A couple of blacksnakes crawling over the lowest charged wire died in the same way. After several years using electric net, however, I had no further such fatalities. I think the animals in the neighborhood came to recognize the fence as a threat to be avoided, and "word got around."

Despite what I said about the low danger of a jolt, it is believed that a shock to head or spinal cord has greater potential for injury. Don't take chances—*cut power to the fence when working close to it.*

The most tragic example of electrical hazard is that of a crawling infant in wet grass who got tangled in a charged net and died. *Never allow young children near the net unsupervised.* Warn visitors, of any age, who are unfamiliar with electric fencing. Attach highly visible electric hazard signs on your fences (see figure 11.6), as required by law in some jurisdictions.

CHAPTER 12

Mobile Shelters

If you day-range your flock, or use temporary electric net fencing anchored on the henhouse to rotate the flock over fresh plots, the birds always return to the same shelter at night. If you pasture them farther afield, however, you will need a mobile shelter of some sort to rotate them to new ground, and to shelter them at night or when it rains. I've seen hundreds of mobile coops, and no two are ever the same. The design you come up with depends on the size of your flock, how you intend to use their services, leftover materials from other projects begging to be used, the nature of your climate and ground—perhaps on how whimsical you happen to be feeling.

The first movable shelter I built was a copy of the classic Polyface design—like the one being framed in figure 12.2. If Joel Salatin's mobile pens can produce tens of thousands of market broilers a year to put money in the bank, surely all of us creative amateurs can come up with shelters that allow our birds continual access to fresh grass while protecting them from opportunists on the prowl.

Designing a Pasture Shelter

The following are some issues to ponder as you plan your mobile shelter project. It could help with your planning to look as well at appendices C and D for design and materials considerations and step-by-step construction of my two most successful all-purpose shelters.

Pasture Pens and Pasture Shelters

Micro-flocks on lawn or pasture are often confined entirely to the shelter, which is moved frequently to new grass. The larger the flock, however, the larger the protected space needed for the birds. Use of electric net fencing gives the birds an extensive area to roam outside the shelter. But if you do not use electric fence, you might provide your mobile shelter with a separate *pas-*

Figure 12.1. My friend Jon Kinnard combined whimsy, utility, and the urge to recycle in this micro-flock mobile shelter. It is entirely self-contained, with feed storage and nest in the compartment under the hinged metal roofing and roosts in the rest of the shelter. Photo courtesy of Deborah Moore.

Figure 12.2. Mobile shelter under construction, based on Joel Salatin's Polyface broiler pen. Note how even a structure as large as 10 by 12 feet can be made with light framing when it includes smart use of diagonal bracing.

ture pen using a set of light wood-frame panels covered with chicken wire, easily locked together using bolts with wing nuts, and just as easily disassembled for moving. Whether you need to attach a cover over the top of the pen will depend on aerial predation where you are.

Cody Leeser, a friend of mine, designed a pasture shelter-and-pen set for her flock of half a dozen layers: She mounted a small shelter (2½ by 3½ feet [76 by 107 cm]) on a landscaper's wagon, complete with roosts, nests, and a ramp. She made a separate 8- by 8-foot (2.5 by 2.5 m) pen, 4 feet tall (1.2 m) and with a cover of wire over the top, and mounted on small wheels for moving. Framed into one side is a narrow opening into which the door of the shelter docks. A nylon strap, shown in figure 12.4, runs through the shelter and attaches to its door, hinged to serve as a ramp. In the morning Cody moves the pen onto fresh grass; wheels the shelter into docking position; then, using the strap for remote-control, lowers the door into ramp position to release the hens into the pen. At dusk they retreat into the shelter on their own, and Cody pulls the ramp back into place with the strap, to guard against unwanted night visitors.

Trade-Offs: Size, Weight, and Stability

The size of the shelter will be determined by the size and nature of the flock it will shelter and its intended use. As said, the first mobile shelter I built was a copy of the classic Polyface model, 10 feet by 12 feet (3 by 3.5 m) (see figure 12.2)—first used to raise fifty comparatively inactive Cornish Cross broilers at a time: Each bird had about 2½ square feet of shelter space. When I later used that same shelter for confined layers, I limited the number of hens to sixteen—7½ square feet (.7 square m) each. Remember that you will be more likely to rotate your birds

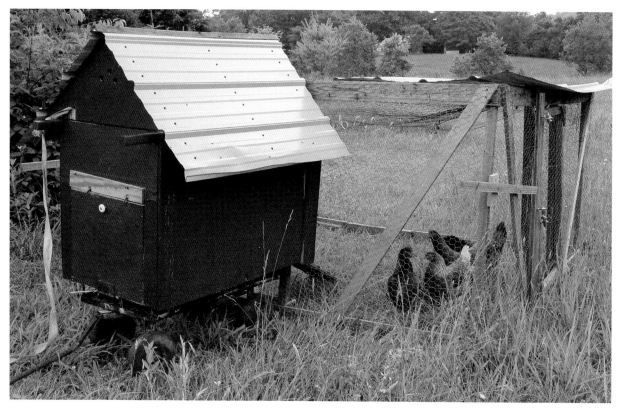

Figure 12.3. Cody Leeser's ingenious design for a small wagon-mounted shelter and a separate wheeled pasture pen. She moves the pen each morning, then docks the shelter onto the pen and releases her hens into it for the day.

to fresh grass as frequently as you should if it's easy to move their shelter. It might make sense to split the flock into two smaller shelters rather than keeping them all in one large one that is more difficult to move. At the low end of the scale, a shelter could be designed as a chicken tractor, holding six to ten tiller chickens and sized to a single garden bed. (See sidebar "Chicken Cruiser" on page 110.)

As with the main coop itself, size has everything to do with whether it will be sleeping quarters only for a flock that is ranging outside during the day or will confine the birds full time.

The heavier a shelter, the more difficult, and possibly the more dangerous, it is to move. On the other hand, the lighter it is the more likely it will be tossed into the next county by a rambunctious wind. Of course, it would be possible to anchor even the lightest shelter to the ground; but again, the more difficult we make

Figure 12.4. A key to Cody's design is the nylon strap that runs through the coop, allowing "remote control" raising and lowering of the ramp on the far end.

Chicken Cruiser

Andy Lee and Pat Foreman introduced the idea of the *chicken tractor* (or as I call it, a *chicken cruiser*)—a small, easily moved chicken shelter sized to fit a single garden bed. A few laying hens till and fertilize the bed while finding free food in the form of worms and slugs and snails, but have no access to adjacent beds.

Since this tractor gets maneuvered in tight spaces and needs to be moved frequently, it is better to make it small and nimble. Be sure to provide enough cover on parts of the sides and top for shelter from blowing rain, and for shade in hot weather.

Figure 12.5. My best-ever garden cruiser–light, nimble, and rock solid.

Figure 12.6. Chickens in a cruiser work a single garden bed while leaving adjacent ones undisturbed. Separate hinged lids–one aluminum roofing, the other wire-on-wood framing–give access to the interior. Recycled milk crates attached to the framing in the covered part of the cruiser serve as nestboxes.

Figure 12.7. Diagonal bracing in this A-frame shelter provides rigidity without excess weight. Note the collar ties, set low enough to serve double duty as roosts.

a move—undoing and redoing a complex anchoring routine—the more inertia will inhibit frequent moves. Shape also plays a part in stability in heavier winds: I have found the boxier-type shelters with a higher profile catch the wind, while hoop or A-frame shapes tend to keep their feet on the ground. (The classic Polyface model, 10 by 12 feet [3 by 3.5 m], is indeed rectangular in shape, but is only 24 inches [60 cm] high and stable even in strong winds.) Materials choices (page 113) have the biggest impact on weight of the shelter.

Diagonal bracing greatly reduces weight of the frame. I framed some of my early shelters in 2 × 4s exclusively, all at right angles—the results were clumsy, inelegant, and balky about moving. I discovered that smaller shelters do not need full 2 × 4 framing even for the bottom rails. I began ripping 2 × 4s to 2¼ inches (5.7 cm) to use as the bottom rails, and using the resulting 1¼-inch (3.2 cm) strips for the verticals and diagonal bracing. Thus the weight of the entire frame was not much more than the bottom rails alone before this modification.

You can also reduce weight by positioning bracing where possible to do double duty as roosts. In a larger A-frame, for example, you will want to include collar ties, the horizontal pieces that tie the rafters together, providing greater rigidity. Position them low enough below the peak to allow use as roosts by two or three hens. Horizontal stringers reinforcing the frame can also be positioned for use as roosts.

A final option for reducing weight is to use mesh wire as much as possible in lieu of solid material, consonant with the need for protection from rain, sun, and sharp chilly winds in part of the shelter. In many shelters wire replaces at least part of the roof, and much of the sides. Use of wire has the further advantage of maximizing airflow and sunlight into the interior.

Wheels

Include wheels in the design of any larger shelter. Instead of installing axles across the entire width of the shelter, I permanently install ½-inch bolts in the bottom rail at each corner, using nuts, flat washers, and lock washers. Pop a wheel on each axle bolt and lock it on with a wing nut and you're ready to roll. (See figure C.15 in appendix C.) The wing nuts make it easy to use a single set of wheels for multiple shelters.

In my first A-frame (shown in figure 12.10), I used 8-inch (20 cm) wheels. If your ground is nice and even, 8-inch wheels might work fine for you. But I found that, on my uneven pasture, wheels of that diameter didn't give enough clearance—the bottom rear rail of the shelter frequently hung up on bumps or tussocks of grass when I moved it.

When I built the improved A-frame I describe in appendix C, I used 10-inch (25 cm) wheels instead. But here's a trick I used for even more clearance: Previously I had mounted the bolt for the wheel squarely in the center of the bottom rail. But now I offset the hole for the bolt by 1 inch toward the bottom of the rail. Aha, the combination of the greater wheel radius and the lower axle hole gave me an additional 2 inches (5 cm) clearance rather than one.

Based on your own needs, you may prefer a smaller clearance—to prevent escape of young birds from inside while moving, for example—but I loved the new ease of rolling my Chicken Ferrari over my bumpy pasture.

The Ultimate Mobile Shelter

The wheel is a great thing—I described on the previous page how it made the moving of even a substantial shelter easy. But you know what? Putting on and taking off a set of four wheels using axle bolts and wing nuts, for a *daily* move—that gets *old*, fast. Not long ago I discovered the *lift kit*, an ingeniously designed and sturdily made hardware set that revolutionizes the process of moving a heavy shelter (up to 400 pounds [180 kg]). The shelter comes up on its wheels, ready to roll, literally within seconds—and another of man's oldest tools, the lever, makes lifting and pulling from the front a breeze.

It would be hard to beat this design for a stand-alone (not protected by electric netting) mobile shelter. See appendix D, "A Robust Mobile Shelter with a Lift Kit," for construction details and use of a lift kit in the design of my best-ever "mobie."

Figure 12.8. The ultimate stand-alone mobie. Use of the superbly crafted lift kit is revolutionary.

If wheels are to be permanently installed, bicycle wheels—or other large wheels looking to be recycled, like the front wheels from an old tractor—make moving over uneven ground easiest of all.

Does Your Shelter Need a Floor?

The whole idea of using a mobile shelter is to give its occupants access to fresh grass, so it usually makes sense to construct a floorless shelter. Some management choices, however, might call for installing a floor. For example, young birds are easier to move with no risk of injury from the rear bottom rail (see page 116) if on a floor. If you do install a floor in your shelter, I recommend using wire or plastic mesh, because droppings will accumulate on a solid floor, requiring frequent clean-out, which is difficult given the tight confines of the interior.

Predators

If the shelter is inside an electric net perimeter, you will not have to worry about predators that might dig under the rails looking for a meal. However, if there are large owls in your neighborhood, close the shelter at night—nocturnal owls hunt on the wing, but also land and walk around looking for prey.

If the shelter is not inside an electric fence, raccoons and dogs may tear a hole in chicken wire—in the case of 2-inch mesh, a raccoon may feed on its victim by tearing it apart right through the wire. If you are designing for such threats, use *½-inch hardware cloth* instead, well secured to the framing. Foil digging predators with a wire mesh floor (2-by-4 inch welded wire allows both access to the grass and protection from digging predators)—or by laying 18-inch (45 cm) panels of chicken wire on light wood framing flat on the ground, entirely around the shelter.

The best option of all is to *wire for defense*: Run a couple of lines of single-strand electric wire around the entire shelter, standing it off from the sides with plastic or porcelain insulators. (See figures D.9 through D.11 in appendix D.) An inexpensive charger powered by a 9-volt battery is sufficient to charge such a small run of wire.

Nests and Other Thoughts

If the shelter will house layers, you should add *nestboxes*, which can be mounted above ground level onto its framing pieces. A hinged door—to shield the nest from rain but give you access from the outside—is a better option than crawling inside to collect eggs. If hens are inclined to roost and poop in the nest, an additional hinged cover to swing into place at night may be in order.

Install a *door* in the shelter even if you rarely use it (such as when the shelter is inside an electric net). Latch the door the night before when you plan to move the shelter from one electric-netted enclosure to another, do a census or selection, or isolate birds for culling.

In smaller, rectangular shelters, often the only door is a hinged *lid* giving access to the interior. Remember that the lid can be popped open by a wind gust, maybe even ripped off the hinges, and install a positive catch for locking it shut. (The country-boy version is a heavy rock set on the lid.)

Even a shelter heavy enough to withstand ordinary winds may flip when a gale blows. When weather predictions here are for winds well beyond the ordinary, I temporarily "nail" my shelters down using an *earth anchor*—essentially, an abbreviated auger screw on the end of a steel rod with an eye hook on its top end. Once the rod is screwed solidly into the earth, I tie or wire one of the bottom rails to its eye hook. Another way to temporarily secure a shelter is to hang a couple of 5-gallon buckets from the framing inside and fill them with water—that's over 80 pounds (36 kg)—using a garden hose. Just empty the buckets after the storm has blown over.

Remember your chickens' need to dust-bathe. Since there is no opportunity for them to do so if constantly on fresh grass, either provide an onboard *dustbox* (as in the shelter described in appendix D) or set one out for them on the pasture anytime no rain is expected.

Most shelters are designed to be used in the warmer parts of the year only. To house birds in the shelter in winter as well, you will need to ensure that at least the part where they sleep is tight against the winter winds, snow, and rain. As noted in chapter 7, however, the shelter should still allow plentiful airflow.

Materials

Mobile shelters have been made in just about every material other than titanium. Your choice of materials depends on which you feel comfortable working with, what might be in your recycle pile, and considerations of weight and climate.

Wood

I am more comfortable working with wood, so all my shelters have had wooden frames, with one exception—a hoop structure based on ½-inch (1.3 cm) solid fiberglass rods as purlins and as arches, anchored into a wooden foundation frame. I don't use pressure-treated wood anywhere on our place. To help prevent rot, I coat all framing pieces in direct contact with the ground with nontoxic

Figure 12.9. Hinged access allows for egg collection from outside the shelter.

sealer (such as marine spar varnish, discussed in appendix D), renewed periodically as needed. (As mentioned in chapter 7, another option is Bora-Care.) Using a highly rot-resistant wood—eastern red cedar in my area—would be an option if you can get it. You might design so that the bottom rails—the parts most subject to rot—can be replaced without taking apart the entire shelter. Or mount the frame on plastic rails. (See figure 14.4.)

When out of service over the winter, a wood-framed shelter should always be set up on blocks.

Plastic

Beginners often think of lightweight 1-inch plastic pipe or the like for framing a shelter. I've never seen one that inspired much confidence—such plastic is fragile and breaks down in sunlight. Heavier plastic pipe (Schedule 40 PVC, for example) is another matter—I've corresponded with flocksters who have used it for shelters that are both sufficiently rugged and easily moved. (See figures 7.1 and 7.2 for examples that could be scaled down for smaller shelters.) I've never used plastic pipe myself.

Metal

Electrical conduit is light and easily shaped. You may see references to its use for framing mobile shelters, but most reports I've read about it have been negative. Both angle iron and rebar—concrete reinforcing rods

Figure 12.10. An 8- by 9-foot A-frame mobile shelter covered with 24-mil woven poly. Ten years old at the time of this photo, it was still going strong.

made of soft iron—make sturdy frames for those with welding skills and equipment.

Some flocksters use cattle panels to frame hoop-style shelters. The standard length of these panels is 16 feet (4.8 m); height varies by species of livestock but in this application likely 52 inches (1.3 m); 6- by 8-inch heavy mesh wire makes a substantial barrier but is flexible enough to bend into curved shapes. Typically, the panels are attached to one side of a wooden frame (or welded onto a metal one); bent in the long dimension into a hoop attached to the frame on the other side; and covered with a tough, flexible, opaque cover. The result has the usual trade-offs among weight, mobility, and stability in the wind but typically has considerably more capacity than shelters framed in other materials.

Covers

If light plastic pipe is a bad idea, such pipe covered with lightweight plastic tarps makes the worst possible combination. Not only do such tarps break down in sunlight—and shred, and blow in the wind—but the combination is so light, even a sneeze will move it. Heavy canvas tarps, tough and weatherproof (like the one in figure 7.1), are a better choice than plastic tarps.

There is one option in plastic covering worth considering, however: 24-mil woven polyethylene—incredibly tough, durable plastic sheeting interwoven with a fiber mesh. It is available in semitranslucent white and various opaque colors, including one that is black on one side, silver on the other.[1] The side you face to the outside depends on whether you need to reflect or gain solar heat to the interior—in my climate, putting the reflective side out is the obvious choice. Did I say *tough*? I once had an 8- by 9-foot (2.5- by 2.7-m) A-frame shelter covered in woven poly sent tumbling by a gust of wind through 30 yards of underbrush. The result was one broken strut (a testament to diagonal bracing) but not a single tear in the poly.

I have used metal roofing for the solid covering on several of my shelters. Aluminum roofing is lighter but more expensive; steel, heavier but cheaper. Steel roofing is available either as plain galvanized, or with a baked-on enamel finish guaranteed for twenty-five years. Though the galvanized is cheaper, the paint you would have to apply to extend service life would over time cost more than the initial investment in the baked finish.

I used baked-enamel steel roofing on the A-frame shelter in figure 12.9. See appendix C for my reasons for choosing metal roofing over 24-mil poly.

Fasteners

Do not use nails when building a mobile shelter—they will work loose as the shelter is jerked about in moving—always use screws instead. I prefer self-tapping types such as coarse-threaded decking screws (galvanized or coated against weather), which unlike conventional wood screws don't require pilot holes and thus save time. (I do drill a pilot hole for a deck screw going into the last 3 inches (8 cm) of a framing piece, to prevent splitting.) Even better are star-drive stainless-steel decking screws. Though more expensive than ordinary decking screws, the star-drive heads are much easier to screw in, and they last longer than the galvanized in prolonged exposure to weather.

Moving the Shelter

Twisted wire or cable, run through a piece of scrap garden hose, makes a convenient pull for moving the shelter. Wire pulls can be permanently attached to both ends of the shelter; or a single pull—with twisted loops at either end that slip into open eye hooks screwed into the bottom rail—can be used on multiple shelters.

When moving a floorless shelter with young or careless birds inside, watch the trailing edge of the bottom frame. Usually the chooks come running as fresh grass is exposed, but those who dither at the rear may get a leg caught between the ground and the moving rail. Actual injuries are rare if you pull slowly and stop to release a hapless bird at the first shriek of distress.

PART THREE

Working Partners

Photo courtesy of Bonnie Long.

CHAPTER 13

Putting Your Flock to Work

The most significant difference between a flock living within the rhythms of the abundant ecology and a flock of pet chickens is enlistment of the birds as partners in the work of food production in the backyard. And what partners they are! They start work early in the morning and keep at it all day, are sharp-eyed and meticulous, and are anything but lazy. And they have a great skill set.

The strategies suggested in this chapter are not games (though they are *fun*)—they are opportunities for accomplishing serious work on the small farm or homestead, work that would otherwise cost the flockster-gardener-orchardist additional time and labor.

But you know something? Maybe this chapter shouldn't be titled "Putting Your Flock to Work." How about "Putting Your Flock to Play"? Think back to the bison, doing amazing ecology- and soil-building work on the prairie. *They* didn't see their great grazing sweeps as work—they were just living the lifestyle that seemed the best fit, the most pleasing. Just looking for good things to eat, looking to make babies. That's the way to think of "putting the flock to work" on the homestead—not as giving our chooks demanding, onerous chores to do but as making possible for them the *lifestyle* they are most comfortable in, the activities they most want to do. Truly, this approach to putting your flock to work, *works*—by making your chickens *happy*.

The Great Work: Soil Fertility

Soil loss . . . is a problem that embarrasses all of our technological pretensions. If soil were all being lost in a huge slab somewhere, that would appeal to the would-be heroes of "science and technology," who might conceivably engineer a glamorous, large, and speedy solution—however many new problems they might cause in doing so. But soil is not usually lost in slabs or heaps of magnificent tonnage. It is lost a little at a time over millions of acres by the careless acts of millions of people. It cannot be saved by heroic feats of gigantic technology but only by millions of small acts and restraints, conditioned by small fidelities, skills, and desires. Soil loss is ultimately a cultural problem; it will be corrected only by cultural solutions.

—Wendell Berry[1]

There is no higher duty on any farm or homestead than caring for and nurturing the soil. Soil care *is* the food independence project. Our soil is our life, and *everything* we do should be in some way an attempt to improve it. A lifetime should suffice.

A troubling but instructive question is this: Why is it that, in natural ecologies the world over, there is a *spontaneous accumulation* of soil depth, quality, and fertility over time; while human practice of agriculture has usually led to *degradation* of the soil—*never so much*

Figure 13.1. No other factor has been as important in transforming our garden soil–from the native clay on the left to the rich, mellow loam on the right–than the many contributions of our hardworking flock.

so as in the era of industrial farming? As we explore the answers to that question, we discover the profound wisdom of imitating natural systems.

Soil Is *Alive*

The soil food web, the complex community of organisms living on and under the soil, is an extraordinary manifestation of the abundant ecology described in chapter 1. Begin with bacteria, up to 13 tons of them in an acre of good prairie soil, an astounding weight considering that half a million bacteria could fit on the period at the end of this sentence. Then add actinomycetes, algae, slime molds, protozoans (maybe ten billion in a square yard of meadow topsoil), and much more—and that's just at the microbial level. Though we think of fungi as "mushrooms" (which are actually the reproductive bodies that produce spores), essentially fungi are meandering underground strands, both invisible (hyphae) and visible (rhizomorphs)—a teaspoon of prairie soil could contain several miles of them. Animal life includes worms—from microscopic up to earthworms 10 feet (3 m) long (well okay, those guys live in Australia)—beneficial nematodes, and other invertebrates (several million in a square yard of garden soil), ground-dwelling insects and spiders, but as well countless other minuscule arthropods that usually go unnoticed—a full census of this kaleidoscopic community fills whole books.

However different the organisms in the soil food web, they all have to *eat*. And what do they eat? Well, *everything*—everything that falls on or dies in the soil becomes *food* for some member of the soil community. Some classes of organisms (prominent among them bacteria) are first at the table, feeding on the more easily decomposed components. The residues of what they cannot break down, and what they excrete, serve as food for the secondary and tertiary decomposers (prominent among them the fungi). Thus the elements and the energy in the parent materials, originally created by photosynthesis in the leaves of plants, passes from one trophic (energy exchange) level to another, and at the end of the process *nothing* leaves the soil. This is the great secret of *spontaneous increase of fertility* in natural soil systems.

At the end of this sublime alchemy, what remains of the original organic matter is *humus*, carbon compounds that resist further breakdown—in particles too small to see, visible only as a darkening of the soil. Though not used directly as a nutrient source, humus content has everything to do with soil quality and fertility: It increases water retention, assists in the chemical bonding of nutrients with plant roots, and improves soil texture. If we as gardeners want to imitate nature's soil-improvement process, the best way to do that is simple: *Feed the soil organic matter to increase its humus content.* And chickens can help!

The presence of so much teeming *life* under our boots is not merely an incidental fact about the makeup of soil. If you have thought of fertility as something you *add to* your soil, consider instead that the complex interplay of species, their ceaseless exchange of energies available in the various stages of organic matter breakdown—*is* soil fertility. And if you have thought of nature as an arena of relentless *competition*, here in the dirt is a marvelous vision indeed: *Everybody is feeding everybody else.*

The concept of soil fertility underlying industrial agriculture is a good deal more static: Soil is a mix of tiny particles weathered out of the parent rock, mixed with air and water—a medium for anchoring the roots of crops as they grow. There is far too little thought given to *feeding the soil*—the idea instead is that, if we *feed the roots of crop plants* with purchased (and, in most cases, industrially processed) nutrients, they will slurp them up and grow profusely, and all will be well.

Chickens are a good deal smarter than the average Big Ag scientist. Their first task when passing over the soil is to offer and scratch in a priceless gift—"holy shit," Gene Logsdon called it.[2] That gift adds minerals needed by plants (prominent among them nitrogen, phosphorus, and potassium, the "NPK" with which industrial agriculture is so obsessed) but also, and of far more importance, supercharges the web of living critters in the soil. Just by pooping, chickens leave the soil more alive, and thus more fertile—more capable of "abundant increase." But chickens can do much, much more to help feed the soil with organic matter, as we will see.

Diversity versus Monoculture

Industrial agriculture grows its crops in huge monocultures—gargantuan fields of nothing but corn, soybeans, or grain—for feeding to gargantuan factory flocks of nothing but layer chickens or nothing but meat chickens. Where in nature do we find such monocultures?

If you've decided to raise chickens, and if you already have a garden (or plan to one day), you're already light-years ahead of monoculture thinking. Do remember that almost everywhere we look in nature, we see complex mixes of both plants *and* animals—a diversity that always tends toward enrichment of the soil underlying all. Though plants start the process, animals participate in the great cycles of reproduction and growth and death and decay and renewal. The small farm or homestead most closely imitating nature, therefore, will be based not only on domesticated plants, but on domesticated animals as well. Any strategy suggested in this book regarding care of the natural-lifestyle flock has a tie-in with increasing diversity—and hence, fertility—in our soil.

Witches' Brew

Perhaps the starkest difference between natural soil systems and those under industrial agriculture is the latter's witches' brew of synthetic chemicals. Some of those chemicals are intended to feed (chemical-salt fertilizers); some, to kill (insecticides, herbicides, fungicides)—all have serious unintended side effects. Chemical fertilizers are highly soluble and readily leach from the root zone into groundwater and natural water systems—where they function more as toxins than as nutrients. The various *-icides* are largely indiscriminate and kill large numbers of nontarget species essential to a healthy ecology.

Wherever applied and for whatever purpose, these chemicals find their way into the soil, and *all* have lethal effects on most classes of soil organisms. That shouldn't surprise us—living organisms have not had to process these compounds during three and a half billion years of life on this planet. Maybe in another billion years soil organisms will adapt to such gifts from Monsanto, Cargill, Dow Chemical, et al., but for the present *there is no more sensible start on a soil fertility program than a Great Vow to avoid all synthetic chemicals. Whatsoever.*

Fortunately, chickens and other poultry can help—not only with direct contributions to soil fertility in lieu of chemical salts, but also by offering alternatives to "going nuclear" to achieve insect, weed, and disease control.

Soil Disturbance

There are occasions of soil disturbance in nature, sometimes on a massive scale. But the beautiful patterns of interactive soil life that result in spontaneous fertility accumulation over time take place *in the absence of major soil disturbance*. Industrial agriculture, by contrast, is based as much as anything else on soil disturbance. Tillage, a frequent ritual on almost all acreage under cultivation, results in the inversion or mixing of the distinct layers in a natural soil profile—and massive disruption of living soil communities. If the next round of tillage comes before the soil food web has healed itself, the soil is on a downward spiral of decreasing structural integrity and fertility.

Chickens, managed creatively, can accomplish many of the tillage chores required on the homestead or small farm, while at the same time preserving soil structure, with net benefits to soil life.

Tiller Chickens

I told in chapter 2 of my disillusionment with a power tiller, based not only on its many annoyances but the realization that in the long term it required *more* time with soil-care chores, not less. But power tilling would be problematic even if its operation were as smooth as cream: The whirling tines are *soil disturbance* in action. To a depth of 10 inches (25 cm) or so, the soil is "blenderized," the natural layering of its zones and its open structure (the result of burrowing earthworms and decay of roots) destroyed. These days my soil is so friable I just prepare seedbeds with a rake, but it is in that condition because, in the early years, I switched from whirling steel to a supremely elegant alternative: tiller chickens.

Cover Crops

I do more cover cropping every year, in all parts of the year—perhaps no other strategy does more to feed the soil and improve its structure. There are cover crops for every season, including winter, their possible applications limited only by your imagination.

Most of the small grains grow well without added fertility in the cool parts of the year, spring and fall. Crucifers, a large ensemble including mustards and rape, have a cleansing effect on soil (suppress soil fungal diseases). Crucifers with fat, deep roots such as turnips and forage radishes have a tremendous loosening effect on compacted soil when they decompose. Buckwheat is frost sensitive but grows fast in summer to smother weeds and keep the soil cool and moist. Sorghum-sudangrass hybrid is a warm-season grass-family cover that gets as tall as corn and produces a tremendous amount of biomass. (Bonus! Ducks and geese *love* it as green forage.) Legumes are especially useful, fixing nitrogen in the soil for following crops—up to hundreds of pounds per acre. Many of them are cool-season crops, though some, such as cowpeas and field peas (also called winter peas—a subspecies of common garden pea), grow well in summer. Some legumes—alfalfa and most of the many clovers—are more long-term crops, ideally left in place a full year for maximum nitrogen fixation.

We think of cover crops as providing a big addition to soil organic matter when they are *tilled in*. But don't forget that the root systems of cover crop plants are far more extensive than the plant parts we see above ground. Roots grow deep into the subsoil or make extensive underground mats of fine rootlets. Alfalfa, for example, pushes its thick taproot 8 to 10 feet (2.5 to 3 m) into ordinary soils, and rye grows a dense mass of fine roots 4 or 5 feet (1.2 to 1.5 m) deep. When the plants die, all those roots decompose in place, not only feeding our buddies in the soil but opening passages through which air enters (essential for soil organisms as well as critters like us above ground) and even the heaviest rain soaks in (rather than washing away over the surface, carrying soil as it goes). Note as well that the most intensely bioactive part of the topsoil is the *rhizosphere*, the surfaces of roots and the area immediately around them. This is where plants feed their microbial allies, with exudates synthesized in the leaves and released through the roots, and where those same allies assist with uptake of mineral nutrients in the other direction. Many processes in the rhizosphere produce soil "glues" that help aggregate tiny particles of soil and humus into larger clumps. Soil aggregation is especially important in heavy clay soil like mine, bricklike when dry and impossibly sticky when wet. Aggregation in effect creates larger particle size, turning clay into a fine loam.

Probably you wonder: Doesn't chicken tilling *disturb* the soil? Happily, chicken-tilling involves plenty of *stirring* of the soil in the top few inches, but no breaking down of soil structure nor inversion of its layers.

Tiller chickens can be the key to making extensive, continuous cover-cropping strategies work. Once I made the marriage between cover crops and tiller chickens, my soil improved more in a few years' time than in all the years before. There was no cover crop that my chickens failed to till in with ease. Even that notoriously tough, tine-fouling cover crop—4-foot-tall rye (1.2 m)—never stood a chance when I sicced tiller chickens on it.

Keep in mind a couple of caveats. Any pass of tiller chickens should be followed by the *immediate* planting of that ground, either to crop plants or another cover crop. Bare ground is always anathema, but we should as well see the tillage pass as *an application of manure*, and should ensure the growth of new plants on the plot that will take up its mineral nutrients and store them for the future—to prevent possible pollution by runoff or leaching. And remember that deposition of *too much manure* for boosting fertility may create problems. (See the sidebar "Too Much of a Good Thing?" on page 134.)

Fighting the Jungle

I expect your garden is a showcase of order, but frenzy is more my style—I routinely have fourteen different projects going at once. The result, inevitably, is heavily weed-grown areas I haven't gotten to. At one time, the more those tough guys dug in their roots, the longer I would put off the assault I call "fighting the jungle." Ah, the joys of tiller chickens—so much easier to just unleash a tiller flock on such a patch and let them "do chicken"!

Remember, if things do come to such a pass in your garden, that weeds are friends, too: Various species specialize in the mineral mix they concentrate and release for use by following crops when chicken-tilled in. I'm sure you know they often have bodacious root systems, impossible for you to tear out, but that yield the multiple benefits discussed in the previous section as they decompose, after the plants have been killed by tiller chickens.

New Garden Ground

Remember why I stressed continual rotation of chickens on pasture? If you leave them in one place *too long*, they will *kill the sod*. Ah, but if you leave them in one place *long enough*, they will *kill the sod*. The promise in that thought should be abundantly clear if you've ever busted your butt wrestling a bucking power tiller, tearing up a tough established sod over compact clay. (*But they said a power tiller makes the work easy!*) When breaking new ground is the next chore on the list, *chicken power* wins, hands down.

At least three times over the years, I have opened up plots of new ground measuring 1,600 square feet (150 square m) or more. Of course, the initial chicken-tilling in of an established sod is just the start of the miracle; afterward the ground is still heavy clay—as compact as it comes (though that would be as true using a power tiller)—but big changes are on the way. Oh, and remember a tiller doesn't poop: Boosting soil fertility starts right away with tiller chickens.

Figure 13.2. A garden expansion begins as multiple pickup loads of horse manure.

I usually follow the birds' initial work with some high-calcium lime—which helps microscopic clay particles *flocculate* (clump into larger soil particles)—and a cover crop, as diverse a mix as the season allows. After a second pass by the chickens to till in the cover, soil structure and organic matter content are much improved, and the ground is ready for its first crop.

Note that it is possible to combine the work of both killing a tough pasture sod and tilling in large quantities of organic matter in the same project. The first time I used tiller chickens to work up new garden ground, I started by hauling multiple pickup loads of horse manure from a nearby farm. After dumping it on the new ground, I allowed it to "ripen" a couple of months—to become rife with earthworms and fat grubs. Then I surrounded the plot with electric net fencing and put a couple dozen chickens to work. Or, from their point of view, invited them to a feast. Within a few weeks they had killed the existing grass sod and tilled in the manure, evenly, over the whole plot. As for me, I had kept busy with other work.

Figure 13.3. The manure is now tilled into the new garden area thanks to these happy chickens. A boost to fertility in a hurry.

Figure 13.4. The first step in establishing a stand of pure alfalfa is removing the established sod cover on the site, courtesy of a flock of chickens doing exactly what they love to do.

While a large, *one-time* application of manure can be a good start on new ground, please do note: *It would be a mistake to apply large amounts of manure to your soil season after season.* Note as well that, to kick off this garden expansion, I could instead have laid down a thick cover of all sorts of decomposables. That, too, after sitting awhile would have come alive with earthworms and other detritivores, and my tiller chickens would have made an equally impressive start on rapidly increasing organic matter.

I once used a flock to till up a much larger area—something like 4,000 square feet (370 square m)—in preparation for sowing a stand of pure alfalfa. Covering an area that size with a deep layer of garden and landscape residues was out of the question, so I just encircled the plot with several connected electric nets, anchored on the henhouse, and released the flock into the fenced area daily. After a couple of months I moved the flock elsewhere and sowed a buckwheat crop. When its seeds ripened in early fall (more free feed!), I returned the flock to till in the buckwheat, then sowed the alfalfa. Imagine the effort required to produce the result shown in figure 13.5 if I had to do all the heavy lifting myself!

Figure 13.5. With the old cover out of the way this fine stand of alfalfa is thriving.

Managing the Tiller Flock

Okay, okay, I hear you! *Gimme the numbers! How many chickens? How long will it take?* The answer, as always, is: It all depends. If the ground to be worked is small, and you have only a few birds to put to work, a chicken tractor parked long enough on the new ground will do the trick; I discuss this approach in greater detail in the next chapter. For a somewhat larger area, you could use the lock-together wire-on-frame panels suggested in chapter 12. Or you could modify the dueling-gardens strategy from chapter 7: Grow your vegetables in an existing garden plot this year, while the chooks work up new ground on an adjacent one; then next year, switch.

I have occasionally used chicken tractors for this work, but most often, as in the previous projects, I've set up a mobile shelter on the new ground and surrounded the shelter and my tiller flock with electric net fencing. I'll describe one project in greater detail for a better idea of "the numbers."

A standard section of electric net fencing is 164 feet (50 m) long. Set it up as a square, 41 feet (12.5 m) each side, and you're ready to till almost 1,700 square feet (160 square m) with chicken power. In one project begun in midsummer, I used three dozen chickens in that space, leaving them for five weeks until they had killed the existing sod cover. After moving the flock elsewhere, I spread high-calcium lime and sowed as diverse a mix of cover crops as I had seeds for: field peas, cowpeas, buckwheat, various crucifers, and small grains. (Though the latter two are not ideal crops for hot, dry weather, in my experience, if you supply enough water in the germination phase, they will do the job.)

The cover crops grew well for five weeks—many cover crops do better in compact soil than more delicate garden crops; and don't forget the boost from that great chicken poop—then I returned the chickens to their work. This time tillage required only two weeks—cover crops are not as tough as established sod grasses. As well, the loosening of the soil already under way, together with enhanced soil moisture in the dense shade of a lush cover, made the chooks' scratching more effective. During this phase, the flockster could have the birds till in as well any amount of organic matter available—compost, garden weeds, grass clippings, autumn leaves.

After moving my tillers back to the main flock in early fall, I sowed a final overwinter mixed cover. When I planted it the following spring the new area had a long way to go to become best garden ground, but it produced nice crops of corn, cucumbers, summer squash, and sorghum in its first season.

To be sure, the process I've described took longer than the single day of backbreaking, nerve-jangling labor that would have been needed to power-till that ground into submission. But didn't someone say patience is a virtue? When opting for tiller chickens, patience is not only good for your soul but good for your soil—the rapid increase in its quality is an order of magnitude beyond anything you can do with a power tiller.

Shredder-Composter Chickens

If we were satisfied with nothing more than substituting chicken poop for chemical fertilizers, we'd be making the mistake of adopting a modified version of the "NPK mentality." When it comes to *feeding the soil with organic matter*, chickens have so much more to offer.

Remember how I advised putting a thick organic duff over the area if you must restrict your flock to a static run? The emphasis there was on responsibly recapturing the fertility in the birds' droppings to prevent runoff pollution. But now let's consider extending that idea to engage hardworking chickens in making compost. *Lots and lots* of compost.

Yes, I'm a lazy old gardener glad to foist off on my chickens any work I can con them into doing for me. So it shouldn't be surprising if I prefer shredder-composter chickens to the unquestionably magical but labor-intensive method of making compost developed by Sir Albert Howard.[3]

In the classic method the "greens" (fresh, moist, nitrogenous)—such as spent crop plants, grass clippings, spoiled hay, and manures—are laid down between layers of "browns" (dry, coarse, carbonaceous), such as straw and autumn leaves. A feeding frenzy ensues, involving some of the same microbes involved in breaking down organic matter in the soil, and before long the conglomerate of organic roughage has turned into dark, crumbly, earth-smelling stuff we intuitively know will be gratefully received by our friends in the soil.

There's just one catch. Imagine assembling 54 cubic feet (1.5 cubic m) of the compost materials mentioned—

Figure 13.6. Shredder-composter chickens on the job. Photo courtesy of Bonnie Long.

that's a heap 3 by 4 by 4½ feet (.9 by 1.2 by 1.4 m), in fact a pretty small beginning for serious homestead composting. That heap could easily weigh a ton, especially if a good portion of it is fresh manure. Now imagine using a manure fork to *turn* that ton—not once but twice during decomposition. If I put my chickens to work instead, making compost is easier for me, and the chickens are happier as well.

The logistics of setting up the composting project are pretty much the same as for managing tiller chickens—just put the materials to be worked in a heap, provide protective fencing and a shelter if needed (which might be your coop with a fenced run), and send in the chickens (see figure 13.6). Using this approach to composting is extremely informal in comparison with Sir Albert–style composting. I just keep dumping any organic debris I generate here or have ready access to locally—spent crop plants, fall leaves, grass clippings, weeds, deadheaded plants from flower beds, cartloads of comfrey. There is *nothing* compostable that leaves our place—it all goes on the compost heap.

The chickens pretty much ignore the heap for a couple of months—until it comes alive with the usual cast of decomposer organisms. Before long, the first place the chooks head when released from the coop in the morning is the heap, ignoring the offering in the feeder until they've had a run at the good stuff.

Remember what I said about metabolites of microbes—vitamins B_{12} and K and immune system enhancers—generated in deep litter? Chickens foraging a compost heap ingest these compounds as well—along with crickets, pill bugs, earthworms, fat white grubs, and more. They eat rhizomorphs that meander through the cellulosic parts of the mix, following them with intense concentration—interesting, given that fungi are known to synthesize numerous natural antibiotics.

How much compost can a flock of working chickens make? For the largest production I remember, I assembled an impressive heap of all our home-generated organic "wastes," intermixed with half a dozen pickup loads of a neighbor's horse manure—enough to cover more than 1,200 square feet (110 square m) to a depth varying from half a foot to several feet. (Again, remember that in retrospect I advise *Go easy on the manure.*) I released my composter flock onto it for almost half a year. When I moved them to work elsewhere, I sowed a mixed cover crop over the compost-in-making—cowpeas and small grains—assuming that a profusion of roots in the mass would assist the decomposition process (the rhizosphere is intensely bioactive, remember) and that

Figure 13.7. Compost by the ton, courtesy of my hard-working flock, is plenty for a 4-inch (10 cm) application on a bed of asparagus, a heavy feeder.

the leguminous cowpeas would fix additional nitrogen in the final compost. The chickens returned to incorporate the cover crop as part of the finished compost. That fall we took about 2 tons of compost off that area for application in one of our two gardens—with a substantial amount left over for use the following spring.

In addition, my flock generates compost by working the deep litter in my 13- by 24-foot (4- by 7.3-m) poultry house, where there is over a ton of decomposing litter at any time. I lay down much of that as a nutrient mulch under nut trees. During the years I used the composting enclosure in a corner of our main garden (described in the next chapter), the composter flock made an additional couple of tons of compost for the greenhouse.

Clearly, even aside from the direct fertility value of their poops, chickens can help provide massive amounts of organic matter for feeding your soil.

Help with Competitor Insects

I assume that, if you are reading this book, it is highly unlikely that you blast your garden or orchard with toxic pesticides, with their devastating consequences for *all* forms of life in and above the soil, and their potential to end up as runoff pollution. I hope you have discovered that doing everything you can to support and increase insect populations in your backyard ecology solves most problems of crop-damaging insects. As said in chapter 1, the solution is not *Kill insects!* but *More insects!*—greater insect diversity, more opportunity for predator and prey species to come into balance. *It really works.*

Figure 13.8. Wasp eating caterpillar. Photograph by istockphoto.com.

As you read the following useful strategies, please do not misunderstand my intent. I am *not* recommending *insecticider poultry* as a better alternative to toxic sprays in a kill-insects strategy. But the fact remains that some insect species present greater challenges than others, and therefore require more active intervention. For example, I have found that Colorado potato beetle is easily controlled by natural means—*if* I hand-pick assiduously when the first beetles show themselves. Their reproductive powers are awesome, and if I ignore the earliest of the "reproducers," they are more than capable of deploying truly out-of-control numbers of progeny. So, beginning when I see the first adult beetle, I hand-pick with great care, two or three times a day, for about ten days; and, as the weather warms, predator species become more active *and no further hand-picking is required*, all the way up to harvest. I have many times, for example, seen *Polistes* wasps flying away with the fat pink larvae of the potato beetle (like the wasp in figure 13.8, preying on a caterpillar.)

Here are some ideas for focusing a little more of your poultry's attention on a few especially determined competitor insects.

Orchard insects and diseases: If you use electric net fencing, you can easily net an entire small orchard and send the flock on patrol. In my little orchard, I have seen leaping guineas snapping codling moths right out of the air. Remember the importance of good *orchard sanitation*—primarily the picking up of dropped fruits—not only to break the life cycle of the codling moth and apple maggot fly, but also to rid the orchard of disease spores that overwinter in dropped fruits. Ah, but why *pick them up*? Why not let the flock take care of that chore? My experience with fallen fruits is that chickens like them, ducks love them, and geese *inhale* them.

Do keep in mind all those poops, however. They're a good thing for soil fertility, right? But in the case of fruit trees, which don't need a lot of nitrogen fertility, you could well end with too much of a good thing if you leave the birds on the orchard full time. I did that one year and had more problems with fire blight in the apples and pears—which are more susceptible to this bacterial blight when growing too rapidly, as can happen with excess added fertility. After that I limited poultry insect patrol in the orchard to spring—when adult insects emerge from the soil and look for sites to lay eggs—and fall, when larvae are looking for places to pupate.

Chestnut weevil: Chestnuts may be the easiest of the nuts to cultivate, with the earliest payoff, but most homesteaders who try growing them find that controlling infestations by the chestnut weevil is difficult, even by chemical means. The best strategy is to break the life cycle—by picking up every nut as soon as it has fallen, before the larva inside has had a chance to emerge and burrow into the soil to pupate. An alternative that doesn't require your constant attention is to range chickens under the trees at nut drop in the fall, and again in early spring, when they either scratch up the pupae or eat the adult weevils as they migrate toward the trunks of the trees. (See figure 5.4.)

Squash bugs: Being heavyweight scratchers, chickens can trash garden crops in no time. However, guineas are not scratchers, if provided with a patch of soft soil outside their shelter for dust-bathing, so keep them in mind if you are troubled by squash bugs. (In the southern and mid-Atlantic regions in which I've gardened—who *isn't* troubled by squash bugs!) I have found guineas *100 percent effective* in preventing squash bug damage in both winter and summer squash. This strategy is easily set up: Move a small mobile shelter to a corner of the squash patch, surround the plot with an electric net perimeter, and plant the squash. Monitor the plants through the early stages of rapid growth; and just about the time the vines start to run and the first fat yellow blossoms appear, you'll see the first of the squash bugs. Move the guineas into the shelter—and like magic, no more squash bugs. Note that, for growing squash at homestead scale, only a few guineas are needed—maybe three, no more than five.

I always had excellent results with guineas in winter squash, though some groups were rough on summer squash. When that was the case, I surrounded the summer squash planting with a close perimeter of 3- or 4-foot (.9- or 1.2-m)chicken wire, staked lightly in place, which kept out the guineas but allowed me to harvest. (Guineas can also be used in tall crops under a lot of insect pressure—trellised or caged cucumbers and pole beans, corn, sorghum.)

Slugs and snails: These *gastropods* have the reputation of being the gardener's bane. I have had fewer problems with them each year, using strategies to increase diversity and provide habitat for as broad an array of species in the garden as possible. Thick organic mulches encourage slugs? Yes, but the mulches also encourage ground beetles and rove beetles, which are major slug predators. Birds (coming by for a bath in our garden pool) love slugs—and so do toads, frogs, snakes, turtles, nematodes, shrews, praying mantises, and firefly larvae. But if you need extra help dealing with gastropods before such natural protections kick in, then—either before planting in the spring or after the last fall crops are harvested—you can set up electric net fence around the entire garden and send your chickens or ducks on slug patrol. During the garden season itself, chooks confined to individual beds inside a "chicken tractor" will eat gastropods—and their eggs—as they till. (I never used ducks in a "tractor," but why not?)

Ticks: With serious tick-borne diseases on the increase, flocksters able to free-range their flocks will be glad to have them on patrol. Sharp-eyed turkeys eat a lot of ticks. It is said that a pair of guineas will keep an acre entirely free of ticks.

Other insect challenges: You will of course discover your own strategies for dealing with competitor insects. Grasshoppers, for example, cause no significant crop damage in my garden, but they might in yours. I've corresponded with a flockster in the south of Portugal, where gardens are plagued by swarming grasshoppers. Most gardeners there find that keeping chickens in or around the garden (with movable fences, for example) greatly reduces grasshopper populations.

Other Work for the Flock

The observant, curious flockster finds new strategies all the time for using a working poultry flock. A successful strategy is one in which "What's in it for me?" intersects with "What's in it for them?"

Weeder Geese

Before the modern agricultural era, geese were widely used for weeding certain crops. Even today, smart flocksters find that weeder geese earn their keep. I have never used geese as weeders myself—and indeed, a diverse mix of crops in a small space mitigates against their use on many typical homesteads. Someone with a market garden might take advantage of their weeding skills, however. An essential would be growing crops compatible with weeding by geese (garlic, strawberries, potatoes, cane berries, herbs, tomatoes, onions, carrots, blueberries, asparagus) separately from incompatible ones (lettuces, cooking greens, and other crops the geese would eat or trample). At the farm scale, geese are extremely effective weeding tobacco, cotton, sugar beets, hops, corn, tree and shrubbery nurseries, vineyards, and plantings of flowers for the floral trade.

Any breed of goose will weed crops, but the smaller Chinese are often preferred for their lighter tread on the beds. Goslings—young geese in their first year—are also preferred for weeding, so weeder geese are usually butchered or sold at the end of the season, in lieu of carrying them through the winter. If a heavier breed is preferred as a better market goose, the more active African would be a better choice than less energetic large breeds like Toulouse and Embden.

Note that success requires more than pointing the geese to the plantings with a cheery *Go get 'em!*—training and knowledgeable management are required to take full advantage of their talents.[4]

Stacking Species

"Stacking" livestock means keeping two or more species that work well together on the same ground. If each utilizes different species of plant and animal life on the area, yield overall increases. The companion species may benefit each other as well.

Geese and sheep pastured on the same ground graze different forage species by preference. Chickens housed under caged rabbits in Polyface Farm's Raken House turn the rabbit pellets and urine into a 12-inch (30 cm) wood chip litter, speeding decomposition and eliminating odor. Gene Logsdon gave chickens access to the space beneath the floor slats of a raised pigpen, where they harvested fly eggs and larvae out of the accumulating manure pack. He also wrote about the work of researchers at the Rodale Institute experimental farm, who built chicken coops over small fish ponds to provide feed for fish—the manure of the poultry, falling through mesh floors—a model that originated in traditional aquaculture practice in Asia.[5]

Poultry on pasture with grazing ruminants—cows, sheep, or goats, for example—break life cycles of internal parasites of the ruminants by eating them and their eggs and larvae in the manure. Muscovy ducks and guineas can be especially useful in the control of liver fluke, eating the snails that serve as a vector in the complex life cycle of this potentially fatal parasite. Since avian biology is so different from that of ruminants, the fowl serve as dead-end hosts—that is, they utilize as high-protein food the parasites of other livestock but are not themselves parasitized by them.

An excellent example of stacking is another Polyface model—following the pastured beef cattle with a big layer flock. Joel waits to move the layers in his eggmobile (a henhouse built on a hay wagon chassis) until the fly maggots in the cow pies are "just right"—that is, the point at which they have produced maximum protein for the hens but have not started hatching. The chickens eat the maggots as they scratch apart the cow pats, and in the process expose the manure to air and sunlight—nature's sanitizers. The hens scatter the fertility over the whole sward, rather than leaving the cow pies concentrated in little hot spots.

CHAPTER 14

Chickens in the Garden

Make no mistake, chickens given free range of the garden would make a mess of things and decimate the lettuce and tomato supply. But for a couple of years we used a "garden flock" in a couple of strategies, with great success.

As with many successes, the first started with a big failure. Ellen and I had gardened here for twenty-four years with virtually no interference from the deer—but then, from one season to the next, Bambi's raids threatened to shut down our gardening entirely. I held the deer at bay temporarily with the same electric net fencing I use to confine my flocks, but knew from research that deer are likely to defeat electric fencing eventually. We decided to bite the bullet and invest the money, time, and effort into erecting a permanent wire fence to keep the deer out—12½-gauge galvanized woven wire, 4-inch mesh, 8½ feet (2.5 m) high. At the foot of the fence we dug in chicken wire to a depth of 10 inches (25 cm) as an additional barrier to ground-level intruders such as groundhogs and rabbits.

I was not pleased, having to expend so much effort to protect my garden—until the fence started a cascade of ideas about new approaches to gardening within its confines. Prominent among those ideas was bringing in a flock of gardening chickens for the entire growing season.

Compost Corner

When we put up our deer fence, I had been successfully using chickens as shredder-composters, described in the last chapter, for a couple of years. Along the far side of the fence was an area of newly enclosed ground that I planned to use as the garden's compost-making area, and it occurred to me that the fence there could serve as one wall of an enclosure for a flock of composting chickens.

Once I had adopted the fence itself as one wall, it was a simple matter to complete an enclosure with three more sides made of 1- by 3-inch (2.5- by 3-cm) wood framing, ripped on my table saw from rough-cut oak fencing boards. I hinged a wire-on-frame door in an opening on one side, 48 inches wide, allowing plenty of clearance to haul compost materials in, or finished compost out, using either wheelbarrow or garden cart. I surrounded the enclosure, including the rear wall with its 4-inch mesh, with 1-inch chicken wire, and attached 2-inch chicken wire over the top of the enclosure as well, to prevent attacks by both aerial predators such as owls, and climbing marauders such as raccoons (see figure 14.1).

As soon as the wire was up, I planted vining hyacinth beans (*Dolichos lablab*) at the base of the entire Compost Corner fence, except the door. By the time stressfully hot weather came on in midsummer, the

Figure 14.1. A working flock turning garden debris into compost in our Compost Corner. Photo courtesy of Bonnie Long.

Figure 14.2. Fast-growing hyacinth bean vines covering the enclosure provide dense shade for the chickens inside. Photo courtesy of Bonnie Long.

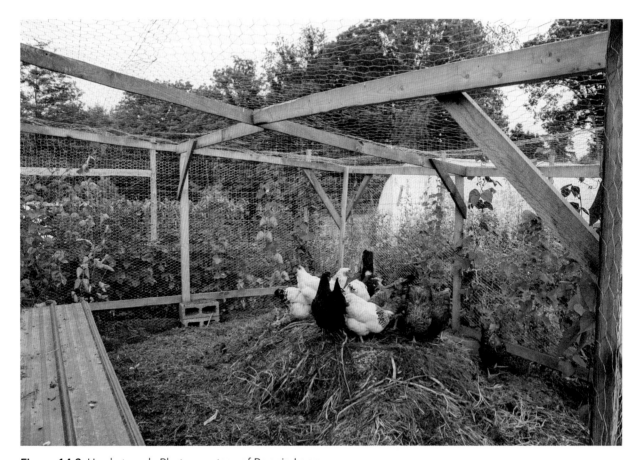

Figure 14.3. Hard at work. Photo courtesy of Bonnie Long.

Figure 14.4. Runners of recycled plastic decking keep frame out of contact with soil and make it easier to move the cruiser.

Figure 14.5. The garden flock is working up a new asparagus bed in the cruiser. Photo courtesy of Bonnie Long.

beans had overgrown not only the sides, but the top as well, creating a dense, cool shade within, which also helped retain moisture in the compost heap (see figure 14.2). (At the end of the season, the spent vines were a big start on the following year's compost heap.)

I split off from the main flock a dedicated subflock to work in the garden, one cock and ten hens. Since the chickens were in residence only in the warmer months, I built a minimalist shelter inside the enclosure, using exterior-grade plywood and metal roofing, to provide sleeping quarters and protection from rain.

We hauled into the enclosure all the organic debris the garden generated—just weeds initially, with soil clinging to their roots to "inoculate" the growing heap with soil microbes, and then spent crop plants as we harvested through the season, plus countless cartloads of oak leaves and of fresh-cut comfrey, an excellent composting plant. As the heap became more active with decomposer organisms—pill bugs, crickets, earthworms, together with the usual hosts of microbes and fungi—the chickens spent more time working it. While doing so, they also worked in the ongoing deposits of their own manure and the residues of the generous amounts of cut green forage—from lawn clippings to chard and mangel leaves to cuttings from small grain and cruciferous cover crops—I threw to them daily (see figure 14.3).

By early fall the chickens had turned the heap of garden debris into about 2 tons of compost, applied in the greenhouse for the winter gardening season.

Chicken Cruiser

As noted in the chapter on mobile shelters, what Andy Lee and Pat Foreman called a "chicken tractor" I call a "chicken cruiser," and I have used several of them over the years to employ chickens for tillage in the garden. Earlier versions hardly deserved the name *cruise*-er, however—they were too large and heavy, clumsy to move within the tight confines of the garden layout—and their frames were too subject to decay. I made changes to past designs for my best-ever garden cruiser—nimble, solid, and durable (see figure 14.5). Here are key design changes:

Size: I knew I would constantly be making tight turns as I maneuvered it from one garden bed to another—so the cruiser had to be small. I made it a mere 8 feet (2.5 m) long, 40 inches (1 m) wide—the width of one of our garden beds—and 26 inches

Too Much of a Good Thing?

Strategies to make soil care and keeping poultry a single integrated project really work: In the years I've raised poultry here at Boxwood, the organic matter content of our soil came to measure, in various tests, between 6.8 and 8.35 percent. (Gardeners in my area start with native organic matter content in the range of 2 to 2.5 percent.)

The same tests, however, have been a reminder of what my mother often told me, and probably yours as well: *It's possible to have too much of a good thing.* I want my reader to be aware that levels of certain minerals can become dangerously high if we're not careful. "Irrational exuberance" with the use of *manures* as fertility amendment is especially problematic.

Manures contain a lot of *phosphorus* (P) and *potassium* (K), both vital to the health of the soil-life community and to growth and productivity of plants. But *too much* of either creates imbalances deleterious to both soil health and crop performance. For example, too much phosphorus depresses the activity of *mycorrhizal fungi*, which form symbiotic relationships that assist transfer of water and soil minerals into plant roots, and help ward off diseases. Reduction of mycorrhizal functions by excess P may in effect make nutrients *less* available to plants, resulting paradoxically in the same slow growth and low yield of plants growing in a soil starved for phosphorus.

Accumulation of dangerously high levels of P and K is more likely in heavy clay soils like mine. Gardeners complain about their clay soils—brick-like when dry, too sticky to work when wet—but the fact is that clay can be among the most fertile of soils. Mine has always tested "Very High" for P and K. Adding lots more with excessive manure applications is not a good thing at all.

I suggest three major guides to manure management to ensure that its addition to your soil will be blessing rather than bane. First, *avoid bringing in additional manure from elsewhere.* The reality is that the feed we buy—and almost all us flocksters

(66 cm) tall. The total area inside the shelter was almost 27 square feet (2.5 square m).

Weight: Another factor favoring maneuverability was shedding as much weight as possible with better materials and design choices. I used metal roofing recycled from previous shelters—some aluminum and some baked-enamel galvanized steel—to cover much of the top and half the sides. The rest of the sides and top I covered with ½-inch hardware cloth, which helped minimize weight and allowed more sunlight into the interior.

Frame: "Strong but light" was the guiding principle for making the frame. Fortunately, I happened to have on hand a few pieces of high-grade, ¾-inch-thick (1.9-cm) clear fir, left over from installation of a new floor on our front porch. Its greater strength, in comparison with common construction-grade lumber, allowed me to use narrower framing pieces, reducing weight. The result, with generous use of *diagonal bracing*, was a light but rock-solid frame easy to maneuver among the garden beds.

Durability: I spent the extra bucks for star-drive, stainless-steel deck screws to fasten the framing pieces, to prevent rusting through years of constant exposure to weather. To keep the bottom rails of the frame out of contact with the highly bioactive garden soil, are at least partially dependent on purchased feed—represents *a gain of P and K over time*, as these minerals from the feed are excreted onto the pasture, in the garden's soil if chickens are working there, and into the deep litter in the henhouse. My biggest mistake in the early years here was seeing the abundant availability of manure from a neighbor's horse boarding and breeding operation as an unalloyed plus to my soil fertility program, heedless of its potential for creating imbalances.

Second, use the excellent strategies I've suggested, especially using tiller chickens for help with weeding and cover cropping, *only in the early stages of your soil-building program*. I found that, after sufficient accumulation of organic matter in my soil, I no longer *needed* significant tillage of my soil, which is friable enough without it. I kill a cover crop by cutting the plants below the crowns with a sharp hoe, then simply rake the bed smooth in preparation for planting—and use the cut plants as a mulch, either on the same bed or elsewhere. Tiller chickens helped get the soil to that level of friability, but are no longer needed now—a good thing, since continued additions, year after year, of the P and K in their droppings could become a case of "wretched excess."

Third, *have a plan* when it comes to manure management. Monitor the changes in your soil through the seasons—every three years at most, every two years if P and K levels are rising. Soil testing may be free through your state extension service, though my preference is for an expert private consultant. (You get what you pay for.) When mineral levels in your garden's soil are at optimum levels, start shifting the chickens' increases elsewhere on the homestead. The larger the pasture area on which you can range your flock, and the smaller the flock size in proportion, the longer the period when manure deposition is likely to be a boost to fertility rather than a turn toward imbalance. I apply deep litter compost to my extensive plantings of nuts (pecans, hickories, chestnuts, hazels), which are greedy for fertility when young. If you have a bit of woodlot, it might benefit as a site for fertility transfers, either in the form of droppings directly from chooks ranging there or as deep litter compost from the henhouse.

I mounted them on two runners cut from recycled plastic decking, 5½ inches (1.4 cm) wide (see figure 14.4). In addition to increasing the frame's resistance to rot, the plastic runners made the cruiser much easier to slide over garden beds.

Access: I allowed for access to the interior, for use when feeding or replenishing the waterer, with a wire-on-frame lid over the section of the top not covered with metal roofing.

Nestbox: I provided the sheltered end of the cruiser with a hinged piece of scrap plywood for easy egg gathering from the straw-lined nestbox mounted on the framing inside.

I put all eleven birds into the cruiser—about 2½ square feet (.23 square m) for each. This turned out to be too crowded, and I began limiting the number of tiller chickens to eight—3⅓ square feet (.3 square m) each. Even that spacing was rather tight, but work assignments in the cruiser were periodic—with frequent rotations back into Compost Corner, which totaled about 225 square feet (21 square m).

The first assignment for my birds in the cruiser was to work up new ground for an extension of our asparagus bed. I had made a start the previous fall with a heavy application of chicken-powered compost and a cover crop. Now I ran the cruiser down the new bed, and the

Figure 14.6. At the end of the season, when many gardens are abandoned for the winter, the garden flock (*in the cruiser, far left*) is still preparing beds for overwinter cover crops.

chooks tilled in the cover crop, one 8-foot (2.5 m) section at a time. On the return up the bed, each day I dumped in large loads of earthworm-rich compost which the birds tilled in while picking out the worms. I then moved the cruiser elsewhere and planted my new asparagus plants.

For the rest of the season, the garden flock alternated between Compost Corner and the cruiser. There was always work to be done on the compost heap, so the assignment at any given time depended on whether there was a garden bed with a cover crop needing to be tilled in.

This cruiser was designed to fit well in a new approach to gardening following installation of the perimeter fence: devotion of fully half our garden beds to cover crops. Specifically, each season every other bed in our garden would be planted to cover crops, and the rest to harvest crops. The following year, the planting of the beds would be reversed. Since a particular cover crop bed might be planted to several covers in succession, there was plenty of work for cruiser chickens to do.

On the last passes of the cruiser over beds in the fall, I laid down large amounts of compost for the chickens to till in while eating the earthworms in it—then reunited the garden chooks with the main flock. Fast-growing fall cover crops sown into the recently tilled beds captured nutrients in the droppings, preventing their leaching to groundwater and "banking" their fertility for the following spring. Thus as winter approached—a time when many gardens are abandoned or, worse, bare—my entire garden was lush with cold-hardy cover crops: clovers and alfalfa, winter peas, small grains, various crucifers. Many of the latter—mustards, turnips, rape, kale—we harvested as needed for the table (and as green fodder for the flock), in lieu of growing them as separate fall garden crops.

Caveats

It turned out Compost Corner had a fatal flaw: Remember the hawk-deterrent chicken wire over the top? I assumed that its 2-inch mesh would never accumulate a snow load, and that no central support was needed under the frame supporting it. (See figure 14.3.) Wrong! Sometimes the snows we get here are wet and clinging—and yes, one snowstorm built up enough load on that 2-inch wire to take down the whole structure. If you want to try such a Compost Corner for a flock of working chickens in your garden, either give the top framing better support than I did, or remove and store the 2-inch chicken wire after moving the garden flock to winter quarters.

Somewhat to my surprise, I never had predator problems with the garden flock tilling beds inside the cruiser. Raccoons are great climbers—but I think the distance established by the fence between the perimeter and the cruiser prevented them from recognizing the chickens in the cruiser as available prey.

Sadly, Compost Corner was another matter. Yes, I had surrounded it—including on the perimeter-fence side—with 1-inch chicken wire, dug well into the ground. But please do remember: *One-inch chicken wire is not sufficient to stop raccoons!* I surmise that a raccoon family came around. Mama tore a hole in the poultry wire (I bet *you* can't do that!) and invited her young ones to "come on in." They were at the age when she was *teaching them to hunt*. Shall I just say: It wasn't pretty.

If I were to set up another such Compost Corner, I would protect it with single-strand electric wiring (as in figure D.11 in appendix D).

The biggest caveat of all is about too-frequent use of working chickens in the garden—and indeed about many of the practices I've described involving introduction of *heavy and repeated applications of manure* of any sort. Without question, it's possible to have "Too Much of a Good Thing," as I describe in the sidebar on page 134.

PART FOUR

Feeding the Small-Scale Flock

Photo courtesy of Bonnie Long.

CHAPTER 15

Thoughts on Feeding

It was hard to decide where in this book to introduce a discussion of feeding. Truly, most of this book is about feeding the flock: Every strategy for making a partnership with them—insect patrol, garden tillage, making compost, raising new chicks using mother hens—is effective mostly because it offers the birds *access to the foods they most want to eat*. But let's look more specifically at the principles that guide best feeding of the farm or homestead flock.

The Three Food Groups

For me the starting question about feeding is not "What's on the label?" or "What is the required ratio of nutrients?" but rather, "What would a chicken eat on her own in the wild?" And while I can't travel to Southeast Asia to observe the eating habits of *Gallus gallus* (the Red Junglefowl), I have a model of natural feeding to contemplate: seeing my grandmother's flock of hens feeding themselves, free-ranging on a small farm. Granny knew little and cared less about "scientific feeding," yet her largely self-sufficient flock kept eggs and chicken 'n' dumplings on her family's table. I spent many hours of my youth observing Granny's busy hens, and found that they ate a diverse mix of three food groups:

Green plants: Though we do not think of chickens as grazers, a small but important percentage of their diet is green plants. Plants contain fiber that helps with efficient excretion; key nutrients such as minerals, vitamins, and proteins; and enzymes—complex proteins that catalyze the body's metabolism. Importantly, plant enzymes assist digestion, enabling more complete utilization of the other nutrients in the chicken's feed.

Seeds and fruits: Chickens avidly seek out seeds and fruits, the structures into which plants concentrate the most food energy as part of their reproductive strategies.

Animal foods: Chickens eat a wide range of animal foods such as insects, worms, slugs, and snails—potent sources of both proteins and fats, which are essential for growth and maintenance, cellular energy, absorption of minerals and fat-soluble vitamins, and functioning of the hormone system. Animal foods are also important sources of beneficial enzymes and minerals—for example, the exoskeletons of most insects contain a lot of calcium, especially important for strong eggshell formation.

What do these diverse natural foods have in common? They are *alive*, and they are *raw*. Best feeding of our flock means maximizing their access to such living, unprocessed foods. It means taking the feeding of our flock into our own hands as much as we can.

The prospect of relying more on natural and self-foraged feeds can bring on a good deal of "feeding anxiety" for many beginning flocksters. We have been conditioned to believe that manufactured feeds are "scientifically balanced" and fear that in our ignorance we cannot possibly get it right and ensure the nutrient balance required by our birds. Perhaps the first feeding principle is thus: *Relax!* To be sure, our chickens need a balanced diet, but tell me what's wrong with this assumption: The more diverse the live, natural feeds our chickens eat, the more *naturally balanced* their diet will be.

Before the era of "scientifically formulated" manufactured feeds, my grandmother's flock and other free-ranging homestead and small-farm flocks filled many of their nutrient needs through their own foraging. Keep in mind, too, that almost all poultry feeding research studies assume confinement feeding as the norm—of *course* they are going to conclude that ensuring the correct ratios of nutrients is a fine balancing act indeed. Regrettably, few or no large feeding studies have been done targeting pastured or free-ranging flocks.[1]

Bag or Bug? The Feeding Spectrum

Think of feeding as a *spectrum*, like the one shown in figure 15.1. At one end the birds' diet is manufactured feeds exclusively, and the flock is totally confined (that is, completely dependent on what we give them to eat); and at the other end, the flock is completely free-range, foraging all their own feeds from biologically diverse ground. Throughout this book, my intention is to help you learn how to move away from the former end of that spectrum toward the latter. Few of us have the option, as my grandmother did, of allowing our flocks to supply almost all their own diet, so our feeding program will be somewhere in the middle of the spectrum. *But every flockster, without exception, can find ways to increase the range of natural feeds available to the flock.* Our feeding program should strive to shift "from bag to bug."

Figure 15.1. The feeding spectrum for poultry ranges from total confinement with 100 percent manufactured feeds to the all-natural free-range approach of Granny's flock. We flocksters are all doing the best we can, looking for creative possibilities to move in the direction of Granny's flock. Illustration by Elara Tanguy.

The more we find ways to give the flock maximum access to live, natural foods, the more we will think of purchased, highly artificial feeds—or even concentrated feeds we make ourselves—as *supplemental*, provided as needed to the extent not enough natural feeds are available.

Feed Costs

Another obvious parameter of the feeding spectrum is that the greatest expenditure on purchased inputs lies at the confinement end, while there are few or no cash costs required at the other end. While most of us are likely to remain in some degree dependent on purchased feeds, there are some practical strategies for cutting feed costs.

Perhaps the starting point for thinking about feeding is getting rid of the notion that all feed for our flock must be *purchased*. Contra that rote assumption, I have used my grandmother's flock as a model of feeding chickens almost *for free*. Now please take an advance peek at the sidebar "From Waste to Resource" on page 330 for the amazing story of how Karl Hammer, owner of Vermont Compost Company (VCC), makes hundreds of tons of high-quality compost with the aid of working flocks of chickens—up to 1,200 at a time, who lay up to fifteen thousand dozen eggs a year for sale in local markets. What do you think *his* feed bill is a year? Answer: *Nothing*—the chooks feed themselves *entirely* out of the biologically active mountains of compost a-making.

If you're intimidated by the sheer scale at which VCC operates to pull off such a "waste to resource" feat, see chapter 19 for options for working similar magic in *your* backyard. Truly, when our basic working model is participation in the give-and-take of the abundant ecology, there is no end to alternatives to the feed bag.

Challenge the Flock to Self-Feed

Even those of us who routinely range our flocks worry a bit, I expect, about whether our birds will find sufficient nutrition on their own; we tend to feed them more from the feed bag than they really need. Don't be afraid to push your flock to work harder at finding their own feed by offering less from the feed bag—some chickens prefer hanging around a full feed trough to scratching and pecking. As VCC's Karl Hammer observed: "The less we do for them, the more they do for themselves." I've received reports from numerous other correspondents as well that the key to more active foraging is being stingier with the feed scoop. Confirm whether your birds are getting enough to eat by occasionally feeling their crops after they have gone to roost: If they are ending the day with full crops, they are getting enough feed however little you're providing.

Practice "challenge feeding": Record current egg production based on purchased feeds. Give the birds maximum available access to natural, self-foraged feeds, and reduce the prepared feed by a measured amount each day. The point at which egg production drops indicates the point at which they really do need the concentrate feed to supplement what they are getting on their own.

Cut Feed Costs with the Hatchet

It is common for us flocksters to enjoy our flocks so much that imperceptibly the flock grows in numbers each year, beyond what's required to meet our production goals. We can save on feed costs by avoiding "flock creep" through constant review of the composition of the flock and vigilant culling.

Stack Species

Among the many benefits of keeping more than one livestock species together is the savings on feed costs, as when sheep and geese share the same pasture, or chickens clean up feed spilled by other animals—or even when chooks pick undigested bits of corn, small grains, and acorns out of the manure of cows or pigs. Protect young fruit trees with wire mesh and use sheep to mow your orchard, then follow with chickens to clean up dropped fruit and any parasites in the sheep manure. Use the Polyface model on a smaller scale by following the family cow with chickens in a mobile pasture shelter or protected by electric net fence. The birds will glean fly maggots out of the cow pies.

Avoid Free-Choice Feeding

Some flocksters advise that chickens do best if fed free-choice—that is, with feed available in the hoppers all the time. I'll tell you why I haven't found that a good idea.

One winter I started seeing enormous numbers of mice in the poultry house anytime I went in at night—the litter and the wire partitions appeared to be boiling with the little rodents when my flashlight beam swept the interior. I took the infestation as a personal challenge and started setting dozens of the common spring-operated snap traps. I lost track of the number of times I caught two mice with one snap—but twice, I caught *three* at a whack! *That's* your basic industrial-strength rodent rodeo.

Despite trapping a couple dozen mice per night, there was no discernible effect whatever on the population. Then I noticed mouse droppings in the feed hopper and realized that keeping a concentrated feed source constantly available was an open invitation to the mice to "be fruitful and multiply." Creating the solution was as simple as creating the problem: I stopped feeding free-choice. Instead, I fed a calculated amount of feed early—no more than the birds would clean up entirely within the morning, leaving no freebies when the mice came exploring at night. The rodent population plummeted. Since that time, there are always a few mice around, but they have never again approached biblical plague proportions since I started denying them a ready food supply.

Other feeding strategies to foil rodent incursions: When possible, feed the flock outside, as far as practicable from the henhouse. When I have no alternative to feeding inside, I put the feed containers into a dustbox like the one in appendix B: The depth of the box and the "lips" around its top perimeter prevent scattering of feed into the litter. Essential to success with this approach to denying rodents their free ride is putting a tight lid over the dustbox, *every night, without fail.*

How *Not* to Reduce Feed Costs

The one way not to reduce costs is to buy cheap, mediocre feeds. If it is true that "you are what you eat," then you are what your chickens eat, too. Buy the highest-quality feed you can afford. If you're stuck with commercial feed of questionable nutritional value, it's all the more important to supplement with vital, fresh, natural—*free!*—feeds like those suggested in following chapters. The difference natural feeds make for the birds' performance is illustrated by the experiences of a poultry breeder I know. During the confinement period of winter and early spring, his early hatches—from eggs of hens eating commercial feeds exclusively—suffer from low hatchability. After he releases his breeders from the winter house and they start feeding on new grass, however, hatch rates jump to 90 percent or better. If it's possible, buy feed grains from a local farmer. The ingredients are likely to be of better quality, and your purchase will certainly boost the local economy and help keep that small farmer in operation. It might even be possible to form a buying club of flocksters in your area to contract with a local farmer to grow your feed grains.

The Far End of the Chicken

Dried poultry manure contains 4 or 5 percent nitrogen, 6 percent phosphorus, and 4 percent potassium. The obvious implication is that some of the expensive feed we offer our flock is being lost in transit out the other end of the chicken. That's the bad news, given the high cost of putting the feed into the near end of the chicken in the first place. But the good news is that, 365 days a year, our birds give our pastures and gardens a *gift*. If we treasure that gift—making sure that it gets deposited directly onto the pasture, or composted in the deep litter—then we will see that part of every feed dollar is an investment in enhanced *fertility* in our gardens and pastures.

When we think of the fertilizer value of manure, we usually think of the mineral values I referred to previously. But remember that the remaining 85 percent or so is *organic matter*, which—as any organic gardener knows—is as much the key to soil fertility as any amount of added minerals.

A Paradigm Shift

I've described Vermont Compost Company's approach to feeding, not primarily to encourage you to seek out food residuals as a feeding resource—though that might be an option on a smaller scale for some. Nor do I imagine that more than a few of my readers have the option of ranging their flocks over a 50-acre farm, as my grandmother did. I gave these examples to suggest a paradigm shift in our thinking about feeding: *All* of us can find ways to increase the flock's access to live, natural foods, and an important way to do that is to find ways to transform what are too often considered "wastes" into feeding resources.

In previous chapters I've described the virtues of more *natural management*, but keep in mind that one of those virtues is lower feed bills: Providing for the *happiness* of the flock—giving them room to roam—also gives them access to self-foraged foods; use of deep litter for best manure management turns out, amazingly, to have positive feeding benefits; and *all* the strategies for putting the flock to work are based on pointing them toward *free food*.

In the following chapters I discuss commercial feeds in more detail and provide more advice on making our own feed mixes as a replacement for purchased feed. Things get more interesting as I point out the many ways to increase access to live foods right in our backyards, gardens, orchards, and woods. I also describe some projects that really bring out the kid in me: cultivating decomposer organisms in "wastes" to produce high-protein feed for the flock.

CHAPTER 16

Purchased Feeds

Many readers of this book understand that the highly artificial foods available in supermarkets, the foods that make up the bulk of our modern diet, are made largely from mass-produced commodity ingredients such as corn and soy, tricked out to masquerade as traditional foods; are created targeting cheapest production and maximum profit rather than nutrition; are processed to the nth degree (with heat, pressure, drying, and other brutalities that destroy the more fragile vitamins and enzymes); and contain chemicals never eaten by *Homo sapiens* throughout the evolution of our species. Many eaters of these modern processed foods would admit, with perhaps a bit of guilt, that it's *convenience* that drives their choice, not a conviction that these foods are more nutritious or better support health.

I make this parallel with how we feed ourselves—or should feed ourselves—to introduce some skepticism about poultry feeds: Why should artificial, highly processed feeds be any better *for our chickens* than artificial, highly processed foods are *for us*? Without that healthy skepticism, we are apt to be suckers for certain conditioned assumptions:

- That poultry feeding is complex and opaque; and unless we want to take an advanced degree in the subject, it is best to leave its formulation to the experts—and buy what they offer us.
- That the experts know what they are doing.
- That manufactured feeds are scientifically balanced, and we cannot possibly achieve the ideal balance when we feed from our own resources and by our own methods.

I question all those assumptions and suggest you put them to the test using this rigorously *scientific* experiment: Put some of that dry, dusty meal from the local farm co-op into a feeder—and throw out beside it a handful of earthworms, crickets, fresh-cut clover, mulberries, and sunflower seeds. If a single chicken even *looks* at the manufactured feed before the flock has cleaned up every morsel of the natural foods, I will eat the remainder of the feed in the bag. Assuming *Gallus gallus* would never have survived long enough to become *domesticus* if she hadn't figured out what is the good stuff to eat—a scrupulously *scientific* assumption—the proposed experiment would suggest sufficient grounds for skepticism about "scientifically formulated" feeds.

The truth is that manufactured chicken feeds are in many ways no different from manufactured people foods:

- Many ingredients are things chickens have never eaten before. Indeed, it's hard to imagine anything farther from the three food groups of a chicken in the wild than feather meal, meal from soybean oil extraction, and stale vegetable oils from fast-food fryers. Such foods are anything but *alive* and *unprocessed*.

- Ingredients are chosen by their market price on the day of purchase, not primarily for their nutritional quality.
- Feeds are highly processed with heat, pressure, and desiccation, with damage to or loss of enzymes and the more fragile vitamins, particularly the fat-soluble ones.
- Feeds contain chemical additives and residues whose full effects—for the chickens, for people who eat their eggs and meat, and for the broader environment into which they eventually diffuse—are not fully known.

If you plan to keep your flock confined, dependent solely on what you feed them, then manufactured feeds might seem a logical choice: As a matter of fact, they are formulated using feeding studies assuming and carried out in precisely those conditions. If, on the other hand, you are convinced about the superiority of natural feeding and management, be aware that large-scale feeding studies do not target such feeding practices—for example, the feeding requirements of ranging flocks. Which is another way of saying that the experts don't know much at all about natural feeding—indeed, are likely to know even less than the curious and conscientious beginner willing to abandon the industrial paradigm of totally confined flocks. The feeding practice of such a beginner will be an ongoing experiment on the cutting edge that leaves the experts far behind.

Three Formulations

Remember what I said in chapter 15 about feeding as a *spectrum*? All of our possible feeding strategies exist somewhere along that spectrum. If you feed your flock commercial feeds exclusively because it is *convenient*, I propose you consider that convenience foods for chickens are not likely to be any better quality than convenience foods for people, and that you explore some of the options for adding in some natural feeds. Many of us, however, must base the feeding program primarily on purchased feeds by *necessity*, whether that necessity arises from limitations of time, or space, or foraging resources. If that is true for you, it's important to learn more about the three formulations of commercial feeds.

These formulations are tailored to the varying needs for the major nutrients—particularly protein and minerals—at different stages of a chicken's growth.

Chick or Starter Feed

Chicks need a higher level of protein in their diet than adults, to support their rapid growth. Any commercial starter feed, therefore, contains about 20 percent protein or more, to be fed the first six weeks or so.

A major issue as you decide what to feed your just-hatched chicks is whether to use "medicated" feeds. Be aware that most chick feeds in the market are indeed "medicated," as discussed in the sidebar "What's in the Bag?" on page 146. But note that qualification, "*most* chick feeds": When I started raising chicks close to forty years ago, there were *no* chick feeds available to me not "medicated" with coccidiostats. I was skeptical of routine feeding of antimicrobials "just in case," which can lead to resistance on the part of bacteria and other microscopic organisms, making them less effective in the long run. It may also leave residues in the eventual eggs and meat, despite standard reassurances to the contrary; and residues must also eventually disperse into the environment, with unknown long-term effects. In lieu of feeding medicated chick feed, I started my chicks on the "stage two" formulation called grower ration or pullet developer, as I describe next. I never had a single loss to coccidiosis.

In the intervening couple of decades, organic feeds (not allowed to contain coccidiostats) have entered the market, though it may not be easy to find them. In a survey I recently made of feeds in a mass-market "farm store," I even found a "non-medicated chick starter" that was not organically certified—though it

What's in the Bag?

It can be an instructive exercise to read the label of one of those colorful packages on a supermarket shelf and realize with astonishment what is actually inside our food. Let's see what we can learn about manufactured *chicken* food by reading the label from a feed bag. This one happens to be a label for "chick starter," provided by a flockster buddy of mine, but you would find bag labels for "grower ration" and "layer feed" much the same, with differing ratios of some ingredients.

Chick Starter: Medicated Crumbles

Aids in the development of active immunity to coccidiosis under conditions of severe exposure to coccidiosis up to 5 weeks of age.

ACTIVE DRUG INGREDIENT
Amprolium 0.0125%

GUARANTEED ANALYSIS
Crude Protein (min) 20%; Crude Fat (min) 3%; Crude Fiber (max) 6%; Salt (min) 0.2% (max) 0.3%

INGREDIENTS
Grain Products, Plant Protein Products, Processed Grain By-Products, Monocalcium Phosphate, Milk Products, Calcium Carbonate, Roughage Products, Choline Chloride, Ferrous Sulfate, Manganous Oxide, Zinc Oxide, Soybean Oil, Niacin Supplement, Sodium Selenite, Vitamin E Supplement, Copper Sulfate, Vitamin A Acetate, d-Calcium Pantothenate, Vitamin D_3 Supplement, Riboflavin Supplement, Vitamin B_{12} Supplement, Menadione Sodium Bisulfite Complex, Folic Acid, Ethylenediamine Dihydriodide, Salt.

FEEDING DIRECTIONS
Start replacement chicks on this feed and offer as the only feed to the chicks until about six weeks of age; at that time, switch to Grower Ration Crumbles.

WARNING
Use as sole source of Amprolium

MANUFACTURED BY [NAME AND LOCATION OF FEED MILL]
163
0251
1347
1116503

Chicks and coccidiosis: The chick starter label states that it contains *amprolium*, a "medication" that kills the microscopic parasites that cause coccidiosis. But who's expecting the "severe exposure to coccidiosis" mentioned on the label? Certainly anyone raising tens of thousands of chicks in a single enclosed space. But if chicks are being "severely exposed" in your brooder, there is something wrong. As mentioned in chapter 6, *Coccidia* are ambient virtually anywhere chickens are found. Chicks that small-scale flocksters raise are routinely exposed to these normal numbers of cocci—and that's a *good* thing: Exposure triggers the chicks' immune systems to develop resistance.

If chicks *need* exposure to coccidia to develop natural immunity, I question the wisdom of routine feeding of coccidiostats such as amprolium to kill them off, whenever and wherever they enter the chicks' systems. And what about the wider effects? I've seen it asserted that coccidiostats don't count as antibiotics, since the latter can be taken strictly as *killing bacteria,* and coccidia are not bacteria. But isn't the distinction something of a dodge? Coccidiostats are fed to *kill microbials*—and we'd all be wise to distrust any practice that has the side effect of releasing compounds into the environment that kill microscopic life, with long-term consequences unknown. I doubt those consequences are likely to be benign, given the strong warning farther down the label against administering amprolium to chicks in addition to that in the ration.

Chick starter ingredients: Exactly what are the ingredients in this bag of feed? The literal answer is: Who knows? The manufacturer is not required to give you that information—and indeed, it would hardly be practical to do so, since the actual ingredients vary constantly, depending on their price on any given day. So long as the manufacturer meets the nutrient percentages listed, the actual stuff making up the feed can be generalized as "Grain Products, Plant Protein Products, Processed Grain By-Products" and the like: whole grains, chief among them corn (wheat in some parts of the country); by-products of the milling of grains for flour and other industrial food products; oilseed meals from extraction of vegetable oils from soybean, sunflower, and canola; cottonseed (cotton is among the crops most heavily sprayed with pesticides). Note the presence of "Milk Products"—any by-products of the processing of dairy products such as whey and powdered skim milk—but the absence of "meat scrap meal" or equivalent, which was standard when I first started feeding chicks long ago. The scare over bovine spongiform encephalopathy (BSE or mad cow disease) led to the omission of meat scrap meals as an ingredient in poultry feeds sold to the small-scale flockster. (They are still used to feed industrial poultry, subject to established processing standards.) Note the extensive list of vitamins and minerals. Some are provided by the actual feed ingredients, but, since those are inadequate to supply the dietary requirements, feeds are artificially supplemented.

Milling date? We'd like for our feed to be *fresh,* so where do we see the milling date on this label? We don't—this is another piece of information the manufacturer prefers that you not have. As a matter of fact, USDA regulations state that the label must carry a date, but the details of the regulations helpfully allow the manufacturer to hide the date in the lines of numerical code at the bottom of the label. The coding records details about the batch from which the feed in this bag originated, information that the mill could retrieve if required. The manufacturer assigns its own code—and when I called the manufacturer asking for help deciphering it, I was told that the date on this one was indicated by the second line: "0251"; that is, this batch of feed was milled on the 251st day of the year. This bag of feed was purchased on October 27, which means it was forty-nine days old on the date of sale. In his excellent compendium on feeding naturally raised poultry, Jeff Mattocks strongly cautions about feeding stale feeds: "Feeds are at the optimum levels for up to 14 days, and are satisfactory up to 45 days after grinding or milling. After 45 days the feed is generally so stale or oxidized that poultry appetite will be severely depressed. Oxidation starts immediately after the grinding or cracking of the grain."[1]

was tucked away on the end of a couple of shelves filled with the default medicated feeds.

To conclude, buy starter feed *without* coccidiostats if you can find it. Or just skip the default "medicated" feed entirely and start your chicks on the second-stage feeds, discussed next.

Grower Ration or Pullet Developer

Requirements for protein decrease as the chicks grow, so the second-stage feed—grower ration or pullet developer—contains less protein than starter feed but more than feed for adult layers, about 16 to 18 percent. It is recommended for feeding growing chickens from six to eighteen weeks of age.

When I was feeding commercial feeds, I always started my brooder chicks on the stage-two feed from the beginning—and boosted dietary protein with crushed hard-boiled egg, raw beef or deer liver, and fish meal. For years Ellen and I maintained a productive flock of laying hens and ate many fine broilers and roasters, all started that way.

Layer Feed

Layers need less protein than growing birds, but they need more minerals. A typical layer feed will therefore contain about 15 or 16 percent protein, and considerably more phosphorus and especially calcium than in chick feeds. Some layer feeds—higher in protein, say 20 percent, and more densely fortified with minerals —are intended to be fed mixed one-to-one with whole feed grains.

An important point to remember about layer feeds is that *they should never be fed to growing chicks.* The additional mineral content of the layer feed can interfere with proper growth of the chicks, particularly development of the reproductive organs. Begin feeding the layer formulation to pullets when they are nineteen or twenty weeks of age, just before onset of lay.

Organic Feeds

Some readers may have access to certified organic feeds. "Organic" as a label has come to be a bit more slippery than it was initially, but in keeping with the National Organic Standards, any feed labeled as certified organic is not medicated. Even if feed is organic, however, make sure it is fresh. Remember that any feed starts to go stale as soon as the feed grains are crushed and should ideally be fed within a month. Check with your source about the date of milling.

The Good News

There's plenty of bad news about commercial feeds. But the good news is that your eggs and table chicken will still be an order of magnitude better than supermarket versions. First, feed sold at your co-op for home flocks, with all its flaws, is likely to be superior in quality to the debased feeds consumed by huge industrial flocks. Second, *every* flockster can feed *some* natural feeds to improve the diet of the flock—with better results if only because the enhanced enzymes in the diet boost the birds' utilization of the bagged feed. So I urge you to explore the many ideas in this book for providing your flock all the natural foods you can within the limits of your own situation; that is, I invite you to a never-ending experiment, which will save you money and improve the diet of your birds, but which is also a kind of play—it's *fun.*

And you won't be troubled with questions like, "Hmm, I wonder if they're getting enough ethylenediamine dihydriodide."

CHAPTER 17

Making Your Own Feeds

Though I now have a tiny flock in comparison to years past, for more than two decades I made *all* the feeds—up to 2 tons per year—for our large productive flocks of chickens, guineas, geese, and ducks. This chapter outlines the basics if you want to make your own feeds, too.

To be sure, making your own feed requires considerably more effort than scooping something out of a bag. So why did I go to the extra effort of making my own feeds? For Ellen and me, it is all about highest quality of everything on our plates, with as few compromises as we can possibly manage. The feed I made, using certified organic grains and supplements exclusively, was fresher and of higher quality than anything I could buy. Because I'm an inveterate tinkerer, making feed was an ongoing experiment in response to new ideas—I was forever fine-tuning my formulations to target specific groups of birds as needed.

In earlier discussions of feeding, I noted that the major characteristic of natural feeds is that they are *alive*. If we stick mostly with whole seeds, we ensure that our prepared feeds are as close to being alive as possible. As someone whimsically observed: "A seed is a tiny plant, in a box, with its lunch." Any one of the feed grains I used could be planted to grow into vigorous plants—indeed, when I needed a cover crop of wheat, oats, or peas, I simply drew the seeds from the feed bin and planted.

Yes, I could have bought premade organic feed mixes from my supplier. However, their feed delivery arrived only once a month—making my own every few days meant fresher feeds. Also, I had found their mixes too finely milled, necessitating having to deal with "the fines" more frequently. (If you remember from chapter 6, these are the feed residues that sift to the bottom of the feeder, becoming stale and moldy if ignored.)

Should you assume, as most people do, that I made my own feeds to save money, let me give you the bad news: Even if you make your own feeds, your table chicken and eggs may well cost you more than the industrial agriculture equivalents sold in the supermarket. That's because *your feed may cost you more*. That's right, even if you make your own feeds from scratch you could well pay more than if you bought them premade from the co-op. Especially in our case—insisting, as we did, on using certified-organic primary ingredients—the eggs and dressed poultry Ellen and I raised cost us *more* than they would have if we had just bought them in the supermarket, or if we had fed our flock with co-op feeds. Only those as uncompromising on food quality as Ellen and I would be dopey enough to accept that bargain. Ah, but remember: "You get what you pay for."

Oops! Speaking of "what you pay for," actually the situation is even worse: If you make your own prepared feeds you will still, as a taxpayer, be *subsidizing the "cheap" chicken and "cheap" eggs in the supermarket*, in effect making a down payment with your tax dollars on industrial foods you are unwilling to feed your own family. Does *that* fact about your ultimate feed costs upset you a bit?

Equipment

When I began making my own feeds, I ground the grains in a hand-crank flour mill—a testament more to my stubbornness than to my good sense. Unless yours is the most micro of micro-flocks, you wouldn't continue making feeds with such a mill for long. When I got serious about making my own feeds—as said, up to a couple of tons a year—I bought a grain mill capable of grinding 25 pounds (11.3 kg) of feed grains in a few minutes, powered by a 1½-horsepower electric motor purchased locally at an electric equipment supply (see figure 17.1).[1] You will also need a feed scoop and a scale for weighing feed ingredients.[2]

Storing Feedstocks

I made my storage bin from plywood left over from an addition to our house. At 5 by 4 by 2½ feet (150 by 120 by 75 cm) it held three-quarters of a ton of feed grains. Three partitions divided the interior into four compartments, one each for wheat, corn, oats, and peas. The floor of the bin was slanted at 45 degrees, for gravity flow through sliding gates mounted on the front of the bin. The slanted floor provided not only more efficient release of the feed grains, but for rotation of the feedstock as the older grains were drawn off below and the newer were dumped in from the top.

Figure 17.1. This heavy-duty mill is what I used to make all my own feeds for twenty years.

A grain bin must be rodent-proof. I never had a problem with mice or rats trying to chew through my bin's ¾-inch plywood. If necessary, they could be thwarted with metal flashing or hardware cloth.

It is harder to guard against two insect opportunists in the feed bin—grain weevils and meal moths—whose eggs come in with the purchased whole grains. Don't assume that, because they'll be eaten by the chooks, their presence is no problem—feeding on the stored grain by these insects degrades its nutritional quality. There are two things you can do to minimize population levels:

- Sprinkle into the bin a cup or so of diatomaceous earth (DE) for every 50-pound (22.7-kg) bag of grains. *Wear a dust mask!* The diatomaceous earth wears holes in the exoskeletons of adult weevils as they move around, leading to dehydration, and the DE particles also coat and smother the larvae, who breathe through their skin. Don't worry, chickens can ingest DE without harm—indeed, it is sometimes administered to livestock to treat internal parasites. Since the weevils and moths are inactive in winter, there is no need to add DE then.
- When you see a lot of weevil or moth activity in the bin, use up your stored grains, then clean the bin thoroughly. I removed the interior partitions and climbed inside with a wire brush and my shop vacuum (always wearing a dust mask), and removed every speck of grain residue, which could contain moth or weevil eggs. Being fanatically thorough in this way, I only had to do this deep cleaning once, at most twice, a year.

Ingredients

I never used refined products such as gluten meal, oilseed meals from oil extraction, nor by-products such as feather meal from slaughterhouses. Except for a few supplements, I made my feeds from whole grains exclusively. Note that I was lucky to live close to an excellent supplier of organic ingredients, New Country Organics (NCO) of Waynesboro, Virginia.[3] Many would-be feed makers I correspond with report terrible frustrations finding some of the feed grains that I routinely bought from NCO, or convincing indifferent feed dealers to special-order them. Finding high-quality ingredients can be a challenging foraging expedition.

Following is a list of ingredients that I used or are used by flocksters of my acquaintance who make their own feeds.

Corn: A carbohydrate ingredient that in typical mixes supplies a major part of energy requirements.

Small grains: I added wheat, oats, sometimes barley, always whole. When available, I used other small grains as well—the greater the diversity of ingredients, the better. The fiber in small grains helps keep the digestive tract efficient.

Wheat: In some regions wheat might be more readily available for use as the main carbohydrate in the feed. If that is true for you, note that you should not feed wheat at greater than 30 percent of the diet, to avoid digestive problems.

Oats and barley: Do not feed these grains at greater than 15 percent of the total diet, either individually or in combination—excess consumption can cause runny droppings.

Other small grains: If you have sources for other small grains—millet, rye (in moderate amounts), triticale (hybrid of wheat and rye), sorghum (also known as milo)—try them in your mixes. If they are good people food or raised for feeding other livestock, you need have no concerns about experimenting with them as poultry feed.

Soybeans: When I first started making my own feeds, I added soybean meal as a by-product of oil extraction to furnish much of the needed protein. When I started getting my ingredients from NCO, they provided whole, roasted soybeans instead, which I cracked when making batches of feed. In response to concern from its customers, however, NCO ceased use of soy altogether in its feeds, substituting field peas instead. Thanks to that substitution by my supply source, I never again fed soy of any sort to my flocks. (See the sidebar "Feeding Legumes" on page 152.)

Peas: After NCO switched to all soy-free feeds, I used their legume substitute for soy, field peas (sometimes called winter peas, a subspecies of common garden pea) in lieu of soy in my mixes. (Field peas have 22 to 25 percent protein; soybeans, from 30 to 50 percent.) But I always had misgivings about that usage as well, given that they had been shipped from as far as 1,000 miles away—how sustainable is that?

Sunflower seeds: Some of my correspondents add sunflower seeds to their mixes, most often the black oilseed type sold to fill wild bird feeders. They provide proteins, fatty acids, vitamins, and minerals, plus dietary fiber to keep things moving through the gastrointestinal system.

Alfalfa: Dried alfalfa can be used as a green addition to feeds in winter. You might start with 100 percent alfalfa pellets (17 percent protein), the kind fed to rabbits and horses—grind them along with the corn and peas, since chickens resist eating the pellets whole. Note, however, that because the pellets are highly processed in comparison with good alfalfa hay, a better option is simply to feed the latter if you can get it. Suspend the hay in a net, and let the chickens pick out the leaf portions. (This strategy has a bonus "entertainment" value for a flock confined to the henhouse in the winter.) Feed in small amounts to prevent accumulation of uneaten excess—with its high nitrogen content, it will quickly heat up in the litter and may generate ammonia.

Flaxseed: At one time, *flax* was something of a buzzword because it boosts omega-3 fatty acids in egg yolks—

Feeding Legumes

At one time farmers grew legumes adapted to their own regions to feed their livestock. Then came the "soybean revolution" of the 1940s, after which the high-protein, widely adapted soybean replaced virtually all other legumes as livestock feed.

More and more people are becoming uneasy about the heavy use of soy in our food supply. Many question the use of soybeans and soy by-products even in animal feeds. Certainly this concern is justified with reference to feeding ruminants —soy feeds have serious deleterious effects in the rumen—though I am less certain about problems of feeding soy to avian species. There is no question that soy is one of the most problematic of all feeding legumes in terms of anti-nutrient factors—components that depress digestion of feeds and growth. Sprouting or heating soybeans neutralizes a good deal of the anti-nutrients—there is disagreement whether either treatment gets rid of all of them. In any case, it is universally recognized that livestock should *never* be fed *raw* soy.

If you want to use soy as a feed ingredient but cannot get whole, roasted beans, the most readily available alternative is likely to be soybean meal as a by-product of extraction of oil, used both for industrial applications and food oil. But note that, unless the oil was expeller-expressed—unlikely—extraction was accomplished using *hexane*, a by-product of petroleum refining. A potent carcinogen, hexane is also highly volatile, and it is claimed that it is entirely driven off by the end of the extraction of the soybean oil, leaving no residues in the meal. Reassured?

Hardly reassuring is the fact that almost all soybeans in today's marketplace are genetically modified.

All legume seeds have anti-nutrient factors—indeed, some of them are highly toxic (sweetpea and vetch), while others are toxic if fed raw (kidney beans)—with soy near the top of the list. However, some legumes are so low in anti-nutrient factors that they *can* be fed raw—field peas, garden peas, lentils, chickpeas, cowpeas, and others. In the case of avian species, anti-nutrient factors in these legumes are likely to be neutralized by the soak in digestive fluids in the crop. More use could be made of these legumes for feeding livestock.

Various subspecies of *Vigna unguiculata* are among the most important food legume crops in many parts of the world. Best known in the United States are cowpea and blackeye pea, with an average protein content of about 24 percent. Flocksters in search of alternatives to soy might seek out local farmers willing to grow cowpeas, blackeyes, or other feed legumes. Home feed makers and small producers of broilers or eggs for market could band together to offer farmers a guaranteed market for such crops. Opportunities might emerge for local production of other feed crops as well, for example corn and small grains, superior in quality to the run-of-the-mill alternatives, perhaps organically certified and almost certainly not genetically engineered.

though so does eating green plants and animal foods such as earthworms and insects. In our modern diets we tend to get far too much omega-6 in proportion to omega-3; thus any way to get these fatty acids into better balance is desirable. I tried adding flaxseed to feeds, but eventually abandoned it and concentrated on boosting omega-3 content in eggs by maximizing our hens' access to natural feeds, year-round. But, if you experiment with it: *Feed the seeds whole*, rather than as purchased ground meal, because flax oil is highly perishable and will go rancid after the seeds are crushed; *do not add too much*, as flax can impart a smell like paint thinner to eggs and meat if fed at greater than 10 percent; and make sure the birds have *plenty of grit*—the seeds are small and hard.

Supplements: In addition to the feed grains that make up the bulk of a feed mix, you might explore some supplementary ingredients for boosting protein or mineral content.

Meat scrap: You may read references to meat scrap meal in discussions of poultry feeds older than thirty years or so. Since the mad cow disease disaster in England in the 1980s, however, there are stringent restrictions on meat scraps in livestock feeds. Though feeding industrial poultry meat scrap meal is permissible if it has met stringent processing standards, manufacturers of poultry feed for the home and small-farm flock do not want to deal with the headaches involved. As a practical matter, you will not see meat scrap meal listed as an ingredient in commercial poultry feeds—nor will you find it available for purchase as a feed supplement.

Fish meal: Menhaden is a species taken in quantity not as a food fish but for turning into a high-protein—60 percent—addition to livestock feeds. It is usually recommended not to add fish meal at greater than 5 percent of a feed mix, to avoid fishy off-flavors in eggs or meat. Most people I know who make their own feeds use fish meal—it's hard to blend feeds that are high enough in protein without it, at least if you want to avoid refined or highly processed alternatives like pure lysine from corn. That is especially true for mixes fed to growing birds, whose protein needs are higher than for mature fowl. (Like flax, fish meal also boosts the omega-3 content of egg yolks.) However, it's important to ask: How sustainable is turning countless tons of fish into feed supplements? Furthermore, though it's a potent source of protein, fish meal is not a fresh, live food, so it will never be as nutritious a food as some of the alternatives discussed in the next two chapters. Note that in the first edition of this book I wrote: "I am especially interested in the possibility of small-farm conversion of organic 'wastes' into high-protein replacements of fish meal. My own cultivation of earthworms and soldier grubs as live protein feeds is driven in large part by their promise as an alternative to fish meal; I try to encourage farmers to give this idea a whirl every chance I get." I'm happy to report that, a decade later, both earthworms and black soldier fly larvae are cultivated at mass scale on organic wastes, and are being used as protein supplements in poultry and livestock feeds.

Crab meal: Dried, crushed shells from commercial processing of crab and lobster meat are a source of protein (about 25 percent or so) and of minerals like calcium. Though it's also a good source of selenium—an essential trace mineral in which some US soils are deficient—crab meal should for that very reason be fed in small amounts: Selenium is one of those vital minerals needed in trace amounts that become toxic at greater concentrations. I limit crab meal to 2 pounds per 100 pounds (.9 kg per 45 kg) of feed.

Cultured dried yeast: This supplement provides not only protein (18 percent) but minerals and vitamins, especially the B complex. It is particularly useful in feeds for waterfowl, whose need for B vitamins, and especially niacin, is greater than that of chickens. Cultured dried yeast contains live cell yeast cultures, which become active in the gut. Together with digestive enzymes in the dried yeast, the live cultures enhance feed digestion.

Probiotics or direct-fed microbials (DFMs): It is possible to add minuscule amounts of live microbial cultures as a supplement to boost the microbes in the gut. Theoretically, doing so would make digestion more efficient. This is another area where the science has been formulated with reference to a seriously flawed paradigm: chickens in high confinement without green forage or live animal foods, eating instead stale feeds based on highly questionable ingredients, chickens with compromised genetics to begin with—*of course* the digestive tracts of such birds need as serious a boost as can be provided. But birds eating natural foods are likely to have healthier, abundant, and diverse populations of intestinal microbes; they need no additional boost. Developing strategies to increase natural feeds, year-round, makes more sense than adding probiotic powders to your feeds.

Mineral and vitamin mix: Just as makers of manufactured feeds supplement with added vitamins and minerals, you can buy a similar supplement for your homemade feeds. The best one I know of is Fertrell's Poultry Nutri-Balancer.[4] Not only is Nutri-Balancer a broad-spectrum mix of minerals and vitamins, it contains live *Lactobacillus* spp. cultures that become active in the gut. Again: Poultry foraging much of their diet have little to no need for supplements. However, the diversity and amounts of natural forages available to your flock may vary considerably in different seasons, so you may choose to add a supplement like Nutri-Balancer as just-in-case insurance, either year-round or during winter's "lean season."

Feeding limestone or aragonite: Even if you are adding a mineral supplement such as Nutri-Balancer, you may choose to increase the calcium available in a layer mix. An easy way to do so is to add aragonite or feeding limestone to the mix. (Both are forms of pulverized limestone—calcium carbonate.)

Kelp meal: Dried seaweed meal from the coast of Iceland or Maine is an excellent natural source of minerals. If you're not using a mineral mix, kelp meal—added to your feeds or offered free-choice in a hopper—can supplement a wide range of minerals as well as several important vitamins and amino acids. For supplementing feeds, I used Thorvin kelp meal, certified as organic and sustainably harvested in Iceland.

Salt: An essential nutrient for chickens, but usually supplied in sufficient quantity in commercial mixes or a supplement such as Nutri-Balancer. If you're not using either, consider adding a high-quality livestock feeding salt, which includes many trace minerals in addition to sodium chloride.

Grit and oystershell: Though I would not add grit and oystershell to feed mixes, remember that grit is essential for processing feed, especially by fowl eating whole grains rather than more finely milled feeds; also, that crushed oystershell can provide the extra calcium layers need for making eggshells. Birds on pasture may find enough grit and mineral on their own. The more confined they are in the winter, however, the more important it is to make sure they have enough by offering granite grit and oystershell. Since both grit and shell are cheap, I provide them free-choice year-round.

Putting Together a Homemade Mix

If you have a curious mind and take time to do a bit of research, you will come to know more about natural feeding than the experts do and can experiment on your own with confidence. And the key word is *experiment*—I tinkered constantly with my feed mixes for twenty years. Indeed, making our own feeds allows a level of *flexibility* unavailable if we are dependent on purchased feeds: We can target a mix precisely to different species, ages, or types (meat or layer) of fowl, to the season of the year, and to the availability or scarcity of foraged foods.

Formulating a Mix

Base your feeds on the greatest diversity of feed grains available to you—the more you rely on whole grains as the foundation of your feeds, the less fine-tuning you must do. For example, I started out using complex formulas to calculate recommended ratios of the three macronutrients (protein, carbohydrate, and fat).

I soon found, however, that when *using mostly whole feed ingredients*, the nutrient ratios came out right if I simply targeted my formulation to the desired protein percentage. For someone using a lot of processed and by-product ingredients, of course, the calculation would be more complicated. As said earlier, for example, it is best not to feed fish meal at greater than 5 percent of the total diet. But when feeding chicks *on range*, I sometimes increased fish meal to 6 percent or even 7.5 percent (as in the feed recipe shown in table 17.1). The foraged green and animal foods the chicks ate in effect decreased the *percentage* of fish meal in the *total* diet.

If you are skilled in the use of electronic spreadsheets, you will find it trivial to do as I did: design a spreadsheet for automatically calculating values such as percent protein and cost as you enter varying amounts of ingredients. (See appendix E for more detail. You will find there as well a link to download a fully functional version of the spreadsheet.)

Premix

If you add half a dozen supplements to your feeds, in small amounts per batch, feed making will be more efficient if you combine them all into a *premix*. That is, weigh out enough fish meal, aragonite, kelp, et cetera to supplement, say, 500 pounds of feed. Mix together thoroughly—just scoop it back and forth between two containers until the mix is well blended—and store in a covered container. Now all the supplements are ready to scoop and weigh as one single ingredient as you make each batch.

Grinding and Mixing

I made feed in 25-pound (11.3-kg) batches. I first ground corn and peas only, setting the grinder to coarse—to crack each corn kernel or pea into several pieces, rather than grinding them into a meal. Then I dumped in amounts of premix and small grains for a 25-pound batch—then ran the next lot of corn and peas through the grinder. When I had layered 50 or 75 pounds (23 or 34 kg) of ingredients in this way, I scooped the feed back and forth three times between two bins to mix thoroughly. (Yes, I've said it before but: Always wear a good dust mask when grinding and mixing feed.)

Sample Recipes

I offer here a couple of feed recipes, but only on the understanding that these are *sample* recipes, based on ingredients available to me. They are merely snapshots of what were for me ever-moving targets. I changed the way I fed as I learned—from my own experience and that of others, from experimentation, from research, from crazy ideas. I hope you will do the same.

Table 17.1 is a feed I used for young growing chicks—and ducklings and goslings—on pasture. The recipe is expressed first in pounds per hundredweight to indicate percentages, then in pounds per 25-pound batch. Note that I added cultured yeast to this recipe because ducklings and goslings need more B vitamins in their diet.

Table 17.2 is typical of the mix I fed laying hens on pasture. This mix contains less fish meal, since adult chickens need less protein, and a lot more aragonite, to increase calcium for eggshell formation. Note that I did not add any cultured yeast to this mix.

Keeping Feed Fresh

The key to freshness in your stored grains is to keep them dry to prevent mold: If storing in large quantities, a wooden bin, well protected from weather, is better than metal, which can be subject to condensation. (I have no experience with storing large quantities in plastic bins.) Again, it is essential as well to *keep weevils and meal moths in check*. Once the seed coat has been ruptured by the chewing of these insects, the carbohydrates in the grain oxidize—that is, become stale—and feed made from such impaired grain is less palatable and harder to digest.

Because the same oxidation processes begin when we crush the seed coat (in my case, of the pea and corn portion—I always added the small grains whole), feed should be made in frequent small batches. Feed should remain at an optimal level of freshness for two weeks—

Table 17.1. Feed for Chicks, Ducklings, and Goslings Sharing the Same Range (18% Protein)

Ingredient	Pounds/100 pounds (percent composition)	Pounds/ 25 pounds
PREMIX		
Aragonite	1	
Nutri-Balancer	2	
Kelp meal	0.5	
Fish meal	7.5	
Crab meal	2	
Cultured yeast	2	
Total premix per 100 lbs.	**15**	
Total premix per 25 lbs.		**3.75**
GROUND/WHOLE PORTION		
Corn	27	6.75
Peas	20	5
Wheat	28	7
Oats	10	2.5
Total	**100**	**25**

Table 17.2. Feed for Layers on Pasture (15% Protein)

Ingredient	Pounds/100 pounds (percent composition)	Pounds/ 25 pounds
PREMIX		
Aragonite	6.5	
Nutri-Balancer	2	
Kelp meal	0.5	
Fish meal	4	
Crab meal	2	
Cultured yeast	0	
Total premix per 100 lbs.	**15**	
Total premix per 25 lbs.		**3.75**
GROUND/WHOLE PORTION		
Corn	27	6.75
Peas	20	5
Wheat	28	7
Oats	10	2.5
Total	**100**	**25**

and should be fed within thirty days, certainly no more than forty-five. I generally made about a week's worth of feed at a time.

Store prepared feed in a rodent-proof bin with a tight lid. I used galvanized trash cans. Yes, I recommended previously against metal bins for *long-term* storage. But I didn't have problems with feed made in small batches—my flocks ate it before mold had time to develop, even if occasionally there was a bit of condensation on the underside of a bin lid.

Alternatives to Grinding

I imagine some readers are wondering, "Hey, wait a minute, *Gallus gallus* didn't get its feed *ground*, did it?" Others may wonder if they must shell out big bucks for a heavy-duty feed mill. Isn't reduced dependence on purchased stuff what this book is all about? Such questions beg for a home-feeding alternative that does not involve grinding.

Often during winter I *sprouted* all the feed grains I fed my flock: corn, peas, wheat, and oats. I fed the sprouts in shallow containers, sprinkling over them small amounts of mixed fish meal, crab meal, kelp meal, and crushed, dried comfrey and nettle "hay." The dry ingredients stuck to the damp sprouted grains, rather than sifting to the bottom, and the birds consumed most of the along-for-the-ride supplements before they finished the last of the sprouted grain.

The biggest drawback to this approach to feeding is that the chickens were not especially fond of the sprouted peas—they ate some of them, but left some untouched. They made more complete use of peas when cracked and mixed into a prepared feed.

CHAPTER 18

Feeding Your Flock from Home Resources

When I review the list of feed ingredients in the previous chapter, something that stands out is the number of ingredients with one *problem* or other: Wheat shouldn't be fed at more than 30 percent; nor barley or oats at more than 15 percent; many legumes can create problems in the diet, some of them serious; fish meal is a high-quality protein supplement with major sustainability issues. Such problems inherent in agricultural products convince me that the key to nutritional balance is not spreadsheets and formulas but the greatest diversity of natural foods. Nothing we can buy in a feed bag can match the breadth and depth of nutrition to be discovered right at home. Is there a dangerous upper limit to crickets in the diet? Probably not, but even if so, free-foraging chickens are unlikely to hit it.

As always, what goes in determines what comes out. Here at Boxwood, we don't need laboratory analyses to prove it—we *know* that the quality of our eggs and dressed poultry improved every year, in step with provision of ever-more-natural foods and self-feeding opportunities.

Home Feeding

I have said that I think of my grandmother's flock as the model for natural chicken feeding. Her chickens were eating their three food groups, whose defining characteristics were that they were *alive* and *raw*. Now reflect as well: They ate almost completely *from home resources*. Increased home feeding—*decreased* dependence on purchased feeds—is a possibility for every flock, wherever located. Why not start experimenting now? I have heard from so many who have done so and who report: "It *works*—I'm saving real money on feed bills."

Fortunately, there are a multitude of the three food groups options on almost any homestead or small farm, and many ways to combine natural feeding strategies. And who would have imagined that saving money on feed bills could be so much fun?

Green Forages

Find ways to provide your flock fresh green plants, the first of their food groups, *daily*. Best of all is keeping them on pasture and other natural range with access to all three

Figure 18.1. Ducks and geese eating comfrey. Photo courtesy of Bonnie Long.

food groups at once—even flocksters on small properties successfully "pasture" their birds on their lawns.

Fertility Plants

I grow comfrey and stinging nettle to feed the soil as mulches or composts, not only for their mineral content but because they are high in nitrogen and are quickly converted by soil and compost heap microbes. They can also be fed to poultry.

Both plants have a bad reputation. Many gardeners fear comfrey as *dispersive*, spreading via millions of seeds, or *invasive*, spreading aggressively via underground runners. In fact, it is neither. It sets flowers—thus supporting pollinators—but does not make viable seeds. And it does not spread—the first patch I planted hasn't moved an inch in thirty years. But it certainly is *persistent*—not surprising given the reserves in its thick, fleshy roots, which grow 8 to 10 feet deep (2.5 to 3 m)—and will not move from where you choose to plant it without a serious fight.[1]

Comfrey is child's play to propagate—once a patch is well established, take crown cuttings and expand your plantings as much as you like. I either cut and carried it to my flocks—geese and ducks especially *love* it (see figure 18.1)—or gave the birds direct but temporary access to patches of it planted in ranging areas (see figure 18.2).

Nettle, however, is both dispersive and invasive and therefore needs vigilant "discipline" to keep it in place. I cut it as soon as it started to flower to prevent seed set, and cut it back at the edges of the patch when its roots got adventuresome. Nettle is high in protein and mineral content.

Figure 18.2. Comfrey grazed by chickens.

Figure 18.3. Mixed cover crop, haute cuisine for a mixed flock. Photo courtesy of Bonnie Long.

Often in the summer I dried some comfrey and nettle "hay" for feeding in winter, crushed and sprinkled over the birds' feed.

Cover Crops

I *love* growing cover crops. If you like jigsaw puzzles or chess, you will, too: Despite the challenges to pulling it off—both in space (a garden that is simultaneously used to grow food crops) and in time (all four seasons)—maximizing cover cropping year-round brings so many improvements to soil fertility and texture that no opportunity to sneak in a cover crop should be missed. Fortunately, cover crops can do double duty as green feeds for the flock. We can cut the greenery to carry to the flock—it quickly regrows—or let the chickens till in the plants as they dine (see figure 18.3).

Note that I no longer bother growing fall crucifers—mustards, winter radishes, raab, kale, rape, turnips—as separate garden crops: I just sow them as fall cover crops, and there's a gracious plenty for everybody—the soil food web, our poultry, and us.

Weeds

Weeds may annoy gardeners and landscapers determined that *nothing* grow on their place they didn't plant, but many wild plants with a mind of their own make valuable contributions to flock nutrition; and some (dandelion, lamb's-quarter, nettle, burdock, yellow dock) are higher in protein than alfalfa. Poultry relish them all.

Take dandelion, toward whose demise millions of dollars are dedicated every year but so nutritious as a cooked or salad green that herbalists extol it as *superfood* and *medicinal*. In addition, it is a *dynamic accumulator*: Its taproot grows into the deep subsoil and mines it of minerals, especially calcium, which it makes available to more shallow-rooted plants. (Both the fertility plants discussed on the previous page—comfrey and nettle—are also dynamic accumulators.)

Of course, such friends may not be equally welcome in all parts of the garden, orchard, and landscape, so remember that weeding chores can furnish valuable green fodder for the flock. The most useful weeds where you live will vary. The few that are toxic vary as well, so familiarize yourself with ones that could be hazards for poultry where you live.

Examples of toxic plants are castor bean (*Ricinus communis*), milkweed (*Asclepias* spp.), immature berries of nightshade (*Solanum nigrum*), oleander (*Nerium oleander*), jimsonweed (*Datura stramonium*), pokeberries (berries of *Phytolacca americana*), and yew (*Taxus* spp.). Such plants are rarely a real threat to chickens, however—most they avoid instinctively. Note that in some cases it is not the plant that is the threat but the mature seeds. For example, hairy vetch (*Vicia villosa*) is an excellent nitrogen-fixing cover crop and its foliage is edible for poultry, but remember that its seeds are toxic. Chickens are unlikely to eat jimsonweed but may eat its seeds. Seeds of both plants have been implicated in actual poisonings of poultry.

Once you've identified the toxic plants to look out for, feel free to experiment with harvests from weeding nontoxic species. Weeds vary in mineral content, so the wider the range of weeds available to the birds, the more likely their mineral intake will be in balance.

In most temperate areas, the following palatable weeds should be common. Many of them make fine people food as well.

- Prickly lettuce (*Lactuca serriola*), likely to show up in shady areas.
- Purslane (*Portulaca oleracea*), rich in omega-3 fatty acids, vitamins, and minerals.
- Dandelion (*Taraxacum officinale*), superfood for chooks as well.
- Lamb's-quarter (*Chenopodium album*), one of whose common names, "fat hen," hints at its utility as poultry feed.
- Yellow dock (*Rumex crispus*), rich in vitamin A, protein, iron, and potassium, though it should not be overfed because of its oxalic acid content.
- Chickweed (*Stellaria media*), how do you suppose it got its name? Common and prolific, both the plant and its seeds are highly nutritious feed.

Sharing the Garden's Bounty

A useful garden crop to share with the chooks is the *beet*, but note that the one species, *Beta vulgaris*, comes in several variations, increasing its utility. We grow the common table beet as food for us, but we also grow the variants *mangel* and *chard* for the flock. Both the latter are tremendously productive, and you can progressively pull off the large lower leaves—all types of poultry *love* them—which only stimulates the plants to vigorous new growth, of more tender chard leaves for us and a mammoth beetroot in the case of the mangel.

We can share green foods from other harvest crops—wrapper leaves and less-than-perfect trimmings from chicories, lettuces, and cabbages; spent harvest plants that are still green such as broccoli and vines of bean, pea, and sweet potato (*not* the foliage of white or "Irish" potato, *Solanum tuberosum*, which is poisonous); and most cull fruits and vegetables—all are excellent fresh foods as well. Some crop plants readily reseed and show up in areas patrolled by the flock, or volunteer in the garden as more "weeds." Sylvetta arugula (*Diplotaxis erucoides*) and purslane are examples in my garden.

Squashes offer useful feeds, from the lurking monster zucchini to the protein-rich seeds scooped from the cavities of winter squash. You might even grow an oilseed type such as Lady Godiva, with its nearly naked seeds, easy for the birds to digest and rich in oils and protein. Large winter squash such as Hubbards store well for fresh feeding deep into winter.

Root crops are rarely utilized as much as they could be. In wartime England and other European countries, potatoes substituted in whole or part for scarce grains for feeding poultry. In *The Resilient Gardener*, Carol Deppe makes a compelling case for potatoes as a major contribution to independent feeding, especially of ducks.[2] Note that—though potatoes are fed raw to ruminant species such as cows—they *must be cooked* for feeding to any monogastric species (poultry, pigs, horses, or people).

Did I mention the "mammoth beetroot" of mangels? I grow 12-pound (5.4 kg) roots—that's on the *small* end of the spectrum—and store them through the entire winter in a protected pit in the ground. They are a welcome source of fresh food for the flock in a part of year when it is scarce and provide great "entertainment" value as the chickens peck them apart.

Every gardener accumulates tail ends of outdated seeds. I like to mix them, perhaps even adding remainders purchased cheaply after the seed-buying season is over, and sow a salad bar for the chickens. It's a bonus if I end up with some prime salad or cooking greens for us as well.

Grass Clippings

If your situation prohibits bringing the flock to the pasture, bring the pasture to them: Lawn clippings—from lawns that have not been treated with toxic chemicals—are excellent fresh forage. Short, rapidly growing grass yields the highest levels of nutrition. Do not feed too large a volume at one time—excess clippings accumulate into an anaerobic, slimy mess. If you are using your birds to work compost heaps, however, you need not worry about feeding too many clippings—the chickens will work whatever they don't eat into the heaps.

Seeds and Fruits

A good deal of the feed we purchase for our flocks is grain, but many wild seeds are equally nutritious and are free for the taking if the birds are given the opportunity to range. As an experiment, I once offered my laying flock, side by side: cracked corn, cracked peas, wheat, and oats (the four main ingredients of the feed mix I made for them); purchased wild bird seeds (seeds of black oilseed sunflower and niger, a type of thistle); buckwheat, clover, and annual rye grass seeds; and seeds stripped off mature pasture grasses. The birds preferred some over others but by the end of the day had eaten almost every last seed.

Easier Grain Crops

Conventional feed grains, the ones grown at massive scale for agricultural markets, are easy crops for us small-scale flocksters to grow. But an attempt to do so at significant scale will entail serious obstacles to harvesting, threshing, and storage. It makes sense to look for simpler strategies for using these crops.

Figure 18.4. Chickens eating mature seeds of sorghum, which is as easy to grow as corn. Photo courtesy of Bonnie Long.

I experimented with several dent corns over the years—Hickory King, Bloody Butcher, Reid's Yellow Dent—as well as Nal-Tel, an ancient landrace flint corn from Central America. I used a simple hand sheller to take the kernels off the cobs—until finding it was easier to break them open by stomping with the heel of my boot. Once the cob was crushed, it was easy for the chooks to peck the kernels free.

I also grow sunflowers—for their beauty and as support for pollinators—and harvest the seed heads just as the seeds ripen. I either throw the ripened heads to the chickens, or string them from rafters in the chicken house to protect from rodents for later feeding. I prefer *black oilseed sunflower* for feeding poultry. For ten years I've developed my own strain of Peredovik, selecting for single rather than multiple heads, large heads with well-filled seed, and resistance to lodging. Wild birds—and squirrels—like sunflower seeds as well; I monitor the ripening heads daily and harvest as soon as I see evidence of raids on them.

Sorghum is closely related to corn. Varieties of *Sorghum bicolor* have been specialized for different purposes: syrup making, broom making, and grain. My chickens love sorghum seeds, whose nutrient profile is similar to corn's (see figure 18.4).

I experimented with eight varieties of the shorter, more dense-headed strains of grain sorghum, commonly called milo, but found them more difficult to grow than the other types, for example more subject to growth of mold in the densely clustered seed heads in rainy spells. And oddly, none of my poultry—chickens, geese, or ducks—seemed more than casually interested in the seed heads of this supposedly more feed-grain type of sorghum. Since then, I have instead grown the taller broomcorn and syrup types, with big open sprays of shiny seeds, highly ornamental. As with the sunflowers, I either cut the ripened seed heads and throw them to the birds or string them up in the poultry house for winter feeding.

Amaranth is a plant I grow for its beauty and its nutritious seeds, hundreds of thousands of them per seed head—extremely tiny, but the Aztecs built an empire on their high-protein nutrition. It readily reseeds—since my original plantings, I have just weeded out the abundant volunteers if in the way, welcomed the ones that fit in, and cut the mature seed heads for feeding to mixed flocks of chickens, geese, and ducks.

There is a sweet spot for the four crops mentioned here—corn, sorghum, sunflowers, and amaranth—when the seeds are ripe but the leaves are still green. If I cut entire plants at this stage and throw them to the birds, the ducks and—with somewhat less eagerness, the geese—eat the leaves down to the stalks.

Triple-Duty Cover Crops

Here's another idea for turning the work of big-boy equipment—harvesting, threshing, winnowing—over to the chickens themselves. If you give cover crops such as small grains, buckwheat, cowpeas, even clover—whose seeds are relished by chickens—enough time, they will ripen a fine crop of feeding grains for the flock. Just let the flock till in the crop while eating the seeds (see figure 18.5). While it is true that the proportion of certain feed grains—oats, barley, and buckwheat come to mind—should not be too high as a regular part of the flock's diet, in my experience a brief spike in these feeds as the birds till in a mature cover crop does not create digestive problems.

No, with less than 3 acres, we couldn't come close to raising all our own feed grains. But if we get a nice feeding boost from seeds of cover crops we grow for soil improvement anyway, that's just gravy, right?

Tree Crops

In more frugal times, farmers released flocks of turkeys (and pigs) into the woods to fatten on the abundance of free feed under oaks, beeches, chinkapins, and persimmons. Planting trees in the backyard with a thought to their contributions to a feeding program is worth considering as well, for the flockster planning long term.

ACORNS

J. Russell Smith pointed out in his classic *Tree Crops* that acorns can be used as feed for poultry, quoting a report from England during World War II of acorns being

Figure 18.5. Chickens in garden cruiser tilling in a mature oat cover crop, its seeds an additional payoff for their work. Photo courtesy of Bonnie Long.

used to replace up to half the feed ration for chickens.[3] Traditionally, farmers fattened turkeys for free on acorns.

Since tannin can depress feed intake and digestive efficiency, sweeter (low-tannin) acorns are better as feed. My white oaks produce large, low-tannin acorns, which in some seasons I have added to feed mixes. Because the acorns are so large, my birds could not eat them whole, so I ran the acorns through my feed grinder. Acorns alone balked at running between the grinder's burrs, but I found that if I mixed them with typical scratch grains such as whole corn, wheat, and oats, the auger pushed them into the burrs more efficiently. I set the mill for coarsest grind—my purpose was not to grind the acorns finely but simply to crack them open. When I offered acorns broken up, garnished with the usual scratch grains, my whole flock—chickens, ducks, and geese—devoured them greedily. A flockster of my acquaintance who doesn't have a grinder smashes acorns inside a stout denim bag she made from an old pair of jeans, using a 5-pound hand sledgehammer.

HAZELS

I grow hazels (filberts), both the larger type (*Corylus avellana*) and the American (*C. americana*). The nuts of the latter are small, and cracking them out is tedious, so in many harvests I have used them as feed instead, setting the burrs of the feed mill at a width to crack the nuts out of the shell without grinding them to a paste. At feeding values like protein 15 percent, high-quality fat 60 percent, flocksters without a mill might even choose to smash the clusters of nuts, same as for chestnuts (next).

American hazel, planted as a thicket, would be great forage for turkeys.

CHESTNUTS

Three Chinese chestnuts I planted on the flock's pasture have produced abundant crops of chestnuts for many years—far more than we can eat, plenty to share with the flock. These large nuts I smash on a rock with a small sledgehammer so the birds can get at the meats—and I expect that my friend's heavy-duty denim bag would be good for smashing them in quantity. All my poultry have loved the fresh nutmeats.

The chestnut trees also provide shade for a flock helping to break the life cycle of the chestnut weevil, as described in chapter 5.

OTHER TREE CROPS

Another shade tree for the flock helps feed them as well: Mulberry trees have a wide range; are easily propagated, precocious, and almost completely carefree; and in an extended season produce an abundance of nutritious dropped fruit that chickens relish, in either full sun or the partial shade of a woods edge.

I gather the wild hickories and black walnuts growing in our bit of woodlot and smash them on a rock for my chickens to eat. Cultivated nuts such as pecans and English walnuts culled for weevils can be crushed and fed to poultry as well.

Live Animal Feeds

As discussed in earlier chapters, our worker chickens get a lot of high-octane feeding value foraging insects, earthworms, and slugs. Be on the lookout for other opportunities to offer live animal feeds. If you, like us, are blessed with an abundance of Japanese beetles, shake them off your grapevines or plum trees into a 5-gallon bucket with some water in the bottom—in the cool of the morning or early evening, when they are less likely to fly. Dump them out to the chickens, and step back out of the feeding frenzy. Ducks are keen on beetles as well—they're like little vacuum cleaners. If you take the time to handpick crop-munchers such as squash bugs, plop them in a bit of water as well—one of my correspondents said her chooks think they're bonbons.

Put down some scrap boards and leave them in place until they accumulate a nice assembly of earthworms, slugs, and pill bugs underneath, then flip them over and let the birds feast. Repeat in a couple of days.

When working in the garden, keep a container handy for the earthworms and fat white grubs that turn up—they're excellent fare for the flock.

There are some animal species we can actively *cultivate* for feeding the flock. These allies offer such a significant boost to the feeding program, I discuss them at length in chapter 19.

Other Protein Feeds

Keeping dairy animals such as cows or goats is like growing zucchini: If you do it, you're going to have a surplus. As well, homemade dairy products such as butter and cheese generate skimmed milk and whey as by-products. These dairy foods make excellent feed—even better if you culture them with natural microbe cultures such as kefir grains before feeding. I expect your chickens would get far more benefit from excess milk fermented with various cultures than from purchased microbial powders or probiotics.

Eggs are high-powered feed. Of course, we are not likely to use many of our precious eggs for feeding; but cracked eggs or those that are too dirty to rescue for table use can be hard-boiled, crushed coarsely by hand, and thrown to the birds. Contrary to what you will hear, feeding eggs in this way will *not* encourage egg eating in the flock. Eggs are especially useful for feeding chicks and young growing birds, whose needs for protein are higher.

Offal—of deer, lambs, goats, or other livestock, or from cleaning fish—makes excellent feed. Liver, if not favored by your family, is yet another protein booster for young growing birds. On slaughter day, if you make the effort to catch the blood as the animal is bleeding out, you can offer it as high-protein feed that chickens relish. I have even experimented with feeding poultry blood to my chickens without ill effect. I have also fed my flock the carcasses of animal competitors who push it a bit too far, such as a rogue groundhog that persistently raids the garden, or the eight squirrels I trapped one year to save my shiitake mushroom production—even the odd roadkill. I just open up the carcass a bit with a hatchet before giving it to the flock. (Note, however, that I found that the flock didn't eat such carcasses entirely. See "Why Not Feed the Carcass Right to the Chickens?" on page 171 for more on this topic.)

Sprouting for Enhanced Winter Feeding

Those of us who have a dormant winter season need strategies for producing fresh green feeds even when the snow flies.

Lots of people like to sprout seeds for the table, on the quite reasonable assumption that the sprouts are more *alive*. Sprouting is a possibility for producing more lively food for the chooks as well, especially in winter. Any of the people-food sprouts are fine; the main concern is that you find seeds in bulk—*untreated!*—to keep the expense down. (Feeding sprouts from health food stores would be prohibitively expensive.) Seeds of sunflowers, alfalfa and clovers, broccoli and other crucifers, buckwheat, radishes, beans, and peas—all these salad bowl sprouts would be fine for your birds. Any grain you feed whole to your flock—corn, peas, any of the small grains—can be sprouted in lieu of adding to a feed mix.

Think of sprouting as value-added home feeding, even if you start with purchased seeds. Sprouting changes nutrient profiles of seeds—an increase in protein, vitamins, and enzymes, though a reduction in carbohydrate content. Of course, you can sprout any time of the year, but I never made the extra effort in the growing season, when the birds enjoyed constant access to fresh green foods.

While any of the seeds mentioned in "Seeds and Fruits" can be sprouted, keep in mind during the discussion of sprouting methods next that only those of cold-hardy plants are appropriate for greening up in trays at low temperatures. Even warm-weather species such as sunflower and buckwheat can be sprouted using my bucket system.

Bucket Method

If you don't require a greened sprout, my bucket method is the simplest for sprouting seeds easily in any quantity needed. I sprout in my basement, where waste heat from the furnace keeps the ambient temperature high enough to sprout any seeds, even warm-weather ones. The concrete floor drains directly to a sump pump, so draining the buckets daily on the floor is no problem.

At its simplest, this method requires five food-grade plastic buckets with loosely fitting lids. One is the *soak bucket*; the other four serve as *draining/sprouting buckets* once I have drilled them with dozens of small holes, in the bottoms and halfway up the sides. The size of the holes is important: They must be large enough for dusty debris to flush through, but small enough to prevent the grains from lodging into them, creating blockages.

The sprouting schedule is as follows:

Day 1: Measure the grains, any mix you like, into the soak bucket and cover with water.

Day 2: Dump the soaked grains into one of the drain buckets and rinse thoroughly; then allow to drain while measuring another batch of grains and covering with water in the soak bucket.

Day 3: Repeat the dumping of the soaked grains into a second drain bucket, rinsing, and starting a new batch of grains in the soak bucket. *Rinse the sprouting grain in the first drain bucket as well.*

Day 4 and thereafter: Now you're rolling. Each day, dump the soak bucket into an empty drain bucket and start a new batch of grains—and each day, *rinse the sprouting grain in all active drain buckets.* The daily rinsing is necessary to prevent growth of molds in the sprouting grain.

Tray/Greening Method

If you have the protected space to sprout grains in growing trays, you can allow sprouts to develop to any green stage you like (see figure 18.6). Indeed, if you time the feeding right, your birds get the benefit of both green feed—the tender shoots that have greened up—and, underneath, the sprouted grain itself, with its enhanced nutrition. I use nursery flats that I have saved from past plant purchases. It is better to start the process by soaking the grains at least overnight, maybe longer. Drain and spread in a thin layer over the bottom of the tray. Cover with straw, shredded leaves, even coarser parts of the litter from the henhouse—any organic material that will help prevent drying. Water daily.

If you're growing cold-hardy types such as small grains, you could green the sprouts in a cold frame. If you are willing to make the additional effort, you could set the trays outside during the day, then bring them inside or into a basement at night. If you have an unheated greenhouse like I do, that's an excellent place to grow trays of sprouts. Hanging the sprouting trays from the side purlins of the greenhouse saves growing space at ground level.

Be an Opportunist

If you stay alert to the open-ended possibilities for feeding your flock naturally, you're going to have lots of fun. And if you are uneasy about feeding your flock exclusively from grain fields far, far away, experimenting based on even your wildest ideas will bring you closer to independence of purchased feeds. To conclude this chapter, I offer a medley of additional ideas I've come across.

Duckweed is a prolific pond plant with a wide range. If you have a pond, experiment with harvesting duckweed and feeding it to your flock. Making this high-protein plant—40 percent or more, dry weight—an important part of a feeding program is more than a theoretical possibility.[4]

One of my correspondents reported that she depends on *browntop millet* and *proso millet*—often planted by hunting aficionados as feeding fields to attract game birds—as major components in the feeding of her flock. Once the plantings have matured their

Figure 18.6. A tray of sprouts from the greenhouse finds a customer.

seeds, she rotates her flock over the field in long swaths outlined with electric net fencing. I experimented with both species myself and found them easy to grow and easy for the birds to self-harvest.

It is not difficult to make *hay* from alfalfa and clover, even lawn clippings, and store for winter feeding.

If you need privacy screens or boundary fencing, it might make more sense to grow *living hedges*, which provide the ecological benefits of edge habitat—shelter and food for birds and other wildlife, enhanced insect diversity including pollinators such as bees, and more—you will never get from a manufactured fence. Figure 18.7 illustrates the possibility of a temporary privacy screen using the annuals sunflower, amaranth, and sorghum, all of which, as described earlier, mean more free food for the flock. If they are suited to your climate, permanent privacy or boundary fences could be based on *goumi, goji,* or *Siberian pea shrub*. I have corresponded with several flocksters who use these tough, productive plants for hedges and get nutritious berries or seeds as a feeding bonus for their flocks.

Remember *stacking of livestock species*, which can yield feeding benefits. One of my correspondents in British Columbia soaks mixed grains and peas for all her livestock, usually for two or three days. When things are busy and the grain has not soaked as long, however, she has observed: "Sometimes the less soaked grain goes right through our milking cow or the pigs. The chickens love this manure and pick through it for this grain. If the chickens miss the whole grain, it grows like a weed in the manure pile. The chickens and pigs love this green crop."

Never doubt that you, too, can make *significant advances* in feeding your flock from home resources. One of my correspondents in far Saskatchewan reported that he cut his monthly feed bill from $20 to $5 when he began feeding largely the peas of Siberian pea shrubs, available for the picking in miles of shelter belts in his area; home-grown alfalfa hay; and chard, fed fresh daily when abundant in summer, and dried, or pureed and frozen in serving-sized portions, for feeding the flock in the winter.

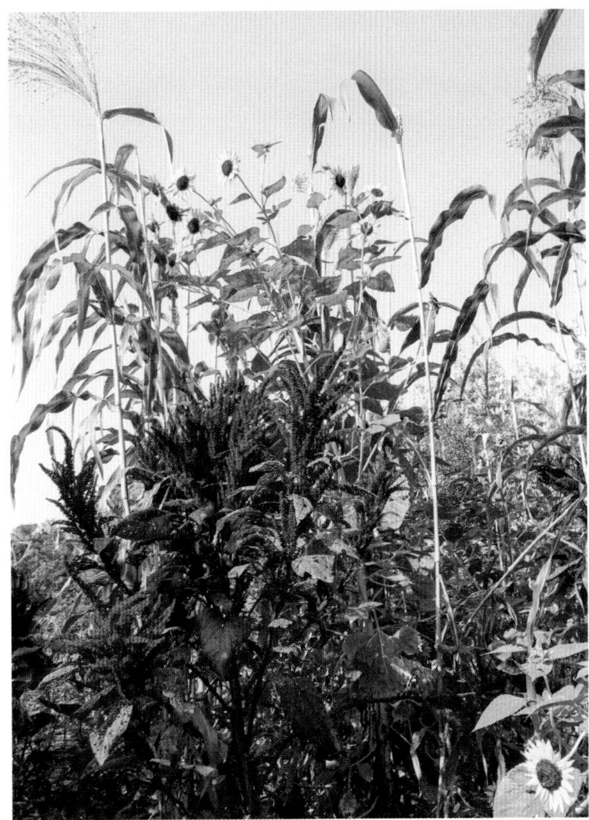

Figure 18.7. Sorghum, amaranth, and sunflower make a privacy screen that is beautiful, supports insect and wild bird diversity, and helps feed the flock.

During our sojourn at Boxwood, we have seen two emergences of Brood II of periodical cicadas. After hiding seventeen years underground, they swarm up in their billions and bedeck every tree and bush like jewels. This old chicken man marveled at those numbers—and their *size* (a couple of inches long, figure 18.8)—and an alert sounded: *Bing-bing-bing: Protein!* Since cicadas are slow and nonchalant, especially in the cool of the morning, it took only minutes to fill a gallon bucket with the creatures to throw to the flock. Indeed, the flock couldn't keep up with this *embarras de richesses*—I made a stash of gallons of them in the freezer. All told, more than a month of gluttony followed the cicadas' emergence.

The point of this story is not *Look for free protein at biblical scale in seventeen years!* but rather *Be an oppor-*

Figure 18.8. Cicada (also known as windfall protein) on catnip.

tunist. Natural-feeding opportunities abound—seek and you will find.

You may have wondered when reading the strategies in this chapter: Wouldn't it be nice if more of them provided natural foods that, like the cicadas, are rich in proteins and high-quality fats? Fortunately, we don't have to wait seventeen years to reap nature's bounty, *we can grow our own*, as we'll see in the next chapter.

CHAPTER 19

Cultivating Recomposers for Poultry Feed

All living things die, and all living things poop. Imagine the state of our abundant ecology absent the work of specialist organisms that decompose dead plants and animals and their leavings. Happily, key species neatly dispose of such residues, including those that may be vectors of disease, such as corpses and excrement.

Waste is nowhere to be found in nature—*all* organic residues become food for other players in the ecology, who are then fed on by others. The end of the process is *humus*—that is, nature's "waste processing" is the key to spontaneous accumulation of fertility over the millennia in natural soil systems.

"Waste" is always a wasted opportunity—for putting to productive use a potential resource. The USDA estimates that 30 to 40 percent of the food produced in the United States is never eaten—and that food waste is the largest single category of throw-away materials sent to our landfills. Despite undoubted progress in recent years in the recycling of various wastes, close to half of landfill discards are still organic materials (uneaten food, yard and landscaping wastes, and paper and cardboard) that could easily be reclaimed through natural decomposition processes. As well, livestock manures from high-confinement production systems are rarely fully recovered to enhance soil fertility. If we see our society as an energy system, failure to reclaim organic "wastes" represents a sluicing of potential energy resources down a black hole with no return. It is especially tragic when the lost resources end up as toxins in the ecology—increased atmospheric methane and carbon dioxide; and pollution of soil, groundwater, streams, and oceans.

Imitation of nature's decomposer organisms—I prefer calling them *recomposers*—offers alternatives for responsibly managing organic "wastes" and preventing their contamination of the ecology. At the homestead and farm levels, such exercises in white magic can offer yet another benefit: free, high-quality feed for our flocks (and other livestock). Imagine that we feed 100 pounds (45.4 kg) of food residuals to a colony of soldier grubs, the voracious larvae of the black soldier fly, discussed in detail later in this chapter. The hungry grubs reduce the food wastes to 5 pounds (2.3 kg) of high-fertility soil amendment, in the process generating 10 pounds (4.5 kg) and possibly up to 20 pounds (9 kg) of grubs as live protein feed for poultry or pigs—in addition to liquid effluent, which can be used like a manure tea to feed crops. There is *absolutely no waste* remaining after this conversion—it has *all* been converted to resource.

In chapter 5 ("Homestead Partners" on page 49) I gave examples of my Icelandic flock reaping feeding benefits from recomposers that break down "wastes." I have described chicken-powered compost heap comes alive with detritivores: shredders such as earthworms, pill bugs, millipedes, crickets, slugs, and snails; predators such as ground beetles and spiders; fungi, whose rhizomorphs chickens love; and fat white grubs of various sizes (larvae of many insect species, preparing for metamorphosis into the adult phase), who play many roles in decomposition ecology but also serve as high-protein fare for the lucky chicken who finds them. Recomposers at the microscopic level include thousands of species of bacteria, protozoans, yeasts, actinomycetes, and more—which provide nutrients such as vitamins K and B_{12}. And remember those hundreds of laying hens who obtain *all* their feed in the form of recomposers in Vermont Compost Company's compost heaps.

For the adventurous, however, there is an option beyond giving the flock access to recomposers that arise naturally in a compost heap: *cultivation* of some decomposer species as live protein feeds. I have worked with three of them: carrion flies, soldier flies, and earthworms. In this chapter, I describe my various approaches to managing these amazing creatures for the benefit of my flock.

Such cultivation projects are another example of finding the perfect middle ground between the fully "natural" and the entirely manipulated: Cultivating detritivores in populations vastly more numerous than *Gallus gallus* would have found in the wild requires levels of intervention and manipulation that leave "all-natural" far behind. Yet there is no part of the process that departs fundamentally from analogs in nature—we add no toxic chemicals nor extravagant amounts of purchased energy—and at the end of the conversions we feed our birds exactly the kinds of foods they would most seek in nature, albeit in amounts considerably more generous than nature would provide. How marvelous that these projects help to solve "the protein problem" in the feeding program—all as a part of "waste" management!

Protein from Thin Air

Many years ago I encountered a method for generating high-protein poultry feed "from thin air," based on the use of kitchen scraps to cultivate larvae of the common housefly. ("Larvae" sounds a bit less repellant than "maggots," yes?) I experimented with the technique, using our sparse kitchen residuals—coffee grounds and tea leaves, vegetable peelings and trimmings. Though my chickens greedily devoured the resulting fly larvae, I knew there could be no serious increase in cultivation of "free protein" based on the modest castoffs from our kitchen. Further research suggested the possibility of working with another fly family—the carrion flies.

Historically, European farmers hung slaughter offal above pens where they were fattening poultry, as well as over ponds stocked with farmed fish. Flies laid their eggs in the offal, the eggs hatched, the larvae fell out and down to the chickens or fish below, and everybody was happy. Were not those tidbits from history abuzz with opportunity? I thought of my buddy Sam Poles, the best trapper in our county, and the beaver carcasses from his nuisance trapping jobs. Weren't they an untapped resource begging to be used? Sam gave me an unbelieving chuckle when I told him my crazy scheme, but promised a steady supply of carcasses.

I drilled dozens of ⅜-inch (1-cm) holes in the bottom, side, and lid of a 7-gallon, food-grade plastic bucket—holes large enough for female flies to enter and lay their eggs, but small enough to prevent the chickens' pecking at the bucket's contents. I then put a beaver carcass into the bucket and suspended it from a tree limb, *inside the flock's electric-fenced enclosure*.

The larvae of common fly species live, feed, and grow in their feeding medium (such as rotting meat), then—when ready to pupate—instinctively find their way to earth and burrow in, ready for metamorphosis into the adult winged form. When fly larvae working the carcass were ready to pupate, bailing out of their feeding medium inside a *suspended* bucket meant a free fall to earth. My sharp-eyed chickens recognized high-value dining on the fly, and snapped them up.

That is the heart of this simple system for generating free protein for the flock out of thin air. Let me address questions you may have and explain some alterations I made in the system.

"But Doesn't It Stink?"

When I loaded that first 7-gallon bucket, I simply dropped in a 35-pound beaver carcass, screwed on the lid, and suspended it inside the ranging area. And yes, during the last few days of "processing," it got pretty ripe! So I made four more buckets as described—these were 5-gallon buckets—and began using an axe to divide each beaver carcass into five roughly equal chunks.

I surrounded the portions inside the buckets with a thick layer of loose organic material such as dry leaves and straw. The female flies had no problem finding their way through this material to lay their eggs, but it helped enormously with damping down odor. There was also less odor because processing time with smaller pieces of carrion was so greatly accelerated, a matter of a few days only. Toward the end of that time, there was a bit of odor when I passed within a couple of yards of a bucket, but it didn't carry to other parts of the property. Still, those in suburban settings, or with a neighbor right by the fence line, would do well to seek alternative sources of free protein.

"What Happens to the Carcass?"

When no more maggots were emerging from the interior of the bucket, I dumped out the contents and kicked the wad of padding apart. A few of the larvae had entered *pupation* (the stage in the insect life cycle during which the larva changes into an adult) in the padding, and the chickens ate those as well. Nothing remained of the beaver but bones, teeth, and a bit of hair. Those shreds, and the residue of the padding, became part of the organic detritus on the pasture on their way back to soil.

"But You're Just Breeding Flies!"

Correction, this project breeds fly *larvae*. Theoretically, it is likely to *reduce* the ambient population of adult flies. Imagine there are one hundred gravid (ready-to-lay) female flies in the area, and my buckets persuade twenty of them to lay their eggs inside. The larvae that hatch from the eggs become high-quality feed for the flock. Result? The potential new generation of local flies has been reduced by 20 percent.

To emphasize: There truly is no *hatch* of adult flies in this project. Yes, some larvae drop to the ground and burrow in below the buckets at night, but the chickens soon learn to scratch them up as they try to enter pupation.

"Why Not Feed the Carcass Right to the Chickens?"

You may wonder why it doesn't make more sense to feed the beaver carcass directly to the chickens. I tried this, and the birds ate some of the muscle meat—not nearly all of it—but not much of the entrails, other than the heart and liver. Fly larvae convert a carcass almost *entirely* to live protein feed, which seems to have even higher feed value than the carcass itself.[1] (I do sometimes *cook* a carcass for feeding directly to the chooks. *Not* in Ellen's kitchen oven!—but as described in the sidebar "A Solar Cooker for More Efficient Feeds" on page 179.)

"Don't the Buckets Draw Invaders?"

True, the odors produced by the carcasses must be detected by eaters of carrion such as possum, raccoon, fox, and dog. But I always hung the buckets inside an area protected by electric fencing, which kept them at bay. In the absence of the fence, defending the system against marauders would be more of a challenge.

As for visitors from the air: A group of buzzards did land to check out that first experimental bucket. But because the bucket had a secure lid, their frustration level visibly mounted as they failed to get at the tantalizing stuff inside. After moping around the bucket a couple of hours, they flew away. I never had a repeat visit by buzzards after I started making modifications to damp down odor.

"What about Disease?"

If no adult flies hatch from carrion buckets, they can hardly be a vector for disease. As for the chance that adult breeding flies visiting the buckets will spread disease, my research convinces me that is unlikely. Recycling of dead animals by fly larvae is a process that goes on in our abundant ecology all the time—fortunately—and my bucket system simply capitalizes on that natural process for its feeding potential.

The question of disease among the *chickens* gets more complicated. On two occasions during the many years I used this system, chickens who ate larvae from the buckets became ill, and a few deaths resulted. In each case the problem was what the old-timers called "limberneck"—botulism poisoning.

On one occasion my buckets contained offal from a neighbor's recently slaughtered chickens that had not been properly starved the day before slaughter. Their crops were filled with feed, which I believe may have soured and supported the growth of *Clostridium botulinum*, the soilborne bacterium that produces botulin toxin. In another case I used a couple of groundhog carcasses that had been left sitting too long before my trapper friend dropped them off at my place—maybe the delay had given the bacterium the opportunity to multiply in the groundhogs' stomach contents that it would not have had, had the carcasses been fresher.

To prevent botulism, I subsequently ensured that entrails to be cycled through the carrion buckets were from birds properly starved prior to slaughter, to entirely clear the gastrointestinal tracts; and requested that Sam donate fresh carcasses *only* to the cause. When I cycled the odd roadkill through my free-protein buckets, I did so only when I knew for a fact it was absolutely fresh. Following those changes, my birds had no further cases of limberneck.

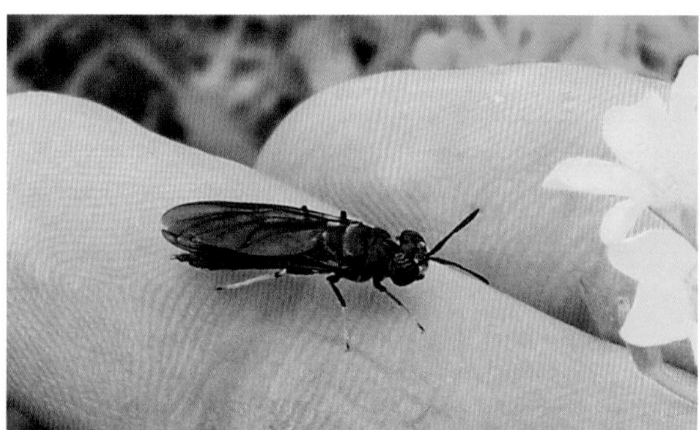

Figure 19.1. Black soldier fly adult. Photo courtesy of blacksoldierflyblog.com.

My carrion fly bucket may not be a feeding model for all flocksters, but I present it as an example of raising a flock in imitation of nature—after all, everything must decompose—and to offer a range of possibilities for converting "wastes" to free feed for the flock. If you ever need to tap into "protein from thin air," this strategy awaits your experimentation.

While you might prefer to cultivate the larvae of the black soldier fly, discussed in the next section, I present my fly maggot buckets for several reasons. Readers far to the north of me will not be able to cultivate soldier fly grubs, whose range ends around Zone 6 or so—whereas carrion flies are active anywhere animals die. In the case of a large donation of raw material—say, a couple of large beaver carcasses at the same time from my trapper friend—the carrion buckets can handle them easily; whereas soldier grubs cannot be fed large quantities of protein-dense feeds. The maggot buckets yield crawl-off protein far more quickly and are much simpler to set up and manage than bins for soldier fly grubs.

An Alliance with the Soldier

The previous section may have reinforced the repugnance for flies most of us grow up with. Houseflies buzz into the house and onto our food, possibly carrying disease-causing pathogens. Horseflies bite. Blowflies lay their eggs in carrion, and the larvae rid the world of dead carcasses—an essential ecological service for which we are grateful, even if we are repelled by the process as "just *too* gross!"

But none of us within its range are either annoyed or repelled by the black soldier fly, *Hermetia illucens*—indeed, it is unlikely most of us ever even notice this innocuous flying insect. Why would we? They look nothing like the flies we find annoying. They do not buzz us or come inside the house. They do not bite. A resting adult looks like a slender black wasp, but without the sting—quite pretty, to my eye (see figure 19.1).

Life Cycle

The stages of development of the soldier fly are a textbook example of the most common insect life cycle: egg, larva, pupa, adult. It starts when the gravid—mated, ready to lay—female finds a mass of succulent vegetable matter or manure and lays her eggs. The eggs hatch in four days to three weeks, depending on environmental conditions, into larvae, feeding busily at one end and expelling undigestible residues out the other.

Under ideal conditions, the larvae mature in ten days (see figure 19.2), ready for metamorphosis. As with many other fly species, however, pupation does not take place within the feeding medium—the prepupal grubs leave it and find a place to burrow into earth. After ten days or so, they emerge as winged adults.

Unlike some other fly species, the adult stage of the soldier fly is exclusively sexual—the winged phase is solely about mating, and, for the female, finding the best place to lay her eggs. They do not feed at all in this phase, which lasts only five to eight days. Death quickly ensues for both male and female adults once mating and egg-laying are complete, and the cycle starts anew.

A Useful Ally

Salient points about the soldier fly's life cycle suggest ways we could make an alliance for mutual benefit: Since the black soldier fly is a specialist in recycling succulent organic residues, we might recruit it in the responsible management of spoiled or unused food, manures, culled fruits and vegetables—reclaiming them as additions to soil fertility. Their high level of feeding activity and rapid growth in the larval stage suggest the concentration of considerable nutrients, and the potential for the use of mature grubs as a protein-rich feed for poultry, pigs, or farmed fish. Since the adults do not feed—indeed, they have no functioning mouthparts—they do not bite, nor do they come buzzing around us or our houses looking for something to eat.

Of special interest is that larval habit of crawling out of the feeding medium when it is time for the grubs to pupate—it suggests the possibility of directing their crawl-off for *self-harvest* into a collection container.

Figure 19.2. Mature grubs ready for pupation. They have just completed crawl-off from the feeding medium in the bin. Photo courtesy of Bonnie Long.

Other facets of soldier fly biology promise a happy working relationship. If there is an inherent problem working with the larval phase of the carrion fly, discussed in the previous section, it is that conditions in the feeding medium tend to be *anaerobic*. Most pathogens, *Clostridium botulinum* among them, thrive in oxygen-starved conditions but do not grow well in oxygenated media. Because the frenzied activity of soldier grubs keeps the feeding medium constantly aerated—and because the adults do not feed at all—this species is not a vector for diseases. Because soldier grubs inhibit development of larvae of all other fly species—including houseflies, fruit flies, and blowflies—cultivating them may actually *reduce* populations of flies with a higher nuisance profile.

The Grub Bin

I cultivated this fascinating recomposer for many years, beginning in 2009. I had noticed the flat, segmented, leathery-skinned soldier grubs in my worm bins long before I even learned what they were. As I became more interested in the possibilities for working with recomposers—especially for their contribution to more self-sufficient poultry feeding—I learned that a lot of creative work had already been done toward making a productive alliance with the soldier fly. Since it was obvious there was a wild population in my backyard, it

Figure 19.3. The original BioPod, a manufactured bin for cultivating soldier grubs. Key features visible here are the two-part lid, the ramp molded into the body of the bin for crawl-off of mature grubs, and a small bucket into which they self-collect. Photo courtesy of Bonnie Long.

was easy to set up a tempting feeding medium—cast-offs from the kitchen—in a bucket in which soldier females could lay, and soon I had an actively feeding colony of grubs. My chickens, both growing birds and adults, snapped up the mature grubs, in preference to all other favored feeds.

As always, my determination to produce more free, natural foods for the flock was only fueled by this first success, and I bought a BioPod, a molded plastic bin designed to accommodate every aspect of soldier fly "lifestyle" (see figure 19.3).

Though the pictured BioPod is no longer available, it is worth examining its design. Remember that the intent is to *capture* a natural process that is going on around us all the time, to *steer* it toward yielding benefits for us—and, in this case, our flock. While single-batch cultivation of soldier grubs in buckets or open compost heaps is possible, efficient "flow-through" production requires a properly designed bin that accommodates

Figure 19.4. Their favorite meal. Photo courtesy of Bonnie Long.

the soldier fly's life cycle. Here in short is another case where a homestead effort leaves behind the strictly "all natural" without succumbing to a Big Boss illusion of being in control.

With that in mind, please note the BioPod's features: Since the most efficient feeds for soldier grubs tend to be high in moisture content, a *drain* with a filter directs liquid effluent into a screw-off collection jar (unseen below the bottom of the bin in figure 19.3). Open space between the two parts of the lid allows both for essential ventilation through the bin and for access by female flies to lay eggs in the feeding medium. Molded into the unit is a *ramp* around the entire inner wall—a migration route out of the feeding medium. The ramp ends in a hole at the top of a *cylinder*, through which the grubs in the *crawl-off* phase drop—following their instinct to find a way to earth and to burrow in for pupation. The positioning of a *collection bucket* beneath the hole, however, results in the migrating grubs' self-collection—how convenient! Most active crawl-off is at night, so the flockster simply grabs the collection bucket in the morning and offers the chickens their favorite meal of the day (see figure 19.4).

Sadly, I found the flimsily made successor to the original BioPod unsatisfactory. My advice is to make your own, in lieu of spending up to $200 for a manufactured bin. The design concepts are now widely understood, and online searches will turn up numerous simple designs requiring only basic handyman skills and tool set.[2]

Managing a Soldier Grub Colony

Cultivation of soldier grubs is easy with proper attention to details. The following are some design and management parameters common to all setups.

RANGE

If you live in climate Zones 10 through 7, there is almost certainly a native black soldier fly population ready to work for you—just set up appropriate feeds in a protected bin that fits their needs, and the gravid females will come. Soldiers may possibly survive in areas colder than Zone 7, though the limits of their range are uncertain. If there is no wild population, you can purchase starter grubs through the mail. How easy it would be to establish a sustaining population—one that reproduces naturally in following years if none exists locally—might depend on how far north of their natural range you live.

BIN

The bin must protect the colony from predation and rain, be readily accessible to gravid females, ensure compatible living conditions, and provide for crawl-off by the mature grubs. It should be in the shade: The high metabolic activity of the grubs generates heat, so any additional heat supplied by direct sunlight could be disastrous for the colony. Otherwise, placement of the bin depends on your own convenience. It does not have to be stuck off in the back forty, as the oxygen-rich conditions in a well-managed colony prevent unpleasant odors.

Protect the interior of the bin, the crawl-off route (see later in this section), and the collection bucket from grub eaters such as raccoons, skunks, and possums—and from dogs, who might make a mess of things just for fun. I have used the salvaged end of a damaged electric net with a battery-powered charger to protect a soldier grub colony under the shade of a mulberry tree.

FEEDS AND FEEDING

Most efficient conversion to biomass occurs in typical food wastes. But if yours is a frugal household such as ours, there is no edible food wasted, other than castoffs such as coffee grounds—which grubs love—tea leaves, peelings, and trimmings. If you have ready access to food scraps from schools, restaurants, and other purveyors of food, that would be an excellent feeding resource for your project. In the absence of such food residuals, any mix of succulent vegetable and fruit matter works well, such as overmature and cull fruits and vegetables. Grubs love the big outer wrapper leaves of cabbages and still-succulent spent broccoli plants. Note that feeding efficiency increases if such garden feeds are cooked. (See the sidebar "A Solar Cooker for More Efficient Feeds" on page 179.)

Some manures make good feed. Low-fiber pig and chicken manures are best—there may be no better way to deal with these manures than using energy-hungry grubs. The larvae will feed on horse, rabbit, and ruminant manures, though the higher fiber content of these manures reduces feeding efficiency—the larvae cannot digest cellulose from plant stems.

Dairy products and meat and fish scraps—including occasional roadkill—make excellent feeds but should be fed only in limited amounts: Grubs will not thrive with too much high-protein feed. The usual recommendation is to limit such foods to 5 or 10 percent of total feed offered. My experience, however—confirmed by a correspondent with a lot of experience cultivating grubs—is that a temporary spike in protein-dense feeds is not a problem.

Do not use as feed materials that are dry, fibrous, high-cellulose, or tough, such as weeds, grass, leaves, stalks, paper, or cardboard.

Grubs are voracious feeders, and the temptation is to "throw it to them." But *overfeeding* creates problems—particularly excess accumulation of undigested (as opposed to undigestible) feed residue in the bottom of the bin, which can unbalance the overall ecology and encourage growth of undesirable organisms. The ideal colony is a pure wriggling mass of feeding grubs, their accumulated fecal residues in a thin layer below, and above, an amount of feed they can clean up entirely in twenty-four hours.

DRAINAGE

Look again at the list of appropriate feeds. Note that all have high moisture content, and remember that grubs do not thrive in anaerobic conditions. Therefore: *Efficient drainage* out of the bin is *essential* and could prove to be your major management challenge. If conditions in the colony do become too wet, either cease feeding for a while or add a moisture-absorbing material such as coir or dry pet food kibble (which will both reduce moisture and feed the grubs).

Note that the effluent from the bin can be diluted and used exactly like a compost tea or manure tea. Some sources advise using it to feed cover crops only, or on vegetable crops that are not in close contact with the ground (corn, pole beans, trellised tomatoes) but not on those that are in close contact (spinach, lettuce, radishes, sprawling cucumbers).

THE CRAWL-OFF

An essential feature of any bin design is provision of some sort of ramp the mature grubs can use to exit the feeding medium. If the incline is no greater than 45 degrees, they will have no problem wriggling up the ramp. The grubs' migration can be further channeled with a gutter that ends over a collection bucket. Most crawl-off is at night—in the morning simply pick up the bucket and throw the grubs to your chickens. Grubs are also relished by guineas, turkeys, and ducks—though my geese never showed the slightest interest in them.

Remember that a continuously productive soldier colony requires a wild population to support egg laying in the bin; and from time to time it helps to scatter a few mature grubs where they can burrow, pupate, and emerge as adults ready to carry on the cycle. If you seed grubs at the end of the growing season in a place that warms up early in spring—such as near the foundation of a south-facing wall or, in my case, inside the greenhouse—adults will emerge earlier in the spring to start the cultivation cycle.

PRODUCTIVITY

A colony of soldier grubs is like a chicken flock: A well-managed colony—a *domesticated* colony—is a lot more productive than a wild one. But what levels of production might we expect? That depends on many factors: bin design, feeds offered, ambient temperatures, and management experience and skills. But as observed earlier, the equation, at whatever level of production, is all positive: The grubs convert the organic "wastes" on which they feed *entirely* to resource. Their work is right at the heart of the abundant ecology, and thus of closing the circle in the self-sufficient farm or homestead.

Do not imagine that soldier grubs have mere hobby potential—they offer the opportunity for conversion of organic wastes at industrial scale. Dr. Paul Olivier, founder of the nonprofit Empowering the Poor with Waste Transformation in Vietnam, made important breakthroughs in the use of black soldier flies for bioconversion. With the goal of turning "trash" (organic wastes creating serious environmental problems) into "treasure" (employment and income for poor people, animal feed and soil fertility amendments for farmers), Dr. Olivier designed systems for soldier grub composting in Vietnam and other developing countries. Academic researchers in North America, such as Sophie St-Hilaire and Craig Sheppard, built on that work to develop grub cultivation systems to manage fish offal from processing of farmed fish, and manures in commercial poultry and swine houses—to yield high-protein feed supplements for various livestock species, including commercially raised carnivorous fish.

Happily, in the decade since this book was first published, conversion of "wastes" at farm and massive industrial scale has brought dried grubs—42 percent protein, 35 percent fat, dry weight—into the market. Not only are whole dried grubs widely available—from my supplier of organic feed, from Tractor Supply, probably from your local feed co-op—but they are showing up as an ingredient in commercial poultry feeds. As production continues to ramp up, dried grubs could potentially replace the unsustainable harvest of hundreds of thousands of tons of ocean fish annually for protein feed supplements.

FLEXIBILITY

H. illucens is enormously adaptive in response to changing environmental conditions. If their food supply runs out, the grubs go dormant until more food is available. During winter they delay maturation for several months before resuming development. This adaptability gives the cultivator great flexibility in managing the colony.

CLEAN-OUT

Because of the enormous reduction in volume of the feedstocks offered to the colony, cleaning out the substrate—the undigestible residue at the bottom of the bin (*frass*), largely the cellulosic portion of plant tissues—need not be frequent. This residue makes a great soil amendment or you can use it to prepare something similar to compost tea.

As in management of deep litter, it is better not to clean out the bin entirely. As grubs feed and grow in a colony, a whole ecology of ally species—bacteria, actinomycetes, and fungi—emerges. When you clean out, retain some of the existing substrate as an inoculant to speed reestablishment of a total, balanced bin ecology.

WINTER

Wild soldier grubs that have gone deep into earth for pupation in the fall survive the freezing of the ground. However, grubs in an active colony will die if they freeze. Since an existing, actively feeding colony would start its work earlier in the spring than a wild one that must start from eggs from newly hatched adults, survival through the winter would be desirable.

My own experiments with keeping a colony alive were not always successful. I routinely kept the bin in the greenhouse until the heavy frosts. But I found that after moving them to the basement, the waste heat from the furnace kept ambient temperature too warm, triggering a heavy crawl-off and even premature metamorphosis of grubs into bewildered-looking adults. My solution was to remove the colony from the bin and return it to the greenhouse, but this time placed into one of the vermicomposting bins dug 16 inches (40 cm) into the earth, protected from freezing. For much of the winter, the grubs remained active, eating all the castoffs from the kitchen such as coffee grounds, tea leaves, and trimmings—but without becoming active enough to complete development into the prepupal stage. In some winters, few grubs survived. In others, a small colony survived the chill for an earliest-possible start on the new cycle of pupation, maturation, and egg laying. My most reliable guarantee of an early start on the season, however, was always the "seeding" of prepupal grubs into the soil of the greenhouse, ensuring emergence of mating adults in earliest spring while frost was still in the ground outside.

Soldier Grubs or Vermicomposting?

Since both soldier grubs and earthworms can be used to convert many of the same organic "wastes" into useful amendments and feeds, which should you choose? Answering that question starts by recognizing the important differences between the two, the most significant being that the population in a well-regulated worm bin is internally self-sustaining, whereas a soldier grub colony requires continual renewal by egg-laying females from an external *wild* population.[3]

Other differences? Grubs digest fresh putrescent matter in a hurry; worms wait until bacteria are consuming decaying matter, then feed on the bacteria. Earthworms, in cooperation with fungal species, can convert cellulosic (high-carbon) materials grubs are unable to digest. Because grubs have a more frenzied level of activity than redworms, it is easier to maintain aerobic conditions in a soldier grub colony.

It is *not* the case that grubs and earthworms can be successfully cultivated in the same bin. However, there is one important way they can work as partners: The *frass* from the grubs—their fecal excretions, plus the nondigested cellulosic residues from feeding—is an excellent addition to worm bins. Studies in Asia demonstrated that redworms grow three to four times faster on the residues from soldier grubs than on food wastes. As for those cellulosic residues, worm bin fungi make short work of them.

Like the role of trailblazer? Give soldier grub cultivation a try. Though enough work has been done to demonstrate the utility of this species, any of us who ally with the soldier are still making it up as we go. I always found grub cultivation a way to remain at play on the homestead, one that brought out the kid in me.

Feed from Worm Bins: The Appelhof Model

Mary Appelhof popularized the idea of *domesticating* earthworms as helpers in the recycling of organic "wastes,"[4] educating her readers about the life cycle of composting worms, optimal conditions needed to cultivate them, and instructions for setting up worm bins. Since that time thousands of enthusiastic recyclers have fed their kitchen and table scraps to partner worms, who turn them into one of the most valuable of all soil fertility amendments—worm castings.

Here, I describe my adaptation of Appelhof's techniques to produce "black gold" for the garden—but with the added twist that I substantially supplement my chickens' protein needs from my bins. In the next section I discuss working with composting worms at larger scale.

A note about terminology: Worm-cultivation projects are sometimes referred to as *vermicomposting*, sometimes as *vermiculture*, depending on whether the main purpose is "making compost" or "growing worms." Since I'm equally interested in doing *both*, I use the terms interchangeably.

The Bins

We have a half-basement with storage shelves, a propane-fired furnace, and a bench under grow lights for garden seedlings and baby salad greens. Since ambient temperature stays in the range of 58° to 65°F (14° to 18°C) year-round, it is an excellent place for worm bins. (Ideal temperatures for earthworms are 70° to 80°F (21° to 27°C), but they thrive in temperatures a little below and a little above that range.)

I use plastic storage tubs for my bins. (No, I'm not big on plastics, but a wooden bin would not last long in contact with highly bioactive worm bedding.) I found space for four 18-gallon bins in out-of-the-way spaces in the basement, and tucked four that were shallower (12-gallon size) under the bottom shelf of my grow bench. The combined volume allows for serious production; and having eight bins is convenient for the sequential feeding described in the next section.

Provision for drainage is essential, as anaerobic conditions in wet bedding are dangerous for earthworms. I drilled something like eighteen ¼-inch holes in the bottom of each bin, then set it on 1-inch blocks in a shallow catch basin.

Oxygen is as essential for worms as for you and me. I drilled ½-inch holes just under the lip of the bin all

A Solar Cooker for More Efficient Feeds

Food scraps are the most efficient foods for both soldier grubs and earthworms. But there is no edible food wasted from my household's table. It occurred to me that I could most closely imitate food scraps by *cooking* the foods I was offering my grubs: outer leaves of cabbages, culled fruits and vegetables, the more succulent cover crops such as rape and mangel leaves, spent crop plants such as pea vines and broccoli stalks.

Of course, it would be a big waste of energy to cook these items daily in the kitchen. But how about using the same source of energy that grew these plant foods in the first place—*solar power*? The solar cooker in figure 19.5 is the one I made for cooking feeds for grubs and worms, increasing feeding efficiency and speeding up production of prepupal grubs, harvestable worms, and castings (earthworm poop).

The cooker is a double box in ¾-inch (1.9-cm) exterior-grade plywood, with a 2-inch (5-cm) space between the bottoms and sides of the two boxes, stuffed with fiberglass insulation. The lidded cooking vessels—covered Granite Ware roasting pans—sit on a steel plate in the bottom. The interior of the cooker, the steel plate, and the outsides of the cooking vessels are all coated with a nontoxic flat black paint appropriate to painting a barbecue grill. Sunlight passes through the glass lid—a window sash—but is trapped inside by the black, insulated sides, cooking the food as the temperature rises. Strips of felt insulation, stapled around the top perimeter of the cooker's sides (after figure 19.5 was photographed), increase retention of heat inside.

In the morning I put the foods to be cooked into the covered roaster pans, seal the cooker with its glass lid—and walk away. Which is to say, the amount of *my* time "cooking" for the grub or worm bins is minimal—it doesn't matter if the cooking process itself takes most of the day. (Do note the absence of added cooking water, which is not needed, and indeed could decrease cooking temperatures.)

I frequently use this cooker as well to cook fresh roadkill and trapper-donated carcasses for my chickens. Smells great when I open the roasting pans!

Figure 19.5. The solar cooker I made for cooking feeds to increase feeding efficiency in both grub and earthworm bins.

around, maybe fourteen in all, and an additional six in the lid. Ventilation this free allows plenty of air exchange through the surface of the bedding, assuming it is properly made and maintained.

More than any other detritivore species I know, earthworms excel at waste management. I know you already appreciate that there is no waste in nature and that what some would consider "wastes" from the garden (culls,

Figure 19.6. Setup of one of my bins starts with drilling drainage holes in the bottom and ventilation holes in the lid and around the bin sides an inch or so below the rim. Bin is set on blocks in a catch basin. Typical bedding to start is 6 inches (15 cm) of torn, soaked, and drained cardboard.

Figure 19.7. *Eisenia fetida*, red wrigglers.

spent crop plants) and landscape (fallen leaves) are in fact feeding and soil-building *resources*. But if you still think of old *newspaper and cardboard* as "wastes," with resource value only if converted in industrial recycling operations, here's the good news: Since setting up my basement vermicomposting project, almost *all* the cardboard and some of the newspaper that come onto our place is recycled right here, in the vermibins.[5]

Setting up a new bin starts with soaked cardboard torn into strips. (It helps with tearing the cardboard if I soak it first.) I drain it thoroughly before putting it into the bin, about 6 inches (15 cm) deep to start. Use of cardboard alone is fine—it is coarser and thus more open to oxygen exchange. Newspaper alone tends to condense into tight clumps that resist breakdown. A mix of the two works well if I go light on the newspaper.

As for moisture level, I'm sure you're familiar with the test for proper hydration in a Goldilocks compost heap: Squeeze a handful. If water runs out, it's too wet—if you can't get out a drop, it's too dry—and if you can squeeze out just a drop or two, it's *just right*. Shredded leaves also make good bedding. Note the "shredded," either by aging in a pile or using a power mower—whole leaves are more likely to compact when wet. Note that all three of these bedding materials are *high carbon* and (except for pure newspaper) maintain an open structure for good *air exchange*.

As I fill the bin I *inoculate* the bedding. The worms I introduce will of course bring with them a host of microorganisms essential to bin ecology, but I want to be sure I have a broad range of microspecies, including *fungi* that will drive breakdown of the high-carbon newspaper and cardboard. So I gather handfuls of well-decomposed material under a garden mulch, some finished compost, and some leaf litter—the old crumbly stuff between the soil surface and the more recently fallen leaves. I sprinkle that lively mix in between each layer of bedding.

Now I add worms. But what *type* of worms? All us gardeners are used to seeing worms pop up when we dig the soil. But those soil-burrowing worms are *not* the ones best used in worm bins. Think instead of a shovelful from inside an aged manure heap. Those "red wrigglers" you see are "manure worms" or "compost worms," a type that specializes in dense deposits of decomposing vegetative matter and manures. *Eisenia fetida* is the species most often used for vermicomposting (see figure 19.7).

That aged manure heap could well be your best source for composting worms for a bin. Or a nearby vermicomposter might share a batch of starter worms. And there are many sources online that supply worms—I have purchased from several, always with good results.[6] It is okay to start with just a few—patience will be rewarded as they reproduce into an efficient working colony. To start off on the run, buy more worms, but be willing to foot the bill—large orders are expensive.

Stocking density is important. Starting with modest numbers is fine, as said, but it is not good to crowd excessive numbers into the bin. "Seed" the bin with about 1 pound (453.6 g) of worms (approximately 1,000) per square foot (.1 square m) of bedding surface, no more than 2 pounds (907 g). The bedding will grow in depth as the worms help decompose their feed and bedding, but you should think of stocking density in terms of *square footage*, since composting worms feed close to the surface of the bin rather than deeper down.

Let newly introduced worms settle in at their new home. Leave them quiet and undisturbed for a day or two before beginning to feed.

Feeds and Feeding

There are other approaches to feeding, but in my eight-bin setup I follow a feeding schedule that lasts eight days, feeding only one of the bins per day. On Day 1, I start at the first bin. I add an amount of feed equal to the weight of worms I put into it. Given that there is a lid on the bin, it's okay to spread the feed over the surface—or to bury the feed shallowly. But remember that earthworms retreat from light, so if there is no lid keeping the bin dark, they will go hungry rather than feed in the light. On the other hand, it's not a good idea to bury the food deeply. Monitoring the feed closely is essential, and that is more easily done if it is near the surface. Even a shallow covering of the feed with bedding will ensure darkness and get the worms eating.

On the second day I feed the second bin in the rotation, again putting in an amount of feed equivalent to the weight of worms I seeded. Now let's pause: When I say "equivalent to" I mean, yes, at least initially, I *weigh* the feed—using a handheld digital scale. Why am I being so exacting? Because *it is essential not to overfeed.* Reread that sentence. *There is no more certain way to kill an earthworm colony than by overfeeding.* Each day that follows, I feed the next bin in the rotation. As a colony grows and I get a feel for its capacity to consume the feed, I no longer use the scale. However, when I come back around to Bin 1 in the rotation (on the eighth day from the beginning), I check to be sure there is *no* feed remaining. If there is residual feed, I pause all feeding a day or two to give the worms time to catch up, resuming only when all traces of the previous feeding have disappeared from Bin 1. Worms who have *less* feed than they can ideally consume simply reduce feeding—they may even enter dormancy for a while. Worms constantly being fed *more* than they can eat—*die.*

What are good feeds to use? Any of the "greens" that would go into a conventional compost heap: kitchen trimmings, coffee grounds, tea leaves; cuttings from comfrey and from cover crops such as alfalfa, rape, mangel (the leaves only); spent but still-green crop plants; culled vegetables and wrapper leaves of cabbages and lettuces—no productive homestead is likely to come up short on feedstuffs for even sizable vermicomposting operations. Remember that these "green" materials are the ones that supply nitrogen in a compost heap—and cause heating as they supply the energy for breaking down the high-carbon components such as leaves and straw. But in a worm bin, care must be taken to prevent heat spiking—and in the worst case, generation of *ammonia*—in the bin. Both are lethal to worms, which have no way to escape the gases. Hence the above precautions about overfeeding of "green" type feeds. (If there are only small amounts of nitrogenous feeds, the worms retreat from any modest heating generated, and migrate back to resume feeding after the mini-spike.)

Unlike nitrogenous feed, high-carbon bedding can be added any time. I usually add a little every time I feed. Remember how I stressed the importance of drainage? Since excess moisture can be particularly challenging when wet feeds such as cooked "green stuff" are used, a little new bedding can be added dry rather than soaked.

I might lay down a *thin* layer of torn newspaper or cardboard, or a sprinkle of coir, and place the wet feed on top of it—or perhaps, mix enough of the dry bedding directly into the feed until excess moisture is absorbed. Let me stress that addition of dry bedding must be in *small amounts*, applied as needed to prevent wet bedding, but never applied as one big addition to correct for excessive moisture in the bin. Judiciously added dry carbon matter reduces drainage from the bin to zero.

What about manures, also often used for making compost, precisely because they speed decomposition with its heating spike? Manures make excellent feed, so long as they are used with caution. As with "green" feeds generally, use in small amounts will not generate excessive heating—though high-nitrogen chicken manure is difficult to use safely. *All manures can be pre-composted and used as feed after a couple of heating cycles.*

Pre-composting is especially useful as well for cooked foods with high moisture content, or in cases where the volume of raw feedstocks is greater than the worm bins can handle. It is important to emphasize that the purpose of pre-composting is to produce a *partially completed compost*. A *finished* compost that has completed breakdown has few remaining nutrients for the worms. But a *coarse* compost that has gone through a couple of heating cycles still has plenty of usable nutrients and will not heat the bin.

Since the only manure available to me is from my chickens, I don't want to lose it as a feeding resource if I can just find a way to prevent its aggressive tendency toward heating. The "coarse compost," partially decomposed concept suggested a feeding approach that works well for me: Remember from the discussion of deep litter in the henhouse that the chooks shred the leaves—and turn in their droppings—as they scratch. Eventually the bottom layer is a slightly moist residue of granulated leaf and manure, with decomposition well advanced. Remember as well, however, my cautions in the sidebar "Too Much of a Good Thing?" on page 134. Two of my eight bins are dedicated to processing well-worked deep litter exclusively, and I spread castings from those bins under young nut trees, *not* in the garden, which at this point I don't want to overload with phosphorus and potassium from the chicken's manure.

I have found those partially decomposed deep-litter "fines" one of the best—and certainly easiest—of all feeds for worm bins. I set aside the coarser litter on top, and scoop up the small-particle material below. When it is time to add feed to one of the dedicated litter bins, I spread some of the "fines" in a *thin* layer (too thick a layer generates ammonia), add a bit of lime and a sprinkle of water, and cover with a layer of new bedding. Results are excellent. I think anyone who keeps both a herd of worms and a flock of chickens, and who uses deep litter for the latter (and *every* smart flockster uses deep litter in the henhouse, right?) will find that aged, well-worked, granulated litter is the most convenient of all "feeds" to use in a worm bin. Don't you just love the marvelous *circularity* at play here: The chooks scratch in their high-nutrient poop, shredding the litter in the process, the granulated residues are fed to worms, the worms are harvested—and fed to chooks. Now we're rolling!

We think of the worms "eating" what we feed, but in fact they do not. *Bacteria* are specialists in decomposing—feeding on—the nitrogenous "green" feeds. They colonize such feeds in unimaginable numbers—and the worms feed on *them*. It would seem a good strategy, then, to speed up bacterial colonization. That can be done by breaking down the tissues of the added feeds before placing them in the bin, either by freezing or cooking. In summer, I use the solar cooker described earlier in this chapter. Since we use a woodstove for heating, in winter I usually cook a pot of green feeds on it when needed. It's easy then as well to set a bucket of green feeds outside to freeze overnight, then thaw in the basement and feed. Do remember to manage bedding moisture carefully if using such cooked green feeds.

Ah, but bacteria don't specialize in high-carbon decomposables, do they? How does a bedding of newspaper, cardboard, or autumn leaves get "eaten"? Remember the "inoculation" of the bin at the start. Sprinkling in those microorganisms from the wild—especially from the old leaf litter—ensured that fungal species would have an active role in the bin's ecology. It is *fungi* that colonize the high-carbon bedding. Most fungal activity remains invisible, though occasionally I see small mushrooms growing from the bedding's surface, typically dainty specimens from the *Coprinus* group. Earthworms eat the *hyphae*, the fungal threads spreading through and feeding on the bedding.

While the range of feeds is broad, there are a few organics that should be kept out of worm bins—onions and garlic; peppers; citrus, pineapple, pawpaw; fat, grease, oils; meat and dairy—in large quantities. Don't make yourself crazy picking out every tiny bit of any of these from the kitchen compost bucket—but avoid them in significant amounts. Large quantities of anything *acidic*, for example pomace from apple or tomato processing, should not be fed, to avoid driving pH too low in the bedding. *Alcohol* is highly toxic to worms—high-carbohydrate foods that can ferment and produce alcohol should not be fed: concentrated starches such as discarded bread and flours; mangels or sugar beets grown as a cover crop (the leaves can be fed, but not the whole roots); potato peelings are okay, but not whole culled potatoes in quantity.

It's a good idea to mix a sprinkle of *lime* into an occasional feeding. Lime helps with pH balance: Worms can tolerate some variation in pH, but keeping the bedding fairly close to neutral is best. Note the recommended "sprinkle"—counting on lime to correct for inattentive management or excessive acidic feeds is a bad idea.

A second reason for a bit of lime is that earthworms—like chickens—need plenty of *calcium* in their diet for reproduction. For that reason I prefer a high-calcium lime (which can be up to 95 percent $CaCO_3$, calcium carbonate) over dolomitic lime (with much higher magnesium content). Dried eggshells are almost pure $CaCO_3$ and can be used instead, the more finely pulverized the better.

Population Growth

Earthworms are hermaphroditic, so it's easy for them to find a mate! Under optimal conditions, therefore, population in the bin can increase rapidly. Look for "cocoons," tiny seedlike pellets that start off pale yellow and turn dark red as four to six embryonic worms develop inside. After an incubation time of four weeks or more, depending on conditions in the bin, white, hairlike hatchlings emerge. Depending, again, on environmental conditions, sexual maturity occurs in about seven to ten weeks—and the cycle repeats, though in this round with vastly increased numbers of partners in the dance. It is theoretically possible that numbers could increase from 100 to 8,200 in twelve weeks.[7] *Be alert for overcrowding*, signaled by the appearance of masses of worms on the sides and the underside of the lid as they try to escape.

As well, increased numbers mean increased pooping—that is, deposition in the bedding of worm castings. Over time the decomposing bedding is replaced entirely by the castings—bin contents become increasingly dark, fine-grained, and dense. As the bedding drives toward pure castings, it is increasingly less compatible with the worms' needs. It should be removed every four months or so and replaced with new bedding.

There are several approaches to the harvesting of castings or the "harvesting" of earthworms, which is just to say: to the (efficient) separation of the two. My preference is a sort of "seesaw" between two halves of

Figure 19.8. As bin contents are replaced by pure castings, it is possible to begin feeding in one side only, drawing most of the worms out of the vermicast in the other.

the bin. When the dense black cast is more evident in the bin than the original bedding, I stop feeding in, say, the far end of the bin and from that point feed in the near end only. The worms in the far end continue feeding until the residues of feed there are exhausted, then migrate into the fresh bedding in the near end with its weekly feed (see figure 19.8). During that time, all the cocoons in the castings side hatch, and the baby worms migrate as well. I scoop out the pure castings in the far end and put fresh bedding there.

If there are still worms in material I remove from the mostly-castings end, I heap it on a tray placed in a spot where there is plenty of light—for example peripheral light from the grow light over the plants on the grow table. The worms retreat from the light. On successive trips to the basement for worm or plant care, I scoop off the top, worm-free layer of castings. In the end all the cast has been removed and there remains only a tight mass of worms, available for starting another bin—or feeding the flock. An alternative way to move worms lingering in the vermicast is to heap it into a large container and place feed on top. As the worms migrate up into the feed, transfer the worm-filled feed into the bins. Repeat until few or no worms remain in the pure castings below.

Black Gold

Castings can be stored for later use in the garden. It is okay to allow them to dry some—they will be easier to work with if less moist—but to ensure viability of their abundant microbial populations, don't let them become powder dry.

Castings are potent as a fertility supplement—you don't need much. I mix a handful at each transplant site when I'm setting out started plants. For transplanting onion starts I make a V-shaped trench and fill the bottom with castings. I use castings as a component of seed-starting mixes as well. In a year's time, my eight-bin system fills several 15-gallon patio tubs with castings.

Earthworms as Chicken Feed

In years past I fed worms straight to flocks of chickens—I just grabbed handfuls of worms in the most "clustered" parts of the bedding in the greenhouse bins I was using at the time (and which I describe later in this chapter) and threw them to the flock. If supply was low, I reserved the worms for clutches of chicks with their mamas. There was no problem whatever with acceptance—they came running and snapped the worms up. But my current flock resists eating fresh, raw *E. fetida*. While it is true that *fetida* is Latin for *stinky*, and earthworms squirt out a fluid that smells bad to potential predators—I don't know why my current flock says *Yuck* when earlier ones said *Yum-yum*. Fortunately I found that if I *dry* the worms, my current flock considers them the *pièce de résistance*.

I know you're thinking *Gimme the numbers!* regarding production of high-protein feed from my basement project. Unfortunately it would be difficult to generate precise statistics. Feeding of worms to the flock is highly variable. For example, if I'm servicing one of the eight bins and its population obviously needs to be thinned, I grab a few handfuls and load them in the drying rack. That may happen in another bin in two days—or not for another week. This is another of those cases of reaching out and grabbing a high-quality feeding opportunity on the fly.

Feed from Worm Bins: Vermiculture at Larger Scale

If you have access to large amounts of organic residues for feedstock, you may be interested in scaling up vermiculture considerably, for a major protein addition in your own feeding program—perhaps even to sell castings to local gardeners, or worms for starting others' vermicomposting projects, or for an ambitious farm-scale soil-building program. Remember—if your feedstocks contain a lot of manure or food residuals from restaurants, schools, and the like—it can be advantageous to pre-compost them before feeding in worm bins. The best source I know of for information about vermicomposting at farm or industrial scale is Rhonda Sherman's *The Worm Farmer's Handbook*.[8]

Vermicomposting in a Greenhouse

My first experiment with vermicomposting on a larger scale developed out of a 3- by 4-foot bin that I constructed in my greenhouse and stocked with redworms ordered through the mail. For bedding I used coir, and for feed, our modest kitchen residues. I dug the bin about a foot deep into the earth floor near the door of our 20- by 48-foot greenhouse. My hope was that, in our mid-Atlantic climate, a worm colony below ground level would not freeze in the winter nor get too hot in the summer. The experience of several years proved the supposition sound: However cold the ambient temperatures—or however hot, in summer—the worms in their earth-protected bin continued to feed and reproduce.

I reserved the modest harvests of worms to feed the young growing birds in my flock, and I observed that those birds feathered out earlier and grew more vigorously. That contribution to the flock's nutrition made me greedy for more. A "worms eat my garbage"-sized bin no longer satisfied my ambitions—I wanted to scale up to worm composting sufficient to provide worms as a significant part of my feeding program.

We dug a trench for "vermibins" 4 feet wide and 40 feet long down the center of the greenhouse. Since

Figure 19.9. The worm bins run 40 feet down the center of the greenhouse. Cross-walls every 8 feet make a total of five 4- by 8-foot bins like this one, 16 inches deep. (Note on the left the 12-inch-wide piece of plywood, stiffened with a 2 × 4, which can be dropped into place along one edge of the wormbin to provide continued access into the greenhouse even while the bin remains open for filling and servicing.)

Figure 19.10. Plywood lids over bins provide easy access down the center of the greenhouse–loss of growing space is minimal.

we needed that central access anyway, little growing space was lost to the new bins. We lined the perimeters of the bin space with 4-inch (10-cm) concrete block, two courses deep, and placed a cross-wall of block every 8 feet (2.5 m). The result was a series of five bins, 4 by 8 feet (1.2 by 2.5 m) each, 16 inches (40 cm) deep (see figure 19.9).

We made lids for the bins from ¾-inch plywood on 2 × 4 framing (see figure 19.10). Since a single 4- by 8-foot lid—one made from a single sheet of plywood—on stout framing would be too heavy to move conveniently, we made each lid a more manageable 4 feet by 4 feet (one sheet of plywood cut in half). The result was two lids covering each 4- by 8-foot bin, creating for management purposes two 4- by 4-foot sections per bin but with no partition between them. Keep that point in mind—it was the key to our harvesting method later.

Note there was no floor other than the packed red Virginia clay with which we are blessed. This is not a problem, given that *E. fetida* has no interest in burrowing into tight soil and that our clay soil drains well despite its compaction. Our block-walled bins never had incursions of worm-loving moles, which are sometimes a problem for worm bins outside. I assume larger worm predators (skunks, possums, racoons) were spooked by the strangeness of the greenhouse interior—in any case, they never came calling.

FEED/BEDDING

In the discussion of my Appelhof-style vermibins earlier in this chapter, I made a distinction between bedding and feeds. In the greenhouse bins I made no such distinction: I filled them exclusively with pure horse manure from a neighbor who breeds and boards horses. Note that qualifier *pure*: If the manure is mixed with hay, straw, or pine shavings, it will heat up, as in a compost heap—a disaster for the worms. I found that pure horse manure did not heat up, or only slightly so, and it was an excellent medium for the worms. They lived in this medium (using it as bedding) while converting it into castings (using it as feed). The fibrous manures of ruminants—sheep, goats, llamas—would also work in my greenhouse model. With proper drainage cow manure, with higher moisture content, should work fine as well. I once toured a market scale compost-making operation based on conversion of tons of pig manure in 100-foot (30 m) worm bins.

Again, please note that the description here, of filling the bins with horse manure, is from past practice, and that there came a point when I quit using horse manure to prevent further accumulation of phosphorus and potassium. If you do not have the problems of excessive mineral accumulation I did, be assured that horse and other manures are excellent for large-scale vermicomposting projects.

SETTING UP THE BINS

After hauling in horse manure by the pickup load from my neighbor's place and loading all five bins, I used a garden hose to adjust the moisture. I aimed for a medium that was neither uncomfortably dry (for an animal whose entire body is covered with a wet skin) nor sopping wet, a condition that would drown the worms.

After waiting a couple of days to ensure there would be no significant heating, I seeded the bins with worms from the small bin I had been cultivating. Starting with such small numbers in relation to the large volume of bedding in the five 4- by 8-foot bins required patience as they multiplied. It was about a year before I could start making harvests.

MANAGING THE BINS

Always before watering, I checked the deeper levels of the bin, not just the surface. Overwatering at the surface could cause a hidden accumulation of excess water deeper down.

The horse manure started out as the "road apples" familiar to attendants at Fourth of July parades. Allowing "processing" to reach its fullest extent would convert it entirely to castings with no remaining worm population, since no animal can live in its own wastes. The trick, therefore, was to furnish an ever-renewed source of food for the worms while separating them from the castings. Remember the "seesaw" approach discussed in the description of my basement project. I used similar techniques in the greenhouse bins.

The manure was reduced in volume as the worms worked it, and there came a point where I could shovel all the bedding from one half of the bin on top of that in the other half—and still have the combined bedding fit under the bin lid.

Remember my noting that a 4- by 8-foot bin could be seen as two 4- by 4-foot sections with no barrier between them? Here's the practical application of that concept: I filled the emptied half with fresh horse manure. Note that it no longer mattered if there was some initial heating in the new material—the worms remained safe in the older, established material and could wait out the heating cycle before starting to work the edges of the new bedding. As they exhausted the old bedding, they migrated entirely into the fresh material, leaving behind pure castings for the garden—and the population was ready for the next swing of the seesaw.

HARVESTING FOR FEEDING

So, what did I prefer to *feed* with harvests from my bins? Feed the soil, with vermicompost? Or feed my flock, with high-protein earthworms? My preference was: Both. At the same time. To pull off that feat, I kept a close eye on the changes in the bedding. I intervened midway along the spectrum noted previously (from "road apples" to pure castings), at what I called "the halfway point": The worms had pulled apart the manure clumps into an even mass with plenty of fiber in evidence, sufficiently broken down for use feeding the soil but still containing plenty of worms. That's what I would scoop up and toss to the flock.

Often in winter I released my birds onto one or the other of two heavily mulched yards for winter feeding, which served in summers as two of my garden areas. I dumped the "halfway" bin material onto the thick mulch, at a different spot each day, in order eventually to spread it over the entire area. Doing the spreading of course were my busy chickens, each in a frenzy to get at the worms before somebody else did. (See the action in figure 2.6.) The vermicompost remained in the deep mulch, for a big boost to soil fertility come spring. No spring tillage, no laborious spreading of compost—when planting time came I just laid out the garden beds, raked them smooth, and the season was off and running.

At other times I would instead dump the wormy bin contents onto the deep litter in my flock's winter quarters. Again, as the chooks went after the worms, they incorporated the vermicompost into the deep litter, increasing its fertility value and microbial populations. At any time thereafter I could remove the litter—something like a mix between a finished compost and a mulch—and lay it down thickly wherever I needed both: under heavy feeders like corn or winter squash, or tilled in where I was preparing a new asparagus bed, or under young nut trees, which I've been told need as much nitrogen as corn.

Winter egg production, both in numbers and in quality, was significantly higher when I was feeding a lot of "halfway" vermicompost, even in times when local flocksters were reporting *zero* production from their hens for up to a month at a time. I also found that

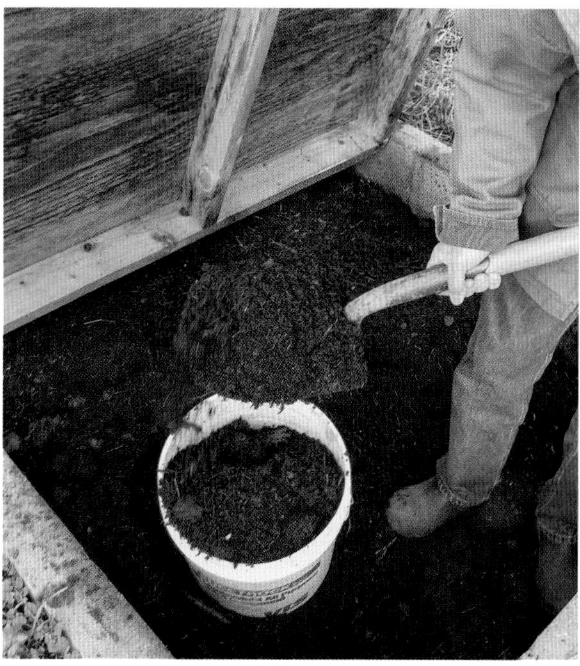

Figure 19.11. Harvesting worms from the bin for feeding.

offering the worms to my hens—together with access to the deep winter mulch, and fresh-cut green forage from the greenhouse and hardy cover crops in the gardens—reduced my use of prepared feed to half what it had been in past winters.

In summer I would sometimes scatter the halfway bedding to feed chickens on the pasture—again, in a different spot each day so the chickens could eventually spread it over the entire sward. If I needed to target the protein boost at younger growing birds, I scattered the wormy bedding inside a *creep feeder*—a feeder permitting entry by smaller birds while excluding the adults—like the one shown in figure 27.9.

AND NOW THE BAD NEWS . . .

My greenhouse vermicomposting project was highly productive for years—*so long as it was based on horse manure for filling the bins.* But remember my discovery that by adding far too much manure-based fertility to my garden soil, I had created excessively high levels of phosphorus and potassium. After I stopped using manure of any sort in my gardens, I experimented with other approaches to bedding and feeding in my greenhouse vermibins—based on cardboard, newspaper, coir, coarse compost, "greens" harvested from cover cropping, and more. I found that I could not with any reasonable amount of labor duplicate the relative ease and simplicity of my previous model: Just haul in horse manure by the pickup load and fill the bins. After three frustrating failures of alternative approaches in the greenhouse bins, I decided the more modest scaled-down basement project described previously made more sense for me. In the two years I have used the new model, I have found it plenty productive for my needs: castings to generously supplement other soil-building strategies in the garden, and a significant contribution to my flock's feeding.

I include the description of my greenhouse project for anyone interested in using a "trench" vermicomposting model protected from both weather extremes and potential marauders of the bins. Bin volume is high in something like my greenhouse setup, which provides about 160 cubic feet (4.5 cubic m) if bedding is 1 foot (30 cm) deep, especially given it is mostly "free space" under the access required through the greenhouse in any case. A conservative estimate is that my bins processed 6 to 8 tons of horse manure a year. Limits to production in a volume this size are more likely to be on access to inputs—whether manures, paper wastes, food residuals, partially decomposed composts produced on-site—than on processing capacity of hungry worms.

Vermiculture at Farm Scale

A more ambitious, farm-scale model might imitate that used on George Sheffield Oliver's grandfather's northern Ohio farm from 1830 to 1890.[9] *All* the added fertility on the farm's 160 acres was in the form of large quantities of earthworm castings from a 50- by 100-foot (15 by 30 m) pit, 2 feet (60 cm) deep, at the center of the barnyard. Each day Oliver's grandfather added the manure from the farm's various livestock to the pit, where it provided ideal habitat and feed for composting earthworms. Periodically Oliver's grandfather scraped the barnyard itself and added its accumulated manures. Aha!—and note that the farm's five hundred chickens ranged the barnyard, scratching up worms from the top layer of the pit.

At the time of spring plowing, Oliver's grandfather removed and set aside the upper layer—with the greatest concentration of worms—and scooped out the deeper layer of worm castings below, which he spread on his fields at the rate of several tons per acre. The reserved top layer, with its teeming established population, was pushed back into the pit, for continued cycling of the farm's manures. Oliver estimated that the worm population in 27 cubic feet (.76 cubic m) of the bin's bedding was about fifty thousand, and that they would convert that cubic yard to castings in one month.

Oliver's grandfather used a tractor with a loader to spread his compost. If I were using his pit model, I would use a manure fork and a wheelbarrow to move wormy compost to my chickens—and let them take care of the spreading and tilling. Feed the soil, feed the flock, beautiful.

PART FIVE

Other Management Issues

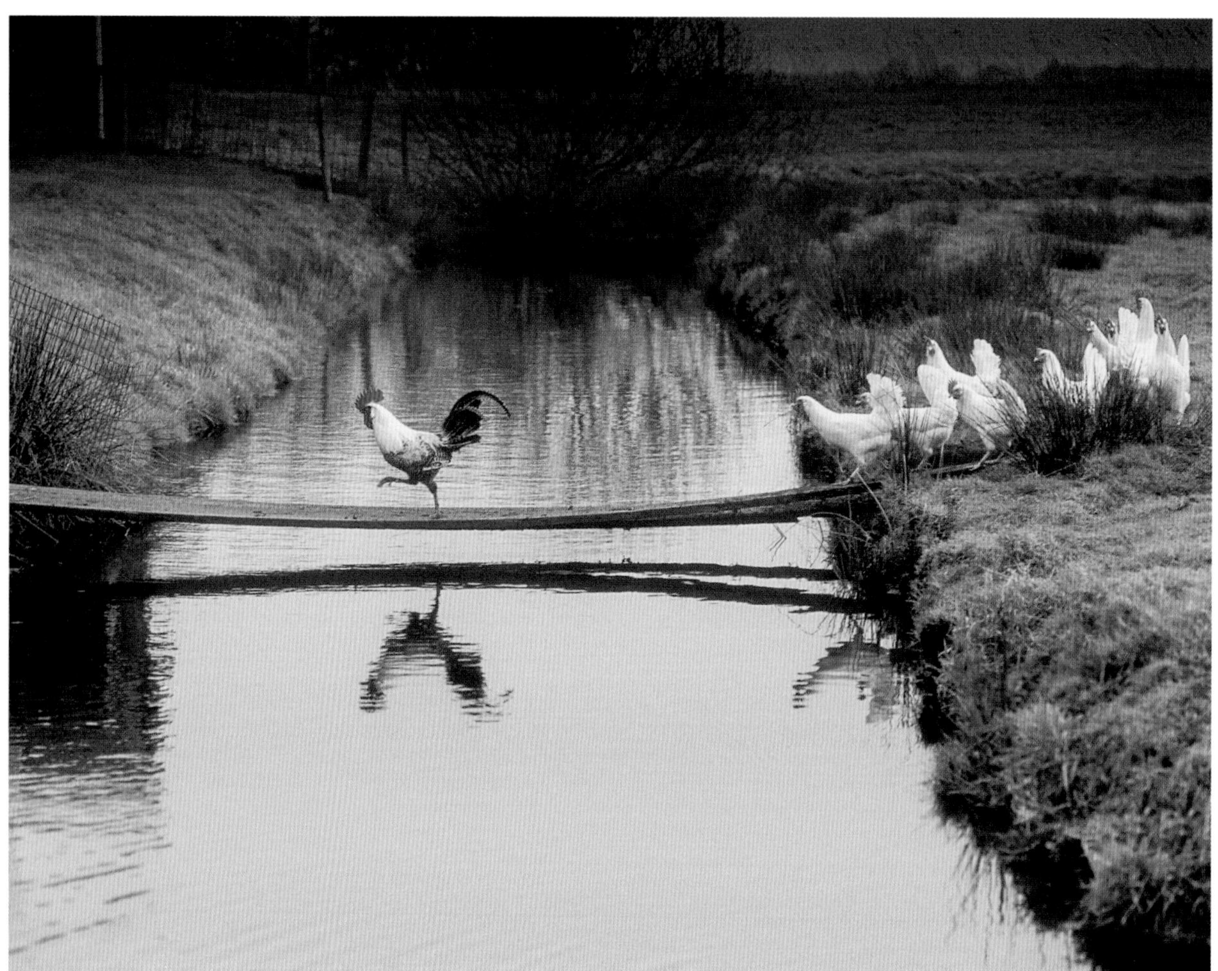

Photo by Auke-Bonne van der Weide.

CHAPTER 20

One Big Happy Family

The homestead or small-farm flock may start as a simple group of a few layer hens, or a group of hens with a cock. Almost inevitably, though, the adventuresome flockster starts a new generation of birds. Or adds an additional cock or more to the flock. Or brings in new adult flock members from elsewhere. Or decides to add other fowl species such as guineas or waterfowl to the backyard menagerie. This chapter discusses the management challenges resulting from mixing it up in these ways.

Working with the Cock(s) in the Flock

The cock has a key role to play in reproduction, and that role is discussed in chapters 3, 25, and 26. Here, let's consider some of the management challenges posed by the cock's role in the flock and his distinctive behaviors.

Sexual Behaviors, Two "Styles"

Ellen and I prefer having at least one cock in our flock—we enjoy the way he completes the flock's social patterns, and how he looks out for his hens. The cock has a reputation for being something of a bully, based on his sometimes-crass manner with the hens: He may approach one without much apparent by-your-leave and mount her in a way that seems more attack than lovemaking. We sympathize with the hen, shaking herself off as if to say, "Well!—glad *that's* over!"

But the cock is not as crass as he seems and in fact is quite solicitous of his flock. Alert to the sky, he gives the alarm if a hawk swoops over, setting off a dash for cover. I have heard stories of cocks attacking dogs or foxes in defense of the flock. And if you want to see the true colors of the cock, throw some special tidbit near him—a cricket or grasshopper. Does he shove the hens out of his way and gobble it himself? Not at all—he calls the hens with a special deep-throated burble used at no other time. It is especially endearing to see him pick up the cricket and beak it to one of his favorite hens. He seems to have the instinctual wisdom to know that the hen needs this nutritional boost to produce the nutrient-dense egg that will carry on their species.

In fact, the more natural style of copulation on the part of the cock is rather different from that described above, too commonly seen today. As with the ancestral Red Junglefowl cock, natural courtship behavior of domesticated cocks involves *dancing* for the hen to encourage her cooperation for mating that is not brutal or rough. Modern breeding, however—focusing on production traits to the exclusion of all else—has produced cocks who have *forgotten how to dance*—Temple Grandin calls them "rapist roosters." Forcing attentions on a hen who has not been properly enticed by dancing sometimes results in copulation violent enough to be injurious.

If the cock in your flock is not a dancer, be on the lookout for *spur injuries* in the hens. In extreme cases, I have seen hens with gaping rents in their sides large enough to show the internal organs. Isolating such hens immediately is imperative—when protected from further injury, their recuperative powers are astounding. Still, I have lost a hen or two to spur wounds, as have others of my acquaintance.

I have not had to do so for years, since I started favoring dancing cocks in breeding and selection, but you may find it necessary to trim the spurs of cocks who injure hens during mating.[1] Use an ordinary pruning shears to remove the spur's sharp point. Don't prune farther down toward the shank, or the bird may bleed badly—shear off just enough to blunt the spur. There will be a little bleeding, and you can use a styptic if you like, but bleeding will not be serious if the cut is properly made.

Aggression: Toward People

In most ways our management of the cock is no different from that of the hen—his needs for feed, water, and access to exercise and dust-bathing are the same. In one way, however, dealing with the males presents a unique challenge: the cock's natural aggression. (True, hens may be aggressive, but in most cases only when they are mothering a clutch of chicks.) Let us consider the problems of aggression—first toward people, then toward one another.

Make no mistake, an attack from a feisty cock can be a painful and stunning experience. He uses both shank and pinion to "flog" leg or torso with a force surprising for an animal so small, and shocking in the sheer violence of an attack that holds *nothing* back. When a sharp, well-aimed spur strikes home, the result can be a lifetime scar, both physical and psychological.

The one time I suffered an attack that serious occurred after I brought into my flock an Old English Game (OEG) cock from elsewhere. I believe he must have been mismanaged at his previous home. In any case, he was Mister Mellow for a couple of days—then made an unprovoked assault as I entered the henhouse. His spurs drew blood on each side of my calf—through the heavy denim of my jeans—and left me badly shaken. (I allowed him a grace period long enough to breed my OEG hens, but then regretfully took him to the chopping block.)

Over the years I have listened in amazement to horror stories about "mean roosters" who terrorized the family for *years*. I can never believe such behavior was tolerated—at my place, any cock who threatens a person earns a short trip to the stew pot. However, it is my belief that examples of serious aggression toward people are almost always the result of mismanagement on the part of the cock's keepers. It is possible to avoid the development of aggression problems, foremost through respect for the bird, and by understanding the instinctual basis for his behavior.

Though we call unwelcome attacks by the cock "aggression," such behavior is more likely *defensive*: He is acting from a deep instinct to *defend the flock*. Once he sees you as a threat, he will attack without hesitation—and believe me, he doesn't care a whit that you are twenty times bigger. The key to good relations, then, is to convince him you are not a threat to his flock.

When you are working with the flock, especially in close quarters, keep your movements quiet and gentle, allowing the cocks, and indeed all the birds, plenty of space so they do not feel pushed or crowded by you. Avoid sudden movements. Don't carry large flat objects such as an empty feed bag or a piece of plywood—the birds see it as a flying predator and react in terror. Avoid having to catch one of the hens in the cock's care, especially in close quarters—perform such chores as applying or examining leg or wing bands at night if possible. If you must get a hand on a hen inside the coop during the day, grab the cock first and settle him somewhere else before proceeding.

I like to offer my cocks special treats by hand from time to time, to encourage trust. If they are with the hens, they will usually approach with them to get the offered treat. If isolated in the breeding pens, they may hang back, uncertain, at first, but eventually will come to get the tidbits from my hand. A crushed hard-boiled egg is a good choice, or a handful of sprouted grain. Soldier fly grubs are like caviar. The more often I take the time to make friends in this fashion, the more mellow the cocks become.

It is especially important that children be taught how to behave around the cock of the flock. We once had visitors whose three-year-old son I allowed into the chicken pen. In his excitement he began running back and forth among the chickens, when suddenly the cock—who had never shown the slightest aggression toward me—jumped up and attacked the boy from behind. I have a neighbor who suspects that his daughter repeatedly teased one of his cocks by rattling a stick on the fence. The resulting tendency to go on the attack whenever a person entered the pen was resolved only when the cock was accorded the place of honor at dinner.

I once had a 10- by 12-foot (3- by 3.5-m) mobile pasture pen that, though designed for broilers, I began using for a laying flock of New Hampshires with a cock. Since there was no access for collecting eggs from outside, I had to crawl into the confined interior to do so. Naturally, the poor cock felt his flock was threatened and reacted accordingly. So intense were his attacks that I had to carry a trash can lid to ward him off as I frantically gathered the eggs. By the time I returned that flock to the poultry house for the winter, the cock's reflexive attacks on me were thoroughly engrained. I regretfully culled him—I *am not* going to be looking over my shoulder around my birds. Still, I felt terrible slaughtering the poor guy, fully aware that his behavior resulted solely from my mismanagement. Having never forgotten that lesson, I now try to anticipate and avoid potential showdowns.

For example, one of my OEG cocks attacked my hand while I was filling the feeder in his breeding pen. My first inclination was to jerk him up by the neck and "teach him who's boss!" Instead, I closed the door to the pen and thought about what had happened. The cock had been no problem whatever as I serviced the pen for a couple of weeks. But I had just moved a couple of breeding hens in with him the day before. Now he had a different sense of what was at stake when I entered that space. I changed the way I serviced the pen—it was as simple as removing the feeder from the pen before filling it—and had no more conflicts with the cock.

Please be on guard against an intensely emotional reaction when challenged by one of your cocks. True, if it comes to a showdown, you may "win"—I have heard people boast of beating a cock into submission. To me that is more admission of defeat than wise management.

Remember, you have more options than he does. Act like it.

Aggression: Toward Each Other

Usually, respectful treatment of the cock in your flock will result in amicable relations. If you have multiple cocks in the flock, however, you must also learn to deal with issues of aggression among themselves.

Cocks have a natural urge to establish their dominance over other males in the flock. How do you respect the expression of that urge and at the same time avoid mayhem? The answer is—it all depends: on the number of cocks, especially in relation to the size of the flock as a whole; the breed(s) of the cocks; the degree of the flock's confinement; and, most especially, the point in the breeding season.

When cocks fight, the intensity of their aggression is strictly a function of how closely matched they are. Understand, however, that by *matched* I refer not to size but to fighting spirit—the will to best the other *at whatever cost*. If the two cocks are *not* closely matched, their fighting will not become serious—after some vigorous sparring, one will conclude he is no match for his opponent and will submit. After that, if the dominant cock flares at him, he runs away shamelessly, and that is the end of the encounter—the top guy does not pursue. If the two are evenly matched, however, they will not back down until one of them is either dead or beaten so near death he *must* accept defeat. We have seen such cocks, utterly humiliated, retreat to a corner, back to the flock and head down, and stay in that position for two days before returning to activity with the flock—the very picture of abject, brokenhearted shame. It is a wrenching sight.

Early on, we tried to intervene—to separate the combatants to help them "get over it." Forget it—if interrupted, they simply pick up the contest where it left off, however long they have been separated—they *must* follow the instinctual imperative to either dominate or submit. Our practice after that realization was to monitor fighting closely; when it became obvious we had a deadly serious fight on our hands, we simply picked the cock we wanted to keep and slaughtered the other. It was just too traumatic allowing the fight to go to its unforgiving conclusion.

The easiest way to have two or three cocks in the flock without serious fighting is to add one cock per season. That is, in year two keep a younger cock from that season's hatch, in addition to the original cock of the flock. The younger cock is almost certain to grow into a position of subordination to the older, without a serious contest to settle the issue. A third can be added the following season, and generally by the next season it's time to cull the oldest cock anyway—most likely before he faces a serious challenge as he ages.

The more cocks you have in the flock, of course, the more likelihood that two or more will decide to duke it out. In any given year, I have kept up to nine cocks in the flock—I got plenty of practice at preventing serious fighting. I found that the more space cocks have, the lower the level of aggression; and conversely, the more confined, the more likelihood of fighting. In the summer there are usually no problems in a flock with several cocks on a large range defined by electric net fencing. In the winter, the need for space to reduce the potential for aggression is one reason I provide a heavily mulched yard on which the flock can spend most of their waking hours.

Figure 20.1. Cuckoo Marans buddies.

If the flock must be confined in winter quarters, I monitor daily for serious aggression. One strategy I use is to provide baffles in the corners of the chicken house as refuges for a subordinate cock who wants to retreat from the attack of a more aggressive one. Fair-sized pieces of scrap plywood, for example, make good baffles—just stand them in a corner, so the corner itself becomes a hidden retreat with narrow access on either side. Once the retreating guy gets out of sight of the dominant one, the aggressor will usually not pursue.

But that is not always the case. I can't stress enough the need to monitor constantly and to be prepared to act immediately as required. For instance, a few weeks after confining the flock to winter housing one year, I noticed that one of the OEG cocks was a little beat up. I set up the baffles, assuming it would solve the problem. It didn't. The next day the beaten cock was in much worse shape. I rushed to prepare the Breeders Annex, where I isolated my breeding pairs and trios for the breeding season; but by the time I had it ready, the beaten cock was in such bad shape that, after a couple days' isolation, he died. I should have been ready for fighting by having already prepared the Annex. Had I done so, I would not have lost one of my breeding cocks.

In any case, the move of the cocks to the Annex stopped the fighting within the day—with the hens out of the equation, what was there to fight about? For a couple of weeks, it was the Peaceable Kingdom. Then one day I found two cocks with bloodied combs. This time my response was timelier: I immediately isolated the cocks in their separate pens inside the Annex.

But there was an interesting thing about that separation. Not having enough available pens, I had to place two of my Cuckoo Marans cocks in the same space. I was pleasantly surprised to find that the two got along amicably, without the slightest trace of aggression toward each other—indeed, they seemed to be great buddies! Clearly a cock's breed has a significant influence on his inclination to fight. My OEG cocks were—not surprisingly—a good deal feistier than a typical dual-purpose cock such as the Marans. (Of course, when I moved in the breeding hens, I separated the two Marans cocks as well.)

Mixing Young Growing Birds with Adults

Most sources advise against mixing age groups in the flock—that is, raising young growing birds with adults. That is good advice in the case of fast-growing meat hybrids, especially Cornish Cross—they are *such* wimps in comparison with sturdier breeds. But in caring for any traditional breeds, I have never had a problem introducing young birds—even those just out of the brooder at four weeks or so—directly into an adult flock. Indeed, in earlier times it was considered good management to expose young growing birds to disease organisms likely to be present among adult chickens, to trigger a healthy immune response in the younger flock members.

If you do introduce a group newly out of the brooder to your adult flock, the more space you can give everybody, the better. Monitor closely the first day or so, but do not be concerned by some hazing of the young ones by the adults—I've never known it to turn vicious. The adults are simply putting the young ones in their place—right at the bottom of the flock hierarchy. Since the young birds accept their submissive place in the pecking order—what other choice do they have?—the older ones do not waste energy repeating their instructions. The younger ones, unable to compete with the older, turn instead to establishing their own dominance-submission patterns, the young cockerels especially strutting and sticking out their chests like boys on a grammar school playground.

If chicks are with a mother hen on pasture with the main flock, you do *not* need to worry about the young ones—the hen will kick butt for anyone foolish enough to mess with her chicks.

If you do run younger birds with adult layers, *it is essential that you not feed the young birds an adult layer mash*. A compromise feeding program is in order, but remember that the layers need a higher mineral content (deleterious to the developing reproductive systems of the growing birds), and the young ones need more protein. My solution has always been to offer plenty of oystershell free-choice, which the layers eat in any amount needed, while the young ones ignore it. To provide a protein boost for the growing birds, feed higher-protein feeds—whether a fish-meal-boosted formulated mix or earthworms or soldier grubs, cultured milk, or excess hard-boiled eggs—inside a creep feeder (described in detail in chapter 27).

Introducing New Flock Members

There may be occasions when you introduce new adult flock members—as replacement stock, perhaps, or to bring new blood into your breeding program. The resulting displays of aggression can make for nervous moments, but it helps to understand that you are seeing a practical and entirely natural behavior, necessary to work new members into the flock's hierarchy. There are equally practical ways we can assist a successful and reasonably smooth transition.

Some flocksters try to short-circuit the barnyard conflict by slipping the newcomers onto the roosts at night, alongside established flock members—the idea being that everybody will wake up in the morning assuming that all the others have always been there. Such a strategy ignores a fundamental reality: The question for a chicken about any other chicken is, *Who are you, in relation to me, in the flock's pecking order?* That question is not answered simply by waking up shoulder to shoulder on the roost.

As discussed earlier, young chickens are relatively easy to work into the hierarchy—their bottom-rung position is a foregone conclusion. In the case of adult newcomers, however, they may well assert their own claims more insistently than birds just out of the brooder ever would. Think of the conflict when the social structure is shaken up and reordered not as *combat* but as *conversation*. The "discussion" sometimes involves a lot of hackle raising and squawking and a little beak stabbing but seldom any real damage before two hens come to an agreement about each's position in the new hierarchy. By the next day they will likely be foraging side by side with no problem.

Chances for an easy reshuffle of the social order will be enhanced if you maximize both the time you can monitor it and the space in which it takes place. Schedule the transition at the beginning of a two-day weekend, perhaps, not in the middle of a hectic week. Ideally, introduce the new flock members in the morning, just after releasing the established flock onto yard or pasture—subordinate members will have more space in which to retreat from a bossier rival. By the time everybody retires for the night, most of the discussions will have been resolved.

Be sure to furnish extra waterers and feeders during the transition, to prevent aggressive competition for these resources.

The potential severity of conflict has a lot to do with how many new birds you bring in. If only a few, more hazing is concentrated on each of the new individuals; if many new birds, the hazing per individual is more diluted. In any case, the worst scenario is an unusually submissive hen destined to inherit the lowest position in the flock—*if* she isn't identified as the weak sister the others ruthlessly zero in on and eliminate. If necessary, use an isolate-in-view strategy (to follow) until the rest of the flock has gotten used to the shy sister.

The default situation for me was typically introduction of new members into a pasture area with an electric net perimeter. If your flock is *either* closely confined—for example, restricted to the coop in winter—*or* completely free-ranging, you will have to modify that strategy. There is much greater potential for serious conflict in a closely confined situation, so it is best to practice what I call isolate-in-view for a few days: Give the newbies their own space in the coop, separated from the others by a wire partition. Everyone sees and gets used to one another, even begins relating socially to a degree, but without being able to get physical. After a couple of days, take away the wire partition, and monitor closely.

In the case of a free-ranging flock, remember that the newcomers have no sense of the new coop and grounds as *home*. Until they have developed that sense, they are lost in the world and are likely to wander away in confusion if allowed to roam. Again, keep the newbies inside their own wire-partitioned section until they are well fixed on the new environs as home—a week or so is probably safe—then release to range with the rest of the flock.

If you use electric fence, before you release the newcomers into the established flock it may be a good idea to clip their wings (as was described in the sidebar "Clipping Wings" on page 103). Otherwise, a new member could fly over the net in panicked retreat from a bossy rival and then wander off, disoriented, and not return.

Introducing new cocks is more problematic than bringing in new hens—the chance of a fight for dominance is high. The cocks of many dual-purpose breeds are more likely to adjust to one another with a minimum of mayhem. A fight to the death may ensue, however, between cocks of more "gamey" breeds such as OEGs and Modern Games, Kraienkoppes, Malays, and Shamos. In this case I always start with isolate-in-view, using ¼- or ½-inch hardware cloth to isolate cocks, who may fight through a larger mesh such as chicken wire. After a week or so, I release and monitor closely. If necessary, I return the new guy to isolation for a few more days. Providing some hideouts on the pasture, such as wooden apple crates tipped on their sides, gives the new cock a place to retreat from an attack by an established cock.

Introducing new flock members is stressful, even when managed well, so it's not a good idea to make the transition if there are additional sources of stress. If there is a sudden heat spike or a storm, for example, delay bringing in the new birds until conditions are more stable.

I limited the previous discussion to introduction of outsiders into a chicken flock. I had only one occasion to introduce an adult new member into a flock of another species: I added a breeding duck from elsewhere to an established small group of Appleyards, two drakes and two ducks. The meeting was pretty much a non-event in comparison with chickens: The three ducks had an amiable bill-to-bill, get-to-know-ya chat, while both drakes looked on in the background, heads cocked at the same angle, appraising the new lady.

Mixing Species

Aside from cocks as a special case, I never had problems keeping different *breeds* of birds together, whether of chickens, ducks, or geese. As for keeping different *species* together, I mixed it up all the time and only rarely had problems.

Guineas with Chickens

If you keep guineas with chickens and there is a conflict of interest, the guineas will win, hands down, *always*. Since the chickens have the good sense to defer to the more assertive guineas, however, the two can be housed and pastured together with few issues. The only problem I've had keeping guineas with chickens comes in the late winter as testosterone levels rise with the approach of the breeding season. The guinea cock then becomes merciless toward the chicken cock. The latter quite sensibly retreats, but the former—unlike another chicken cock—relentlessly pursues, exhausting him and pulling out his tail feathers. One spring I *hobbled* the male guineas to stop their harassment! Keeping the guineas and chickens in separate areas during this period would also be a possibility, though I dislike having to service multiple groups. I've also simply slaughtered the guineas after the winter squash harvest—my main use of guineas was for control of squash bug, as described in chapter 13—hoping to be able to bring in adult replacement guineas by squash-blossom the next year.

Turkeys with Chickens

I have too little experience to address this question with any expertise, though I've raised a few turkeys started elsewhere with my chickens. The big question is not a clash of personalities but *blackhead*, a disease caused by a parasitic protozoan. Though they are not usually seriously affected by the parasite, chickens can pass it to turkeys, among whom it is usually fatal. In the case of *small* flocks, some flocksters report that keeping the two together works okay; for others, there's simply no way around the blackhead threat.

Waterfowl and Gallinaceous Fowl

In my experience, keeping ducks and geese with any of the gallinaceous fowl works fine, so long as you take care to provide a drown-proof waterer that serves the needs of both classes as described in chapter 9. I particularly value the more complete use of feed resources I get, keeping chickens and waterfowl together. For example, chickens eat the seed heads of sunflower and sorghum, but ducks and geese also eat the leaves, right down to the stalk—and those of corn and the sorghum-sudangrass hybrid I grow as a cover crop. Chickens will not eat the stiffly awned seed heads of mature rye, but ducks and geese do so with relish—while the chickens till in the rye in preparation for another planting.

CHAPTER 21

Protecting Your Flock from Predators

I said in chapter 1 that a basic characteristic of the abundant ecology is the degree of *cooperation* among its countless players. But there is plenty of *competition* as well, with the ultimate competition being *predation*. It is essential for us all to reflect on the critical role of predation in shaping and guarding the abundance of the ecology.

In a word, predators guarantee that *nobody gets a free ride*. Absent checks on their abounding fecundity, many species run riot, decimating resources and imperiling

Figure 21.1. This beautiful barn owl carrying its rodent kill shows us where thoughts about predators properly start: We *need* them! Photo by Bob Brewer via Unsplash.

whole ecologies. It is *predators* who constantly reduce their numbers, who keep them anchored in their place in a finely balanced whole.

The threat of out-of-control populations is not trivial. Think of just one catastrophic example: Rabbits, introduced to Australia as a food animal in 1788, inevitably escaped their cages; and in the 1850s were deliberately released for sport hunting—onto a whole continent *devoid of their natural predators*. And you know what they say about "breed like rabbits": Their populations exploded over the landscape. Their ecological impact was devastating—they may be the primary factor in loss of species in Australia, and their denuding of landscapes causes serious soil erosion. Agricultural losses are, literally, incalculable. Despite desperate attempts at control—shooting, trapping, poisons, destruction of warrens, fencing, and more recently biological measures (infecting rabbits with bacterial or viral diseases and releasing them to the wild)—the rabbit population remains today Australia's unsolvable problem.

Let me suggest, as we begin a discussion of predators, that our most anxious question not be *What if they get at my chickens?* But rather *What would we do without them?* Imagine rabbit and rodent and crop-damaging insects, not as minor annoyance but at numbers great enough to wreak havoc. Now thoughts of predators can, properly, start with the realization: *We need those guys!*

Figure 21.2. Least weasel: Tiny but lethal. Photo by istockphoto.com.

Close Encounters at Boxwood

Ah, but—however much we respect and honor their good work—most neighboring predators like a chicken dinner as much as we do, a fact that comes home with a jolt when our flock suffers its first hit by a predator. Unfortunately, despite best efforts to anticipate specific threats, most of us go through some on-the-job training, adjusting management as we are instructed by predators. The following stories are illustrative only, to get you thinking about the challenges and the need for nimble adaptation on the run in a fluid, ever-changing predation environment. At Boxwood, we've always seen a success by a predator as a lesson about a weakness in our management. The good news is that making management changes, as instructed by each lesson in turn, eliminated losses to predators almost entirely.

Mayhem in Miniature

I hope your first predator experience is not as dreadful as ours. We started our first batch of chickens—twenty-six New Hampshire chicks received through the mail—in a temporary brooder in the shop. We moved the birds at four weeks to the coop, where they enjoyed having more space and access to the outdoors. One morning a week later Ellen opened the door to a grisly scene of scattered red-plumed bodies, a few of them with some of the carcass eaten, all of them with grievous neck wounds.

With a heavy heart I gathered the torn little bodies, having no clue as to what had made the injuries to the necks, nor how such a ferocious killer had gotten into the tightly secured coop. Suddenly a head popped up from behind the feeder—a tiny creature no larger than a rat. I had never seen anything like it but in the first glance recognized it as the cause of the mayhem.

Grabbing a shovel, I managed to kill the nimble creature, and picked it up to examine it. It had a pelt like silk—and a mouth filled with needlelike teeth. Here was a marvel fashioned to do one thing superbly well: to kill. Using its head as a template, I tested all the openings up under the rafters—and each one big enough to admit that head I subsequently blocked off with a scrap of wood, nailed securely in place.

When I had set up the coop, I anticipated predators, of course. But I was thinking raccoon, possum, fox. If *weasel* entered my mind at all, I pictured something larger, not remotely as small as what I now held in my hand: a *least weasel*, smallest of the weasel tribe—and indeed, at 7 inches (18 cm) or so and a weight of less than 2 ounces (57 g), the smallest of all true carnivores.

Such a tiny animal could only eat so much, of course. The reason Ellen and I found so many bodies is that the weasel's attack reflex is triggered by movement of prey, not by hunger: As long as chicks had scampered about in terror, it rushed in to kill—for a total of nineteen out of our twenty-six young chickens.

Do note: Attacks by least weasels on poultry are rare. They prey mostly on rodents.

The Pack Is Back

One advantage the flockster has is that most predators are nocturnal—thus, if we religiously shut the flock at night in a secure shelter, our losses should be minimal. A major exception, however, is *dogs*, who could be your major predator threat, not only because they are active during the day but—in contrast to most other predators, who hunt alone—dogs may hunt as a pack, to devastating effect.

Our second predator experience occurred one season when I raised fifty broilers in a Salatin-style mobile pen, and the birds were close to butchering size. One morning I went out to find chicken carcasses flung like rag dolls over the pasture. In the early hours, two dogs from a nearby neighbor had both dug under the bottom rail of the pen and torn a hole in the poultry wire to get at the chickens. These dogs were well fed at home, so their assault on my birds was strictly fun and games.

I locked the dogs in a shed and called the county's animal-control officer. Though their owners paid me for my losses—as required to get the dogs out of the pound—no stack of dollars could compensate for the two-month effort I had put into providing a cache of dressed poultry for the freezer.

Be aware that many jurisdictions, both state and local, support the owner of livestock threatened by dogs—laws in Virginia, for example, even give a landowner the right to kill harassing dogs "running at large." Study applicable laws carefully, should it become necessary to quote them chapter and verse to owners with a hazy grasp of their responsibilities. And remember those laws apply to you as well, if you have dogs of your own.

Masked Marauder

I guess I'm a slow learner, because I continued to use the unmodified broiler pen to raise batches of meat birds. My second lesson in its vulnerability came one night when a raccoon tore a hole in the poultry wire enclosing part of the pen and made off with eleven young broilers. This time I took remedial action, by *wiring for defense*: I mounted single-strand electric wires around the outside of the pen, one at nose level and another higher up. Charging such a short run of wire required only a light-duty fence energizer powered by a 9-volt battery. My electric defense was thus entirely self-contained, with only the energizer's ground rod requiring pulling up and resetting when it was time to move the pen. Until I quit using that mobile pen—when I switched to electric net fencing—it never suffered another loss to a predator.

Raccoons typically eat insects and grubs, fruits and nuts, and fish.

A Sly Fellow

I'm not spooked by snakes, but it gives me a start when I reach into a nest while gathering eggs—and find a blacksnake coiled inside! Fortunately that has been a rare occurrence. Be aware that snakes will eat eggs, as well as young chicks. Or ducklings: Three of mine once went missing when I was brooding a batch. Under a piece of scrap plywood I had foolishly left on the litter,

Figure 21.3. Raccoon: tough, wily, and smart—a competitor worthy of respect. Photo courtesy of Emmanuel Keller.

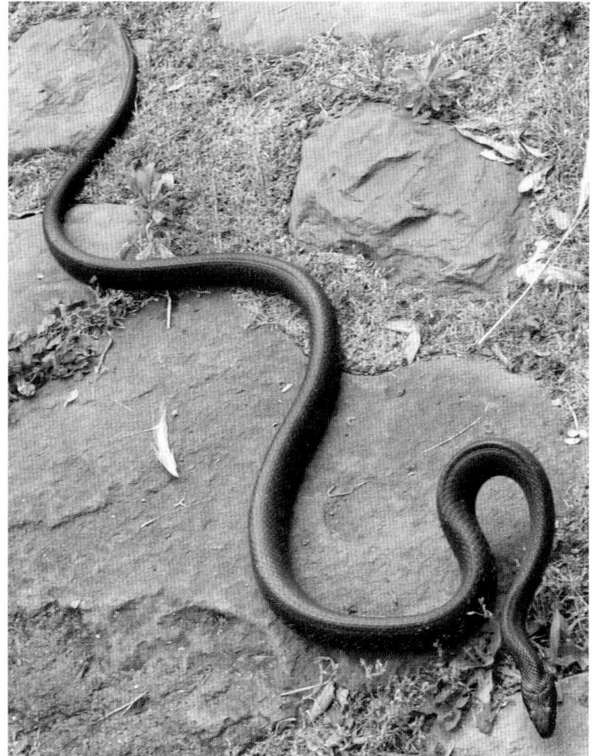

Figure 21.4. Blacksnake: has a taste for mice.

I found a large and very satisfied blacksnake, with three bulges down its length leaving no doubt what had happened to my ducklings.

Since I leave my poultry house open during the day, there is no way to prevent a snake from wandering in. Thus the best way to discourage its doing so is to decrease the appeal of the henhouse—that is, to prevent a *rodent* population inside: If there are rodents in the coop, a snake may come on the prowl for them and then, always the opportunist, grab a snack of egg or hatchling if available. It helps as well to be more diligent than I was about leaving anything about as a place for a snake to hide. You might check the rafter plates as well: If a snake is using that as a place to enter or laze about after a meal, there may be streaks of chalk-white snake poop down the wall to give away its hiding place.

We value the role of blacksnakes at our place, so my practice was always to carry them out of the henhouse and release them elsewhere on the property. But, following the success of my efforts to reduce the mouse population in the coop almost to zero, incursions of blacksnakes have been extremely rare. (See "Avoid Free-Choice Feeding" on page 142.)

Among the blacksnake's favorite foods are rats, mice, rabbits—and other snakes.

Leprechaun

We always love seeing the local leprechaun, the red fox, on our walks through the surrounding pastureland—and in our backyard. Though foxes have a reputation as the quintessential chicken thief, for decades they took not a single one from us. A combination of three strategies allowed us to enjoy the beauty of the foxes passing through our property even as we served ourselves a second helping of roast chicken: Digging the poultry wire 8 inches (20 cm) into the soil at the foot of the fence, when we were using a static run; shutting in the flock at night; and later, routine use of electric net fence.

Ah, but then came that smart vixen! Remember that about foxes—if you don't outsmart them, you lose. The one who outsmarted us was feeding kits, and supported that effort by raids on every flockster in the neighbor-

Figure 21.5. Red fox: "Greatest destroyer of mice." Photo by istockphoto.com.

hood. Smart? She would calculate a line of entry that started with jumping from a bank onto the top of a board fence, run thirty yards along the top board, then sail into a pastured flock with a gay "Hi y'all!" *And*—she figured out she could *jump* an electric net in lieu of getting herself a snootful of pain. *Not* good—she took eight of our chickens of all ages before we could figure out a response.[1]

We were compensated to some extent later in the season, when Mama Vixen showed up in our backyard with three almost-grown young ones, every day for two weeks running, this time daintily eating the dropped bounty from two fruiting mulberries.

My *Fieldbook of Natural History* says of the red fox: "Probably the world's greatest destroyer of mice."

Aerial Predators

Though electric net fencing can deter most predators on the ground, it does nothing to stop aerial predators—hawks, owls, eagles, and other raptors.

A surefire way to deter aerial predation is to net the ranging flock entirely, with some sort of netting above the flock as well as at ground level. For example, I have often kept young birds without a mother in an enclosed shelter with wire on top until they grew large enough to be less vulnerable. But the larger the flock and the larger the plot they forage, the less practical the netting option. An interesting alternative is not to net the area overhead, but instead to string monofilament fishing line here and there in that airspace—40- to 50-pound test. Make the lines as tight as you can, in random runs among trees, fence posts, corners of buildings, or stakes. Some flocksters using this strategy believe the lines more effectively spook raptors flying into the space if they are invisible; others increase visibility by tying on lightweight streamers such as strips of foil or flagging tape.

The effectiveness of this spiderweb strategy was demonstrated by the experience of a member of the American Pastured Poultry Producers Association (APPPA), who reported: "I was having problems losing juvenile turkeys to an owl, and tried tying fishing line from the top of the pen out to the posts of the electric net. (I ran about 10 or so lines.) It worked great. I'd untie the lines from the net once a week when I moved the turkeys, which was a little annoying, but not that bad. I continued using the fishing lines about a month till the turkeys got big enough that the owl gave up." However, another APPPA-er reported that while monofilament line, strung lavishly above her flock, was completely effective at first, eventually hawks learned how to get past it.

Experiment to find what works best for you. As for me, I used this strategy for the first time last winter (no streamers) and found it completely deterrent, even though I frequently saw a red-tail giving my chooks the eye—in a time when several confirmed kills by this individual were reported in the neighborhood.

Happily, in the case of hawks, adults learn that full-sized chickens are not easy prey and focus on small mammals like chipmunks and voles instead. Thus I frequently see hawks flying over on the hunt, but attacks on the flock are rare. It is the juvenile raptor—something like a teenage boy with more bravado than smarts—who is apt to make the most frequent and determined attempts on the flock. Almost all the airborne attacks here have been from juvenile Cooper's hawks.

Several kills by the same individual usually reveal its modus operandi. For example, I had several losses one summer, all in the early hours before I got out to feed. The defense was as simple as it was effective: I started leaving the chickens shut in their house until the morning was well begun. By the time I released them, the disappointed raptor had flown elsewhere looking for a meal.

There was a time when I routinely left my mobile pasture shelter open at night, since it was protected inside an electric net perimeter. But then I had a couple of losses to a barred owl—who will land and walk about on the ground seeking prey—and began shutting up the pasture shelter until I went out in the morning to feed.

Unlike chickens, who readily come in to roost with the approach of dusk, ducks and geese may prefer to stay outdoors even at night once they are well feathered. As a result, waterfowl may be especially vulnerable to raptors who hit in the night or in the dawn hours. I therefore *train* them to return to the shelter at night. It takes herding them into the shelter every night for a week to ten days—a bit of a pain but not *nearly* as bad as trying to herd chickens—to get them accustomed to going inside on their own when night falls.

Be aware that some flock members may be quite capable of seeing to their own defense. A mother hen with chicks has no fear whatever of the taloned monster from the sky. Once as I was doing the morning feeding, a Cooper's hawk, impudently indifferent to me, made a dive on a chick right by my boot. Its mother, an Old English Game hen, responded in outrage before the hawk could even make contact—and kicked its butt right over the electric fence.

We were once looking out over the pasture at lunch when a hawk dived after a chick on a pasture plot with not only several mother hens, but half a dozen geese as well. Did they all retreat squawking and honking in terror? They did not—both geese and mama hens *converged* to face the intruder with an emphatic *Bring it*

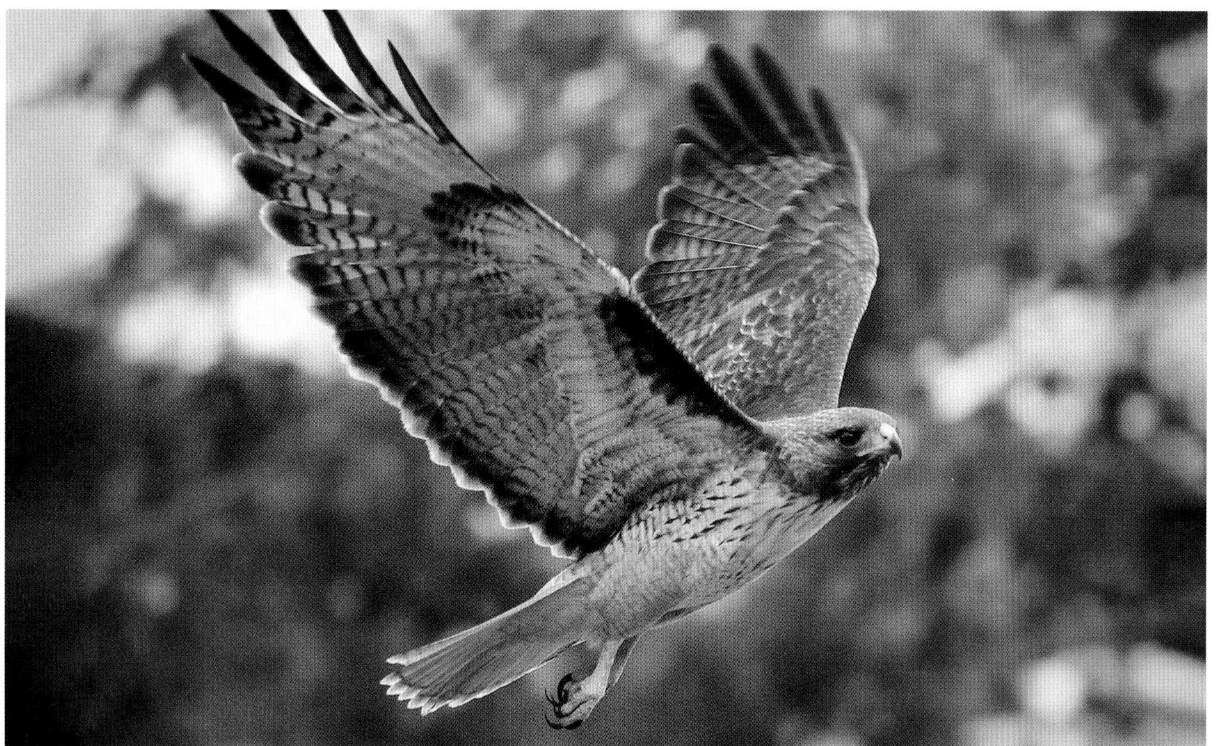

Figure 21.6. Red-tail hawk: beauty on the wing. Photo by istockphoto.com.

on! The hawk made an equally emphatic U-turn in mid-dive, having decided it had urgent business elsewhere.

Other strategies: Be sure your birds have cover to which they can retreat—the coop or pasture shelter, shrubs, fruit or shade trees—to escape attack. Flocks on pasture with larger livestock such as sheep seem to be less subject to aerial predation. Visual distractions that move in the wind—aluminum pie plates, discarded CDs, old clothing, toy pinwheels—may deter aerial predators, who are typically cautious. Switch the locations of the whirligigs from time to time.

Though attacks out of the blue are unnerving, as a matter of fact even the simple expedients I've mentioned can be quite effective. I've received numerous reports from proactive flocksters that they manage to avoid losses to raptors for years at a stretch.

The most important foods for barred owl are mice and other rodents; for red-tail hawks the same, plus rabbits, insects, and snakes.

The Ball Is in Your Court

I have related some of our predator stories not on the assumption they can prevent all your own potential losses to predators—perhaps there is no more "your mileage may vary" aspect of husbandry than predation—but to get you thinking about the nature of predation in your area and some commonsense ways of thwarting attacks. There are predators I didn't mention because we haven't had a problem with them—but you might. *Rats* can decimate chicks in a brooder. *Coyotes*, *bobcats*, *minks*, and *wolves* can cut great swaths through the flock in a successful incursion. *Possums* and *skunks* are mostly egg eaters but will kill chickens as well, especially when they are asleep on the roost at night. *Cats*, both domestic and feral, kill young growing birds and attack and injure older ones—and like dogs, may be active either day or night. *Crows*, *ravens*, and *black vultures* are not primarily predators, but they are opportunists. We once had crows descend on a weakened gosling who had gotten tangled in the electric net fencing. After they discovered what a tasty treat it was, they began swarming healthy goslings and ducklings. I had to return the young waterfowl to the safety of the coop until they grew large enough to be immune to the crows.

A few further thoughts about preventing attacks: I have emphasized electric net fencing to deter predators. If you plan to use a static fence instead, be sure to sink the bottom edge of the fence wire about 8 inches (20 cm) or so into the ground when you install it. Digging predators such as foxes are not stupid: If they find their strenuous effort to dig their way into your flock thwarted, they will not continue wasting energy on the project. I do *not* recommend you put your trust in purchased, high-tech gizmos advertised to have an almost magical deterrent effect on predators: Electronic devices that emit ultrasonic sound or vibrations, flashing lights, and chemical odors have little proven effectiveness—save your money.[2] Since local predation is so variable, your most important first step could be consulting with experienced flocksters in your area about predators to look out for and their methods for coping. (For me that's my buddy Sam Poles: See his useful info in the sidebar "Reading a Kill.")

Some flocksters find that guardian dogs and other guardian animals provide excellent protection for their flocks. Properly trained dogs deter aerial predators as well as those on the ground. Donkeys and llamas can protect flocks on pasture. Both are low maintenance, grazing as they do the same green sward as the birds—and both are especially ill-tempered toward canids: foxes, dogs, coyotes.

Your first loss to a predator will be upsetting, but don't waste time blaming yourself. Instead, understand that the ball is in your court, and that *you must learn and adjust*. I get so frustrated with flocksters who report despairingly, "I can't keep chickens—they keep getting eaten by foxes!" when they have made no changes in management to thwart demonstrated lines of attack. Such slackness is a serious failure in one's duty to the flock. I wonder why they don't take up growing petunias instead.

The good news is that, if you learn from and respond to each lesson from predators, losses will be rare and become more so. The pattern at our place has been: a

Reading a Kill

An effective response to an attack on the flock requires determining the predator involved. Each has its own modus operandi, and in some cases we can tell who made the hit by telltale characteristics of the attack—by "reading" the kill. If you know someone in your area who has plenty of experience with local predators, pick their brain for tips on preventing attacks. If your flock is attacked, perhaps your more knowledgeable friend can read the scene for clues about the perpetrator. For me that friend is Sam Poles, the best trapper in my county, with a vast store of animal lore that has helped me solve predator problems over the years. Of course, the predators you will have to deal with may be different from the ones challenging the flock at Sam's place or mine. But Sam's observations point out some of the signs to look for following an attack.

A hit by a fox may be the easiest to read. A fox will make a single hit—not multiple kills—and will not eat the kill on-site. It will carry it away to eat elsewhere. An odd thing I've seen with fox kills over the years: The fox will stop every 90 to 120 feet (27 to 37 m). I guess it shakes the carcass roughly at those points, making sure it's dead—anyway, the resulting trail will be a single feather here and there, then a cluster of feathers every 120 feet or so, all the way to where it had its meal. In the spring, when the vixen is denning up to raise her pups, the trail of feathers will lead to the underground den. Other times, the trail will lead to a thicket or a crevice among big boulders or similar hidden place where the fox felt safe to let down its guard and eat.

A fox is a digger, but it may have to return to where it is digging under a fence several days in a row to complete a hole big enough to get inside the fence. If it's a fox who's been digging, the dirt will be in a neat, concentrated pile—because the fox has been kicking the dirt back between its hind legs. If the dirt is scattered all around, the digging was by a groundhog. If a fox hits an underground barrier, like dug-in chicken wire, it won't keep trying to dig an entrance hole.

A mink or weasel will eat its kill on-site, starting with the head or neck. They will return to a kill the next day and then eat other parts of the carcass. Either mink or weasel will kill by reflex, as long as it sees prey on the move—so they may leave behind a lot of dead chickens, many with no parts eaten at all. Both these predators are long and skinny—even the larger mink can get through an opening as small as an inch and a half. Hardware cloth might be a better choice than chicken wire to keep these guys out of the chicken coop, especially in the case of the least weasel—really tiny but quite a killer.

A possum will often stay close by after a kill—I've found them on a rafter or even in a corner of the barn, sleeping it off after a nice meal. They don't carry the carcass away and will return to the kill next day. A possum is not a fast, aggressive predator like a dog or fox—if he can get into the coop at night, he takes advantage of the fact that the chickens are "zombied out" on their roosts. A possum will start eating on the head, or will eat in through the anus to get the internal organs, which it prefers to muscle meat.

Raccoons eat their kill right away, on-site. Most of the year they will just make a single kill, and they like the meatier parts—they will start eating the breast meat first. They will return to the kill next day to continue eating. In the late spring or early summer their attacks can be harder to distinguish from a dog's—I think the mother is teaching the young ones how to hunt at that time, so they may kill multiple chickens for practice that they do not eat.

Dogs kill just for fun. A dog who is well fed at home will probably not eat any of his kills—but he can sure leave a lot of dead bodies behind. Dogs may attack day or night.

If you find a young bird who is dead and looks slimed—covered all over with the kind of shiny dried slime that slugs and snails leave behind—that means a snake tried to swallow it but found it too big and gave up.

A bear can be a headache for the chicken keeper, but usually because it's trying to get at the feed rather than the chickens. I once had a bear pay my flock a visit four or five times in two weeks. It ignored the chickens but each time emptied the range feeder. I use electric net fence but I guess the bear didn't hit it with his nose—he just walked right over it. If you're in bear country, it's best not to leave a feeder out with a big reserve in it—it can draw a bear like a magnet.

A hawk will leave a big splash of feathers where it hit a chicken. It prefers to fly away with its prey if it's small enough and eat it up on the branch of a tree, where it feels secure. But a lot of times it can't lift a chicken carcass, so it will eat on-site. It will start with the neck and crop, and return to the kill next day to continue eating. How serious the predation threat is from hawks depends more on species than size. In my area, for example, the Cooper's hawk is more likely to be a problem for the chicken flock than the red-tail, a much bigger hawk.

The worst predator of all is the rogue predator—the one who figures out he has an easy meal ticket at your place and keeps coming back for more. So as soon as you have a hit, you need to figure out who the predator is and take steps to keep him away from your flock. Occasionally that might mean you have to take him out to keep him from coming back.

—Sam Poles

hit from an unexpected direction, an adjustment in our defenses, and never a repeat loss to that particular modus operandi. Note that the heavy hits I recounted were all in the early years of keeping poultry here—losses since then have been rare. And our experience is not unusual: Most of my correspondents in the APPPA—who manage larger flocks than us homesteaders, and whose livelihoods *depend on* keeping their birds safe—report they rarely have economically significant losses to predators.

The Best Offense

Note that I have said nothing about *killing predators* as a strategy for protecting the flock. My opposition to such a strategy should be obvious in my argument for their irreplaceable role in a balanced ecology. But there is an eminently practical reason for not "going on the warpath" against predators: It is likely to be an exercise in futility. True, we terminate the threat from an individual predator by killing it—but only for a while. It is a law of nature that, when the established occupant of a territory dies, another claimant quickly moves in to fill the space. To prevent predation in the long term, therefore, *the best offense is a good defense*—changes in management as required to keep the flock safe in the face of ambient predator populations.

Tell me: Which predator would you *prefer* nearby—one that has learned it cannot get past your defenses (whose nose perhaps has hit your electric fence), or a newcomer who is certain to probe your defenses and may find a vulnerability (may approach your fence when a fallen branch has shorted it out)?

Good Neighbors

This chapter ends with the plea I made at its beginning: Seeing to the health—which is to say the balance—of our local ecology is *everybody's* job; failure to do our part is both myopic and stupid.

I have noted that APPPA members report they make a profit while protecting their flocks from heavy losses to predators—through management practices ranging from defensive fencing and frequent moves of the flock to good shelter design and use of guardian animals—*especially* guardian dogs. In a video produced by APPPA, one notes: "We haven't lost a single animal to predation in the four years since we started using guardian dogs." Importantly, since APPPA members rely so much on fencing, most are careful to ensure constant provision of "wildlife corridors" on their farms—through which all forms of wildlife, including predators, can move, find mates, and thrive. If our goal is first and foremost improvement of our land, another member observes in the video: "When the land improves you see a lot more wildlife, from pollinators to predators—and that's a *good* thing."[3]

Joel Salatin has for years been my own mentor in just such an approach to dealing with predators. Counterintuitively—since he pastures hundreds of layers and thousands of broilers—Joel goes to considerable lengths to make sure there are large swaths of viable habitat for predators scattered over Polyface's 1,000 acres: brushy and wooded areas, numerous ponds, tall trees for perching raptors. Joel knows that Polyface *needs* its predators, lest it become overrun with rodents and rabbits.

Joel knows as well that some predators blunt the threat of others, as in: *The enemy of my enemy is my friend*. If Polyface is generous with habitat for hawks and owls, they ensure that minks and weasels seeking dinner out in the pastured flock become dinner themselves.

While working with predator populations for a more balanced ecology is simply good sense, there is no need to be a silly sentimentalist. Occasionally we, like Joel, may encounter what he calls a "rogue predator"—one that gets around his defenses and makes repeated culinary trips to his flock. In response to such an individual—as opposed to a population—Joel has no qualms about going out on the pasture at night with a strong light and a scoped rifle to protect his birds. Be aware, though, of state and federal laws that protect some predators. Such laws are not arbitrary—they are based on the indispensable ecological services of predator species. Raptors especially are protected, and killing them may bring severe penalties.

Since I've told you that, in the heat of the moment, I did indeed kill the weasel in my first encounter with a predator, I'll end by telling you about another encounter that illustrates the challenge of protecting the flock while respecting and honoring the predator. One summer it was clear that an aerial predator was making repeated attempts on my flock. There were no successful strikes, but at least once and sometimes several times a day the flock would make a squawking dash to the henhouse. Following one of those panicked retreats, I heard a chicken, obviously in distress, near the edge of our patch of woods. Grabbing a garden stake—"Okay, I'm gonna *whack* that guy!"—I ran toward the hen's shrieks in some underbrush. Suddenly the red-shouldered hawk who had made the attack bounded up out of the brush and landed on a fallen branch—we were eye to eye at three feet. I stared transfixed—I could no more have struck out with the stake than I could have struck my grandmother. It spread its wings and lifted into flight, leaving me spellbound by this glimpse of transcendent beauty—and wishing it well.

CHAPTER 22

Helping Your Flock Stay Healthy

The birthright of all living things is health.

—Sir Albert Howard

This is a good-news, bad-news chapter. Might as well get the bad news out of the way first: Regrettably, I am *not* the guy to help you much with curing a sick chicken. The abundant good news is Sir Albert's observation above. We tend to become far too concerned—regarding both our chickens' health and our own—about the myriad things that can *go wrong* and pay too little attention to the reality that *the natural state of a species that has been honed by natural selection is near-perfect health.* That is why this chapter is titled "Helping Your Flock Stay Healthy" rather than "Dealing with Health Problems": The best path to flock health is making sure that everything we do for our birds—proper housing and feeding and provision of opportunities to work—supports rather than compromises their naturally robust good health.

Health Crises at Boxwood

Since the reader may find my thoughts on dealing with health issues a bit cavalier at best, and callous at worst, let me review the actual health issues Ellen and I have coped with in almost four decades of homestead poultry husbandry—while raising seven fowl species, dozens of breeds, and thousands of individuals. When I advise that you can trust in the natural good health of domestic fowl, and should do all you can to support it—instead of buying magic-bullet cures for sickness—that advice is borne out in our actual experience here. The following list is complete, as best I can remember.[1]

Spur wounds: For a while we found that hens were being badly injured by the cock's spurs as he vigorously clasped their sides during mating. I described this incident and the way I eventually resolved the problem on pages 191–92 in chapter 20.

Botulism: There were a couple of outbreaks of botulism poisoning—what the old-timers called limberneck—associated with my infamous maggot buckets, described in "Protein from Thin Air" on page 170. It is possible to have a problem with limberneck as well if you leave spilled feed on the ground where it can get wet and serve as a medium for growth of *Clostridium botulinum.*

Eye infections: One year when the deep litter was unusually damp in the henhouse, our birds had a lot of eye infections, a few of which resulted in blindness or death. We afterward took care not to let the litter get damp, even in the wettest seasons, and had no repeats of such infections.

External parasites: During that same damp-litter episode, the birds were unable to dust-bathe properly, and some of them became seriously infested with mites and lice. Since that time, we've provided a dustbox (as described in appendix B) as an always-available backup to ensure access to dust-bathing. When our chickens can take care of this essential hygiene themselves, we have no problems with external parasites.

Gapeworms: As I describe in the sidebar "But What About Gapeworms?" on page 210, there were a couple of episodes of gapeworm infestation. This is not a common parasite, and an infestation is unlikely if you rotate your flock frequently and practice good range sanitation.

Cannibalism: Injurious pecking of individuals by other flock members can be a problem in highly stressed flocks—debeaking in large-scale industrial flocks is routine for exactly that reason. Our only experience of cannibalism arose solely out of stress during an unseasonable heat spike. Our solution was to ensure our birds would not be unduly stressed even in unexpected weather extremes.

Broken leg: Once when I was moving a mobile pen, a six-week-old broiler was injured when it got caught under the bottom rail in the rear. When I examined it I found that one thigh bone was broken—imagine holding a pencil in your hands and snapping it in two. It was an unusually busy time, and I told the poor guy, "Sorry, pal, you're on your own with this one"—and walked away. Now this was a Cornish Cross, mind you—one of the most compromised examples of chickenkind on the planet. Despite that, I found that he was getting around on the pasture better every day. Long before slaughter six weeks later, I could no longer pick him out of the crowd. (I tell this story to illustrate the enormous capacity of chickens to recuperate *from injury*. But keep in mind this happened in a flock of lackadaisical Cornish Cross. A flock of more vigorous chickens would most likely have eliminated the injured bird, without an isolation period to recover.)

JCOS: The losses from health problems described previously were all in the early years. Since then, the only losses we typically have—one or two a year or so, out of a flock of dozens—are to a condition I call JCOS (Just Crapped Out Syndrome): I walk into the poultry house or onto the pasture and find a bird lying dead. No drawn-out crisis, no anxious interventions, just *Bye-bye*. Or I might find that a bird, though still living, is seriously ill. I take such a bird to the chopping block immediately—if it has an infection, I do not want contagion in the flock. (And no, I do not eat any bird who may have any sort of illness.)

When I reflect on the brevity of this list of problems some thoughts about poultry health come to mind, and I share those reflections in the rest of this chapter.

"It's the Management, Stupid"

That is the message to myself I keep tacked to the wall in my henhouse. Almost all health issues we have had to deal with are amenable to prevention by management changes. Every practice recommended in this book is intended to give the birds the most natural and thus most healthful life possible.

Remember what was said in the chapter on best manure management with deep litter, the alchemy that transmutes lead (the potential for the poops to be a vector for disease) to gold (good health that is supported by what the chickens find to eat in a mature litter). Make sure the litter stays dry, to prevent growth of molds. High levels of ammonia lead to conjunctivitis—also known as ammonia blindness—and can increase the chance of infectious bronchitis as well. Prevent ammonia generation by adding plenty of fresh high-carbon material *before* it becomes apparent—if you can smell it, there is already too much for the chickens' respiratory health.

Maximize the flow of fresh air through the henhouse. Assuming you protect them from direct cold winds, your winter flock is far more likely to become ill from damp, stuffy conditions than from the cold.

"But What about Gapeworms?"

More and more, sensible poultry people these days recognize that *of course* our flocks will be healthier if they're out in the sunlight, exercising and foraging on pasture or other natural ranges. Earthworms are one of those high-value, live animal foods that pastured flocks will most certainly be eating, especially on pastures that are improving in quality every year.

Yet we're likely to see stern warnings about the possibility of chickens' getting parasitized from eating earthworms, which—along with slugs and snails—can be intermediate hosts for *gapeworms* (a parasite of the *throat*, not the gut). Infestation is marked by gasping, sneezing, coughing, choking, convulsive shaking of the head, and death. Sounds pretty scary.

And is. The first time several of my chooks were infested with gapeworm and started wheezing and desperately shaking their heads, I was terrified. After some panicked research—during which one of the most seriously infested hens died—I began immediate treatment (continue) and saw marked improvement of the ailing birds within days. I subsequently had a single episode of gapeworm infestation, not as severe because that time I recognized the problem and took immediate action.

After the gapeworm infestations, I regretfully told my flock, "No more ranging for *you*," right? I did not. Instead I did further research on gapeworm, focusing on environmental factors that could trigger an infestation. After all, I had ranged my birds on our pasture for years, without a worry that the flock might eat earthworms and snails and slugs—indeed, was glad to know they were doing so.

Three things stood out. The infestation followed a long spell of frequent rain—the ground everywhere was soggy. The rain had pushed a spurt of growth—the sward was taller and shaggier than normal. Both factors increased moisture in the soil, making it more compatible to the life cycle of gapeworm. And there was a big muddy spot of bare ground in front of the entrance to the henhouse. I mowed the pasture short (having read that sunlight on the "infected" ground can kill worm eggs), moved the flock immediately to fresh ground, and covered the bare spot with a heavy mulch.

As for treating the active infestation: I am shy of using powerful medications that are rough on the birds' metabolisms. I also fear—despite reassurances to the contrary—that such treatments may leave chemical residues in eggs and meat. Instead, I placed a rush order for an herbal preparation made in Britain called Verm-X.[2] While awaiting delivery, I added raw apple cider vinegar and crushed garlic to the flock's drinking water and mixed small amounts of diatomaceous earth into their daily feed. When the Verm-X arrived, I added it to drinking water as directed to treat the entire flock, and—in the cases of the chooks with the worst symptoms—injected the liquid Verm-X directly into the throat with a syringe. Recovery of symptomatic birds followed within days.

Prevention is always better than cure. The best practice for prevention of gapeworm is *frequent rotation* through your various ranging areas. But monitor for the characteristic symptoms of this parasite and treat immediately on the rarest of occasions (in my experience) if you must.

Chickens are subject to several respiratory infections, such as coryza, Newcastle disease, and infectious bronchitis—as well as adverse reactions to breathing *Aspergillus* molds—but it is highly unlikely that small, uncrowded flocks with abundant air exchange over well-maintained deep litter will suffer from any of them.

We don't have problems with cannibalism, and neither will you if your birds are well fed and have plenty of space. It is a good idea as well, especially if they're confined to the coop in the winter, to hang up various kinds of edible treats for the birds to peck apart—providing both good nourishment and entertainment. (For suggestions of appealing treats, see "Managing Winter Laying" on page 216.)

I've discussed the threat of rodents to the feed supply. Rodents can also carry diseases that affect poultry—among them fowl cholera, chronic respiratory disease, and infectious bronchitis. Few poultry coops are likely to be entirely free of rodents, but take all measures you can to minimize their presence.

What about the role of wild birds in the spread of poultry diseases? Though we hear much of this possibility every time there is a big avian influenza (AI) scare, the chances of infection of your farm or homestead flock from wild avian species are remote. Though wild birds inside the henhouse should not be ignored, as carriers of disease they are mostly a threat only in the huge industrial poultry houses. But those places are *incubators* of disease because of the crowding and filth.

An excellent illustration in my region occurred many years ago, when an outbreak of AI led to a panic in the poultry industry in Virginia's Shenandoah Valley—millions of broilers and turkeys were bulldozed, a more straightforward word for "depopulating" flocks of thousands in a last-ditch effort to prevent the spread of devastating infectious disease. I made no changes whatever in the care of my flocks because of the AI scare; neither did two sustainable farmers of my acquaintance in the Shenandoah, Joel Salatin and Timothy Shell, with flocks far larger than mine. They were *already* managing in ways most likely to keep their flocks disease-free, and sailed through the crisis with no problems.

External parasites—lice and mites—weaken chickens by feeding on their blood. Standard advice for getting rid of them is treating with an "approved, medicated" (read "toxic") powder. As noted previously, though, chickens with adequate opportunities to dust-bathe almost always take care of the problem entirely on their own. Note, however, that an external parasite not prevented by dust-bathing is scaly leg mite, a microscopic mite that burrows into the skin between the scales on the shank, causing the scales to thicken and crust over. You can prevent them on your birds by adding red or yellow cedar chips to the nestboxes. (To treat, see page 213.)

As for internal parasites—various roundworms (nematodes), flatworms, protozoans, and particularly the coccidia that cause coccidiosis—rotating the flock over good pasture as noted previously is the best prevention. A completely confined flock, or one kept continually on the same ground, is more likely to suffer from internal parasites.

For maximum health of ranging flocks, be sure they always have access to shade in summer, and plentiful access to fresh water. And keep an eye out for foreign objects the chickens might mistake for grit and ingest, such as screws, small nails, pieces of glass. Such items if ingested can pierce the crop or wear a hole in the gizzard, with lethal results.

Illness versus Injury

The recuperative powers of a chicken following even quite serious *injury* are astounding. They are extraordinarily capable not only of healing themselves, as did the young broiler with the broken thigh bone, but also of resisting septic infection while doing so, as did the hen who recovered from the gaping spur wound.

On the other hand, where infectious *illness* is concerned, chickens pretty much have two settings: "On" and "Off." Once a chicken has become ill, the chances of recovery—while not impossible—are so low it makes more sense to cull immediately. Which is to say: "Sometimes the best medicine is the hatchet."

The Best Medicine

If you want to treat a sick chicken and try to nurse it back to health, I wish you success. Certainly, if you have a flock of half a dozen hens, the loss of one will be a proportionately greater loss than in a flock of dozens. But the theme of this book is husbandry of the productive small-scale flock, and therefore using poultry as a serious partner for production of your own food; that is, management decisions look to the future as well as the present.

Seen in that perspective, my decision to cull a diseased bird from the flock immediately is hardly as callous as the reader may at first assume. My concern is not only to provide as healthful and enjoyable a life as I can for today's flock but to help ensure the well-being of all its future members as well. Especially for someone who breeds his own stock as I have so often done, is it kindness or cruelty to rescue a sick bird and thus pass on her genes to future progeny?

What Is Disease, Anyway?

If we understand disease as *caused* by germs, then buying some miracle drug to kill the germs makes good sense. And that is plausible enough, until we reflect on the fact that the infectious agents associated with coccidiosis or Marek's disease are universally ambient—that is, present anywhere chickens are—like the virus for the common cold among *Homo sapiens*. So, if they *cause* the infections they're associated with, seems to me all the chickens would be *dead*.

It makes more sense to me to think of disease as an *imbalance* in the organism, an imbalance that changes the default state—near-perfect health—and gives the disease agent its opportunity to proliferate. So the problem is not the germ but a failure of the conditions that support the *default* abundant good health of *Gallus*—problems with the quality of feed, undue stress, or chronic unsanitary housing conditions: damp, dark, with accumulations of poop unable to enter the natural cycles of return.

One way of expressing this idea is the proverbial "three keys to success," in this case: *prevention*, *prevention*, and *prevention*. But I'm not entirely comfortable with that summary. I don't want to be obsessed with *prevention*—of *disease*. My primary goal is really *promotion*: promotion of the conditions that support the natural abounding good health of our chickens. The distinction is between focusing on *keeping bad things from happening* and *helping good things happen*. The three keys to success then become: keen observation of the birds ("The best medicine is the eye of the flockster"); the robust good health and natural immunity of the birds themselves; and husbandry practices that support rather than compromise natural good health.

It's a truism that we should "treat the disease, not the symptom." But if disease *is* the symptom, our challenge is to treat the true *cause*—that is, whatever in the environment or our management is getting in the way of the natural good health of our flock.

Treating Illness or Injuries

I started this chapter with some "bad news"—here's some more: It is common for beginning flocksters to think first of taking a sick chicken to the veterinarian. If you are so inclined, give your vet a try. But in my experience local veterinarians have no interest and little expertise in dealing with chickens. The experts are likely to be poultry scientists in your state's extension office or ag schools, but their focus is on commercial flocks. If it seems that your flock has a general infection, as opposed to the rare unexplainable death, you should seek their help, through your local extension agent.

As for trying yourself to cure a chicken with an infectious disease (which, truly, is not likely), it is imperative to isolate it immediately—do not run the risk of contagion spreading in the flock. Further, the bird itself may have little chance of recovery if left with the others: Chickens have an instinct to eliminate sick or weakened members from the flock, and they can be ruthless and utterly unsentimental about doing so. Isolation gives the bird's natural recuperative powers the best chance to come into play. If it does heal, return it to the flock under close supervision to make sure it is not unduly harassed while reestablishing its position in the flock hierarchy.

If a bird is injured, it should also be removed to its own quiet spot to recover. Yes, I know I said earlier I left a young broiler with a broken thigh to manage on his own. I was lucky in that case—such luck would *not* be the norm. Other flock members will zero in on an open wound especially. If a wound is not too serious, however, I sometimes smear it with a thick salve strongly scented with balsam of fir, and leave the bird with the flock. The dark salve hides the color of blood in the wound, and the scent repels curious fellow members of the flock.

A wound can be cleaned with hydrogen peroxide. I have never applied any sort of bandage to wounds and have never had an injured but otherwise healthy bird succumb to septic infection.

Dust-bathing, as mentioned, prevents infestations of external parasites. If on rare occasions you do have to take temporary action yourself to rid chickens of lice or mites, diatomaceous earth—applied liberally by hand—has always worked well for me. I have also used finely powdered garden lime. *Wear a good dust mask.* As said before, dust-bathing will not prevent scaly leg mites. Still, on the rare occasion when my chickens have been infested with them, I've found them easy to treat: I make a mix of mineral oil, Vaseline, tea tree oil, and oil of oregano, and scrub it vigorously on the entire shank and foot, using an old toothbrush and wearing disposable latex gloves. Repeat after a few days.

And don't worry about the precise recipe. It would be sufficient just to coat the shanks liberally with Vaseline or mineral oil, thereby smothering the mites—the antimicrobial qualities of the tea tree oil or oil of oregano just make for a stronger assault on the entrenched critters. Make up a thick, oily goo of the mineral oil and Vaseline and add enough of one or both oils—a dropperful or two—to strongly scent it, and your mix will be completely effective. These mites are not difficult to kill.

Not long after the second application, the unsightly old scales slough off, and the hens' shanks shine with pristine new skin.

Assisting Natural Immunity

Far from trying to put our flocks in a bubble to isolate them from disease organisms regarded as "the enemy," we should recognize the benefits of exposure to disease organisms in the environment. For example, you could prevent coccidiosis in the flock with vaccination; or by feeding coccidiostats through the entire brooding period; or by ridding the environment of coccidia entirely by cleaning out brooders between batches, sterilizing, and fumigating. A natural means of dealing with the threat, though, is simply to *expose chicks to coccidia in the environment*. While it's true that massive exposure can overwhelm the chicks' immune systems (especially if they are weakened by stress), exposure to levels of coccidia normally present where chickens have been raised for any length of time actually stimulates resistance in the growing chicks. Just as children's immune systems become stronger by being *challenged* by germs in the environment—as opposed to an attempt to keep them in a bubble—so chicks' immune systems develop resistance to cocci by exposure to them. Chicks "learn" resistance in the same way by exposure to other pathogens as well—those of Newcastle disease, chronic respiratory disease, infectious bronchitis, infectious coryza, infectious bursal disease.

Be prepared to encounter advice to avoid mixing age groups in your flock to prevent infectious diseases. When were age groups *not* typically mixed in past eras of more-natural home and farmstead poultry keeping? So long as the adults are healthy, we could think of exposure of chicks to pathogens normally ambient among the adults as "natural vaccination."

I say again, unapologetically, that I never doctor a chicken who seems sick with an infectious illness. The fact that an individual succumbed to disease, while her sisters receiving the same care and facing the same environmental challenges did not, is the best possible reason for culling her from the flock. It is easy to say we plan to manage the flock "naturally," but natural management means adopting nature's sternest measures where infectious disease is concerned. Disease is nature's hoe for weeding out weaker individuals—meaning that over time more individuals in the species will be robust and naturally resistant to disease.

A good illustration of this "bio-logic" is Marek's disease, universally present where chickens are raised. Apparently *Gallus* has already made progress developing complete immunity to this viral disease—some individuals have an inherited immunity, though others are susceptible and will succumb if under stress or heavy exposure. We can help the advance of natural immunity to Marek's in *Gallus* by culling all susceptible individuals who develop the infection. Looked at in this way, vaccination of whole flocks might be the *worst* choice: It confers immunity to *all* potential breeders and thus masks the carriers of immunity and of susceptibility alike, making it less likely that the flock as an ongoing entity will increase natural immunity over time. To be sure, there may be exceptions to the previous bio-logic: You may need a vaccination program, for example, if you exhibit your birds; if serious diseases have occurred on your place, either under your management or that of a previous owner; or if there is a concentrated potential disease incubator close by, such as an industrial chicken operation. *In no case should you vaccinate against diseases you don't have good reason to think are threats to your flock.* With that caveat, it seems to me that the more we use vaccines, antibiotics, or other artificial means to prop up whole flocks that do not have natural immunity, or are too stressed to express it—the more we ensure that future flocks will exhibit weaker immune systems and *require* such supports.

CHAPTER 23

Managing Your Flock in Winter

Will winter for your flock be a time of inactivity and stressful confinement? Is it necessary to say good-bye to opportunities for natural feeding until new growth starts next spring? Or for getting some work done in preparation for next year's garden? Sure, your hens will lay fewer eggs for you in winter, but are there ways to encourage them to lay not so *many* fewer? Might winter be just the time for shaping your flock to fit your plans for the coming breeding season? Read on.

Culling for Winter

Without the availability of as much free food as during the green season, feeding nonessential members of the flock in winter doesn't make sense—costs for purchased feeds will be higher, and amounts of natural feeds available on the small farm or homestead will be reduced per flock member. The approach of winter is thus the time to cull the flock down to its mission-capable minimum. Core size in our flock has always been determined by my plans for the coming breeding season; the number of hens I estimated would keep us and a few friends supplied with eggs, factoring in the usual winter decline in laying; the aging of layers who would not be as productive next year; and the number of reliable broodies needed to hatch new stock the following spring. Thus in my final fall culling, I butchered all males I would not be using as breeders in the next breeding season, older females of declining productivity, and any individuals with undesirable flaws. Typically I retained about three dozen chickens (two dozen layers and about six each breeding cocks and reliable broodies) and—when I kept waterfowl—three ducks with two drakes, and two geese with one gander.

Unless you simply practice a "two years and out" cull of your laying flock, you will need to discover "who's laying and who's lying" among your hens. One option is *nest-trapping*—the use of nestboxes with doors that fall or swing shut when hens enter to lay. (See chapter 26 and appendix A for more information and construction details.)

There are also physical clues that help to determine productivity of individual hens—all a result of increased hormone levels associated with egg laying. If you do not use trapnests to determine layer productivity, it is worthwhile to learn to read these signs.

- The vent will be large, oval, soft, and moist (as opposed to dry, puckered, and round).
- The greatly increased size of the ovaries will round and extend the abdomen between the tip of the keel bone (breastbone) and the tips of the two pubic bones (rising to sharp points on either side of the vent). When palpated, the abdomen will be large

and soft. The distance between the pubic and keel bones should be about four fingers, at least three, pressed down flat and side by side.

- The pubic bones of a productive hen will be wider apart and flexible to accommodate passage of the egg. You should be able to fit at least two and ideally three fingers between the tips of the pubic bones.

Those three are the surest signs that a hen is laying well. Other indicators that are not as reliable but are worth noting: Comb and wattles in a high-producing hen will be larger, softer, brighter, more pliable (as opposed to small and pale)—in a single-comb breed, the comb may even lop over to one side. The nonlaying hen will actually look nicer than the productive one: Since the latter is putting more resource into making eggs, her plumage will look more worn and disheveled and her shank more bleached—the nonlayer's plumage will be sleeker and more pristine, and her shank more intensely colored.

If you are serving a market for eggs, the calculation about culling is more complex. You might delay your heavy culling of older hens until the spring to keep winter egg production up, though in that case you invest more feed per dozen eggs produced. On the other hand, if you sell through farmer's markets, most of which are closed in the winter, the off-season might be the best time to cull unproductive hens. If you supply eggs through a CSA (community-supported agriculture), your customers should understand that the weekly "basket" you furnish depends on what is currently being produced.

Managing Winter Laying

Any sudden or extreme weather fluctuation can cause a temporary dip in egg production—a heat wave or a storm, for example. But there is one fluctuation that brings a decrease in laying for an entire season—winter.

All breeds reduce egg production in winter, some more than others. But remember, it isn't natural for the hen to lay *at all* in winter—domestic hens lay in winter only because selective breeding has unmoored their ovulation from a seasonal reproductive cycle—so we should be grateful for *any* eggs our hens give us in their off season.

Reduction in rate of lay in winter has two causes: The *molt* involves the replacement of every feather on the hen's body by shedding all the old feathers and growing new ones, as discussed in previous chapters. Since feathers are almost pure protein, replacement of her entire set requires a *lot* of resource for the hen—no wonder she takes a break from laying to accomplish the task.

The other reason rate of lay declines in winter is the decrease in day length: For the undomesticated *Gallus* hen, egg-laying was exclusively about reproduction—the decreased photoperiod must still be to some degree nature's signal that winter is not the sensible time to hatch chicks. Selective breeding has partially defeated that inhibition, but it still manifests as reduction in laying.

Not surprisingly, techniques have evolved to further manipulate the reproductive biology of hens to induce them to lay more in winter. The practice of controlled or forced molting is practiced on most industrial laying flocks. The birds are starved for a period ranging from five to twenty-one days, either by complete withholding of food or by the feeding of nutritionally deficient feeds. Water may be withheld for briefer periods as well—together with light manipulation and feeding of drugs, hormones, and metals such as dietary aluminum and zinc. This program of "shock and awe" in the layer house has the intended effect of reducing body weight of the hens by 25 to 35 percent, which induces a sudden and short-term molt and an earlier return to laying. Not only is the procedure highly stressful, but it induces a greater susceptibility to the spread of salmonella in the force-molted flocks and an increased risk of contamination in the supermarket egg supply.

Though there are home versions of forcing molt that are not as extreme as in commercial flocks, to my mind there is no place in small-scale poultry husbandry for the inducement of such high levels of stress. On the other hand, increasing the amount of protein in the diet will increase the speed at which the hen can regrow her feathers and return to normal laying. We noted improved rate of lay in our flock in winters when we supplemented protein by feeding from the vermicomposting bins, or when they had access to a deeply bedded winter feeding yard rife with detritivores (see later in this chapter).

The other approach to increasing rate of lay over winter is the manipulation of apparent day length through use of artificial lights in the henhouse. That is, if the hen's circadian rhythms respond to the decreasing day length of winter by decreasing egg production, we can use supplemental lighting to fool her system into thinking that longer days have returned.

I do not use artificial lighting in winter—it seems ungenerous to push the hens toward greater production when they have perfectly good reasons for cutting back. However, if your layers are well fed and in good condition, a program of supplemental lighting to increase production will not harm your flock. Note you do not need much light to do the trick—a 25- or 40-watt bulb should be sufficient in most small-scale henhouses. One of my correspondents uses a 10-foot (3 m) string of the tiny Christmas icicle lights for softer and more uniform lighting in his coop.

Note as well that the supplemental lighting should be set to a regular schedule, not turned on and off randomly when you think about it. The usual recommendation is to provide artificial light to bring the total apparent day length, natural and artificial light combined, to fourteen hours. The most convenient way to do so is to put the artificial light on a timer. The best strategy is *not* timing the additional lighting in the evening hours—the abrupt darkness of a timer that shuts off after dark catches the hens unawares, and unable to get onto the roosts. Instead, set the timer so that the light comes on in the wee hours of the morning.

Though chickens are hardy in temperatures far below freezing, low temperatures can create some problems in the henhouse. If eggs remain in the nest too long in low temperatures, they will freeze, then crack as contents expand. It takes temperatures lower than the freezing point of water to freeze eggs—28°F (−2°C) or lower. In my experience even at temperatures in the low 20s, eggs do not readily freeze in the daytime. In any case, collect eggs more frequently in winter, especially if temperatures dip into the teens, and make certain none remain in the nests overnight. If an egg does freeze, you can thaw and cook it right away—either for your own table or as a special treat for your pets.

As discussed in chapter 9, your flock needs water in the winter as much as in the summer. If your hens are deprived of water by freezing of the waterer, egg production will decline drastically.

Though I offer supplemental oystershell and granite grit free-choice in the summer, doing so is mainly insurance that the flock will have plenty of calcium for eggshell formation and grit for the gizzard. I don't remember my grandmother offering her flock either, and they took care of these needs with what they picked up on their own, roaming a 50-acre farm. For the flock confined to winter housing, however, it is essential to offer grit and shell (or another calcium boost such as aragonite or feeding limestone). A hopper mounted on the wall high enough to keep it free of litter scratched by the chickens is the most convenient way to offer these supplements for self-serve intake.

Give some thought as well to the mental health of the flock, if confined to winter quarters. Chickens get bored when confined to a relatively small space. Entertainment is to be had first in a fluffy deep litter in which they can scratch to their hearts' content. The birds also enjoy pecking at special treats hung up in the henhouse—mangels, cabbages, alfalfa or barley hay in a mesh net, bunches of nettles, and saved seed heads of small grains, sunflowers, or sorghum.

Winter Feeding Strategies

For most of us, winter weather is the biggest challenge to keeping an array of live, natural feeds available to our flocks. Some simple strategies can help. *Sprouting* represents a kind of value-added feeding we can practice to increase the enzyme, vitamin, and protein content of any seeds we grow or buy to feed whole. In my area *dandelion* and *yellow dock* stay green deeper into the chill than any other forage weeds—as long as I can get a spading fork into the ground, I dig these plants by the roots and throw them by the bucketful to the winter flock. They make short work of the green tops, and the chickens eat part of the dandelion roots as well, though not the tough roots of the dock. The roots of both species get turned under the litter by the chickens' scratching, where they sprout new growth—similar to the forcing of Belgian endive from roots in a cellar. When they surface again, the chooks have second helpings. Since our greenhouse—at 20 by 48 feet (6 by 14.6 m)—has more space than needed for winter greens for Ellen and me, I grow *cut-and-come-again greens* for the flock. And a final strategy: For me, winter feeding includes some of the thriving populations of *earthworms* from my vermicomposting bins—for higher egg rate and quality than would be typical during winter.

Remember that *fat hens do not lay well,* so ensure your hens do not get too fat. Fresh green forages and plenty of exercise are great antidotes to getting too fat in winter. Be careful especially not to feed too much corn, which is high in carbohydrate. Feeding some oats—higher in fiber and lower in carbohydrate—is a good alternative, especially if the grains are scattered over the litter to encourage more active scratching.

It is natural and desirable that hens put on some fat in preparation for winter. But it is important to learn how much is just right. Examine hens by hand, keying on the pubic bones, beneath which are the major fat deposits. Pinch one of the protruding pubic bones: If the fat surrounding the bone is half an inch thick or more, the hen is far too fat. If it is extremely thin—if you feel essentially just the bone itself—she is far too thin. If the fat layer around the bone is an eighth to a quarter of an inch thick, the hen has enough fat for good condition but not so much as to interfere with laying.

The Garden's Overwinter Cover Crops

The cold-hardy cover crops I grow through every winter—crucifers, small grains, peas—make good cut

Figure 23.1. Cold-hardy cover crops over the entire garden can furnish fresh green forage every day of our mid-Atlantic winter.

'n' carry green fodder. I cut daily, starting with the crucifers, most of which succumb when the ground freezes solid. In the bleak days of midwinter, I harvest the grain grasses, wheat and especially rye (not oats, however, as they winterkill as well when the ground freezes). For a while, cuttings come out of the "bank" of garden greenery in place when falling temperatures and shorter days shut down active regrowth. But as days lengthen following the solstice, the cover crops resume growth. My winter forage strategy works for us partly because we don't get a lot of snow. If you live in a zone with a lot of snow, of course, you will have more challenges reaching cold-hardy forages buried beneath the snow, even if they survive the low temperatures.

The Winter Feeding/Exercise Yard

There is no better provision for the winter flock than a heavily mulched winter yard.

Use organic residues on your place or that you can get elsewhere—anything a compost heap would eat. Capture the fertility in the birds' manure while preventing its runoff as pollutants to groundwater. If the mulch is heavy enough to prevent freezing, the continued activity of animal life such as earthworms and slugs at the soil line remains a protein opportunity for the flock.

As discussed in chapter 2, in some winters the feeding yard for my mixed flock was on one of two garden spaces: behind the greenhouse or in front of the henhouse, both deeply covered with organic litter. At the end of the winter, with the garden already covered with made-in-place mulch/compost, I was ready to start the gardening season at full stride.

Some flocksters have mobile shelters large enough to simply park over winter on a garden space where they lay in a deep litter and let the birds do the advance work on soil preparation for the coming year. A sizable hoophouse can also be assembled for the overwinter flock on a garden or new-ground space and disassembled and stored during the growing season.

Figure 23.2. A deeply mulched yard gives a winter flock access to live animal foods at the unfrozen soil line.

A Winter Salad Bar

Leave room for serendipity in your flock management. One winter I exhausted all the sources of fresh greenery discussed previously to feed my ducks and geese and was desperate for more. Suddenly I noticed a large patch of lush green winter rye near the main poultry house. The area had worn bare because of poor management on my part, and I had sowed it to rye simply because I hate leaving ground bare over the winter. The rye had established well, then gone dormant—though not actively growing, it was thick and green. It was the work of an hour or so to set an electric net fence around the patch and move my two breeding trios of geese and ducks onto it. Because the stocking density was so low, I was able to keep the waterfowl on the rye full time until it began growing again in the late winter. I moved the birds elsewhere until the rye matured, then returned the ducks and geese, along with the chickens, to the area to till in its cover. The chickens showed no interest in the mature seed heads—gobbled down greedily enough by the waterfowl—but busily tilled in the plants themselves and prepared the ground for a planting of buckwheat.

One success calls for another, so the next fall I sowed that same ground again as a winter salad bar. This time, however, I sowed barley and wheat in addition to rye. I also sowed a mix of crucifers—mustards, turnips, rape (which all have a "cleansing effect" for fungal diseases, and offer excellent grazing)—and alfalfa. During the winter, whenever there was insufficient fresh greenery from my garden cover crop beds, I released my waterfowl breeders onto that patch to graze. I never put the chickens on it full time but allowed them a taste late in the day, when the coming of dusk summoned them back inside the henhouse before their scratching damaged the winter cover.

The following spring the alfalfa grew slowly, but without serious competition from weeds, in the shade of the more rapidly growing grains as a nurse crop. Once the flock harvested the mature grains, however, the alfalfa came on strong in the additional sunlight. From that stand of pure alfalfa I made a little alfalfa hay for feeding the flock next winter. The alfalfa of course fixed nitrogen in the soil, enriching it for any planting to follow.

A Preseason Sweep of the Garden

The actively growing garden is not compatible with free-ranging chickens. However, chickens can do useful prep work in the garden during the preseason.

If parts of the garden have been under a protective mulch for the winter, the birds get a big protein boost while knocking back the slug and snail population for months to come, and shred the mulch, softened by winter rains, to speed its decomposition to feed the soil.

Other parts of the garden will have been under cover crops for the winter—grains, crucifers, or peas. Less cold-hardy species will have winterkilled and collapsed into the same sort of protective mulch—but one the flockster-gardener didn't have to collect and haul into place. Cover crops that survived the winter serve up green forage as the chooks till them into the soil in preparation for planting.

CHAPTER 24

Other Domestic Fowl

My most enjoyable experiences raising poultry were the years when I kept mixed flocks. Keeping species in addition to chickens meant a broader array of resource use and work skills on the homestead, cooperative warding off of threats from the sky—and new delights at the table.

Throughout the book, I have referred to other fowl species when their needs or contributions differ from that of their chicken cousins. Now let's look at them in more detail.

Waterfowl

If you have "wild water" on your property, keeping waterfowl is a natural choice. But if you don't, it is still possible to keep them happy. There are four *species* of domestic waterfowl, so I will start with observations of this group as a class, then consider them separately.

Waterfowl need plentiful access to *water*. If you do not have a pond, at least give them an opportunity to bathe—to rid themselves of external parasites (as chickens do by dust-bathing)—and for pure *joie de vivre*. You'll know what I mean when you see them taking their first exuberant splash. I provide bathing for my waterfowl, using a stock watering tank, even in the middle of winter. As a far-distant third best, you *must* at least provide waterfowl water deep enough to dip their heads and rinse eyes and nostrils. Note as well that waterfowl may choke on feeds, especially fine-textured commercial feeds, if they do not have simultaneous access to water.

If you keep waterfowl only, their housing can be quite simple. Indeed, if winter temperatures do not drop below 0°F (−18°C), a windbreak of straw or hay bales may be adequate—or an open-front, roofed shed can be used. The same basic principles of housing apply as for chickens: a lot of airflow, deep litter to absorb the droppings, and protection from predators. Allow adequate space in the housing, depending on the size of the birds—always remembering that "adequate" depends greatly on whether the birds just sleep inside the shelter or are confined to it full time, perhaps in winter. For geese, for example, I would advise a *minimum* of 10 square feet (1 square m) per bird in the former case, 20 square feet (1.9 square m) in the latter.

I kept my waterfowl in our main poultry house with the chickens. That worked fine for me, so long as I kept the watering of ducks and geese outside, regardless of the season. In winter I released all my birds, chickens and waterfowl alike, onto winter feeding/exercise yards.

In contrast with wild Canada geese and mallard ducks, who amaze us with their migratory feats, domestic geese and true ducks—bred for greater size and weight—are not capable fliers and are thus easy to confine, even with a fence only a couple of feet high. A possible exception is Muscovies—some flocksters report their Muscovies are forever flying over fences, even roosting in trees at night. I have never had that problem with my Muscovies. True, they proved a little more capable of flight than mallard-type ducks, but they almost never flew over my 42-inch-tall (1.1 m) electric net fencing. Just as with rogue chicken fliers, you can clip wings of flighty Muscovies to encourage them to stay where you want them.

When handling chickens, it is common to catch or carry them by the legs. You must *never* do either with waterfowl of any type—*their legs are easily injured*. To catch waterfowl, I either use a poultry net on a long handle—like a larger version of a butterfly net—or, if in close quarters, simply grab the bird briefly by the neck until I can gather it with hands around the wing shoulders, and tuck it under an arm. Watch those claws—especially in the case of Muscovies, they are *sharp*.

Culinarily, all the waterfowl have the same gifts to offer. Their flavor is rich and delicious. Their fat is easily extracted and as easily rendered for a superior cooking fat. Egg production varies widely among the species and will be noted in the following sections.

Essential for breeding waterfowl is *sexing* (distinguishing the genders)—more or less difficult, depending on species and breed. *Vent sexing* (visual examination of the genitalia), if properly done, gives certain gender identification. Remember that the heavier breeds of both geese and ducks may require sufficient water to swim in for successful mating—the male may be incapable of mounting the female on the ground.

Broodiness among true ducks and geese varies a lot. Broodiness is likely to be strong in most Muscovy ducks (females), however, so they can be used as foster mothers where broody skills are lacking in other waterfowl females. The trait is also likely to be strong in bantam breeds of mallard-type ducks.

So long as they shield the broody females from wind and rain, there is no problem with placing nests outside—after all, wild waterfowl nest in the late winter or early spring when it is still cold. Set up the nests well before the laying season begins, to accustom the females to laying in the nests in which they will brood (see figure 24.1). In contrast to many broody hens, waterfowl broodies cannot be moved once broodiness commences without *breaking them up* (that is, disrupting the "mindset" conducive to a prolonged stay on the nest until their eggs hatch). Onset of broodiness is signaled by the appearance in the nest of a lining of feathers, which the female pulls from her breast, and by her staying on the nest full time. Access to bathing during incubation is a plus—the broody female who bathes during a nest break will take some water back to the nest on her feathers, helping to ensure proper incubation humidity.

In my experience all the waterfowl species are easy and fun to raise. They also make full use of good pasture—I have even grazed my waterfowl flock on four lawn areas around our house. Why mow?

Figure 24.1. I made this outdoor shelter, which contains separate nests for two broody geese. It is inside an area protected by electric net fence, with overwintered rye as a cover crop. Photo courtesy of Bonnie Long.

Geese

Domestic geese comprise two different species, descended from separate wild ancestors. Embdens, Pilgrims, Romans, Toulouse, Pomeranians, and other common breeds descended from the European Greylag (*Anser anser*); while Chinese and African geese descended from the Asian Swan Goose (*A. cygnoides*) (see figure 24.2). An identifying characteristic of the latter two breeds is the large, forward-inclining knob that develops where the upper bill meets the skull, shown close-up in figure 24.3. (Two feral species—the Canada and the Egyptian—have also been domesticated and are kept by some hobbyists.)

Geese are not kept for their eggs—their laying season is short—but the eggs are perfectly edible. Like duck eggs they are especially prized for baking. The best fit for geese in a small-scale flock is as meat birds (and sources of cooking fat) who grow without large feed inputs as long as they have plenty of opportunity to range. They may eat the odd insect now and again, but geese are largely grazers that grow well on good grass and other plant foods: broad-leaved weeds, berries and fallen fruits, seed heads of forage plants. Indeed, once past the brooder phase, they can thrive on such feeds alone—though they will fatten better in the fall if fed grain as well.

Sexing geese is essential if you want to breed them. In most breeds, gander (male) and goose (female) are colored the same; and the secondary characteristics of size, carriage, posture of the neck, and other distinctions often recommended for sexing are *not* reliable. A partial exception is sexing by reference to knob size in Chinese and Africans—the gander's knob grows earlier and larger than the goose's—but that difference is no help in sexing goslings. Even in adults there is so much variation among strains and individuals that

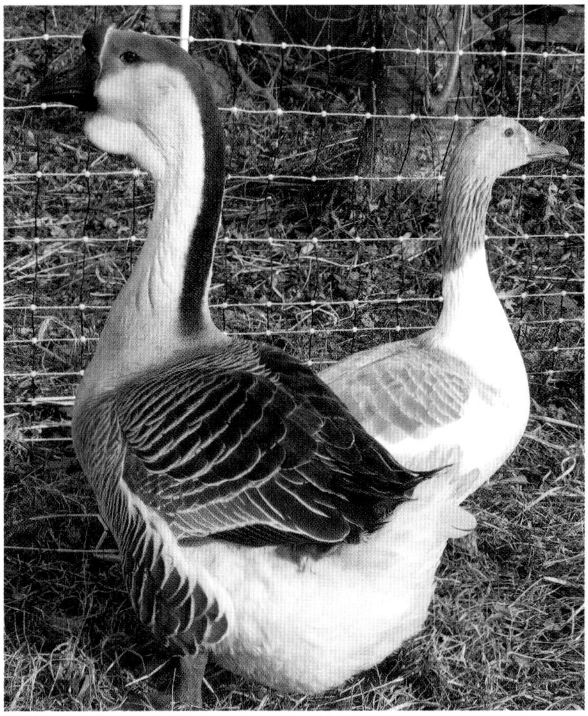

Figure 24.2. Domestic geese descended from the European Greylag are represented by the Buff Pomeranian on the *right*; those from the Asian Swan Goose, by the African on the *left*.

Figure 24.3. Two African ganders.

Figure 24.4. Pilgrim goose family. Pilgrim ganders are almost all white; geese, a soft gray. Even the goslings are autosexing–note the darker female gosling between the two lighter males. Ganders participate fully in parenting. Photo courtesy of Bonnie Long.

knob size will not yield certain gender identification in all cases. Complete certainty is possible only with *vent sexing*—the complete eversion of the cloaca to expose the genitalia. Do note that qualification *complete*. If you fully evert the cloaca of a male, what pops out—an odd corkscrew-shaped penis with little rubbery knobs—makes identification certain: It's a boy! But there is a little "gotcha": If you do not fully evert—and it takes some skill and expertise to do so—the penis will lurk hidden inside the cloaca and give much the same appearance as the genitals of a female.

Note that vent sexing any waterfowl is stressful at best and may even injure the bird if handled clumsily. Learn this skill from someone experienced if possible; otherwise, carefully study directions for this procedure in a reliable guidebook.[1]

A few breeds of geese are autosexing or dimorphic—the goose and gander are distinctly colored, even at hatch. The most available of the autosexing breeds is the Pilgrim. (Other dimorphic breeds exist—West of England, Shetland, Normandy, and Cotton Patch—but are rare.) Do note that the autosexing trait remains useful *only* when breeders take pains to maintain breed purity. I once found, to my dismay, that some cull Pilgrims I was slaughtering—white and blue-eyed, supposedly dead ringers for ganders—in fact revealed tiny egg clusters in place of testicles on evisceration. Crossing breeds is common in commercial goose breeding, and an Embden × Pilgrim cross would be especially likely—the most probable explanation for Pilgrims who were not reliably autosexing—so be certain your breeding-stock source conscientiously maintains breed purity if you want to raise autosexing geese.

I have found geese the most challenging of all the domestic fowl to breed. It took several years of frustration before I finally had a bit of luck. As in wild geese, the gander helps with parenting. Indeed, I found in my Pilgrim family—a gander and two geese—that they *all* shared equally the parenting of three goslings that one of the geese had hatched (see figure 24.4).

Geese are the most long-lived of all domestic fowl. A neighbor of ours had an African gander—Peepie, as known to everyone in our village—who lived to be thirty-five. I read of a goose in England who, at her death at 104 years—in a farm accident, mind you, *not* by natural causes—was still laying, hatching, and rearing a clutch of goslings each spring. Average life span is twenty years or more.

Geese can be pretty talkative, so they may not be a good choice for flocksters who need to minimize noise from their flock. On the other hand, that very trait may be welcome if you would like alert watchfowl on your place.

We have raised Pilgrim, Chinese, African, Pomeranian (Buff and Gray Saddleback), Roman, and Embden geese. All have been fast-growing and easy to raise. Geese are also chock-full of personality and great fun. Be warned if you adopt them into your productive farm or homestead flock, though—those very qualities make slaughter day more wrenching than is the case with any other domestic fowl.

Ducks

All true ducks share the same ancestor—the wild mallard (*Anas platyrhynchos*)—though as noted previously they have been bred for greater size and weight and have largely lost the ability to fly.

You may prefer to raise ducks primarily for eggs. As with chickens, selection for laying has resulted in ducks that lay much of the year, not just during a narrow breeding season. You may be surprised to know that the most productive laying ducks—especially Campbells and Runners—lay more eggs than the best layer breeds of chickens, from 250 to 325 eggs per year or more. It may be useful to know as well: A small subset of people are intolerant of chicken eggs, and experience various digestive problems after eating them. But many of those people find they can eat duck eggs without difficulties.

Heavier breeds such as Aylesbury, Pekin (the fastest growing of all duck breeds), and Rouen lay many fewer eggs and are raised primarily for meat.

As with chickens, there are dual-purpose breeds that perform well as both layer and meat birds—Saxony, Swedish, Orpington, and Magpie. My favorite of all dual-purpose mallard breeds is the Silver Appleyard—or simply Appleyard—which offers the best combination of beauty, egg laying, and fast growth to good slaughter size. While not in the same league with Muscovy or bantam-breed ducks as broodies, Appleyard females are more likely than most of the heavier breeds of mallard ducks to retain the broody trait and make good mothers.

Sexing adult mallard ducks is typically easy. Among many breeds bill color is different in ducks from that in drakes. The drake is noticeably brighter and showier among particolored breeds than the more subdued duck. Except during molt, the drake has distinctively curled feathers on the tail, which are lacking in the female. Vocalization is distinctive as well—the male's is a lower-pitched, throaty *queeg-queeg*, while the female's is the brassier *quack-quack* we commonly associate with ducks.

Sexing ducklings can be more challenging. In some breeds differences in bill color between duckling and drakeling are definitive. Vent sexing as for geese is recommended if you need to be certain of gender.

Figure 24.5. Many Appleyard ducks retain the broody trait and make good mothers. Photo courtesy of Bonnie Long.

Figure 24.6. Appleyards are among the most beautiful of all ducks. The female's coloration is more subdued, while the two drakes exhibit the green head and more showy plumage typical of many breeds of mallard or "true duck" males. Look closely to see the curled tail feathers of the drakes, another key to his gender. Photo courtesy of Bonnie Long.

Ducks will make good use of all the green forage you can give them, but as grazers they are not in the same class with geese. They are, however, voracious eaters of worms and insects and can be particularly useful for controlling slugs and snails.

Muscovies

Though referred to as "ducks," Muscovies have an entirely different wild ancestor, *Cairina moschata*—considered by some more closely related to geese than to true ducks. They can mate with mallard-type ducks, though the offspring are likely to be sterile, like mules. In southwest France, Muscovies and Pekin ducks are crossed to produce prized *Moulard* ducks. Though they are comfortable in water for brief periods, their plumage is less water repellent than that of mallard ducks, so their needs for shelter in extreme weather are greater. Though like geese they are active grazers, like mallard-type ducks they also eat an abundance of animal foods such as insects, worms, and slugs. One of my correspondents told me of flocks of between fifty and a hundred Muscovies on her parents' farm who earned their living entirely through their own efforts in the growing season and were only fed in winter.

Flocksters concerned about the noise level of their flock should note that, in contrast with both geese and ducks, Muscovies are almost entirely mute.

Muscovies are a prized meat breed. Egg production is moderate—somewhere between 60 and 180 eggs per year depending on management and climate. The females—as said, among the best natural mothers of all waterfowl types—may brood more than one clutch of ducklings in a season. If I had to choose a single rugged, self-sufficient, easily reproducing waterfowl type, my choice would probably be the Muscovy.

Sexing of adult Muscovies is no problem, not only because of differences in size—the adult drake is almost twice the weight of the duck—but also the greater size

Figure 24.7. Muscovies are not true ducks. Like geese, they are excellent grazers who make good use of pasture. The drake is considerably larger than the duck, with larger face patches.

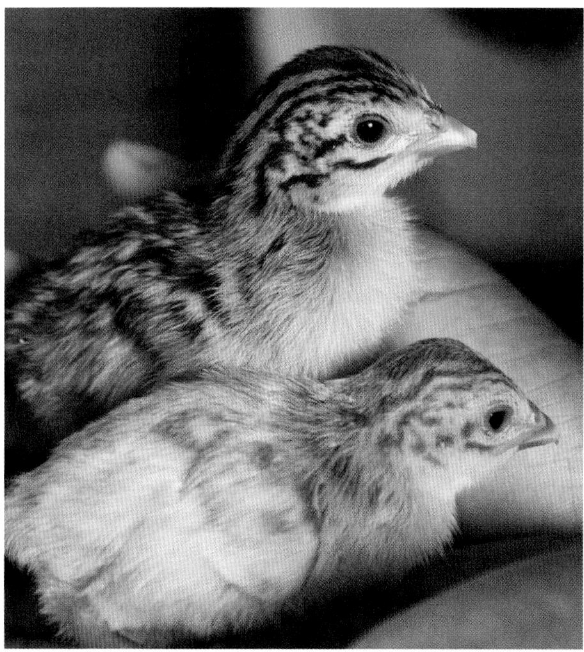

Figure 24.8. Guinea keets. Photo courtesy of Bonnie Long.

of the bare skin patches or *caruncles* on the head and face of the drake. Ducklings must be vent sexed for complete accuracy.

Guineas

Guineas or guineafowl evolved in the central and western plains of Africa and are thought to have been domesticated four thousand years ago by the Egyptians. Both the Greeks and the Romans prized them as table fowl, and the Romans carried them into their expanding empire.

Guineas go feral more easily than perhaps any other domestic fowl, and may need encouragement to fix on the poultry coop as home—as opposed to setting up housekeeping in the woods, roosting in trees. A flockster friend of mine set an open platform on a pole for his few guineas to roost on, thereby accommodating their desire to roost well off the ground while keeping them close by. A further refinement would be to put a roof and wire mesh sides over the platform, complete with a door you could close at night if you wanted to get a hand on the guineas.

Of course, it is possible to "go with" their feral capabilities: I visited a farmer not far from me who assembled an enclosure of metal dog kennel panels, including a panel over the top. He kept the water and feed troughs for his large flock of guineas inside the kennel—but otherwise allowed them to "live wild." Any time he needed to catch some, he just shut the enclosure's door while they were eating. Or sometimes he just "harvested" a couple for dinner with a .22 rifle.

Guinea hens have a reputation as terrible mothers—they may carelessly allow their babies (keets) to get wet in the early-morning dew, chill, and die. Others report success using guinea hens to hatch and raise keets—and that they often nest and rear the young communally. I have always played it safe and set guinea eggs under broody chicken hens. It is said that guineas grow up tamer if hatched and brooded by a chicken mother.

Guineas ranging outside hide a nest so well it is extremely difficult to find. If you're lucky you may see a cock guinea standing guard while the hen is laying her egg, giving a clue to where the nest is hidden. When you do find it, leave some fake eggs when you gather the eggs—otherwise, the guineas will make an even more securely hidden nest elsewhere. Laying is seasonal, from around mid-May to mid-September here.

Guinea eggs are excellent. As with any smaller egg, the yolks are larger in proportion to the whites, making for richer omelets and scrambled eggs. The eggs are wedge-shaped in profile, pointed on one end and bluntly rounded on the other, and are softly banded in earthen colors or finely speckled—a basket of them is a striking centerpiece on the table. They are extremely hard-shelled, and fun for practical jokes: An unsuspecting guest asked to help break the eggs for omelets *will* conclude that you have substituted ceramic eggs for the real thing!

"Guinea hen" is considered a gourmet item in some quarters. The meat is all dark with a hint of game in the flavor. Younger guineas are usually preferred for the table. I do not find guinea any better on the plate than our pasture-raised, traditional-breed chickens—both are superior by orders of magnitude to supermarket chicken.

Cocks and hens are colored the same, making gender selection challenging. You may read advice about sexing adult guineas by comparisons of body size, size of the "helmet," and the like, but such physical characteristics vary widely enough to make them unreliable for determining gender. Still, there are two foolproof ways to sex guineas:

Vocalization: It is said that the distinctive two-note call of the guinea is *buck-wheat! buck-wheat!* or *come-back! come-back!* (or, according to my father, *pot-rack! pot-rack!*); but actually that two-note call is the call of the guinea *hen*. The cock's cry is a harsh staccato shriek, *chi-chi-chi-chi-chi!* Be aware, though, that guinea hens will make the same cry as the male if alarmed; so the trick is to listen for distinguishing vocalization when the flock is relaxed and on routine patrol.

Figure 24.9. Sharp-eyed guinea on the prowl.

Tidbitting: If you throw a special treat such as a cricket to your guineas, guinea cocks exhibit the same gentlemanly behavior as chicken cocks: Rather than gobbling up the special treat himself, the guinea cock calls one of the hens to come and eat it instead, maybe even passing it to her with his beak.

The caveat to both these methods for sexing your guinea is that now you have to catch it!

If you're looking for watchfowl, guineas are hard to beat. However, I have raised some groups of guineas who had something to say only when alarmed, while others kept up a shrieking chatter *all* the time. You'll have to decide for yourself whether noisy guineas fit into your situation.

Guineas are among the best of all fowl for insect control. I have read that a pair will keep an acre entirely free of ticks. As I described in chapter 13, I use guineas for control of squash bug, nemesis to squash family crops in my area.

Guineas can become major players on the landscape when grasshoppers swarm to biblical-plague proportions. A correspondent in the Midwest reported seeing a flock of a hundred guineas in a line, taking out great swaths of grasshoppers down a field, then reversing to take out a swath in the other direction. An aficionada from Texas got her start with guineas when she faced a similar infestation. Since that time, she has more than once had the experience of seeing the landscape around her as 360 degrees of brown, stripped clean to dirt by grasshoppers, with her own property a guinea-protected green emerald in the center.

Turkeys

Turkeys are a fun addition to the small farm or homestead. A friend's big tom turkey follows the family around like a dog, gobbling and displaying importantly, endlessly entertaining (see figure 24.10).

Figure 24.10. Narragansett male turkey displaying.

Figure 24.11. Standard Bronze hen turkey.

Turkeys also have a lot to offer in terms of homestead self-sufficiency—I've seen how my same friend's turkeys glean insects and other free feed on his place.

Unfortunately, I have had dreadful luck trying to raise day-old poults bought through the mail. On my first attempt I lost every last one I ordered. On the second I did a little better, with fifteen surviving out of an attempted nineteen. So I can only be honest with the reader that my sad experience bears out what a lot of flocksters find—that turkeys are fragile and die readily in the brooder phase, though they are tough as nails after that. (The same is true when starting guineas as purchased keets.) I would bet that poults would do better with a real mother—maybe a good broody chicken hen as a foster mother—than with any poor effort I can make in a brooder.

Once into the tough-as-nails phase, turkeys can roost outside in even the harshest weather, with just enough cover to keep the rain and sharpest winds off them. Remember how vulnerable to predators they are when asleep, and give them adequate protection at night.

A major question for the small-scale flockster considering turkeys, particularly one who wants to keep chickens as well, is *blackhead*, a usually fatal disease. Blackhead is caused by a parasitic protozoan, *Histomonas meleagridis*, likely to be present almost anywhere chickens are raised. Symptoms among turkeys include lethargy, retracted neck, drooping wings, suppressed appetite, and, especially, sulfur-colored runny droppings (sometimes dry, solid black droppings with waxy yellowish streaks). The head may darken—though usually it does not, so the common name is something of a misnomer.

Chickens themselves are highly resistant to the parasite and rarely suffer ill effects from it—though may succumb to it if chronically stressed. However, chickens are *carriers*—that is, they help perpetuate the complicated life cycle of the protozoan and pass it on to turkeys, who are highly susceptible. For most flocksters, therefore, turkeys and chickens don't mix. If you raise both, keep them strictly separated, and run the turkeys only on ground chickens have not ranged. I have heard from a few flocksters with *small* flocks that they keep turkeys and chickens together without problems of blackhead. If you start with a clean slate where blackhead is concerned, you may be able to keep it that way. The key is whether your chickens have ingested cecal tonsil worms that transmit the *H. meleagridis* protozoan—noninfected chickens on clean land cannot infect turkeys.

PART SIX

Breeding the Small-Scale Flock

CHAPTER 25

Trapnest and Hatchet: Breeding Essentials

Here is a conundrum: For best health, self-sufficiency, and *contentment*, chickens should follow a lifestyle that most closely mimics that of *Gallus gallus*. After all, their "lifestyle" adaptations are what honed *G. gallus* into the state of perfection we admire today. But perfect adaptation to what? To life in the *wild*. To perfection as a *wild* species. Those tough, resilient, and resourceful chickens foraging the edges of the jungle were not, in comparison to modern breeds, especially *productive*. As mentioned previously, the hens laid eggs only during the breeding season. Their carcasses were lean and mean, rather than plump and tender. It was the *breeding* of *G. gallus*, following the alliance between that wild species and *Homo sapiens*, that heightened production over the millennia. To practice improvement breeding is to strive for (in the best case, *perfect*) adaptation to *domestication*.

What are the keys to breeding for the retention of—indeed, the improvement of—*productivity* in future generations of our flocks? Again, consider *Gallus,* the Natural Chicken: In the crucible of natural selection—most especially, assaults by numerous highly capable predators—only the fittest of the fit survived to leave behind progeny. When we breed our own, in this most interventionist of all phases of flock management, it is *we* who determine "the best of the best" in our flock, and who gets to mate with whom. *We* take the place of the predator, eliminating from the flock not only the poor performers but the great middle range of the so-so. However complicated the process of improvement breeding may at first seem, it comes down largely to two simple practices: *selection* and *culling*.

Improvement Breeding: What's It All About?

What does *productive* mean, anyway? If we define *production* narrowly—for example, how much meat and eggs grace our tables in exchange for the time, effort, and expense required for our flocks' care—then we could hardly find better models of breeding for production than those in the large-scale poultry industry. Imagine a fast-growing meat hybrid ready to slaughter in as little as forty-four days to fill supermarket coolers. That's the Cornish Cross, a wonder of the modern world produced in multibillions annually. (A child growing at a comparable rate would weigh 349 pounds [158 kg] at age two.) Industrial layer hens are equally impressive, coming into lay at sixteen weeks old and laying more than three hundred eggs a year. But—*oops!*—these miracles of the breeder's art are bred to fit a production model that is unsanitary, wasteful, polluting, and destructive of agricultural soils; and they require higher inputs of protein in their feeds to fuel that level of performance—and a steady diet of interventionist medications just to survive. Hardly the genetics we look for in the sturdy small-scale flock.

Breeders for exhibition do equally brilliant breeding work. I enjoy attending a big poultry show, but I know most birds I see are more works of art, sketched in DNA, than anything I would want in the rough-and-tumble of my backyard. Many exhibition breeders focus not on production traits but on the fine points of color and pattern, comb and carriage—and win ribbons for such refinements, even if their grand champion hens lay only a couple dozen eggs a year. I have encountered competition breeders as well who have no compunction against crossing in "a little of this and a little of that" to produce their finely honed specimens, even if the result is a muddying of the water in the gene pool.

I hasten to add that my hat is off to competition breeders who conscientiously breed for the *production traits* for which their breeds were originally valued. A good example is Don Schrider, poultry writer and breeder, who wins ribbons with his show-quality Light Brown Leghorns and Dark Brown Leghorns, even while breeding to maximize egg production.

Most of us flocksters depend on commercial hatcheries to do the breeding required to stock our flocks with productive birds. Just how carefully hatcheries breed for production traits we have no way of knowing, however. What is highly likely is that hatchery birds have been bred in a production environment far removed from the conditions of a working homestead flock. And what is certain is that formerly productive breeds often decline in productivity when mass-bred. Good examples are the Delaware and the New Hampshire, breeds formerly valued both as reliable layers of large brown eggs and as table fowl. Indeed, performance in the latter role was so outstanding that each had its turn as the foundation of the broiler industry—before "the Cornish Cross revolution" swept aside all competitor broiler breeds. Until recently, the Delaware declined greatly as a meat fowl and layer and was of interest only on the show circuit. The New Hampshires on offer at most hatcheries should more properly be termed "Production Reds," far removed from Andrew Christie's meaty original, or from the higher-laying strain developed by Clarence Newcomer in the 1940s.

However well or sloppily poultry breeding is being done, by whatever supplier, one thing is certain: None are breeding to *your* specific climate and backyard resources, tailored to *your* goals, using *your* management system. At one time it was common for small farmers and even backyarders to breed their own stock—and they were rewarded, like gardeners who save their own seed, with flocks ever more finely attuned to their own unique situations. Readers who have taken to heart the central theme of this book will not be satisfied with poultry that is *productive* narrowly defined—the payoff in dozens and pounds of eggs and meat. They will wonder as well: How can our birds most efficiently furnish those eggs and meat while at the same time hustling up a lot of their own grub? What useful work can they accomplish if we provide them a life as close as we can manage to that of *G. gallus*? How do we ensure that our flock will be healthier and more resilient with each generation?

It is disturbing to realize that traditional and historic breeds are being lost—as in *forever*. Breeds are not like precious jewels, put in a vault for safekeeping and brought out when desired; they can be conserved only as living animals repeating the cycle of growth, reproduction, and death. But in the long run they will not be conserved by well-intentioned efforts to save them as zoo specimens or a library of DNA. The key to the conservation work encouraged by The Livestock Conservancy (TLC) is its recognition that breeds will be conserved only to the extent the *economic* traits for which they were originally developed are valued and used. A sense of altruism about saving heritage breeds may inspire us to take up their breeding, but only a self-interested *What's in it for me?* will carry the work into the future. And even those who do not choose to breed their own stock can help in the work of conservation and breed improvement—by seeking out and buying from breeders dedicated to those goals rather than from mass-market sources.

It is unsettling to recognize that there is no guarantee that anyone is breeding to maintain poultry breeds at their historic peaks of productivity. And it is a certainty that no one is breeding stock to best fit *our* homesteads and farms. Don't these two bleak realities suggest an obvious strategy: *to breed our own*? And if we do so, wouldn't it feel good to know that we are helping conserve the priceless genetics of the historic breeds? We never know when we'll need them.

As an example, historic breeds such as Old English Game were the forerunners of numerous modern breeds. It would be foolish to assume that the only breeds we need to maintain are those of immediate use in today's circumstances. Who can say when we will need to draw upon the widest range of poultry genetics in response to new challenges and changed circumstances? An example from agricultural crops illustrates that this is not merely a theoretical concern: By 1970 the hybrid varieties that made up the bulk of the American corn crop came from an increasingly narrow genetic base. That year a devastating epidemic of southern corn leaf blight wiped out 15 percent of the US corn crop—50 to 100 percent in some areas—resulting in at least a billion-dollar loss to the agricultural economy. The solution was simple: Introduce resistant genes from older varieties of corn into breeding programs. But imagine the dimensions of the catastrophe if no one had taken the trouble to conserve those earlier varieties.

Bad News, Good News

If your head spins when you hear about alleles, genes, autosomes, homologous chromosomes—be assured mine does, too. Fortunately, you do not need a degree in genetics to understand sound poultry breeding. As in many things agricultural, the best direction for progress could be a giant step backward. Savvy farmers of an earlier time discovered and used techniques for improving their livestock and poultry, and we can still learn from them today. That is what TLC did in 2006 and 2007: Based on principles well established in agricultural texts of the first half of the twentieth century, TLC carried out a breed-improvement project with the Buckeye—an outstanding dual-purpose breed developed in Ohio at the end of the 1890s, which in recent decades had been sadly neglected as a production breed. Indeed, by the time TLC's breeding project began, the Buckeye was almost extinct—there could hardly have been a better test case of improvement breeding. Guided by those earlier selection and breeding criteria, the project, under breeder Don Schrider's direction, achieved in just three years an average increase of 1 pound (453.6 g) live weight at slaughter and a decrease in growth to that weight from twenty weeks down to sixteen. Impressive gains indeed, and an illustration of what is possible in a serious breed-improvement program.

A more recent example is the successful restoration breeding of the much-neglected Delaware by Will Morrow of Whitmore Farm in Maryland. Using the principles of TLC's Buckeye program to identify and select his best performers, he brought his Delawares back up to their standard weights (as defined by the American Poultry Association) in two generations (see figure 25.1).

While the bad news is the sad neglect of many production breeds, the good news is that the field is thus wide open for homestead and small-farm breeders to restore some of the traditional breeds to serious levels of utility—that is, to make them productive enough to efficiently supply the family or small local markets. Within neglected traditional breeds, there is substantial leeway for big gains in productivity in a relatively short time—*if* we understand how to select for the traits we want.

Note especially, for example, that in both Will Morrow's work with the Delaware and in Don Schrider's with the Buckeye, weight gains did not require sacrifice of good egg production levels. To be sure, there are genetic limits on possible gains—at a certain point, for example, continued gains in meat qualities may come at

Figure 25.1. A pair of Will Morrow's Delawares. Photo courtesy of Will Morrow.

the expense of laying capacity, and vice versa. In breeds whose productive traits have been neglected, however, gains in both may be possible.

Start your breeding project with the best stock you can find. If your only source is hatchery stock, work up from there—as Will Morrow did in his restoration work with the Delaware. The better your starter stock, however, the better the results, and the less time and effort required to overcome generations of neglect. Get in touch with conscientious breeders through TLC, the Society for the Preservation of Poultry Antiquities, and the Sustainable Poultry Network. (All are listed in appendix J.)

Thinking through Your Breeding Project

Improvement breeding is of course about breeding superior performers. But isn't it empowering and liberating to realize that it's not some anonymous expert who determines what *superior* means? However rudimentary your knowledge at the beginning, *you* are the expert about selection criteria and breeding goals for your flock.

You might decide you'd like to improve laying performance in one of the traditional breeds, such as Leghorns, Anconas, or Hamburgs, that have been developed to maximize egg production, with some sacrifice of meat characteristics. But if your aim is a good meat bird, it still makes sense to strive to improve egg production as well as meat qualities. After all, a breeding program with a lower average rate of lay requires more hens to provide the number of eggs needed for hatching, resulting in lower efficiency in the breeding project. This incentive toward improving *both* laying and meat qualities may explain historically why so many preferred farm breeds have been dual purpose. It may be that production breeding is most of all about putting the *dual* back into dual purpose.

If you're breeding for meat qualities, which is superior—a bird that grows faster to slaughter weight or one that has better flavor? If you're growing to sell, pricing constraints in a market that demands plump carcasses and is less discriminating about flavor may dictate the former choice. In that case it is probably best to work with breeds that were important in the production of market chicken before the Cornish Cross became the standard broiler choice—for example, Plymouth Rock, New Hampshire, and Delaware.

If you're not constrained to make maximum slaughter weight your priority, dedication to producing truly gourmet poultry might incline you to work with breeds that are slower growing but have traditionally been valued as superior table fowl—Dorking, La Flèche, Crèvecoeur, Houdan.

As for layers, which is superior in terms of your goals and needs—a hen with early onset of lay or one with higher production long term? One with better annual production, or who produces fewer but larger eggs, or who holds her production better in the winter? If you are taken with the beauty of dark-shelled or speckled eggs, are you willing to accept lower production if necessary to favor that trait?

Remember the importance of *breeding the all-around bird*. Don't focus exclusively on production narrowly defined—rate of growth, fleshing, egg production—because productive capacity depends on a high level of health and vigor in the bird overall. We are not breeding drumsticks or cartons of eggs, but birds with a high state of fitness in every aspect. Traits such as rate of growth, disease resistance, early onset of lay, fertility, and hatch rate are controlled by interrelated sets of genes. It is better to choose breeders with a good balance of such traits, in preference to those who excel at a single production trait alone.

It is especially important to eliminate as a breeder any bird who has a weak immune system, as demonstrated by suffering illness of any sort. The chicken's immune response is naturally robust—our breeding efforts should keep it so.

So long as breeders of whatever age, male or female, are meeting the requirements of the breeding program, they may be retained as breeders, to foster longevity (which is almost entirely neglected in commercial breeding) and more extended productivity in future generations. Genes for longevity are associated with genes for vigor and health, so they should be among the most important of our selection criteria. Hens that maintain good egg production as they age are especially valuable for use as breeders.

As much as you can, try to select for high levels of *foraging skills*. If you notice that a particular individual is a real go-getter when it comes to hustling its own grub, you might favor it as a breeder even if he isn't quite as heavy, or even if she doesn't lay quite as many eggs, compared to birds that hang around the feed trough.

In a flock valued as working partners, and maybe almost as members of the family, we want our birds to "fit in"—and that means selecting for *temperament* as well. Only you can define ideal temperament in general, but *all* of us would do well to cull any birds—especially cocks—displaying unprovoked aggression toward humans.

Don't forget the *brooding* instinct as an important selection criterion—whether in your project that means reducing it or increasing it—or, as discussed later in chapter 27, *improving* it. Clearly the trait is heritable—some breeds retain it to a high degree, others almost not at all—even though the specific genes for this complex behavior are unknown. Most commercial breeding seeks to eliminate broodiness among hens, and that could be a reasonable choice for you as well, if you want to maximize egg production and plan to hatch with incubators. But for more self-reliance in your homestead or farm flock, you might well decide you want to encourage the broody trait instead. Of course, a flock featuring broodies exclusively would mean no egg production at all during the hatching season. An elegant solution might be the breeding of both a world-class broody breed such as the OEG to furnish a reliable working-broody subflock and a layer breed such as Light Brown Leghorn to keep egg production up. Expression of broodiness would then be equally crucial for selecting breeders in both breeds: Among the OEGs any hen who failed as a broody would be culled; among the Leghorns broodiness would be selected against by culling broody hens.

I noted in chapter 5 an equally elegant solution: Keeping a breed such as Icelandics with enough reliable broodiness among the hens to cover hatching needs, but with a significant portion of hens having no interest in motherhood who continue laying uninterrupted.

When selecting among broody hens, further refinements are possible. I used natural mothers exclusively for hatching and preferred starting all my new stock in one big wave in the spring. My selection, therefore, favored hens who became broody early, and who did not return to broodiness after their spring clutch. You might prefer a more random expression of broodiness, to spread rearing of new chicks over the green season; or favor hens who brood two or even three clutches of eggs in the season, if you have a lot of hatching to do. As for winter broodiness, it makes no biological sense to me—I culled any hen who went broody in winter.

For years I experimented with crossing OEG cocks onto proven broodies of larger-bodied breeds. The daughter offspring, my "Boxwood Broodies," almost always expressed the broody instinct; and, being larger than OEG hens, they incubated more eggs per clutch. It would be possible to continue selecting from such initial crosses to generate not only new strains but even a new breed of hardworking broodies. (See "Something New Under the Sun" on page 251.)

Mating Systems for Genetic Diversity

A breeding project succeeds in the long term only if it maximizes *genetic diversity* among the offspring, which is to say: We will not achieve improved strains of stock through haphazard, free-for-all matings. Though many assume that reproduction in nature is based on haphazard mating, that is a misconception—many social and sexual behaviors in the wild promote matings that are not random at all but follow patterns that serve at least two critical functions.

First, the gene pool of any population must remain as *diverse* as possible if it is to respond to changes in its environment—whether from increased predation or disease, food scarcity, or changing climate—by adapting as required to thrive in the new circumstances. The more diverse the array of genetic combinations in the population, the greater the chance the needed adaptations are already latent—and that their possessors will enjoy greater reproductive success and lead the way into the future. For improvement breeding, likewise, the greater the genetic diversity we maintain in the flock, the greater the chance that the traits we're targeting—faster growth, greater disease resistance, better egg production in winter, broodiness, more docile temperament—will express in the offspring and be available for selection.

Second, a population must avoid *inbreeding depression* resulting from excess breeding of individuals too closely related to each other. Simplifying for brevity: Closely related individuals are more likely than unrelated ones to share the same recessive gene for a negative trait. If both the male and the female share the same recessive gene, then that trait will be expressed in the offspring. Thus, indiscriminate breeding of too-closely-related individuals generation after generation has a strong tendency toward inbreeding depression—the expression of more and more negative recessive traits, leading to an increase in deformities and a decline in health, vigor, productivity, and reproductive success.

How to choose who mates whom is a complicated question, and certainly closely related matings *can* be used with care to achieve particular goals (as in well-managed linebreeding, which often involves frequent parent-to-progeny matings). The long-term thrust of the mating pattern, however, must be toward the prevention of inbreeding depression. Strictest interpretation of "too-closely-related" prohibits first-degree matings—those between parent and son or daughter, and those between siblings and half-siblings.

Since breeding populations in home and farm flocks are typically small, it is critically important to manage matings so that—despite the limited numbers of breeders available—they mimic patterns of natural mating that avoid the breeding of too-closely-related individuals while maximizing overall genetic diversity in the flock. Fortunately a number of *mating systems* are available to achieve these goals, with the best choice among them dependent on questions of breeding flock size, practicality, management style, and comfort level with record keeping.

The Rest: A Range of Options

If the breeding flock is large, *flock mating* is the simplest of all: Just allow free-for-all mating. But keep in mind the "large flock" caveat: A breeding population of 200 is considered the minimum if eventual inbreeding is to be avoided; 250 to 300 would be better. Proportion of males should be about 10 percent. Record keeping is minimal. Flock mating is likely to be the system used in big hatcheries and industrial poultry operations.

Pedigree mating (or *stud mating*) features the recording of every single mating, with the result that both the dam and the sire are known for every individual in the breeder flock. This method is favored by competition breeders for tracking highly specific strengths and known flaws in every single bird in the project. Impressively rapid improvements in specifically targeted traits are possible, though the method wouldn't serve as well for targeting the broad range of utilitarian traits desired in homestead or small-farm breeding. Great care must be taken if using this system to guard against a "creep" toward inbreeding. And yes, the complex record keeping required would far exceed my patience.

Most mating systems at scales smaller than commercial flocks rely on periodically bringing in breeding stock from outside sources, most easily cocks. One system is *out-breeding* or *out-crossing*, bringing in breeders from entirely unrelated outside strains, to rejuvenate flocks starting to suffer the effects of inbreeding. Another is *out and out breeding*, bringing in cocks from multiple outside sources as new blood. A third is *flock sourcing*, bringing in new stock periodically from a single outside source. All these options require that you trust the supplier to be as conscientious a breeder as you are—and that their breeding practices and goals are compatible with your own. Note that relying on outside sources to prevent inbreeding depression makes your project *dependent* on those sources; and that, the more unusual your breed, the more logistically challenging such outside-sourcing becomes.

The Best: Clan Mating

Suppose you weren't happy with that sort of dependency, that you wanted to maintain as closed a flock as you could manage—while maintaining genetic diversity—with a level of record keeping that supports rigor in the breeding program but would not drive you crazy with excessive detail. Fortunately there is *clan mating*, the method I've used the most—and enjoyed using the most. (It is also known as *family mating*—and as *spiral mating*, for its rotational mating pattern). Yes, I'm being facetious when I name it "the best"—*your* best mating system will be the one that *works best for you*. But for me, clan mating is the ideal lower-input system that ensures genetic diversity in the size breeder flock I can manage.

The first step in clan mating is separating breeding stock into *clans* or *families*. Most sources recommend three as the minimum number of clans (see figure 25.2). Thus the smallest number of breeders at the start is six: one pair, cock and hen, in each of the three clans. If you have sufficient access to good stock, you could of course start with larger numbers of breeding birds in each clan. Whatever number you start with, you will probably have lots more breeders in future seasons. During a typical breeding season, each of my Icie clans typically numbered a dozen or so hens, penned with one cock—a ratio, in a lighter weight breed like the Icelandic, virtually guaranteeing 100 percent fertility in the hatching eggs. (As noted in the next chapter, each clan typically had *two* breeding cocks, though only one would be in service in the breeding pen at a time, with the other in waiting elsewhere.) Keep in mind that *every bird* must be assigned to a clan. So if you (subsequently) bring in stock from another breeder—whether as hatching eggs, just-hatched chicks, or adults—each bird must be assigned to a clan. Assignments may be made for some specific purpose or entirely randomly—but remain in effect for the rest of the bird's life.

How you assign birds to clans initially is up to you, but don't worry if you distrust your ability to judge who is "best" to mate with whom—it is fine *at the beginning* simply to make clan assignments randomly. The exception: Do *not* place two birds into the same clan if you know they are either siblings or half-siblings, or are parent and progeny.

Could you work with *more* than three clans? You could. The main intent with the clan system is avoidance of too-closely-related matings, and reduction of the possibility over the long term of inbreeding depression. I've read that a closed three-clan flock should guard against inbreeding for twenty years before requiring the bringing in of new blood from the outside. The more clans, the slower would be the cumulative creep toward inbreeding depression—theoretically that time frame would stretch out to a century or more with five clans. But, especially in a breeding project at the smaller end of the scale, a five-clan system would entail far more time, management, and infrastructure input than would be practicable. On the other hand, managing a three-clan system is within reasonably easy reach—even at the scale of my Icelandics project (as I describe in detail in chapter 26).

Figure 25.2. With the clan mating system, start by separating breeding stock into at least three lines or "clans." Illustration by Elara Tanguy.

Figure 25.3. In the first breeding season only, breed cock to hen in the same clan. Illustration by Elara Tanguy.

Figure 25.4. Assign all chicks to the clan of their mother. Illustration by Elara Tanguy.

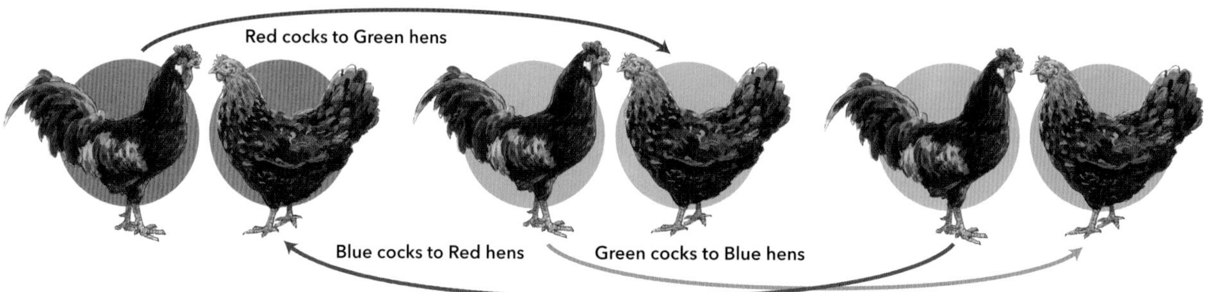

Figure 25.5. In the second breeding season, breed cocks to hens of "the next clan over." This remains the rule for all future breeding seasons. Illustration by Elara Tanguy.

You can give your clans creative names if you like (Hotstuff, Gogetter, and Mojo), or just One, Two, and Three. Naming the clans by reference to color of wing band or leg band (both commonly used for individual identification in the flock) is a practical choice: *All* individuals of the Red clan will have red bands, for example, while *no* members of other clans will have a red band. In my Icelandic breeding project my clans were Red, Green, and Blue. Once a bird is *assigned* to a clan, it remains in that clan *for life*.

Breeding season one: Assume the initial clan assignments have been made—a minimum of one cock and one hen per clan—and it is time for the first breeding season. In this first season *and this season only*, simply breed Red cock to Red hen, Green cock to Green hen, and Blue cock to Blue hen (see figure 25.3). Understand that mating *within* the assigned clans is the correct choice *just this once*, assuming that the birds in each clan are not first-degree related. But note that *never again*, in any future breeding season, should any cock mate any hen of the same clan.

As chicks hatch, assign them—whether pullet or cockerel—to the clan of their mother. That assignment seems obvious now, since both dam and sire of a chick are of the same clan. That will *not* be true in future breeding seasons, however, so it's worth emphasizing: *Assign all chicks to the clan of their mother* (see figure 25.4).

Breeding season two—and ever after: In the second breeding season, *and in all future breeding seasons*, breed cocks to hens "one clan over." Breed Red cocks to Green hens, Green cocks to Blue hens, and Blue cocks to Red hens (see figure 25.5).

Now we see why I emphasized assignment of every chick to the clan of its mother: The progeny of Green hens have Red sires, but in this matriarchal system all chicks are assigned to the clan of their mother (figure 25.6).

Mating systems can be confusing, but the two rules at the heart of clan mating are elegantly simple:

Rule 1: Each mating season, mate cocks to hens of "the next clan over."

Rule 2: Assign every chick to be in the same clan as its mother, permanently marking each chick to keep assignments crystal clear. (I describe how to keep track of clan assignments in the next section of this chapter.)

If these two rules are applied each generation for the next hundred years, there will never be a single mating of two siblings or half-siblings, the type matings that if too often repeated are most likely to bring on

Figure 25.6. In the second breeding season, and all future seasons, continue assigning all chicks to the clan of their mother. Illustration by Elara Tanguy.

inbreeding. (Nor will there ever be a father–daughter or mother–son mating *if* cocks are used for a single mating season only.)

Please do note: In my discussion of clan mating here and in the following chapter, I refer only to its use in improvement breeding of *chickens*. With a couple of caveats, clan mating can be used to breed most other domestic fowl as well. The major exception is geese, who pair-bond for life and are not amenable to its rotational breeding pattern. Some turkey hens develop a strong preference for a particular tom. Though they eventually accept mating from a different tom, the time lag involved might well not fit the needs of the breeding season.

Improvement Breeding in Practice

Enough theory. How is the breeding project managed? How do we keep track of who's who in the flock? How do we select the chosen few for breeding? How can record keeping and data manipulation help?

Identification of Breeders

Even the simplest of breeding schemes will require that you keep records about breeders, if not as individuals then at least by group—family, year of hatch, whatever. There are several ways to assign identities to your chickens, both temporary and permanent.

TOE PUNCHING

You can start when chicks are young—up to three weeks old—with *toe punching*. There are two webs between the forward-facing toes of a chicken's foot, four webs in all on a bird's two feet. These webs are a "coding space" available to us if we imagine, say, numbering them left to right 1 – 2 – 3 – 4 (see figure 25.7) and consider that we have the option of making multiple punches to identify a bird. I'm no mathematician, but even in this simplest of all identification options, using multiple punches would enable the coding of a lot of data.

Figure 25.8 shows how I used single punches to assign chicks permanently to their mother's clan in my clan mating system of Red, Green, and Blue: Any bird

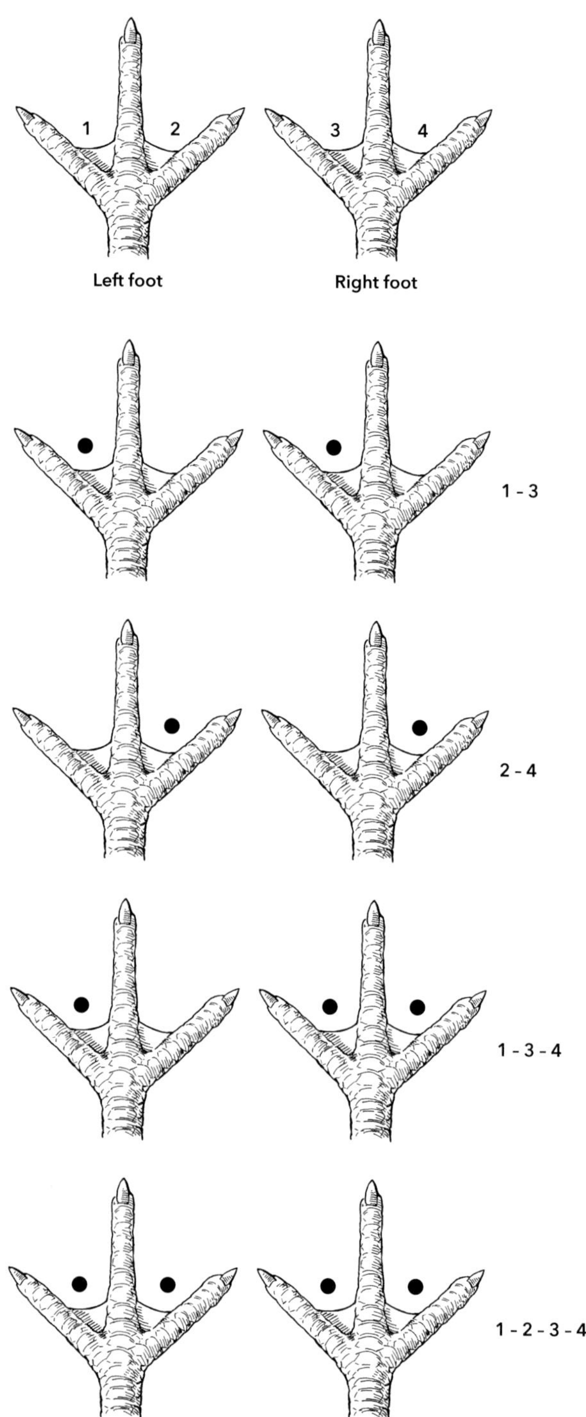

Figure 25.7. Toe punching can record identification information using the webbing between toes as the "coding space." Multiple punches increase coding capacity. Illustration by Elara Tanguy.

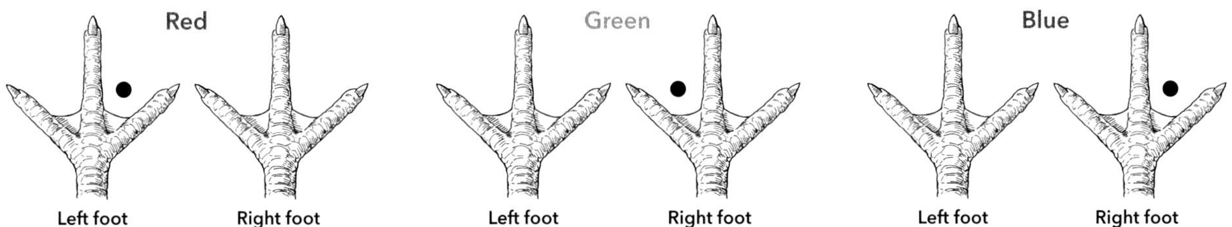

Figure 25.8. Toe punches are a reliable way to indicate—permanently—in which of three clans a bird belongs. Illustration by Elara Tanguy.

of any age or gender showing a punch at position 2 was in the Red clan; if at 3, Green clan; and if at 4, Blue clan. The remaining three spaces allowed for additional coding—for year of hatch; to indicate sourcing from another breeder, whether as chick or hatching egg; or other attribute.

The tool required is like a ticket punch, but smaller, closer to nail clippers in size. Punch one or more holes in the webbing between the toes. Make sure that the hole is clean—that is, that the punched-out tissue is not still attached on one side—otherwise, the tissue may regrow, as normal webbing showing no punch. I always

Toe Punching

Note two things about the toe punch chart in figure 25.9. First, there is nothing official about the coding positions shown. So long as you are consistent, you can pattern your codes as you wish.

Second, the chart shows there are sixteen codes possible, but only if *leaving both feet without punches* is *also* considered an ID code. However, I avoid "both feet blank" as a possible code. If punched webbings of a bird regrow, as occasionally happens, that bird will seem to be part of the "both feet blank" group. If "both feet blank" is not allowed, a bird who has regrown its webbing is an obvious unknown—it will not be misidentified with another group. Even if you have solved the regrowth problem by notching, there's always the possibility that you missed a chick when marking groups for identification.

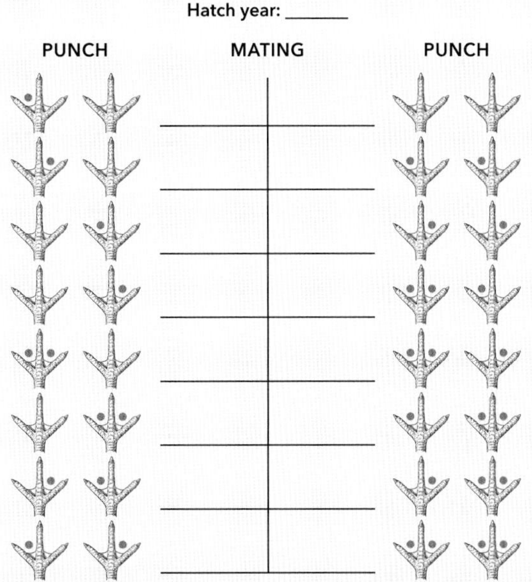

Figure 25.9. Toe punch guide. Illustration by Elara Tanguy.

re-punch the holes after a couple of weeks to be sure there will be no regrowth, and that even at maturity there will always be a hole (or holes) in the webbing, one-eighth inch or a bit larger in diameter.

It is possible to avoid the problem of regrown toe webbing by *notching* instead of punching, using a small surgical shears—resulting in visibly shorter webbing at that position that will never regrow. As well, there is no hole to catch or fill with dirt as the chicken scratches.

Is toe punching painful for the chick? I can only assume that it is. Might that argue for another option, such as the plastic spiral bands I describe next? Not so fast. It would be illogical to focus solely on this question of causing pain, while ignoring the fact that handling the chick is itself stressful. If you opt for spiral bands, you will have to change them several times as the bird grows—over the course of its growth to full size, multiple handlings might well result in more total stress than two brief toe punchings.

LEG BANDS

Identification rings and bands of various sorts are available from poultry supply houses. All can be frustrating to use. Numbered *metal bands* make it possible to identify specific individuals, but I have found them clumsy to put on and difficult to remove, creating stress for the bird in either case. As well, you would want to apply them only after the bird has finished growing and the shank is full-sized.

Plastic *spiral bands* allow for coding both by color and by the leg on which it is placed. They are available in eleven sizes and as many colors. The spirals allow for identifying *groups* within a flock, but, unless numbers of birds to be identified are quite low, not for unique identification of individuals.

Snap-on plastic *bandettes* are both colored and numbered, enabling unique identification of every bird in the flock. They are available in five colors, in lots of 25, numbered in sequence from 1 to 200, keyed to shank size of the more common breeds (see figures 25.10 and 25.11). After the expanded bandette is placed on the bird's shank, it springs back into its coiled shape.

Figure 25.12 illustrates use of single leg bands in a three-clan flock: The color of the band, always on the right leg in this flock, indicates clan—Red, Green, or Blue—and its number is unique to that individual in the breeding records. You can apply a second band to capture more information. For example, a yellow band might be placed on the left legs of all birds in the flock hatched in a particular year. Both those pieces of information—clan and year of hatch—are obvious even at a distance (see figure 25.13). If you approach quietly and slowly, you may be able to read the numbers on the bands as well, without having to catch the bird. And since each yellow band's number is also unique in any

Figure 25.10. Plastic bandettes are available in sequentially numbered lots, in distinctive colors and different sizes.

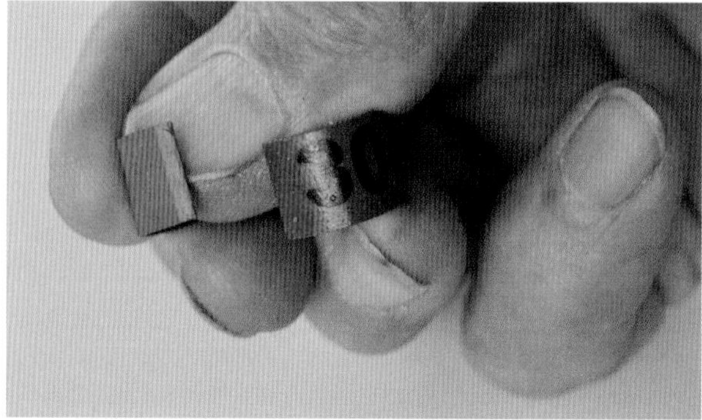

Figure 25.11. Expand a bandette between fingers and thumb.

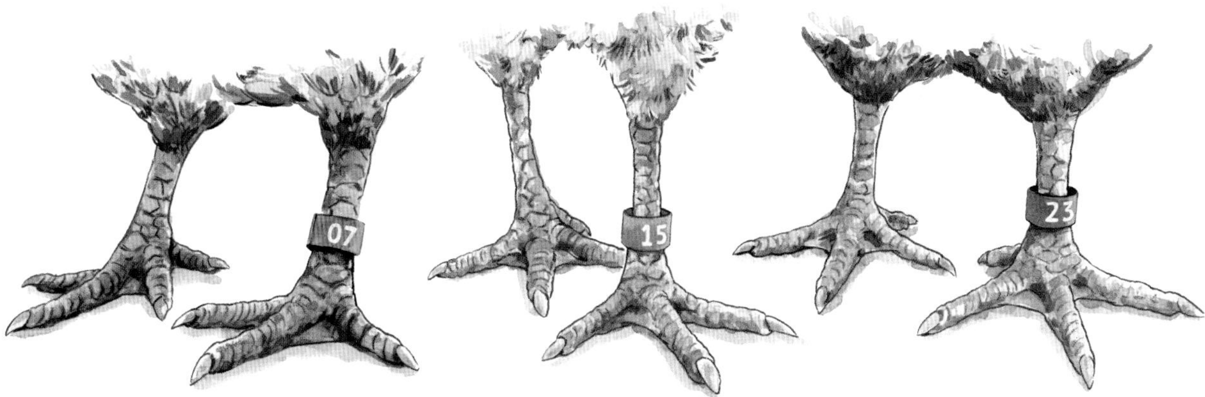

Figure 25.12. In this example, leg bandette color indicates clan membership, while the combination of number and color is unique in a group of up to 200 birds. Illustration by Elara Tanguy.

Figure 25.13. The blue bandette on this hen's right leg identifies her as in the Blue clan; yellow on the left is a code for her year of hatch, 2013.

flock that numbers in the dozens rather than hundreds, double-banding makes the ID doubly unique, which could come in handy if one band were lost.

An inherent problem with both types of plastic leg bands is that they must be changed for larger sizes as the bird grows: If left on too long, a band can cut into the flesh of the growing leg. Monitoring and changing of bands makes them rather high-maintenance with a sizable flock. I chose to wait until my chickens reached adulthood before banding them. Up to that point the essential identification was clan of the bird, which was shown by the toe punch. I was interested in uniquely identifying individuals only after most of the season's culling had been done.

Even if placed only on adults, spiral bands and bandettes sometimes work free of the leg or, worse, move up over the hock, constricting the lower thigh and resulting in crippling and pain if not removed.

Another drawback is that the numbers on the bandettes may wear off over time or get badly obscured by grime. They are easily replaced, though breeding records must then be updated as well.

WING BANDS

Despite the problems noted in the previous section, numbered leg bands worked okay for me until I started raising Icelandics. Guess what—some Icie hens learned how to *pull them off*. Replacing the pulled-off bands, not to mention having to update my records multiple times, became irksome. The final straw came when four broody hens in the Green clan each removed *both* their bandettes—a catastrophic data loss for tracking performance of those broodies in my records. I switched entirely to wing bands, which I now recommend over other options. Like bandettes, they are available in multiple colors with sequenced numbering. Unlike bandettes, if properly attached they remain permanently secure and, under a covering of feathers, clean and easy to read. Their disadvantage in comparison to bandettes is that wing bands can only be read with the bird in hand.

There are several styles of wing bands (though "wing tags" would be a more accurate term). The ones I use are Jiffy 893, packaged in numerical sequence on cardboard sleeves (see figure 25.15). They are available from numerous poultry supply vendors; I bought mine from National Band and Tag Company. There are several wing band designs—be sure the applicator you buy is designed specifically for the one you have chosen.

Two short videos on National Band's website (https://nationalband.com/jiffy-893/) illustrate proper attachment of Jiffy 893 bands in detail. Do check them out—"A picture is worth a thousand words"—but here are a few tips for applying this type of wing band.

1. Pluck out the wing feathers at the spot where the band will be affixed.
2. Remove the band from the sleeve.
3. Load into the pliers-like applicator. (See figure 25.14.)
4. Position the applicator so that the sharp point at the end of the upper half of the band is directly above the plucked area of the wing (see figure 25.16). Clamp down on the applicator's handles to drive the point through the wing's webbing and then crimp the band securely in place.
5. Test the band (as described on the next page) to ensure that it is now permanently attached.

I'm going to be honest: I have had my frustrations applying wing bands. As with many techniques, "practice makes perfect"—I become more proficient with each tagging session. You will as well. At the start, focus closely on the smallest detail. Here are a few pointers:

- I am right-handed, and it feels more natural to band the bird's right wing, as shown in figure 25.17. Band the left wing instead if that is your preference.
- Care must be taken to remove *every tiny feather* at the point of attachment. Even the smallest feather will deflect the narrow aluminum point, the band will not be properly crimped, and attachment will not be secure. As noted previously, I do not band my chickens until they are close to maturity. If you band chicks that are in the downy phase, the fine down will not deflect the point like feathers do, so no plucking is needed.

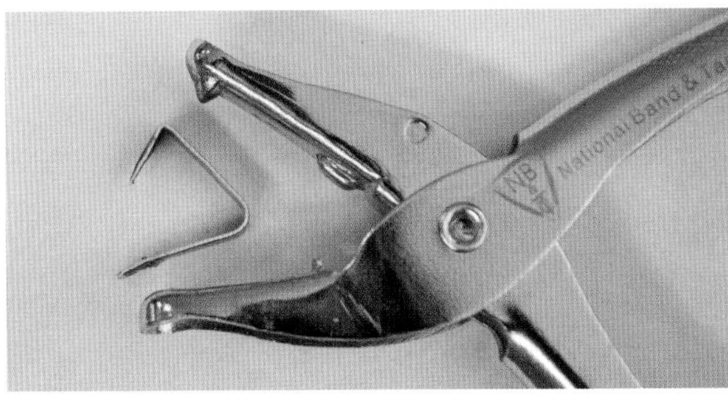

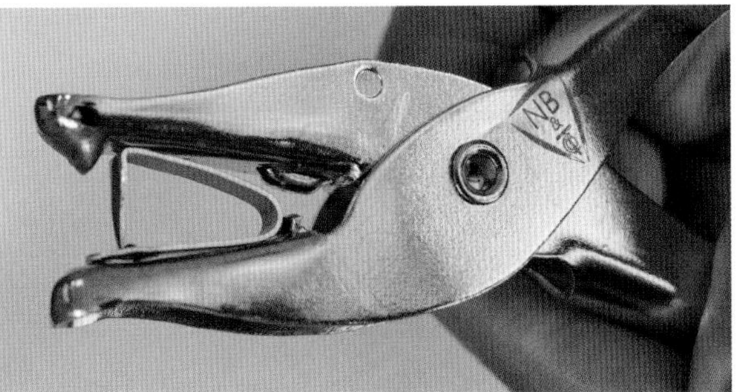

Figure 25.14. *Left*: Loading the expanded wing band into the applicator. *Right*: Once the sharp point at the end of the upper side has pierced the wing's webbing, further pressure on the handles will drive home the point and crimp it tight, locking it in place.

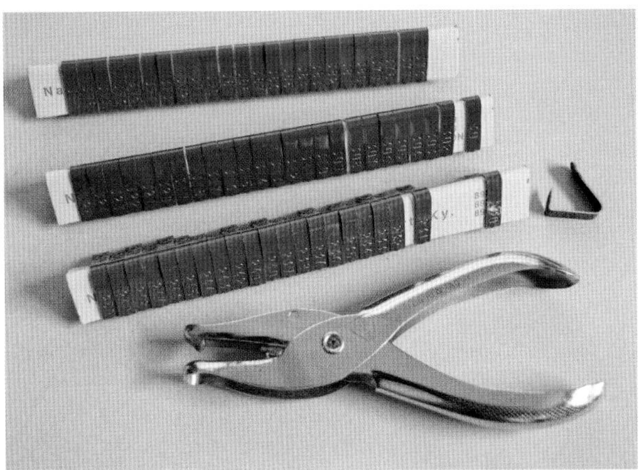

Figure 25.15. Wing bands with applicator.

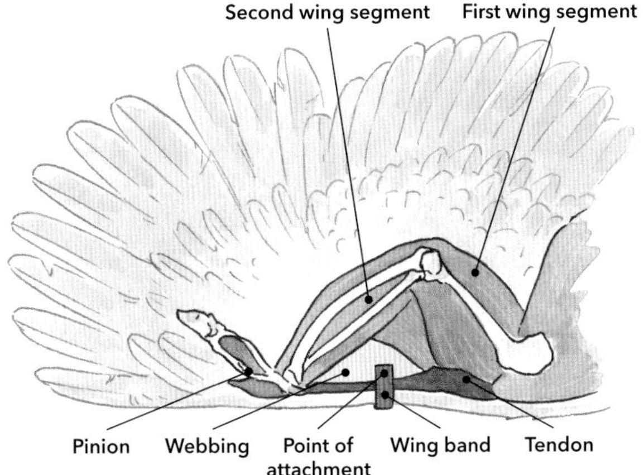

Figure 25.16. Attachment point for the wing band (viewed from above the wing). Illustration by Elara Tanguy.

- When starting out, I found it helpful to have a partner hold the bird, its head toward me, with the webbing of the wing fully expanded and the tendon along its edge stretched taut. I could fine-tune techniques for piercing the webbing and crimping the band's point into place.
- When you clamp down on the applicator handles, the band must be precisely aligned. If it is, the end of the band will crimp tightly in place, and the completed band will never come off.
- To achieve that perfect alignment: The bottom half of the band must be centered on and absolutely flat against the lower jaw of the applicator, with the button seated into the round depression at the end of the lower jaw. If you find it impossible to maintain this precise alignment as you move the applicator into position for crimping, first *manually* force the point of the expanded band through the wing webbing, and *only then* load the band into the jaws of the applicator and clamp down.
- I always test the completed band to ensure it has properly crimped. If I pull on the upper and lower halves of the band, using thumbnails and fingernails of each hand, I cannot pull open a properly crimped band. If it is not well crimped, the two sides of the band will pull apart. In that case I pull the point out of the hole in the webbing and have another go.

- Of course, I *record the color and number* of the attached band for each bird as I go, and *enter that information* in my flock-identification spreadsheet immediately after I finish banding.

Selection of Breeders

Whichever mating system you choose, improvement breeding starts with *selection* of breeders. The following are some of the selection practices used in previous, more hands-on farming eras, which TLC's Buckeye project and its offshoots emphasize.

TRACKING WEIGHT

Both rate of growth and mature size are key production factors—track both prior to selection. Note that the two are not always positively correlated, however. Both are important, so it is desirable to use as breeders both those individuals with a good rate of growth and those who attain a large adult weight.

To measure progress between generations, weigh *at the same age* when evaluating prospective breeders. TLC recommends weighing at eight weeks and again at sixteen. Will Morrow found that individuals with superior weight performance at earlier stages tend to be superior in later stages as well. He weighs at twelve weeks only, an acceptable slaughter age for his Delawares, and thus combines breeder selection with production of broilers for his market.

Figure 25.17 illustrates my technique for keeping a bird quiet while I weigh it in the scoop of a hanging

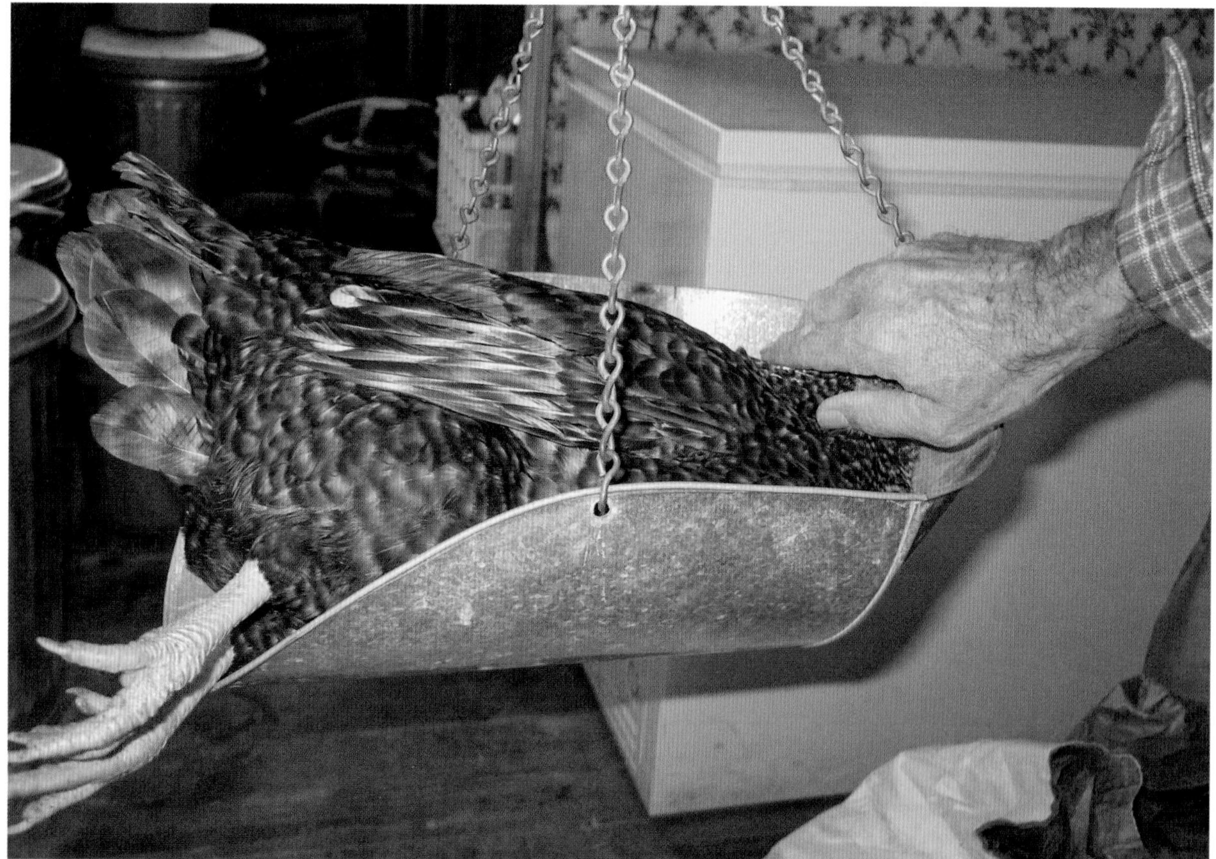

Figure 25.17. The key to an accurate weight is to avoid touching the scoop as you closely hood the bird's eyes with your hand while simultaneously reading the scale.

scale. I lay the bird in the scoop and, just before releasing its feet from my left hand, hood its head with my right—cupping my hand closely but *not quite touching* over its eyes. Having her vision blocked keeps her quiescent just long enough—before she can react and try to escape, I read the weight on the scale and again grasp her feet with my left hand.

SELECTING FOR EGG PRODUCTION

Certain physical traits related to high egg production can be determined through hand selection: space between the pubic bones, distance between the pubic bones and the tip of the keel, a soft pliable abdomen, and other structural indicators of large internal capacity for the ovaries and other reproductive organs.

Observing and recording molting characteristics aids selection of high-producing hens, too. Hens who do not begin to molt until late fall or early winter are generally those who have laid best during the laying season. Those who molt quickly and get back to the business of laying will produce more eggs in a year's time.

Remember that the male transmits genetics for egg production as well: A cock that had a productive dam tends to pass the trait to his daughters. Cocks should be hand selected using the same conformation criteria used for selecting productive hens. (Find more information about selection for breeding in the four listed organizations in appendix J. Especially useful is TLC's "Selecting for Egg Production.")

SELECTING FOR MEAT PRODUCTION

Judging capacity for meat production is not just a matter of weighing the growing birds at a given age or checking fleshing of the breast. Certain *structural* characteristics are necessary to realize a bird's potential for putting on muscle. Heart girth (circumference of the body at the level of the heart); flatness, length, and breadth of back; depth of body; straightness and length of keel—all are related to allowance of room for a large digestive system and hence capacity for growth. For producing good table fowl, therefore, hand selection is also essential when choosing breeders.

A couple of other traits correlate less intuitively to good meat qualities: The width of the skull is related to the structural traits and thus is an indicator of productive capacity. Also, birds with wider feathers tend to grow faster: Wider feathers retain body heat better than narrow ones, thus more energy from the bird's food is available for growth. (Again, there is more information on selection and breeding to be had from the four organizations in appendix J. Especially useful is TLC's "Selecting for Meat Qualities and Rate of Growth.")

NEST-TRAPPING TO GATHER DATA

For many utilitarian traits we might target in a breeding project, "*who laid that egg?*" is the most important question to answer, because some traits cannot be tracked without knowing *which hen laid which egg.*

For example, TLC's excellent guide "Selection for Egg Production" advises hands-on inspection for "soft, enlarged combs and wattles; wide, moist vents; generous space between pelvic bones, and between pelvic arch and keel; and expansive, soft, pliable abdomens." Well and good, hens with "wide, moist vents," for example, are the ones in active production. But is a specific "moist, wide" hen laying five eggs per week, or six? Is her average egg weight 57 grams, or 64? Knowing such answers with certainty is essential if we are selecting layers as *breeders*, as laying hens whose daughters might lay even more eggs of larger size. And that requires *trapping the hen in the nest with the egg she has laid*, so we can identify her and mark her egg for our records before releasing her.

Targeting other egg traits might also be important in a breeding project, for example intensity of shell color in breeds such as Ameraucana or Marans that are known for unusual color. And what about egg shape? Or shell texture, thickness, and integrity (absence of flaws that weaken the shell)? Like rate of lay and average size, such eggshell qualities are heritable, so breeding for production of more-desirable eggs in future years requires removing the layers of less-desirable eggs from our breeding program now.

I cannot overemphasize: If you are serious about your breeding project, use *trapnests* as your most essential selection tool. The topic is so important I discuss it in great depth in the next chapter.

Use the Hatchet!

To select *for* the best few means selecting *against* the average many. Which is to say, an essential for breeding success is *culling ruthlessly*. As Nature does. Never forget that it was ruthless culling—by environmental factors, by microbes, by critters with curved fangs and long talons—that honed *Gallus gallus* to a state of high fitness: robust, resilient, adaptable, with near-perfect health. Our challenge is to reproduce that state of near perfection in the *domesticated* version. We do so by taking on the role of Nature's cullers—through generous use of the hatchet.

Will Morrow's Delaware breeding project illustrates what it means to "cull rigorously." He started with 250 hatching eggs from a large hatchery, then culled *to the top 10 percent* of the first breeding year's hatch. Since that time, he hatches about 200 Delaware chicks each year as potential breeders and selects the top 5 percent of cockerels, since fewer cocks are needed for breeding, and the top 20 percent of pullets.

Though inevitably there are variations in practice, it's a good idea to stick as closely as you can to "the 10 percent rule": *Keep only the top 10 percent as breeders each year.* Combined with the rule about maintaining genetic diversity, however, this really means: Keep the top 10 percent of each *mating* during the season.

To clarify, imagine there are three groups of breeders in your breeding program—however selected and using whatever mating system you have chosen. You hatch 40 chicks from each group, for a total of 120 for the breeding season. The 10 percent rule suggests you should retain only the best 12 as breeders. But the need to maintain maximum genetic diversity suggests that you not simply select the top performers from the 120 offspring, but instead select the best from each *mating*; that is, 4 each from Group A, Group B, and Group C.

Culling rates of 90 percent or so may seem logistically difficult in a small-scale setting. But consider: You can forget about the time and effort some flocksters spend raising big batches of "meat birds" for the freezer—the culls from your breeding project can furnish all your family's meat chicken. Sharing a breeding project with a poultry friend nearby may help achieve the high culling rates: Manage the birds as one flock in terms of selection and breeding, but split their actual care between you.

Essentials of Selection

Culling starts at hatch. *Always* cull a chick who is weak or struggling to make a good start. Cull all runts. Examine the growing birds by hand and cull for structural defects as soon as they manifest: wry tail (permanently held at an abnormal angle because of a deformity at the end of the spine), crooked keel, crossed beak, malformed feet or wings. (Yes, an unpleasant task, but to do it most humanely: Hold the chick's body in the palm of one hand, its head between thumb and forefinger of the other, and—gruesome, yes, be forewarned—quickly pop the head entirely free from the neck. Death is instantaneous.)

Without exception, never "doctor" your potential breeders for infectious disease of any kind. Look around: The sick bird's brothers and sisters are doing *just fine*, in exactly the same conditions. Who are the best candidates as breeders? And if compassion inclines otherwise, ask yourself: Do you *really* want to burden future generations with a demonstrated immunity weakness? In your *compassionate* concern for *their* robust good health, use the best medicine of all for infection—the hatchet.

While selecting breeders for good size and rate of growth, don't choose as breeders birds who are inordinately fast growing or larger than their peers—such birds often have a tendency toward immune system weakness.

Remember that selection of the best breeders is followed by selection of the best eggs to keep for hatching. Choose large eggs for incubation, since genes for large egg size are highly heritable—and larger chicks make a more robust and vigorous beginning. Avoid setting the smaller eggs of pullets when they start laying, at about eighteen to twenty-two weeks—wait until they are twenty-six to twenty-eight weeks old and their eggs are larger. Note, however, that abnormally large eggs should not be selected—they may have problems with hatchability. Don't set eggs of odd shape or surface texture—ridged, lopsided, chalky shell, calcium bumps—these traits are highly heritable. Even if the flaw is only a temporary glitch in the system, hatchability of such eggs is reduced.

The focus should be on improving *production traits* in the early stages of a breeding project. Once the flock's productivity is making good progress, secondary traits such as desired feather and eggshell color can be selected for as well.

Minimum standards for retention as breeders should become more stringent each year. In the ideal case the standard this year will be what the best achieved last year.

Something New under the Sun

There is a critical need for conservation breeding to protect our precious heritage of useful poultry breeds from the past. But Joel Salatin, in his *The Sheer Ecstasy of Being a Lunatic Farmer*, makes the excellent point that we shouldn't be concerned only about saving the heritage breeds, many of which were bred in Europe. We should as well be breeding our own uniquely American, region-specific strains and breeds. Should you care to take up Joel's challenge, know that the task certainly can be done. Nettie Metcalf developed her Buckeye, suited to her harsh Ohio winters, out of crosses among Barred Plymouth Rocks, Buff Cochins, Rhode Island Reds, and black-breasted red games. Other American breeds, developed out of assorted crosses to produce breeds attuned to their local environment, include Plymouth Rocks, Dominiques, Wyandottes, Rhode Island Reds, Jersey Giants, and more.

If there are a couple of breeds you like a lot but you would really prefer one that combines the virtues of both, seems to me you're almost ready for takeoff. If you cross the two, the first (or F1) generation may well reward you with higher average performance than that of the parents—it's called *hybrid vigor*. The next (F2) generation, however, is likely to be all over the map genetically, with as many underperformers as stars. Choose the individuals who come closest to your ideal. Proceeding by selection and breeding as suggested in this chapter, you should eventually end up with your own unique breed—your gift to the world. And *eventually* needn't mean so very long. I believe Nettie Metcalf's development of her Buckeye took seven or eight years. My friend Tim Shell developed his Corndel, a cross between Delaware and a proprietary strain of Cornish Cross, into a reasonably stable genetic package for use as a pastured broiler in about three years.

I have not myself brought such a project to fruition. A few years ago, however, I made a start on one. I've described how much I liked Icelandics as a do-it-all breed. But the performance of some of my Icie broodies was well below par, and I thought I could improve it through a mix-'em-up breeding project. I also wondered whether I could keep their self-sufficiency virtues while nudging a little toward a more traditional dual-purpose breed—a bit larger, a few more eggs in the nest. I made an ambitious start, but regrettably, personal developments that intervened after my first full breeding season required termination of the project.

I considered my Icelandics as the "base" for the new mix. I brought in stock—hatching eggs, chicks, adults—for an additional twelve breeds, all of them with long histories as home and small-farm flocks. Eight of my sources were private breeders who impressed me as conscientious conservation breeders. Four of the breeds I ordered from a commercial source dedicated to some of the older traditional breeds. Layer breeds I brought in were Hamburg, Ancona, and Light Brown Leghorn; meat breeds, Australorp, Salmon Faverolles, New Hampshire, Barred Plymouth Rock, Buckeye, Buff Orpington, Red Dorking, and Dominique; and breeds to ensure a strain of broodiness, my Icelandics on hand plus Old English Game. (The latter I planned to breed in with caution, with the hope I could enhance average broody performance—but avoid the 100-percent broodiness and lack of winter laying in OEG hens and the heightened aggressiveness among OEG cocks.) My goal was a robust, healthy, active go-getter that would keep feed costs down while yielding reasonably good meat and egg production, with a mix among hens of high-performing broodies and layers with no interest in setting.

The Buff Orpingtons, sadly, turned out to be a cautionary lesson: They were huge, gorgeous—but their breeder had focused on those traits exclusively, with no thought whatever to basic soundness or production. Weak, disinclined to forage, with an extremely low rate of lay and nagging health issues, they went straight to the slaughter table, with never a turn in the breeding pens.

Many flocksters might deride my project as merely "breeding a bunch of mongrels." But although the project began with a grand sweep of cross-breeding, I carried it out from the beginning with the exactitude and sound principles discussed in this chapter: Assigning every breeder, of whatever breed or origin, to one of three clans, and strictly controlling all matings using the rotational patterns of clan mating; and great care in selection, both for desirable traits and for ruthless exclusion of flaws and weaknesses, keeping only 10 or 20 percent of hatches as new breeders. Had I been able to carry it forward, the result truly could have been "something new under the sun."

Should you be inclined to attempt to "blaze new trails" in such a project, some advice and a promise: Focus on utilitarian traits and forget narrowing the "cosmetic" traits to an all-alike sameness. So long as the birds you breed, generation after generation, more closely approach your own ideal for exactly the traits you are targeting, consider their variability a bonus—something to enjoy and admire—and value the greater genetic depth that kaleidoscope implies.

And my promise: However much you raise poultry for their utilitarian offerings, I bet you do so for less tangible reasons as well. I guarantee you'll find a something-new breeding project exciting, challenging, and fun; that it will add a new intimacy with your evolving flock and with the abundant ecology you share; and that in the best case you will feel a deep pride if you pass on your "something new" as a gift to the future.

CHAPTER 26

Managing the Breeding Season

Please don't let anything said in the previous chapter make you think that breeding your own stock need be a big, complicated project. It's pretty straightforward for the cock and the hen, that's for sure. In this chapter I discuss a few of the practicalities of organizing the breeding season, however casual or refined the breeding project.

Progress in a breeding project depends on keeping good records. How can we make decisions about breeders in the current breeding season if we don't remember how they performed last year? Because record keeping is so important, I also go into detail about a topic that some may find intimidating—working with electronic spreadsheets. Don't worry, if I learned how to do it, you can, too.

Throughout this chapter, keep in mind that descriptions of my practices are illustrative only—you will find your own best ways to manage the breeding season.

The Mysteries of Fertilization

We are familiar enough with the biological intricacies of reproduction in our own species but may be a bit clueless about the ways in which avian reproduction differs. Yet an understanding of these fundamentals is essential if we are to avoid some common mistakes producing the eggs we intend to hatch.

The female of a *wild* bird species starts laying eggs only after she has found and mated with a male. Laying eggs for her is always and exclusively about reproducing her species—a wild bird never lays eggs that are not fertile.

In the case of the chicken, however, selective breeding has resulted in hens whose egg production is no longer restricted to a seasonal reproductive cycle. (For the domestic hen, ovulation is more analogous to that of a woman, who ovulates periodically whether or not she is sexually active.) The hen lays her egg every day or so, in *readiness* for reproduction, regardless of whether there is a cock in the flock whose attentions would fertilize the egg. I emphasize this point because I am so often asked, "Do I have to have a rooster in my flock?" If you prefer, for whatever reasons, to have hens only in your flock, be assured they will still lay eggs—though, of course, the eggs will not be fertile. For fertile hatching eggs we must ensure the hens have sufficient exposure to an active, virile cock.

What constitutes *sufficient*? A general recommendation is eight to twelve hens per cock for a near guarantee of 100 percent fertility in the eggs—with the lower end of the range more appropriate for heavier breeds, the higher end for lighter ones. But a ratio of up to twenty-five hens per cock will supply eggs with a high fertility rate. Thus the hen-to-cock ratio you choose for the flock may vary, depending on how important it is to ensure 100 percent fertility in the eggs.

An interesting difference between human and avian reproduction is the term of viability of sperm in the female's reproductive organs. In humans the viability of sperm inside a woman's body—the window of opportunity for fertilization—is limited, as short as one day, rarely more than three to five. But in the hen, the cock's sperm live much longer—and continue to fertilize eggs as they develop in the hen's oviduct over the course of many days.

Just how long do sperm survive inside the hen? It is said that sperm remain viable longer in some breeds than in others. But imagine we have a group of hens who have been frequently mated. We isolate them, with no further exposure to a cock, and incubate a batch of eggs each day thereafter, tracking the resulting hatches on a graph. The first clutches of eggs would likely hatch at a rate approaching 100 percent. At a week to ten days, we would see a falling curve on the graph, representing the decrease in fertilization in the eggs as sperm inside the hens died. At some point, maybe about three weeks, the hatch rate would fall to zero—indicating the death of the last sperm in the hens' oviducts and complete absence of fertilization.

This is not merely a theoretical exercise. Suppose we want to be completely certain of the parentage of the chicks we are hatching—that is, certain that fertilization comes *only* from the cocks we have isolated with our hens and not from more free-for-all matings that occurred in the general flock *before* we isolated the breeders. How long do we wait after isolation to be certain that sperm from previous matings have now died, and that fertilization can result only from the parent birds in our breeding pens? Again, it depends on how important is absolute purity in the breeding program. Consider the example of Dominique hens who have been mating regularly with Australorp cocks but are then removed to breeding pens, where they are mated solely by Dominique cocks. The longer the time since the last breeding by the Australorp cocks, the fewer and less active will be the sperm from those encounters, while there will be vastly greater numbers of more vigorous sperm from the more recent matings with the Dominique cocks. It becomes ever more likely that in the competition to fertilize the germ cells in the developing egg yolks, the Dominique sperm will win. The ratio of pure Dominique chicks to Australorp × Dominique chicks would track as a rising curve on a graph, until the former represents 100 percent of every hatch.

In practical terms, I usually waited a minimum of ten days, preferably two weeks, after isolating my breeders before collecting hatching eggs.

Trapnests: Fundamental to Success

The use of trapnests is fundamental to successfully managing the breeding season. Yes, several periods of nest-trapping should be scheduled throughout the year. But whenever scheduled, each test period is truly a part of "managing the breeding season," since its priceless data will guide management of the *next* breeding season. Once the breeders have been isolated for the new breeding season, the trapnests should be used full time: No data source is as essential to the success of the breeding season than the jottings in pencil on eggs collected from trapnests.

Trapnest Designs

The basic concept of a trapnest is simple: A door of some sort is added to the front of a regular nestbox, and that door can be "set" in an open position to allow the hen to enter. As she does so, her settling-in ritual triggers some sort of release that causes the door to snap or drop or swing into blocking position. She is now trapped until the breeder releases her, after checking her unique identification and noting it on the egg (in pencil, *not* marker pen) with the date.

An online search will turn up numerous design options, some of them rather complicated, including some from earlier agricultural eras at backyardchickens .com.[1] Here, let's keep it simple and consider three basic designs: one based on purchased wire fronts and two others featuring doors that either swing or drop into the closed position.

PURCHASED TRAPNEST FRONTS

When a hen pushes against the bottom of a purchased wire frame door as she enters the nest, she triggers the wire front to snap into the closed position (see figure 26.1). I found that my smaller Icelandic hens did not exert sufficient force to push the wire front out of its locked position as they scooted underneath it. Results with larger, heavier hens might well be better, but a success rate of about 50 percent made wire front doors essentially useless with my Icie hens, and I cannot recommend them.

Note, if you do opt for the wire fronts, that few suppliers of poultry equipment still offer them for sale to buyers at the small-farm and backyarder scale—probably because demand is so small. Indeed, as of this writing I can find only one source (https://www.eggcartons online.com) for sales of wire frame trapnest fronts (in two sizes) at the smaller end of the market. However, as far as I can tell, the manufacturer, Kuhl Corporation, will fill small online orders (https://catalog.kuhlcorp .com, key in "trap nest front" as search terms), though they mainly supply the large-scale needs of the poultry industry and setting up an account is required.

SWINGING DOOR DESIGN

The essence of the swinging door design is a door hinged to the top of the nestbox's entrance, suspended out of the way as the hen enters, which is triggered by her settling-in ritual and swings down. Because the door is slightly longer than the height of the nestbox opening, its swing is stopped by the retainer strip along the base of the opening. The hen is trapped—until you release her after recording her achievement.

Figure 26.1. These three nests have two different types of doors: purchased wire fronts and a homemade wooden door. The wire front of the center nest is in the "set" position. To enter the nest, the hen must push against the lower edge of the wire frame, which triggers it to lock behind her in the shut position (*far left*), trapping her inside.

While this trapnest design—an ordinary nestbox with a hinged door in front—could hardly be simpler, a closer look at figure 26.2 suggests a couple of caveats. Note in the left drawing how deeply into the nest the hen must enter before contact with the door's edge triggers its release. Might she decide to settle right in the center of the nest, but in the process knock free the door to swing down and smack her? The obvious solution—a deeper nest—would add to the size of the nest, and a multiple-nest unit (like the one in figure 26.1) would be heavier, would take up more space in the coop, and would be much clumsier to mount on the wall.

The right drawing in figure 26.2 shows two modifications I made to this basic design to solve these potential problems. First I added a sort of "vestibule" in front of the retainer strip. Because the vestibule is empty, it directs the hen deeper into the nest—that is, toward the nesting material beyond, the only sensible place to lay her egg. Only then does she trigger release of the door. Second, I cut the plywood door in half and hinged the halves together so the door could fold in the middle—it has effectively half the length in the set position, but extends to full length as it swings into place after its release is triggered.

The nest on the right in figure 26.1 has the modified hinged door (shown in its "set" position). Note as well the "approach perch" in figure 26.1. Such a perch should be added in front of any wall-mounted trapnest: The danger of premature triggering of the door is minimized if the hen is able to hop or fly up and land on the perch, in preparation for a more sedate entry into the nest.

Now look at figure 26.3. The hinged door has opened into full blocking position as it dropped. Note the strips that have been added on either side of the nest front to act as stops that prevent the hen's pushing open the door and escaping the nest.

As for propping the door open in preparation for its timely triggering, figure 26.5 shows three options: a prop stick, a hook, and a notched pivot. The *prop stick* is fashioned from two dowels of different diameters: A large dowel is cut to the width of the nest, and the smaller dowel is glued into a hole drilled near one end (see figure 26.4). A *hook* can be made from stiff wire, the upper end bent to hook into a screw eye or open screw hook secured in the ceiling of the nest, the lower end bent to support the edge of the door at its center. A freely swinging *notched pivot* that supports the corner of the door is easy to make from a scrap piece of wood. Simply cut a

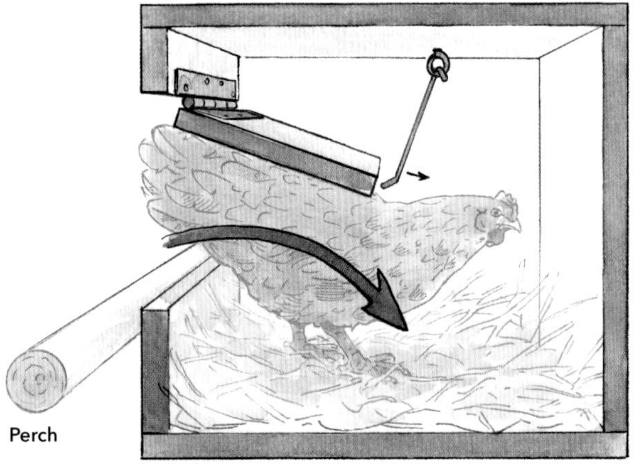

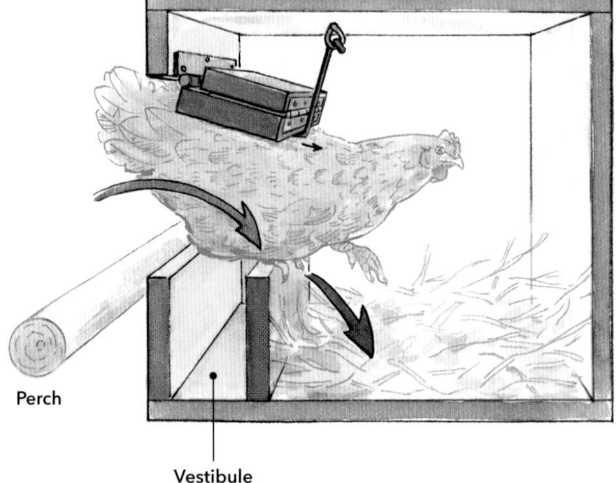

Figure 26.2. The left drawing illustrates the core concept of the swinging-door design, which becomes more practical to build and install if the door is cut in half and hinged, to open into full blocking position as it drops, as shown in the right illustration. Illustration by Elara Tanguy.

Figure 26.3. The hinged door after it has opened into full blocking position as it dropped.

Figure 26.4. Prop stick made from two dowels.

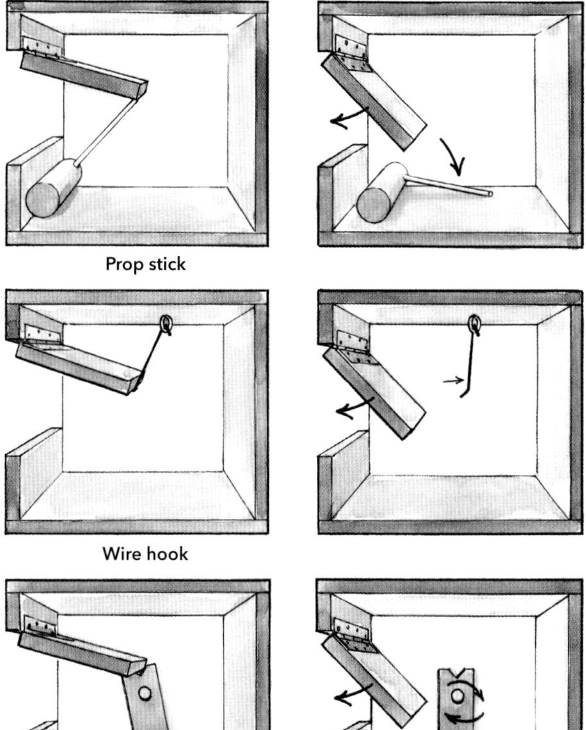

Figure 26.5. Three designs for a release that holds the door suspended until nudged by the hen. Illustration by Elara Tanguy.

small notch in one end, and then use a bolt to attach the pivot piece loosely to the wall of the trapnest.

FALLING DOOR DESIGN

There are also design variations for a door that *drops* (rather than swings) into place when triggered. I use an original design that emerged during one of my father's visits. Father always *insisted* we have "a project" to work on when he and my mother visited. (*Oh, why don't we just take it easy?* was not even to be considered!) One time, I proposed making a set of trapnests. After mulling over a couple of designs I showed him, he concluded he didn't much like either and proposed, "Why don't we do it the way Granddaddy used to put together his rabbit boxes?" I remembered the live rabbit traps ("rabbit gums," the old-timers called them) my grandfather used for trapping rabbits in his fields and woods. The trap was a long rectangular box made of scrap lumber, enclosed on all sides except the front. A door was suspended above a slot in the top at the front end, and connected via string and a dowel pivoting on a notched support to a baited trigger stick in back. Nibbling by the rabbit on the bait knocked loose the trigger stick, releasing the door to fall and block the rabbit's escape. Despite some skepticism about turning a rabbit trap into a trapnest for hens, I agreed to give it a try.

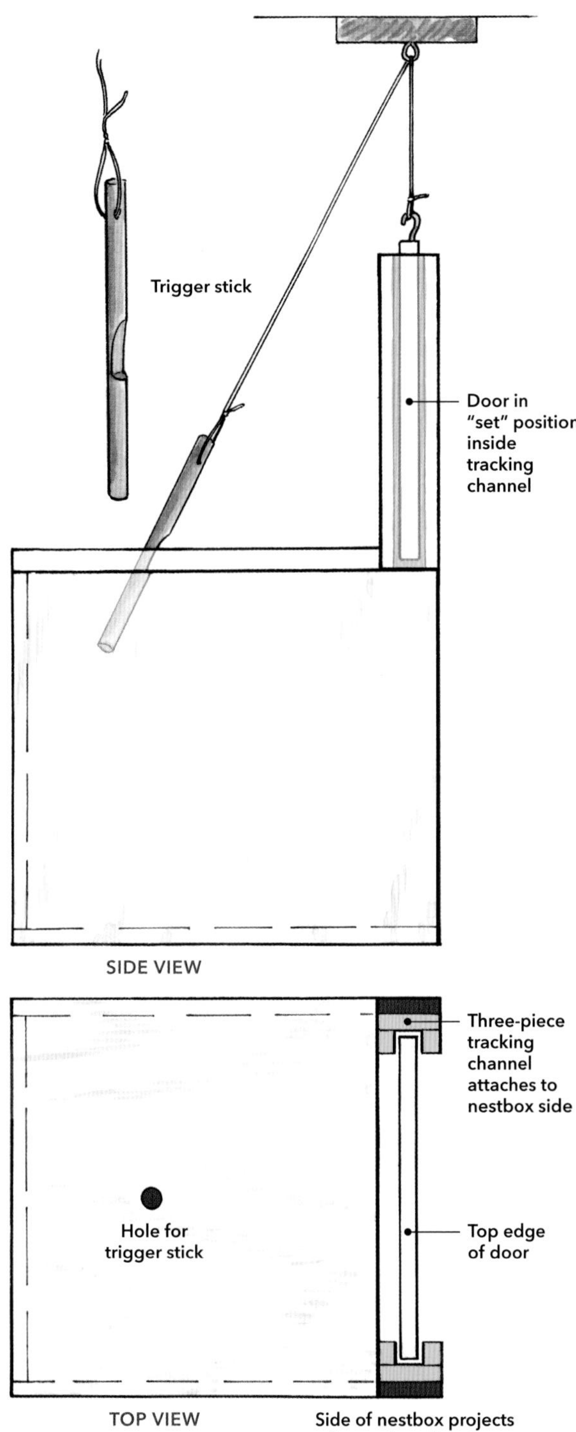

Figure 26.6. The most effective trapnest I've used, hands down. Its step-by-step construction is described in appendix A. Illustration by Elara Tanguy.

Figure 26.6 shows our adaptation of the core design: The door hangs in a slotted track, suspended by a string that passes through an open screw hook overhead and is tied to a notched trigger stick on the other end. The trigger is not baited, of course, but the hen knocks it loose as she settles in to lay, releasing the door to fall.

As it turned out, this design proved to be the best I've used, performing at an ever so slightly less than 100 percent success rate. I share its step-by-step construction in appendix A. And about that "slightly less than 100 percent" performance? Because of a design flaw, on rare occasions—maybe in one or two falls out of a hundred—the door would drop slightly off-kilter and jam at the top of the track. If you build this trapnest, therefore, be sure to include the two modifications noted in appendix A. If you do, I guarantee you will never have a single failure of this design however long you use it.

Using Trapnests

Nest-trapping hens may not be possible in your breeding project, since it is time consuming, but it does enable far more targeted selection of breeders. If maintenance of winter production is important in the breeding program, for example, tracking of laying in winter by individual hens is essential to breeder selection.

Nest-trapping is not a task to do all year-round, of course—that would be wildly impractical, given the intensity of monitoring required. Instead, schedule a period of several days at a strategic point in the laying cycle—"strategic" being determined by the specific laying traits you are targeting. Trapping in summer would track egg production at its peak, while trapping in the darkest days of winter would reveal breeders most likely to hatch daughters with best winter production. If early onset of lay is important, test an age cohort. (Note: A *cohort* is a group of individuals in the flock with something in common. In this case—comparative evaluation of earliest onset of lay—the cohort would be defined as layers who all hatched the same day. In other contexts I will discuss, the cohort referred to may be the *year* of hatch—or perhaps the cohort of *proven broodies* among the total population of hens in the

Other Thoughts on Design

I've discussed several trapnest designs—and there are many others you could consider. Whatever you choose, however, there are some basics to keep in mind for any design.

- You may need to adjust the sizing of various elements of a trapnest, depending on the size hens you will be trapping. There is a good deal of leeway, but don't expect a bantam hen to trigger door release in a trapnest sized for Wyandottes.
- Hens may become nest-shy if they experience unpleasant surprises in a trapnest. Premature triggering of the door must be prevented—it could whack the hen while she's still entering the nest, or even rudely "slam the door in her face." I've mentioned the role of an approach perch to lessen the chance of premature release. As well, be sure that whatever release mechanism you use locks the door in its open position securely, and requires a significant nudge from the hen to trigger. Damp the force of the door's closing impact by rigging some sort of "bumper," such as scraps of discarded inner tube stapled in place at impact points.
- Make sure there is plenty of ventilation in the sides, bottom, and tops of trapnests. A hen trapped for a quarter hour in summer will appreciate good airflow.
- Because trapnests will be used as ordinary nestboxes most of the year, design their doors to be easy to remove for storage and then to reinstall.

flock.) The age cohort test would start at, say, twenty weeks and continue until the last pullet in the cohort has laid her first egg. Trapping in late winter and in early spring identifies which hens lay best during those shorter days—they are likely to be the best layers the rest of the year as well.

It's worth emphasizing a point made earlier: However we choose to schedule nest-trapping tests during the year, once the breeding season begins the breeder hens should be trapped full time—until the last of the hatching eggs are set. During this time the breeder makes selection decisions about every single egg laid.

Modern poultry breeding has ignored the question of longevity, to say nothing of maintaining good egg production in older hens. But suppose you want to breed for greater longevity (genetically related to greater vigor, health, and hardiness, by the way) while keeping a higher level of productivity as hens age. Nest-trapping to the rescue: If Hen 106 and Hen 141 trap at the same level of production but the former is a year older than the latter—who is the obvious candidate for breeder?

Be sure to have enough trapnests in proportion to the number of layers you are testing—one trapnest for approximately every six hens. The integrity of the data generated depends on trapping for *every single egg* laid in the test period—if urgent hens blocked from the nests lay in the litter of the pen floor, you have lost irretrievable bits of your selection data.

Avoid nest-trapping to track egg production during the molt. Because the hen is putting so much of her resources into replacing feathers, and individual hens molt on different schedules, laying data during this period (typically early or mid-fall through early winter) are not valid.

Set aside a period of perhaps five days when you will be able to check the nests frequently—as often as every quarter hour during the hours when hens are most likely to lay. Testing for one or two days is not useful: Remember that even the best-laying hen has a periodic "off" day when she lays no egg while returning to the beginning of her diurnal laying cycle. Imagine again our two hens with identical average rates of lay. In a test of only two days, one of those days may be the off day for Hen 106 so she would lay only one egg, whereas Hen 141 is at the point in her cycle to lay an egg both days. In this too-short test period, Hen 141 seems *twice as productive* as Hen 106, even though their average production is the same. A five-day test is less likely to yield such distorted results.

The best strategy for averaging out the vagaries of hens' laying cycles is to schedule two or more tests in an overall testing period, say four or five days each and three weeks apart.

Since nest-trapping requires a major commitment of time during the testing periods, it could be dismissed as a drag on efficiency. In fact, nest-trapping can *increase* overall efficiency. Imagine you have twenty-five hens in your breeding program. You don't nest-trap, but it's obvious that one hen lays a chronically small egg; one, lopsided eggs; and three others, eggs with shells that are thin or have wrinkles or fracture lines or unsightly calcium bumps. Knowing these traits are heritable, you always discard these eggs when selecting for incubation. Problem solved regarding selection against these undesirable traits, but . . . How many breeding hens do you have again? Twenty-five? I don't think so: You have *twenty*—plus five freeloaders on your breeding program! Why not use a set of trapnests to eliminate the freeloaders and *boost efficiency*?

Isolating and Managing Breeders

In the simplest scenario—a breeding project with a single cock, a group of hens not requiring division, and incubating and brooding artificially—the breeding season can be scheduled to the breeder's preference. If there are multiple cocks in the breeding flock, or the mating system requires separation of breeders, or you rely on some of your breeder hens *also* to serve as mothers—things get more complicated. By way of illustration, I describe next management of breeders in a typical breeding season in my yearslong, three-clan Icelandic breeding project.

It makes sense to keep a flock together for as much of the year as possible—why service separate pens unnecessarily? I started isolating breeders in mid to late March. I kept six cocks in the flock, two each in three clans, and their rising levels of testosterone started the clock running. Although the level of aggressiveness among Icelandic males was not excessive, as winter ended and the first green of spring appeared, competition for access to the hens spiked. Suddenly, cocks who had accepted subordinate positions in the hierarchy with good grace were more likely to mount serious challenges. And they were more inclined to play for keeps. As a practical matter, then, I isolated "the boys" as soon as I began seeing conflict beyond routine sparring.

Remember the basic layout of "The Chicken Hilton," our main henhouse, divided internally into three main sections (see figure 7.9 in the sidebar "The Chicken Hilton," page 72). I hung two doors between sections A, B, and C, and separated the hens by clan, typically about a dozen each: Red hens in the left section, Green hens center, and Blue hens right. (To be precise: I did not hang additional doors to make subsections D and E.) Recall that in the first season breeding my Icelandics, it was acceptable to breed cock to hen of the same clan. In all subsequent breeding seasons, however, I followed the rotational pattern at the heart of clan mating: I put one of the Red cocks with the Green hens, one of the Green cocks with the Blue hens, and one of the Blue cocks with the Red hens. As for the extra Red, Green, and Blue cocks, I moved them to the pens in the far end of my greenhouse. They got along fine there—with no hens around, what was there to fight about?

Because I prefer not to confine my birds too closely, I set up ranging areas for all of them, including one yard inside electric net fencing behind the greenhouse for the "off duty" cocks. If you look closely at the graphic illustration of the henhouse in figure 7.9, it will be clear how easy it was to set up three electric-netted enclosures that kept the clan mating groups isolated from each other even when they were ranging outside: with Red having access through the exterior door at the front of the left section; Green, through the pop-hole in the rear of center; and Blue, through the exterior door at the front of the right section. (That left the exterior door in the center section for my own access into the henhouse.) In those four ranging yards, the cocks and hens enjoyed all the benefits of exercise, fresh air, and sunshine as they had before separation. Each ranging area contained debris fields of decomposing organic litter or large compost heaps to ensure continued access during the breeding season to live, nutrient-dense feeds—pill bugs, slugs, grubs, and earthworms—for maximum vitality in the hatching eggs.

Each of the breeding sections inside the henhouse was furnished with a set of trapnests. Now more than ever, with the breeding season unfolding day by day, it was essential to know precisely *which breeder hen laid which potential hatching egg*.

I did not put the trapnest doors in place until I wanted to begin nest-trapping. And when was that? The earlier discussion—about the prolonged term of sperm viability in relation to ensuring purity of matings following isolation of breeders—implies that I would start nest-trapping for egg selection about two weeks before setting hatching eggs under the first hens that settle into *broodiness* (undergo the hormonal changes inclining them toward a long spell of uninterrupted incubation). But in practice it was a good idea to begin nest-trapping a week or two *before* that. Why? Once a hen went broody, she no longer laid any eggs. Consider Blær, my best broody and, even in advancing age, one of my best layers—not surprisingly, I very much wanted to set *her* eggs. But I knew she would reliably be among the first hens to "go broody." Thus, in order to set *her* eggs, I had to have *already collected* them (marked with ID and date and set aside in readiness), *after* the required two-week period following breeder isolation, *before* she entered broodiness.

(Oh, and by the way, please don't think I go all goo-goo naming my individual chickens—the majority are just numbers. But the relationship with my broodies becomes over time rather intimate. While Blær herself came to me with a name—given to her by her Icelandic-American breeder—in other cases it's not so much that I give a broody a name as that suddenly she just *has* a name.)

The "reserve" cocks got their turn in the breeding sections as well: Every three days or so I would switch the cocks, again taking care to pen reserve Red cock with the Green hens; reserve Green cock with Blue hens; and reserve Blue cock with Red hens. The three cocks who had been in service would then be in the greenhouse "bachelor" pen, enjoying each other's company as buddies with no hens around.

Collecting Hatching Eggs

A common misunderstanding is that a broody hen will only set her own eggs. A wild mother bird, of course, sets a nest containing her eggs only. In contrast, the chicken broody will happily set any clutch of eggs you choose to put under her—I have even set duck eggs under chicken broodies. Indeed, one of our bantam hens once hatched a goose egg! The eggs you choose to set can be from another source—from a fellow flockster, from a commercial hatchery—or from any birds in your flock from whom you want progeny.

Bear in mind that a wild mother bird lays an egg a day in her nest, and each egg remains viable for many days at ambient temperatures. That is our key to proper handling of hatching eggs: Store them in an egg carton at *room temperature*—not beside a heat register, nor in the full light of the sun. Store gallinaceous eggs—those of guineas, turkeys, and chickens—narrow-end down, since the air cell inside the egg is at the other end, which is where we want it for now.

The eggs should be turned a bit daily: Doing so imitates the female bird, who turns the eggs in her clutch as she lays another each day (and who later, when she is incubating them, will turn them several times a day). Turning the eggs prevents the germ cells (before incubation) and the embryos (during incubation) from sticking to the eggshell. As well, turning helps keep the all-important yolk in the most protected position, in the center of the egg, surrounded by the albumen. We do not have to duplicate entirely the broody bird's turning of her eggs, a gentle daily tilting will do: Just prop one end of the holding carton on a block of scrap 2 × 4 or the like; next day, change the prop to the other end.

Orientation of waterfowl eggs is not critical for storage up to a week. If holding longer than that, store them on their sides and rotate them a half turn daily—clockwise one day, counterclockwise the next.

When I released a hen from the trapnest, I penciled on her egg the date and her band number and clan color (R, G, or B). I placed the eggs into separate cartons, labeled by clan. When a carton was full, I rotated out the oldest eggs each day. Eggs rotated out, at most a few days old, were still perfectly edible.

Using this strategy, I always had on hand a minimum of a dozen freshest eggs from each clan, ready to be set without delay as soon as the next hen went broody.

If an egg was dirty, I did not keep it for hatching. However, I would use eggs that had a trace of smear or stain for hatching. In my experience that was no problem for the development of the eggs. I emphasize this point because it is sometimes recommended to wash hatching eggs with a special solution. I never washed hatching eggs. The wet coating on an egg when it is expelled by the hen—called the bloom—not only helps lubricate the egg to ease its passage but provides a protective barrier against bacteria. Washing eggs may reduce the time they can be safely stored, and I believe could decrease protection of the developing embryo.

How long can you store eggs and be sure they are likely to retain hatchability? At least as long as a hen who has hidden a nest would take to assemble a clutch of eggs before setting—up to two weeks. Conservatively, if you store hatching eggs under the conditions described earlier, they should all remain hatchable for ten days. After that time there will be a falling-off of hatchability.

However, if older or less-than-ideal eggs are for some reason the only ones available to you, know that many a clutch of eggs three or even four weeks old have yielded surprisingly good results. A friend of mine had sold all her eggs—and then her prize cock was killed by a fox. In desperation she asked her customers if they still had any of the fertile eggs she had recently sold them. Some did, and donated them to the cause. Hatch rates turned out to be excellent, and my friend's breeding line was saved.

My Record-Keeping Spreadsheet

Only you can say what constitutes helpful records for your breeding project, including what traits should be tracked and, especially, what degree of exacting detail is personally tolerable.

The simplest records can be jottings in a notebook, but these days electronic spreadsheets are available to many flocksters. If a spreadsheet is used in a static way—as a place to note numbers of eggs laid or weights at slaughter—it is just an alternate form of a paper notebook. A spreadsheet becomes considerably more useful if we take advantage of its capabilities for *manipulating* the data entered as a guide to breeding choices. When combined with the results of well-organized nest-trapping, a spreadsheet becomes a powerful selection tool indeed.

Let me illustrate by sharing with you a spreadsheet I designed for selection in my Icelandic breeding flock. I hope it demonstrates the range, granularity of detail, and capacity for data manipulation possible when using a calculating spreadsheet as your record-keeping tool of choice.

You might say that such a spreadsheet allows a breeder to "ask questions" of their compiled data. I had an extraordinary range of questions I wanted to ask, starting with the obvious ones that anybody breeding a layer flock would ask: What is the rate of lay of this hen—this potential *breeder* hen? What is the average

Table 26.1. Layer Performance Score Sheet

Age	Flock ID	TEST PERIOD 1		TEST PERIOD 2		Average Score*	Total Eggs Laid	Weighted Rate of Lay†	Weighted by Age‡	Broody?	Status
		Score	Eggs laid	Score	Eggs laid						
2012	R102	0.73	4	0.70	4	0.71	8	5.7	10.3	Yes	Good broody!
2013	B103	1.17	3	1.28	3	1.23	6	7.4	11.8		Good older layer
2013	R103	1.06	3	1.18	3	1.12	6	6.7	10.8	Yes	Good broody!
2014	G104	1.04	3	1.00	3	1.02	6	6.1	8.6		Good older layer
2014	G106	1.24	2	1.20	2	1.22	4	4.9	6.8	Yes	Good broody!
2014	B104	0.69	1	0.49	1	0.59	2	1.2	1.7		*Cull*
2015	G108		0	1.07	2	1.07	2	2.1	2.6	Yes	Good broody!
2015	R115	1.15	2	1.13	2	1.14	4	4.6	5.5		Give to Gary
2016	B129	0.50	1	0.49	1	0.49	2	1.0	1.0		*Cull*
2016	B135	1.38	2		0	1.38	2	2.8	2.8		Give to Gary
2016	B140	1.08	1	1.02	2	1.05	3	3.2	3.2		Give to Gary
2016	R135	1.14	1	1.28	5	1.21	6	7.3	7.3	Yes	Good broody!
2016	G128	1.08	3	0.86	4	0.97	7	6.8	6.8		Good 2nd year layer
2016	G131	0.96	3	0.95	5	0.96	8	7.7	7.7		Good 2nd year layer
2016	G144	1.20	4	Setting!		Setting!			Setting!	Yes	Good broody!
2016	R158		0	1.07	2	1.07	2	2.1	2.1	Yes	Good broody!
2017	R122	1.32	2	1.21	3	1.27	5	6.3	6.3		Good 1st year layer
2017	R123		0	1.23	2	1.23	2	2.5	2.5		Give to Gary
2017	R124	1.18	3	1.11	3	1.14	6	6.9	6.9		Good 1st year layer
2017	G122		0		0	0	0	0	0		*Cull*
2017	B111	1.24	3	1.17	3	1.21	6	7.2	7.2		Good 1st year layer
2017	B113	0.79	1	1.36	3	1.07	4	4.3	4.3		Give to Gary
2017	B117		0	1.48	2	1.48	2	3.0	3.0		Give to Gary
2017	B119	1.38	2	1.13	3	1.25	5	6.3	6.3		Good 1st year layer
2017	B147		0	1.27	3	1.27	3	3.8	3.8		Give to Gary
2017	B148	1.33	3	1.24	2	1.29	5	6.4	6.4		Good 1st year layer
2017	R174		0	1.29	3	1.29	3	3.9	3.9		Give to Gary
2017	R177	1.33	2	0.99	2	1.16	4	4.6	4.6		Good 1st year layer
2017	R182		0		0	0	0	0	0		*Cull*
2017	R188	1.28	2	1.19	3	1.24	5	0	0		Good 1st year layer
2017	G165	1.36	3	1.09	3	1.23	6	7.4	7.4		Good 1st year layer
2017	G167	1.12	3	1.13	2	1.13	5	5.6	5.6		Good 1st year layer
2017	G173	0.99	1	1.21	3	1.10	4	4.4	4.4		Give to Gary
2017	G177	1.36	2	1.25	3	1.31	5	6.5	6.5		Good 1st year layer
2017	G178	1.06	1	1.25	2	1.15	3	3.5	3.5		Give to Gary
2017	G180	1.17	3	1.24	2	1.20	5	6.0	6.0		Good 1st year layer
2017	G182	0.80	1	0.80	1	0.80	2	1.6	1.6		*Cull*
2017	B154	1.35	2	1.20	2	1.28	4	5.1	5.1		Good 1st year layer
2017	B156	1.29	2	1.28	3	1.28	5	6.4	6.4		Good 1st year layer
2017	B157	1.35	1	1.15	1	1.25	2	2.5	2.5		Give to Gary

* Average Score is an average of the scores for the individual test periods 1 and 2.
† The Weighted Rate of Lay score also accounts for factors such as egg shape and shell quality, not simply number of eggs laid.
‡ The Weighted by Age score also factors in a hen's age, giving higher scores to hens who maintain good egg production as they grow older.

size of her eggs? Simple enough: Set up the trapnests and record her performance day by day. But how does her laying performance vary through the seasons? That question can only be answered by multiple tests spaced throughout the year. Some hens might lay impressively in spring, some might excel in summer, and some may have learned the trick of laying well even in the cold and dark of winter. But only if I test my layers in all three seasons can I discover which hens perform best in all of them—that is, which are supremely desirable as my very best breeders. And, since I never practiced a rote two-years-and-out cull of my layers, I wanted very much to know about a hen's rate of production as she aged. Since all hens lay fewer eggs as they get older, however, I needed to judge the hen's rate of lay in her advancing years *within her age cohort*—that is, in relation to other hens of the same age in the flock. To me the most valuable hen in the flock—and thus the ideal candidate as breeder—is the one who continues to lay well even as she advances into vigorous old age.

I also wanted to address some "nonproduction" laying questions that were seriously "bugging" me. A major disappointment with my Icelandics was the realization that egg shape and shell quality had been sadly neglected in breeding flocks "back up the line" from whom my stock descended. You may well think I'm simply being persnickety to make a big deal of egg shape and shell issues. But even at the aesthetic level I beg to disagree. The classic egg is broader at one end, tapering elegantly at the other, its surface smooth and flawless. I observe more and more lumpy shapes in eggs from many sources—tell me why I *shouldn't* want the original instead, that object of quietly stated beauty, gracing my kitchen counter? But such characteristics are in fact not just aesthetic—a lumpy, misshapen egg with a too-thin shell (see figure 26.7), to say nothing of one with a fault line—is a weaker egg, hence inherently less fitted to successful hatching. An egg that is too-round or too-evenly-oval? The classic shape is important to somebody handling a lot of hatching eggs, which should be stored with the air cell up—that is, in the broad, not the pointed, end. If the egg doesn't taper, it means taking time to candle eggs before placing them in holding cartons, and who wants to do that? (Refer to "Candling the Eggs" on page 276 if you need more information on candling.) Correction of such flaws was a major goal of my breeding project from the beginning, and a major focus in the spreadsheet I designed.

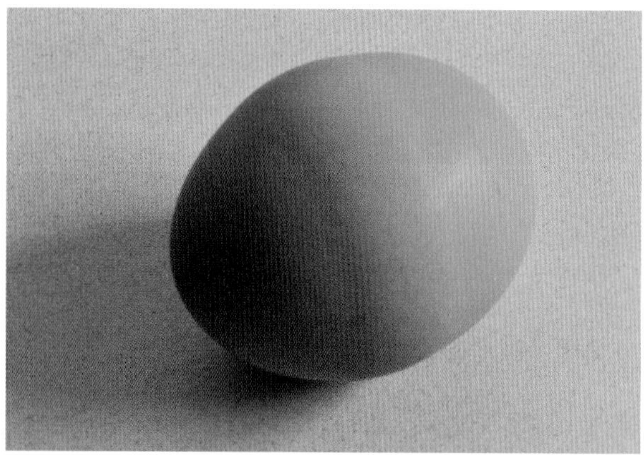

Figure 26.7. This egg illustrates poor shape and shell quality. Note that some eggs laid by my Icelandic hens were far more compromised than this one.

I set up the spreadsheet with a separate row for each hen in the flock, identified by wing band color and number, with hens divided on the sheet into cohorts by age (all hens hatched in 2016 in a distinct group; all hatched in 2017, in the next). During a given test period I nest-trapped *every* hen in the flock and entered data about her production in her individual row on the sheet.

Not surprisingly, I wanted to track the total number of eggs laid by a hen in the test period, as well as the average size of her eggs. Since I value maintenance of good egg production as a hen ages, however, it wasn't good enough simply to record an aging hen's rate of lay in absolute numbers: Of *course* a two-year-old hen will lay better than a five-year-old. But, if that five-year-old lays twice as many eggs as any other hen *her same age*, she may well be more desirable as a breeder than a high-producing two-year-old. So I designed the spreadsheet to *weight* a hen's laying performance in comparison with that of all other hens *in her age cohort*.

Finally, regarding the flawed shape and shell qualities occurring all too frequently in eggs from my

Icelandic flock, I wanted to identify which specific hens laid poorly, so-so, or well—in comparison to the shape and shell characteristics of the ideal egg. I designed cells in the spreadsheet to evaluate every egg, minutely, for such traits.

Table 26.1 shows an example of the type of score sheet for layer performance that can be generated using my spreadsheet. For a larger-scale, more detailed view, please see appendix F for more information about the design and use of my "Layer Performance Spreadsheet." You will find there as well an online link from which you can download a fully functional electronic version.

While the spreadsheet may seem at first complex and difficult to follow, understand emphatically its real-world utility in my breeding program: Using it to target not only gross egg production but also egg shape and shell quality, I found marked improvement in the latter after only two years. Use of the spreadsheet identified hens with better winter production, and those especially valuable breeders who maintained good production as they aged.

At the End of the Season

As said, I preferred to have six cocks on hand at the beginning of the breeding season, for both increased genetic diversity and as "insurance." (Lose one Green cock and you still have one left for breeding. *If* you have two Green cocks.) Once the breeding season was over in mid to late spring, however, and I was no longer separating the breeders, six cocks was too many in a flock of two to three dozen hens. I would typically cull one mature cock from each clan immediately, retaining one of each clan. As the new cockerels grew and I made my selection for two replacement cocks in each clan, I culled the remaining older cock as well. Note how managing replacement of breeding cocks in this way ensured that *both* cocks doing breeding service in any given year *hatched* in the previous year. So long as first-year cocks *only* are used in a clan-breeding rotation, there will never be parent–progeny matings.

Mature cocks are even tougher than old stew hens, and at our place they are reserved for the same stockpot.

CHAPTER 27

Working with Mother Hens

It has been decades since I started routinely moving new chicks with their mother directly from the hatching nest to the pasture, my confidence that Mama would not fail them increasing with each new hatching season. So it was that one morning a few years ago I moved a just-hatched clutch of chicks with their mother, Blær, best broody in my flock, to the grassed enclosure in front of the Broody Annex. Around the middle of the day, off Ellen and I went to a party.

As we were leaving the party in late afternoon, a storm unexpectedly blew through. The skies opened, with rain of biblical proportions—a neighbor's gauge recorded more than an inch in twenty minutes. I was terrified for the new chicks. I knew from experience that Blær would not at first "remember" the built shelter as the place to bring her chicks in a rain, that she would get back into that habit only after I carried her and the clutch into the coop the first couple of nights. On this first day she would do what any *Gallus gallus* mother would have done: She would hunker down in place, sheltering the chicks under breast and wing. And while I knew a good hen provides an amazing degree of protection against extremes, for these chicks it was a stark "dry or die"—and there was *no way* Blær could keep them dry in rain this torrential.

I ran straight from the car to the enclosure, looking frantically for Blær in the lashing rain. I found her in the tall grass, soaking wet but serene as the Buddha, tight to the ground as a barnacle. I picked her up and found her chicks—*fluffy* dry.

How miraculous was that moment! It explains why hatching with natural mothers is simply a given for me. I have written a lot about backyard chickens as the *domesticated* Red Junglefowl, but as I held the drenched Blær, I touched the indomitable spirit, the motherly valor, the purity of devotion of *Gallus gallus* herself. I hope you will get to experience the rare beauty of such moments as well—and to share them with your children.

Why Not Let Mama Do It?

A regular part of the green season at our place is watching a Carolina wren in her rituals of motherhood—building a nest, incubating her eggs, and caring for her chicks until ready to fly off into the joys and hazards of the wide world, beginning the cycle anew. Seeing her at this great work is a special joy.

When we need to raise chicks, why not let our hens imitate the mother wren and hatch their own young? There are great advantages to recruiting a mother hen for this work, perhaps the most important being that she is a lot smarter at it than we are. No incubator manual, no instructions for tending mail-order chicks in a homemade brooder, will put you anywhere near her level of expertise. There are few joys on the homestead to match seeing a hen with her clutch of chicks. A mother hen saves us the multiple expenses of incubators, heating elements, and thermostats—the chicks she provides us are essentially free. Her eggs' embryos continue growing even during prolonged electrical outages. After the chicks hatch, she reduces the cost of feeding them by teaching them to forage natural, nutrient-dense foods of a quality superior to anything we can offer from a bag.

I have found, over many years giving advice about working with natural mothers, that the most common mistakes arise almost exclusively from a lack of understanding of the details of avian reproductive biology. So let's begin with the mother wren's spring rituals—mating, nesting, caring for her young—to understand the natural behaviors in play. Then we may better shape strategies for working with mother hens.

The Mother Wren

Please understand that the following discussion of nesting behaviors in this wild species—the Carolina wren—is greatly generalized and does not address the many variations in avian reproductive behaviors. I am not writing a zoological treatise, just trying to develop some general comparisons to better understand broodiness in chickens and other domesticated fowl.

Seasonality, Mating, and Building the Nest

The time of year when the wren's mind turns to love and motherhood is not haphazard—it is strongly keyed to the opening of the green season, when chicks can be

Figure 27.1. *Gallus gallus* mother (in my backyard).

fed from the newly abundant insect life. A female may hatch only a single spring brood, or may hatch a second in the summer, but she will never nest in any other season of the year.

Increasing day length and warmer temperatures trigger nesting and mating behaviors. The female finds a mate—her eggs will not be fertile without the participation of a male—who usually has already started building a nest. Note that the nest is small and well hidden, the wren's best provision for thwarting hungry predators. Though some bird species nest communally, the wren wants to be secluded, secure from intrusion by other nesting females, who are also looking for a place to hatch their babies.

Laying a Clutch of Eggs

Mating and completion of the nest usher in hormonal changes that bring on the production of eggs in the female's oviduct. The wren typically lays one egg per day, until she has assembled her clutch. Do understand three essential things about the egg-laying period: First, unlike in our own species—in which sperm remain alive only a limited time in the female's body—in avian species the male's sperm remain viable and mobile for much longer. Thus *the sperm from a single mating continue to fertilize multiple eggs*, as they are grown sequentially in the oviduct and laid over the course of many days.

Second, the wren lays *only the number of eggs she can completely cover and keep warm*, and no more—she has the instinctual wisdom to know that embryos in eggs not adequately covered against the chill will die. Once she senses she has assembled as large a clutch of eggs as she can effectively incubate, hormonal changes shut down further egg production.

The final critical point is this: During the egg-laying period, the wren comes to the nest, lays her egg, and flies away—she does not remain on the eggs, which thus remain at ambient temperatures. Inside, the fertilized cells from which embryos will grow are poised and waiting, but dormant. The brief warming of eggs already in the nest, while the wren is laying the next egg, does not trigger embryo growth. *It is only the sustained warming of the eggs to the mother's body temperature* that will do so.

Incubating

Only after the clutch is fully assembled, and egg production has ceased, does the mother wren settle into the nest and remain on it full time. Only at this time does the uninterrupted body heat of the mother cause the fertilized cells in the egg yolks, *regardless of when the eggs were laid*, to begin growth as embryos. All at the same time. All on the same schedule for hatch.

It is worth emphasizing: *No additional eggs are laid after the mother begins incubation.* Embryos in eggs subsequently laid in the clutch would be on a different schedule for hatch—a disaster.

Once the embryos begin to grow, they cannot revert to dormancy—they must remain at incubation temperature and grow rapidly. They die if left too long without the mother's body heat. The mother makes only brief outings from the nest, for food or water. She also leaves to poop, to avoid fouling the nest. Such brief forays from the nest do not chill the embryos sufficiently to kill them. Indeed, among some avian species—waterfowl, for example—a brief daily cooling of the eggs as the female takes a nest break is apparently necessary for proper development of the embryos.

Once the mother wren is settled into incubation, traumatic disturbance—say, by something she sees as a predator—is likely to break up her nesting behaviors, wrenching her out of the broody state of mind and causing her to abandon the nest. It makes more sense, perhaps, to leave the compromised nest and start the entire process elsewhere, rather than risk return of the predator and loss of the current clutch after yet more time and effort.

Hatching

Hatch occurs after about fifteen days. The sounds and movements under the mother as the chicks heroically break out of the shell call the wren's mind into the next phase: nurture of the chicks. She will not wait indefinitely on a final egg that for some reason fails to hatch, nor will she make any efforts for a chick that is weak and struggling after hatch. She focuses all her energies on the vigorous chicks.

Nurturing the Young

During the following couple of weeks, the mother serves three vital roles for her rapidly growing young. Assisted by the male, she feeds them rich natural foods, mostly insects. She ensures they stay warm enough, since the flimsy down with which they hatch does not provide sufficient insulation for the chicks to maintain body heat. They do not chill dangerously while the mother is away on routine flights to catch insects, and she covers them in the nest at night, as well as during wet, windy, or especially chilly conditions, when the chicks need some on-the-spot warming. And of course, any mother bird does her best to protect her babies from predators.

Defensive behaviors vary among avian species. I don't think wrens attack threatening predators—their main defense is the cleverness with which they hide their nests. Some mother birds, especially ground-nesting species such as grouse, put on a distraction display—floundering away from the nest, faking a broken wing—to draw a nosy predator away. Some species will attack a predator outright—we have observed an entire local community of barn swallows gang up on crows and blue jays flying too close to a nest with swallow nestlings. A mother Canada goose will fearlessly attack a fox who is trying to get at her goslings.

The high-payoff insect food from the tireless mother brings on an amazingly fast rate of growth in the chicks. The season is short, and the chicks are vulnerable to predators in the nest—snakes, rodents, other bird species—so they must grow up and start living on their own in a hurry. By the time they are well feathered enough to maintain body temperature on their own, and are strong enough to fly, their hardworking mother has enabled them to do something truly extraordinary: In one swoop they start flying, feeding themselves, and roosting in the wide world, without retreat to a nest. The mother has made a heroic effort getting them to this point, but she now breaks the maternal bonds and resumes a life focused on her own needs.

The Mother Hen

Understanding natural brooding behaviors among wild birds prepares us to anticipate challenges and to find solutions when working with broody hens. The important thing to appreciate from the beginning is the degree to which chickens, and other domesticated fowl, have moved away from strictly natural behaviors. In some ways, as an avian species, chicken nesting behaviors parallel those of the wren—and in some ways they have changed considerably in the alliance with *Homo sapiens*. Fortunately, both the behaviors that are still natural *and* those that are much altered can be used to advantage when working with broodies, so long as we are clear about the nature of a given behavior. The following discussion focuses on comparisons and contrasts with nondomesticated avian nesting behaviors, to reveal effective management strategies for broody hens.

The Great Forgetting

The first question to ask about a hen you want to use for hatching chicks is this: Does she "remember" the intricate lore of incubating, hatching, and nurturing chicks? The question I am asked most frequently about using natural mothers is: "But my hens don't go broody—how can I make them do so?" The simple fact is that *modern breeding has deliberately and dramatically suppressed instincts for natural reproduction in most breeds.*

In the era of mass production of chicks using artificial incubation, broodiness is considered not merely an unnecessary nuisance but an economic calamity. After all, a hen who has gone broody and is determined to be a mother—like our broody wren above—ceases laying eggs. Thus a major component of modern breeding has been a rigorous selection against the broody trait—that is, the elimination from breeding flocks of hens with a strong mothering instinct, in favor of hens who happily lay their egg per day, having forgotten that doing so has any relation to reproducing the species.

Broody Breeds

If you're looking for reliable broody hens to hatch chicks for you, don't play guessing games with hens of hit-or-miss breeds like Buff Orpington or Brahma. While they usually make excellent mothers if they do decide to brood, you can't count on their doing so. Here are three options for broodies almost guaranteed to work for you, starting with my personal favorite.

Figure 27.2. Old English Game hen, possibly the best chicken mama on the planet.

Though *Old English Games* have a thousand-year history as a utilitarian farm fowl, cocks of this breed have been exploited in the cockfighting pits, which may partially explain why hatcheries typically do not offer OEGs for sale. They may be nervous about being tainted by an assumed support of cockfighting (though many of them do offer bantam versions of Old English). As well, breeding management of any game breed would be complicated by the heightened level of aggression among cocks. Your best bet for finding good OEGs is to join the Society for the Preservation of Poultry Antiquities (SPPA). Membership puts you in touch with breeders of older breeds, from whom you may be able to get stock.[1]

There are a couple of other options, if smaller broody hens are adequate to your needs. The

Shouldn't it *bother* us that we consider the ideal chicken one who has forgotten how to reproduce her kind? In any case, the effort to eliminate broodiness as an inherited trait has been extraordinarily, though hardly surprisingly, successful: If we make going broody a capital offense, it doesn't take long for the chickens to get the point.

Hens of most modern breeds, therefore, cannot be relied on as mothers. While it is true that hens of some breeds—Cochins, Brahmas, and Buff Orpingtons come to mind—have a reputation for being more likely to brood than others, trying to find good broodies among them is hit or miss at best. Even when the trait is present, it may be more or less intact. Some hens, for example, will go broody in the nest, set with great determination for a couple of weeks, then abandon the enterprise and return to daily activity in the flock. Or perhaps a hen will successfully complete incubation, then prove somewhat clueless about caring for the chicks.

To be assured of hens with sound mothering instincts, it is best to revert to the older, historic breeds of chickens, in whom the broody trait is more the norm than the exception. An outstanding example is the OEG. Other game breed hens—Asils, Malays, Shamos, Kraienkoppes—are also likely to have strong instincts to be mothers. Dorkings, a breed originating in the ancient Roman Empire, may or may not brood well. Bantams as a class are likely to retain the broody trait, though because of their small size, they are more limited in the number of eggs that can be set per hen.

Nankin, one of the most ancient of bantam breeds, was widely used in England from some time before 1500, for hatching not only farm chickens but also pheasants and other game birds. *Silkies* are another bantam breed favored as mother hens. Silkie hens have a reputation as extremely devoted mothers—I've been told by those who use them: "They don't want to do anything *but* set!" A Silkie hen may assemble her next clutch of eggs even as she is completing the rearing of her current clutch of chicks.

If you're adventurous, consider an experiment cited by Charles Darwin in his *The Variation of Animals and Plants Under Domestication* (1868): Two decidedly nonsetting "Mediterranean class" breeds were crossed, and all the daughter offspring of the cross expressed the broody trait. I find it comforting that, despite the intensity of our manipulations, our domesticated fowl retain just under the surface their native wisdom, ready to reassert itself with a single roll of the genetic dice.

While it is generally a good idea to maintain the purity of historically important breeds, Darwin's observation suggests a case where breeding of crosses could be a good thing. For many years I experimented with crossing OEG cocks onto proven broodies of larger breeds in my flock, such as New Hampshire and Cuckoo Marans, with the goal of developing broody hens who could cover more eggs per clutch than the smaller Old English, but who also retained the same level of broody skills and motherly devotion.

I had a lot of fun with my "Boxwood Broody" crosses, a motley assortment for color, pattern, comb, and shank—selection of "keepers" was based not on any particular conformation but on how well a hen performed as a broody. As hoped, the larger cross hens were able to hatch larger clutches, though a high level of broody skills always trumped body size. I discuss this experiment in detail in the "Breeding for the Broody Trait" section on page 283.

Sadly, even among these petite breeds, the prejudice against broodiness is strong and is being selected against by some breeders.

The Brooding Season

As said, there is a pronounced seasonality to nesting behavior in the female wren: She goes broody in the spring, at the same time all the other wrens in the area make nests and incubate eggs. A particular female may or may not brood an additional clutch later in the summer, but she will never do so in the fall or winter.

Seasonality is one of the ways in which chickens' mothering instincts differ most sharply from strictly natural behaviors. Hens with the inclination will begin going broody as late winter turns into early spring. But there is no predetermined time for onset of broodiness—some hens may wait a month or even longer before making it obvious they want to be mothers. Another trait with a big range of variability is whether the hen returns to broodiness later in the season—many will not, though some will hatch a second or even a third clutch of chicks.

A hen may even take it into her head to set a clutch of eggs in winter. Going broody in winter is rare, but you will occasionally find a hen setting a clutch of eggs in an out-of-the-way corner, indifferent to the bleakness of the winter landscape outside—a testament, I suppose, to the degree to which the domestic chicken has loosed her moorings to natural seasonality. Whether to let a hen who goes broody in winter hatch her clutch is up to you. My own sense is that brooding so much at variance with the nature of the season violates biological boundaries—fragile, highly vulnerable chicks simply have no place in the iron grip of winter, however dedicated the mother. At the least, I break up a winter broody, and I might even cull her outright.

Spotting Broodiness

"But how do I know a hen has gone broody?" Unlike the wren, the chicken broody does not usually make a new nest for incubating her eggs. There can be exceptions—as when a hen who is completely free-ranging hides a nest and assembles a clutch of eggs, with no intervention from you. But typically, *the hen will go broody in the same nest where she—along with her sisters in the flock—normally lays her eggs.*

Thus a further contrast with natural behavior: The chicken broody will enter true broodiness when the requisite hormonal changes occur, irrespective of whether she is in fact setting any eggs. (During the hatching season, I leave a fake egg or two in all my nests, about which more later, to give a hen something to settle on if she is tending toward broodiness.)

A hen who has "gone broody" will remain in the egg nest full time. Don't assume that a hen is broody simply because she is spending a lot of time on the nest—some hens linger a long time when tending to the business of egg laying. But the broody hen does not leave the nest to go to roost with her sisters at night. She takes on a deeply settled, Zen-like intensity that is impossible to describe but highly symptomatic once you have seen it (see figure 27.3). An exploratory hand will draw forth a flattening of the hen in the nest, a fierce raising of the hackle feathers, and an outraged *Skraaaawk!* That hand may receive a peck, usually a mere token of vigilance against the predator—though on a couple of occasions an especially indignant peck has drawn blood from my own hand, especially in the case of guinea hens.

Once you have determined a hen is indeed fully broody, *you must remove her from the egg-laying nest.* Far too many beginners try to leave the broody hen in the regular egg nest she has chosen as the place to hatch chicks. But the hen may leave the nest to relieve herself, then return to the adjacent nestbox by mistake, leaving her incubating eggs to chill and die. Or other hens may enter the nest to lay their own eggs. Even if you have

Figure 27.3. Zen-like intensity.

marked the eggs in the clutch so you can distinguish them from eggs laid by other hens, sooner or later all that jostling will break an egg, coating the clutch with goo. Since incubating eggs "breathe"—via essential oxygen exchange through the shell, which is porous at the microscopic level—a coating of broken egg can smother the growing embryo. Trust me on this one: I've been there. It doesn't work. Don't do it.

Isolating the Broody

Let's be clear: Truly natural behavior would mean that the broody hen would not tolerate being moved into a different space; the change would break up her broody inclination. But chicken broodies are likely to remain broody despite being moved. This is not to say that every broody hen will tolerate it: You will occasionally encounter a hen who will not settle anywhere else than in the egg nest she has chosen for incubation. But because chicken broodies as a group will tolerate this level of management intrusiveness, I *insist* on it; that is, I define a good broody as one who works with me, by submitting to the move, in our mutual endeavor to hatch her chicks.

Do note that broody females of other domestic fowl species (ducks and geese, guineas and turkeys) retain the more natural behavior in this regard: *They cannot be moved, after thoroughly settling, without breaking up broodiness.* With such broodies it is necessary to prepare ahead of time a nest for each female who might go broody. Once an individual becomes broody, add some sort of physical barrier to prevent other females from entering her nest or otherwise disturbing her.

When you move the broody hen, you must do so with as little intrusiveness as possible. *Move her at night only*, when she is in the trancelike state that is chicken sleep. Move her to an isolated place of her own, physically separated from the other flock members, with a previously prepared nest. Initially, provide some fake eggs on which she can settle in the new nest: You can purchase plastic or, less commonly, wooden or ceramic eggs. You could also use golf balls, or anything loosely suggestive of "egg" in shape and size.

Figure 27.4. Broody box, in a unit of six mounted on the wall. Note feed and water free-choice, room to take a nest break, and easy-clean wire mesh floor that increases ventilation.

As for the isolation area or *broody box*—any out-of-the-way corner will do. The area should provide enough space for a nest, access to feed and water, and room to stretch and poop. Like the mother wren, a good broody has the instinct not to foul the nest as long as she has enough space. Use whatever you have on hand—cardboard, scrap plywood, wire screens—to establish a physical barrier around the broody box that will exclude other hens from entering the nest.

If you do a lot of hatching with broody hens, install permanent broody boxes. I prefer mounting them on the wall, so no floor space is lost for the rest of the flock. I have a set of six so mounted, made of leftover plywood from a friend's construction project (see figure 27.4). Each box should be at least 30 inches (76 cm) wide by 24 inches (60 cm) deep and 16 inches (40 cm) tall. *Use wire mesh*—½-inch hardware cloth is best—rather than solid floors. Wire increases ventilation through the unit and is easy to clean—just use a scraper to push the poops through. And it doesn't accumulate an inch of dust in the off-season like a solid floor (the cleaning chore from hell).

As for the nest, you can use cut-down cardboard boxes—which get composted at the end of the brooding season—or scrap lumber or plywood to make sides about 4 inches tall to retain the nest materials: burlap, straw, wood shavings, and the like.

It is not unusual for a hen to be somewhat agitated the day after being transferred to the broody box, moving about restlessly, as if trying to find her way back to her chosen nest, perhaps squawking indignantly. Most often, however, she will settle on the fake eggs in the new nest by the end of the day. If she fails to do so, you might leave her in the broody box an additional day. If she still fails to settle, she is unlikely to do so. Then you must decide what to do about her. You might return her to the main flock and see if she becomes broody again—especially in the case of a first-timer, she may settle with less resistance on a second attempt.

Because a hen who goes broody is not laying eggs for me, I take seriously her offer to work as a mother. If she fails for any reason to carry through, I'm inclined to cull her to the stew pot. Such a strategy may seem drastic, but it leaves me with a subflock of dependable working broodies. On the other hand, I *never* cull a good working broody. If she continues to perform well as a mother each season, she retains an honored place in our flock, until she dies a natural death—or I pass her on to someone else who wants to work with mother hens.

Breaking Up a Broody Hen

A key goal in the previous section was avoiding breaking up the broody hen with our interventions. But suppose a hen enters broodiness at a time not convenient to the demands of the season or our breeding plans. You will be amazed at how persistently a broody will set the nest—sometimes for weeks on end. Is there a way to encourage her to get over it and resume egg laying and normal life in the flock?

An excellent way to break up a broody hen is to do exactly as you would if you wanted her to incubate: Isolate her in a broody box with feed and water and a wire floor—but in this case without a shred of nesting material. Keep her there—typically a week or so—until she lays an egg. Then return her to the main flock.

Another effective strategy is to isolate her with an active young cock, whose constant attentions get her mind running in less motherly directions.

If I am maintaining separate flocks—during the clan mating rotations described in the last chapter, for example—the simplest method for breaking up unwanted broodiness is to remove the hen from her chosen nest and put her in another clan's pen. (Remember, I'm using trapnests at that time, and recording eggs by individual hen, so there is no danger an egg from the "outside" hen will be set accidentally.) Generally this transition—not only into a different physical space but into an unaccustomed social hierarchy—breaks up the hen immediately. Should she try to brood in the new flock quarters, I immediately return her to the other flock. Theoretically, I would keep bouncing her back and forth until she forgets about incubation, but in practice the one additional bounce is all that is needed. (There was one exception—a hen who remained persistently and insistently broody for weeks, whatever I did to break her up. In the end I culled her as the only solution.)

Setting the Clutch of Eggs

Once the broody hen is thoroughly settled, you are ready to set the hatching eggs from your chosen breeders, collected as discussed in the last chapter. *Working only at night*, remove the fake eggs from under the setting hen and substitute your chosen eggs. Here's another big contrast with natural behaviors: Unlike the broody wren, who incubates her own eggs in a nest she built herself—the broody hen sets a nest you provide and place her in, on eggs you choose to put under her.

How many eggs should be set? The exact number is a function of the body size of the hen, the size of the eggs you are setting, and, to a lesser extent, the point in the season. Remember what was said before about the mother wren's instinctual understanding when "enough is enough" regarding clutch size: *She will lay only the number of eggs she can completely cover and keep warm.* You must develop the same sort of instinct for clutch size. For eggs from standard-breed chickens, you might set only five or six under a bantam broody, but nine or ten under an OEG, and up to fifteen under a big matronly Wyandotte, Brahma, or Buff Orpington. Perhaps twenty guinea eggs could be set under a standard chicken broody, perhaps only three duck eggs under a bantam. If in doubt, err on the side of setting slightly fewer. If it is still early in the season and nights are cold, set fewer—the hen's body heat must offset lower ambient temperatures to maintain incubation temperature in the clutch.

I cannot stress enough the importance of not going for broke when determining clutch size, assuming you might as well risk a few eggs that don't make it, for the sake of maximizing the number of eggs that do hatch. Not so fast—that strategy is more likely to *minimize* the hatch. During incubation, the hen moves the eggs around in the clutch (to ensure that all embryos grow at the same rate). If she is setting more eggs than she can completely cover, there is a chill zone around the edges of the clutch in which embryos fail to maintain incubation temperature, and die. When the hen moves those eggs into the center of the clutch and eggs with viable embryos from the center into the overexposed edges, those embryos may also chill and die. The result is a potential loss in the clutch greater than that of the inadequately covered eggs alone.

Once you put the eggs to be hatched under the hen, all the embryos start growing at the same time, and thus are on the same schedule for hatch. Obviously, it would be a mistake to add eggs after the initial setting of the clutch, since those embryos would be on a delayed hatch schedule.

When you set the clutch of eggs under the hen, mark on your calendar the expected hatch date. But I'm going to give you a tip I haven't come across anywhere else: Any source of information on the subject I know will tell you that incubation of chicken eggs takes twenty-one days—as indeed it does, in an artificial incubator. In my experience, however, hatching of eggs under natural mothers is as likely to occur in twenty days as in twenty-one. Make it standard practice to staple on the door of every broody box a note that specifies: the broody inside (name, flock number in your records, or other identification); how many eggs she is setting and of what breed or parentage; and earliest possible hatch date, twenty days out. (On rare occasions I've seen a clutch take a day or two *more than* twenty-one, which would not happen in an incubator.)

Keep in mind, however, that the hen is not tracking incubation period with a calendar. Her setting behavior will end not after a counted number of days but when the sounds and movements of chicks hatching under her signal the end of incubation. Only then will she begin the next phase, mothering the new chicks. This fact is particularly useful if you use chicken broodies to hatch eggs of species with a longer incubation period (guinea, turkey, and mallard-type duck, twenty-eight days; goose, twenty-nine to thirty-one days; Muscovy, thirty-five to thirty-seven days)—a good chicken broody will make no objection to remaining on the nest for the additional time. Similarly, if you need to schedule a hatch to your own convenience, you can keep a settled broody fixed on a clutch of fake eggs a week, maybe up to two, before you place hatching eggs under her—she will set the additional three weeks with marvelous patience.

Managing the Broody Hen

Once you have set eggs to hatch under the broody, she will do the rest. Check on her unobtrusively each day—she will not be disturbed as long as your movements are quiet and gentle. Refill her waterer as needed, and provide feed free-choice. Most broodies eat some feed each day, but it is not unusual for a broody to have little interest in food during this period. Do not be concerned. The broody may lose up to a third of her weight during incubation, so be sure all hens you use for setting are in good condition and well fleshed. As for feed, I've found that if I change to a leaner feed for broodies—coarsely cracked corn and small grains—there is less chance they will have loose, diarrhea-like poops, and the broody boxes remain cleaner.

Some broodies like to leave the broody box occasionally, while others never do so even if given the chance. If a hen makes it obvious she would like to leave the box for a quick outing, I generally allow her to do so. She typically emits an explosive poop of an odd, distinctive smell, then maybe takes a quick dust bath and returns on her own to the broody box, since she instinctively knows the eggs must not cool too much. If she fails to return—say, by mistakenly getting into one of the egg nests to continue setting—the embryos in the cooling eggs will die. If I allow a broody off the nest, it is only when I am caring for the general flock, and I make certain the broody is back on her nest when I leave the area.

Candling the Eggs

It is a good idea to *candle* the eggs midway through the incubation period. Work at night, in full darkness, beside the broody's nest. Remove the eggs from the nest, and working quickly, shine a light through the egg. A flashlight will get the job done, though a stronger *candling light*, homemade or purchased, gives a better view of the inside. I recommend the Brinsea OvaView High Intensity egg candler.[2] A growing embryo will show as a small pulsing mass at the center of a spiderweb of red supply veins. Keep examining eggs until you are sure you recognize a living embryo with its support system. Then it will be obvious when you find a nonliving egg—one with only a yolk showing, or a dark mass. Such eggs should be discarded immediately. There are visual guides online to tracking embryo development by candling eggs.[3]

Figure 27.5. Hatch day.

It is tempting to skip the chore of candling, on the assumption that "it'll all come out in the wash" come hatch day. And frankly, you can usually get away without candling in a typical clutch. But remember, a nonviable egg is a rotten egg, and the putrefaction inside generates gases that can sometimes cause it to explode. Not only is the resultant smell not to be believed, but the remaining eggs get covered with a thick coating of goo, which may block the necessary gas exchange through the shell, in effect smothering the embryos. The contents of the exploded egg carry a heavy load of nasty bacteria as well, which can penetrate the pores of the other shells. You should candle instead.

Hatch Day

Be prepared for hatch day. Remember that chicks become active an amazingly short time after hatch. If the nest in the broody box has sides that could serve as a barrier (such as the one in figure 27.4), just before the expected hatch, place a little straw around the nest to serve as a ramp for a chick who falls out to climb back in. Even in moderate temperatures, *a new chick who cannot get back under its mother will chill and die.*

Check progress without being too intrusive. With most broodies you can slip a hand gently under the hen and feel the eggs. If you feel a crack in one of them, pull it out and examine it. The first stage of hatching is *pipping*—the chick cracks open a little hole from the inside. (At this point, if you hold the egg up to an ear and tap with a fingernail, you hear the chick peeping inside. Kids *love* this.) Later the first crack extends around the entire shell, which breaks open into two halves, the wet, exhausted chick sprawled between. After an hour, the chick will be dry and fluffy and as said surprisingly active (see figure 27.5). During the day you can remove the empty eggshells from the nest as more chicks hatch.

Although the embryos all start development at the same time, their rate of growth varies sufficiently that the first chick may hatch out of its shell sixteen hours or so earlier than its slowest sibling. The hen has the wisdom to know that she must not leave the nest early, and waits patiently for the last chick to hatch. Remember that, just before hatch, the chick absorbs the remaining contents of the egg yolk for a "reserve supply" of nutrients in its system. Thus the early arrivals in the clutch can wait awhile on their slower siblings before requiring feed or water.

Awhile means easily forty-eight hours, and up to seventy-two. I've never seen a clutch of chicks take anywhere near the latter, though one clutch of Muscovy ducklings took two full days to hatch. Do note, however, that this "pause" period in the young hatchlings' life ends as soon as they have their first water and feed. From that point they require regular access to feed and water—there is no going back to the grace period when they could do without.

In practice I typically wait until the morning following hatch day for the last chicks to break free of their shells. Any egg showing no sign of pipping at this point is unlikely to hatch. If you shake it, you may hear a liquid gurgle inside—proof of a nonviable egg. Even if there is pipping that has not progressed, if you tap on the egg and hear no peep, it is clear the chick died attempting to hatch. Such failed eggs should be removed from the nest and the hen should be encouraged to leave and start caring for her chicks.

Sometimes a chick is unable to break free of the shell on its own. While it's tempting to intervene and help it out, this apparent kindness is ill advised. Breaking out of its shell is difficult for the chick, but that difficulty itself is nature's first challenge for the new life. If it isn't strong enough to meet the challenge, and you give it a boost it would not otherwise have had, it is likely to start life weak and struggling. Certainly, if you are a breeder, you would not want to pass its lack of vigor to offspring. Better to let it make that first big step or fall on its own. Like the hen, you should focus your efforts on the vigorous chicks.

Hope, Supermama Extraordinaire

My most spectacular success grafting chicks onto a willing mother occurred with a White Jersey Giant hen named Hope, who came off the nest with only six chicks of her own. Shortly after hatch, I received forty-two chicks in the mail and put them in an adjacent section of the poultry house, set up as a brooder. Believe me when I say that Hope *asked* to be mother to those chicks, and, when her request finally penetrated my thick skull and I opened the door between the sections, Hope rushed in and began busily mothering them.

But that is not the end of the story. That night I tried grafting ten purchased goslings onto a broody goose. The goose was willing, but the goslings wouldn't fix on her (wouldn't recognize her as mother) and kept wandering off through the dew-wet grass. By morning goslings were going down like dominoes. In desperation I scooped up the remaining seven and prayerfully offered them to Hope. She didn't even *blink*.

For weeks thereafter the center of the action in the henhouse was Hope's kaleidoscopic clutch—a tumble of colors, sizes, and shapes during the day; a tight happy clump spilling out from under Hope in a corner at night. And Hope was the happiest hen alive.

Broody Hens as Foster Mothers

If my description of the advantages of a broody hen over the artificial brooder sounds good to you, you might conclude it would be a good idea to give purchased day-old chicks to a broody hen to mother. Will she accept? Maybe. Most of my attempts to *graft* purchased day-old chicks onto a broody have been successful. However, never assume success is certain, and be prepared to brood the chicks yourself if the hen doesn't cooperate. The hen should have been on the nest a couple of weeks—she is unlikely to accept a graft if she has been broody only a few days. But you can hold a willing broody on her nest with fake eggs for four or even five weeks, until your purchased chicks arrive. Remember, the hen is not counting off days on a mental calendar—rather, she moves on to the next phase when she hears live chicks under her body.

To make a graft you should again work only at night. Remove the fake eggs and slip the chicks, who have been kept quiet in their shipping carton through the day, under the hen. Check on them later that night, and again at first light. Chances are excellent the hen will be delighted to welcome "her" new babies into the world. I have only a couple of times had a hen reject grafted chicks—but in the worst case, the hen killed a few of the "intruders." Monitor closely. Be prepared to intervene.

As a rare exception, I have made a daytime graft onto a broody hen—that is, placed in her nest chicks just received in the mail without waiting for nightfall. After all, the chicks have had a long and stressful trip, and I want them in the foster mother's tender loving care as soon as possible. But a daytime graft is chancy, something I attempt *only with a seasoned broody I am confident is rock-solid as a willing mother.*

An Alternative to a Mother Hen

An elderly neighbor Ellen and I got to know soon after we moved to Boxwood told us that, when she was growing up, her mother provided a custom hatching service for the village. People would furnish hatching eggs from their hens and pay Mrs. Green a fee to hatch them out. She set them behind her woodstove and fiddled with them as needed to keep the temperature in the right range. If she could pull it off with a woodstove, you should be able to do so with an *incubator*. For detailed directions and tips from my friend Don Schrider on that subject, see appendix H.

The New Family

My practice is to move the new family directly from the hatching nest to the foraging range (see figure 27.7). If that seems drastic to you, consider this: The first time I did so, the mother hen went onto the pasture in the last week of March with nine chicks right out of the nest. Daytime temperatures ranged from 40° to 45°F (4° to 7°C), with winds up to 25 mph. Water froze in the waterers at night. In contrast with pampered chicks in a brooder, these chicks were out foraging in those conditions from day one, enjoying superior natural foods found by the hen, enjoying an on-the-spot warming session with Mama whenever needed (see figure 27.6). She didn't lose a one.

A good broody is completely fearless in defense of her young. One summer, when I had a lot of chicks in the flock, a Cooper's hawk began making hits, becoming more brazen with each successful meal. As I was feeding one morning, the hawk made a dive for a chick belonging to one of my OEG hens. Just before contact, the mother hit the hawk—a shrieking flurry of wing and stabbing beak and outrage. The Cooper's suddenly concluded that dinner was not nearly as important as

Figure 27.6. On-the-spot warming–and a little loving. Photo courtesy of Bonnie Long.

Figure 27.7. Foraging with Mama from day one. *Bottom right* photo courtesy of Bonnie Long.

Figure 27.8. Hen with chicks in a special mobile shelter I call a halfway house. Photo courtesy of Bonnie Long.

Figure 27.9. Creep feeder when needed, additional mobile shelter at night.

pressing business somewhere else—*anywhere* else. The hen maintained her attack as the hawk made a desperate scramble over the fence and finally managed to break free and land on a tree branch, shaking and stunned.

That same fierceness is directed at any member of the flock so foolish as to show a lack of respect for a broody's babies—ensuring that she and her chicks can run with the main flock on the pasture from their first day.

A problem may arise when two mothers come off the nest at the same time and begin fighting. My solution was a "halfway house" for new families: a low pasture shelter divided into two sections by a wire mesh partition of ½-inch hardware cloth (see figure 27.8). If there was only one new family, I left the doors open—they sheltered in it at night, or during the day if it rained. If there were two new families coming onto the pasture at the same time, I placed each in a section of the halfway shelter. The two hens could see and interact with each other through the mesh divider but could not fight. After a few days, I released both groups to range. Usually the hens now accepted each other's presence and got along peaceably. On rare occasions there would still be a determined fight, with neither hen willing to back down. If it became obvious that a continued standoff would result in serious injury—loss of an eye, for example—I returned one of them to the main flock and allowed the other to adopt her chicks. It was sad, denying the hen who had put so much effort into making a family the opportunity to carry through to completion, but sometimes that was the best solution. I had a few occasions when an especially aggressive mother viciously attacked the *chicks* of another. That behavior goes beyond tolerable levels of aggression—I always culled such a broody.

I don't want to overemphasize the possibility that mother hens may fight. Non-game-type broodies are inclined to be more laid back. Indeed, we had numerous cases in which two mothers formed little communes, the chicks running to either indiscriminately for help finding something to eat or for a warming session, and all settling down for the night in one cozy clump.

If new chicks are on the pasture with the general flock, there are special considerations for feeding. Chicks should not have feeds, whether purchased or homemade, appropriate to layers—that is, supplemented with extra minerals, especially calcium. Feed a compromise mix everybody can eat—a purchased grower ration or pullet developer or an equivalent homemade mix at 16 percent protein. Extra calcium for the laying hens should be oyster shell, free-choice,

Does the Cock Take Part in Parenting?

Anyone who sees a family of geese is impressed with the devotion of the gander as he helps parent the goslings. Does a cock take part in raising chicks? The answer is largely *No*—it is the hen exclusively who sees to the nurture of chicks. However, over the years we have seen a couple of partial modifications to that pattern, endearing because they were so unusual.

When I've had a single pair of OEGs in a pen in the Breeders Annex—one hen and one cock—I've seen on several occasions that, when the hen went broody, the cock settled down in the straw facing her, an air about him of reverent solicitude, almost awe.

One spring one of the young OEG cocks developed an apparently strong family relationship with a mother and her chicks: Wherever they foraged on the pasture, he was right there with them, a behavior I had never seen before (nor since).

Even if a cock exhibits no parental interest in the chicks, he is respectful of them. I have many times seen a hen flare at another hen's chicks—but I have never seen a cock do so.

Figure 27.10. This young Old English Game cock became part of the family after the hen moved out to pasture with her chicks. Photo courtesy of Bonnie Long.

Table 27.1. Broody Performance Score Sheet

Hen ID	**G144**	**R124**	**R103**	**G108**	**R102**	**B148**	**R122**
Baseline Incubation Score*	10	10	10	10	10		
+Early broodiness	4	4	4	4	4	3	2
+Prior broodiness	1		4	2	5		
–"Flirting"	–1				–1		–1
–Panic behavior						–4	
–Excessive agitation							–4
–Refuses broody box							
–Nervous Nellie							
–Poops in nest	–1	1	1	–2	1		
+/–Egg viability at candling	–2	0			–1		Note
+/–Hatch success	1	1		1	1		
+/–Other	–1		7		2	Left nest!	
Incubation Score	**11**	**16**	**26**	**15**	**21**	**Culled**	**Culled**
Baseline mothering score†	10	10	10	10	10		
+/–Mama Mojo			2	1			
–Fails to nurture	–2				2		
–Fails to protect							
–Fails to teach							
–Aggressive to hens							
–Aggressive to chicks							
+/–Other	2	2			2		
Mothering Score	**10**	**12**	**12**	**11**	**14**		
Combined Score	**21**	**28**	**38**	**26**	**35**		

* A hen receives a basic score of 10 so long as she completes the incubation of a clutch of eggs.
† A hen receives a starting score of 10 if she demonstrates basic competence at mothering her chicks.

which the chicks ignore. If the compromise mix seems short on protein (at 16 percent) for fast-growing chicks, remember that the mother hen helps her chicks find live foods that boost their total protein.

If you do want to assure higher protein for the chicks, consider use of a *creep feeder*. The simplest design is a length of wire mesh, leftover perhaps from a fencing project, shaped as a circular or square enclosure, 12 to 18 inches (30 to 45 cm) tall, with a wired-on top of the same material. I have used both 2 × 4 welded wire and 4 × 4 woven wire, depending on which age (size) birds I want to give access while excluding the adults.

I have also used a modification of the minimalist shelter in figure 27.8 to serve as both creep feeder *and* shelter at night or in a rain. Instead of cladding the doors with ½-inch hardware cloth, I affixed slats on them 2⅝ inches (6.67 cm) apart (see figure 27.9). Note that an opening this narrow was designed to exclude my rather small OEG and Icelandic hens. If you have larger breeds exclusively, you could increase that spacing a little. I offered high-protein supplemental feedings for the chicks—crushed hard-boiled eggs, earthworms from the vermicomposting project, Japanese beetles, soldier grubs—inside the creep feeder. I opened the creep feeder at night as an additional shelter. (Do note, however, that I used that creep feeder *only* inside a perimeter of electric net fencing: A door clad with spaced slats would not prevent attacks by predators.)

As said earlier, when the mother wren has readied her brood for the wide world, she breaks her bond

G167	B111	R158	R177	B154	R135	B156	G144 2nd Brood
		10		10	10	10	10
2	1	0	0	0	−1	−1	2nd brood
		1			1		2
−1	−2	−1	−1			−1	
−1			−2				
−1	−2		−1	−1			
					−1		
Note							
Left nest!	Left nest!	5	−4	4	1	2	2
Culled	**Culled**	**15**	**Culled**	**13**	**10**	**10**	**14**
		10		10	10	10	10
		−1		2		2	−1
		−1		2	2	2	4
		8		**14**	**12**	**14**	**13**
		23		**27**	**22**	**24**	**27**

with them and goes her own way. Mother hens break the bond with their young as well, when they are ready to thrive on their own. But the exact point at which a mother breaks the bond varies enormously from one hen to another. Some go their own way early on, roosting with the other hens at night and foraging for themselves. Others maintain the bond much longer—in extreme cases until the chicks are half grown. Chicks whose own mothers have returned to their usual place in the flock, but who still feel the need for a little loving, join the broods of more motherly hens for a while. So long as a mother does not abandon her young before they are ready—fully feathered, robust, and able to forage for themselves—I don't care when she chooses to break the bond.

Breeding for the Broody Trait

I have noted several times the outstanding mothering skills of OEG hens—every one I raised over a couple of decades was certain to go broody and, if allowed to set, was equally certain to be a devoted and fiercely protective mother. I also explained why I chose to raise Icelandics instead—100 percent broodiness is not an unalloyed blessing, and loss of winter egg production is a serious loss indeed.

I found that, yes, enough of my Icelandic hens would go broody to provide all the chicks I needed in a hatching season. In contrast to OEG hens, however, I found the brooding and mothering skills of Icie hens all over the map: The best were on a par with Old English broodies, but many who went broody proved to have disappointing skills, from so-so to abject failure. I suspect that breeders "back up the line" in Iceland had been selectively fighting broodiness in their hens, reducing the presence of the trait in the landrace as a whole, and producing in some hens who retained it a state of confusion.

I designed a spreadsheet to track broody performance, which I'm glad to share with you. Table 27.1 is a sample of the kind of information the spreadsheet can capture. See appendix G for a larger, more detailed view and for tips on using the "Broody Performance Spreadsheet" and an online link for downloading a fully functional version of the spreadsheet that you can adapt to your own uses. The spreadsheet was, first, an aid to memory—choices of broodies to use this year, especially for must-not-fail hatches, depended on precise knowledge of how they performed the year before—and the year before that. As well, tracking broody skills enabled *breeding for the broody trait*—that is, selection of hens likely to enhance its reliability and intensity in their daughters. If you hatch with natural mothers, but find performance of your broodies not always satisfactory, I offer this spreadsheet as a selection tool for its improvement in future generations. And even if you prefer to keep notes by hand, please do flip to the appendix, because you'll also find there plenty of helpful information and ideas for recordkeeping regarding broodiness and motherly performance, no matter what format you use for your records.

Do not doubt the usefulness of a spreadsheet to track performance of broodies. The one I designed demonstrated that in my flock high-performing broodies tended indeed to produce daughters with excellent mothering skills. It revealed as well something about my Icie broodies that astounded me, which I would never have discovered any other way: I found that, year after year, some broodies hatched their clutches at a rate at or close to 100 percent. Others hatched every time at 73 or 82 percent. (If those figures seem odd: I always set eleven eggs per clutch under Icelandic hens. A hatch of 8 = 73 percent; of 9 = 82 percent; of 10 = 91 percent.)

When I checked them in their broody boxes they looked *exactly the same*: settled deeply into that Zen-like intensity, focused, resolute. Yet clutch after clutch, they hatched out at remarkably consistent rates, whether stellar or mediocre. You can imagine I found *that* information extremely useful when selecting for "breeding broodies."

Analysis of broody performance using spreadsheets is an individual choice, not a necessity. Whether or not you decide to keep performance records, I hope that you have the chance to give a willing broody hen a try—for a joy and intimacy experienced more intensely than in any other part of husbandry. If you have children, it is especially gratifying to be able to share with them this miracle of new life and mother love.

PART SEVEN

Poultry for the Table

Photo courtesy of Mike Focazio.

CHAPTER 28

Butchering Poultry

Butchering skills are essential for anyone serious about keeping an ongoing homestead flock. Whether you hatch your own stock or buy in day-olds straight run, you soon have a large surplus of males for whom there is no long-term place in the flock. Even if you keep layer hens only, they will cease or greatly decline in egg production long before the end of their natural lives. Maintaining them "on welfare" is a fine option if you are keeping pet chickens but hardly a practical choice for those whose flock is part of a productive homestead or farm.

Newbie flock keepers often try to find someone else to do their culling for them. I can't tell you the number of times I've received a call out of the blue, a beginner flockster on the other end of the line who started their first flock blithely assuming that *of course* it would be easy to find someone to butcher their birds for a fee. Reality sets in as I assure them that, while dressing out birds for my own table is good work, doing so for someone else would be drudgery—and I don't do drudgery, at any price. "Of course, you're welcome to bring your birds and join me next time I slaughter," I always assure them. "I'll be glad to give you some pointers."

As I write, we are emerging from the Great Isolation of the COVID-19 pandemic last year. I don't know why, but the enforced isolation seemed to nudge many first-timers toward setting up small backyard flocks. Who knows, if most of those in the big surge continue keeping their flocks as things get back to normal, maybe budding entrepreneurs soon will start seeing a money-making opportunity—using mobile processing units that come to *you*, for on-site slaughter and cleanup, leaving neat packages for the freezer behind. But for now, most flocksters find that individuals interested in providing such a service are few and far between, and they are stuck with doing their own butchering.

It is easiest to learn butchering skills working with someone more experienced. In lieu of that better option, I hope you will find the following guide useful. Read it thoroughly, study the pictures, and don't be discouraged if there are points that do not make sense. When you have your bird on the worktable, you will recognize key anatomical features from the pictures, and obscure points from the text will become clear. It may help to work with a partner who assists with point-by-point reference to the guide, while you do the hands-on work—then switch. Good luck![1]

Culling Strategies

The idea of slaughtering their own birds often makes beginning flocksters nervous, so it is important to understand that culling is an essential tool for effective

A Note about Slaughter

Most of us are uneasy at the prospect of slaughtering for the table poultry that we have so intimately nurtured. As we should be—I cannot imagine being complacent about the killing of a beautiful animal for food. As with any moral issue, however, the answers we should distrust most are the easy ones. We need to consider the question of slaughtering in its context. That context at its broadest is how we eat, and therefore "how the world is used."

Reflect first on the origins and nature of *domestication*. The common notion that domestication was initiated and controlled by humans from the beginning is probably a misunderstanding. In a sense certain opportunist species *chose* domestication by moving into ecological niches created by human activities, especially the practice of agriculture. The accommodation that emerged was one based on *mutual* benefit. The result has been that domesticated species have achieved vastly greater reproductive success, in comparison with their wild cousins; and that most domesticated species are dependent on their partnership with humans and would likely not survive outside it.[2]

If we accept that domestication is a relationship of mutual benefits, however, some consequences inevitably follow. Of greatest significance for poultry husbandry is that domestic avian species reproduce equal numbers of males and females. In a domesticated flock, however, keeping all the males that hatch is unmanageable—culling down to just a few is a necessity. You may choose to keep hens only in your flock—but remember that your purchase of pullets-only chicks implicates you directly in the "euthanizing" of excess cockerel chicks by the hundreds of thousands. You may choose never to cull old, no-longer-productive hens in your flock, but those who supply you with chicks will most certainly not make that choice—that is not a viable economic option for them. Again, you are implicated in the killing of their retired breeders if you buy their chicks.

flock management. One year, for example, we had a large group of Silver Penciled Wyandotte cockerels that caused such an uproar Ellen started calling them "the Mafia"—they were constantly sparring, with *everybody*, and mounting the poor pullets and hens nonstop. I took them to the slaughter table en masse, and the chicken run was the Peaceable Kingdom once more.

I like to hatch with natural mothers, but occasionally a hen who has gone broody fails to carry through. Such a hen is failing me as a layer, failing me as a mother, so she serves quite nicely in the stew pot.

Many flocksters like to raise big batches of meat birds for the freezer. I tried to move away from that strategy early on, and to rely instead on the needed culling throughout the year to put meat chicken on the table—call it a just-in-time strategy that has the side benefits of reducing the amount of plastic packaging waste I generate and the energy required to store dressed birds in the freezer.

Progressive culling of birds of all ages opens up culinary adventures unavailable to those dependent on supermarket chicken: young, tender birds just past the fully feathered stage (sometimes called *poussin*, from the French word for "chick"); larger but still tender fryer or broiler size (what I call "spring chicken"), excellent for sautéing, baking, or broiling; older cockerels,

To my mind the conclusion is obvious: We can choose either to keep a domestic flock or not as we wish. But if we do so, there is no escaping our implication in the killing of chickens—*the entire enterprise, keeping domesticated chickens in any fashion, is not a possibility without it.*

But killing to eat is unavoidable, even if we eat nothing but tofu. In our current agricultural practice, the growing of grains and soybeans, supposedly the basis of a "cruelty-free" diet, causes enormous suffering of living beings of all sorts—from soil organisms killed by excessive tillage or agricultural chemicals, to birds and small mammals and amphibians who perish under a rampant agriculture that leaves no room for natural habitats, that "plows to the ditch banks."

It is intuitively obvious to me that eating is an intensely moral act. And I respect the eating choices others make. But I emphatically *do not* cede the moral high ground to anyone eating a plants-only diet because that supposedly means less suffering for living beings. I despise the industrial CAFO (concentrated animal feeding operation) model as a moral abomination, with its reckless disregard for the quality of life of billions of living beings and for the well-being not only of creatures with whom we share the ecology but of ourselves as well. I take my cue from the fact that this abundant ecology thrives because it is everywhere a mix of communities of both plants *and* animals. Keeping *both* in the homestead and small farm as well—what better key could we find to a regenerative and sustainable (which is to say moral) agriculture?

My respect and gratitude to my chickens is as profound as toward the microbes that create soil fertility in my garden, the bees that pollinate my crops, and the decomposer organisms that keep my world clean and sweet rather than a wasteland of putrid corpses. Their contribution to my nourishment and health is life itself. My response can only be a sense of *personal* indebtedness to the chicken on my plate, which leads to a sense of gratitude and duty, not only to my flock but to the totality of the Creation of which it is a Gift.

though less than a year old, for braised dishes such as *coq au vin*; old cull hens, with meat that is flavorful if cooked slowly on low heat but whose main payoff is the fabulous broth they make. Mature cull cocks are also more appropriate to making broth.

Getting Ready

Some elements of preparation, such as the level of tooling up, will vary according to the frequency you slaughter and the number of birds you typically process. Other preparation is the same whatever the scale of the operation.

Starving the birds: I strongly recommend isolating birds to be slaughtered overnight without feed. Do provide water free-choice. The brief starving of the birds clears the gastrointestinal tract, making for easier, more sanitary butchering. (*It is important to note* that "starving the birds" is not limited to simply *withholding feed*: If the birds to be slaughtered are left free to wander over yard, pasture, or even deep litter, the tidbits they find will keep residues in the digestive tract and the cloaca.)

Chill: Do think of freshly dressed poultry as highly perishable, and be prepared to chill the carcasses as you complete them. You can ice them down in a cooler, or pop them into the refrigerator after each bird is finished. I arrange space in the kitchen refrigerator ahead of time and chill the birds, covered loosely with waxed paper, until I'm ready to complete processing.

Butchering waterfowl: This guide focuses on butchering chickens. Butchering geese and ducks is basically the same anatomically, but there are thousands more feathers in these species. You pay your dues when you dress waterfowl!

Use it all: You honor the bird who has made such a contribution to your homestead by utilizing it to the maximum extent possible, minimizing the parts you define as "waste." Learn to make stock from what I call the spare parts. Learn to love liver.

Setup and Equipment

The key operations are: killing the bird, scalding, plucking, and eviscerating. Setup requires, at a minimum, a scalder of some sort, a worktable, cutting tools, and running water. You can work indoors if you like, but I prefer to work outside, setting my worktable in the sun if the day is cool and in the shade of a big white oak if it is hot. I use a 15-gallon fiberglass scalder heated by a thermostatically controlled electric element—like that in an electric water heater. I encase my scalder in 2-inch foam insulation for greater efficiency. You could use instead a large enameled canner, like the one in figure 28.9, on the stovetop or a portable burner. I also use a mechanical plucker driven by an electric motor that features a drum into which are set many stiff rubber "fingers" that slap the feathers off the bird as the drum rotates. The expense of a mechanical plucker—or the effort and time to make one—may only be justified if you process a lot of birds, but a plucker does speed up the operation considerably. You will be more efficient working at a table of a comfortable height, which can be made entirely from scrap or recycled materials. A stainless-steel work surface is best, for sanitation and ease of cleaning.

My father and I made my worktable from scrap lumber, a double stainless-steel sink from a friend's kitchen renovation, and a single sink with drain board I picked up at a junkyard for four bucks. (Still giving best service, decades later.) To accommodate friends who join me for slaughter day, I made insets for the sink wells to provide additional work surfaces: small synthetic cutting boards screwed onto wooden feet (see figure 28.2).

Note the supply hose hooked onto the leg of the table. The pistol-grip sprayer is just the thing for that quick splash of water whenever needed. I spray frequently—for

Figure 28.1. My basic butchering equipment–scalder, plucker, and worktable.

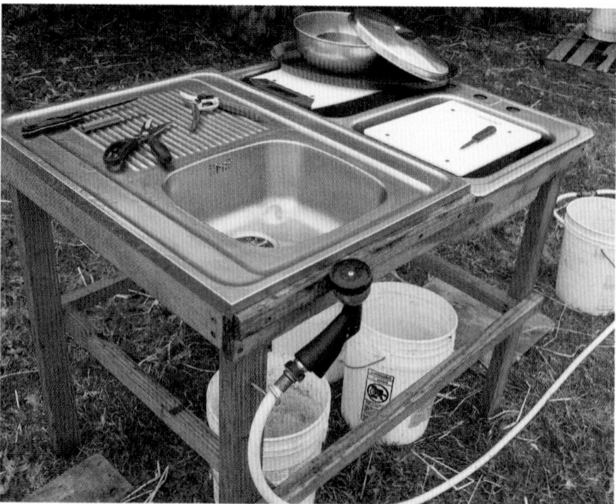

Figure 28.2. A worktable with essential accessories is the key to efficiency. Photo courtesy of Mike Focazio.

better sanitation, to prevent drying of the skin, and to keep flies and yellow jackets from getting too interested.

I like to have on hand a tray for carrying dressed carcasses into the house and a stainless-steel bowl with a lid to hold the usable innards until I get them inside. And of course, the homestead revolves around 5-gallon buckets—I use them under the table to catch the rinse water, so the area does not become a muddy mess, and position them on either side of each workstation, to catch feathers and offal.

About Cutting Tools and Knife Technique

Do yourself a favor and invest in good cutting tools—yes, the good ones are expensive. One of the greatest frustrations in my butchering workshops is the wretched cutlery many beginners bring to the work: Knives that are badly designed, that will not take or hold a keen edge, are clumsy and fatiguing to use. Keep cutting tools sharp—remember that a dull tool is more dangerous than a sharp one, because of the greater force required to cut with it.

You will discover your own preferences for cutting tools. I prefer two knives—one with a thin, flexible, 3-inch (8-cm) blade for more delicate cuts, and the other with a stiff, heavier, 6-inch (15-cm) blade for more hefty, resistant cuts. I also recommend a good pair of shears, for cutting off the neck while butchering, and later for cutting up carcasses in the kitchen. Poultry shears vary tremendously in quality, and I have broken at least half a dozen over the years. After eventually breaking the spring on the best model I ever found (the black-handled "Soft Touch" shears in figure 28.3), I lost patience and bought a Felco No. 2 pruner (red handles). I don't expect I'll ever break that one.

Let me emphasize two points about knife technique, based on the most common mistakes I see beginners make. *Never use the point of the knife when cutting.* All the cuts you need to make are *slicing* cuts, some of them rather delicate—to avoid piercing the entrails—so always use the edge of the blade. And keep it sharp!

You may have wondered about my preference for a stainless-steel work surface, having expected a chopping-board surface designed for contact with the blade. Using my methods, *the blade need never contact the work surface.* Either make downward cuts from above, so the carcass itself prevents contact of blade and work surface, or pull on the part to be cut, using the weight of the carcass to create tension, and make your cut against that tension—rather than sawing or chopping down onto the work surface as when using a chopping board.

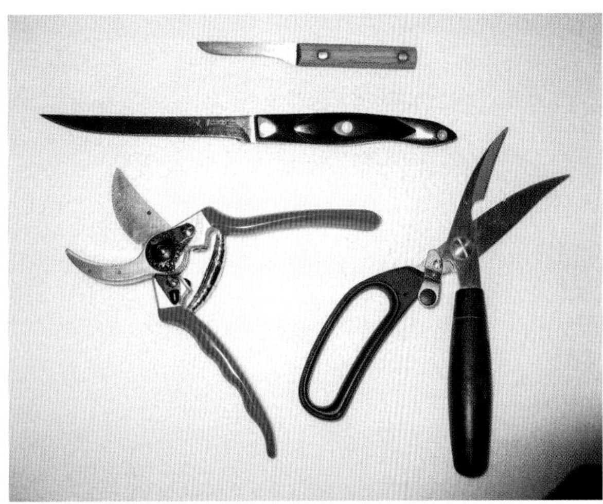

Figure 28.3. You will learn by doing which cutting tools work best for you. Don't skimp on quality: Good tools are a joy to use–mediocre ones are anything but.

Killing the Bird

Emotionally and psychologically, the killing of the bird is typically the most difficult part of the process: You will want to do the job as quickly and humanely as possible. I use the three methods described next, depending on the age, size, and species of the bird I am butchering.

Note that, whatever method you choose, *it is essential that the bird bleed out completely.* The dressed bird will not keep as well, nor taste as good, if the blood remains in the muscle tissue. And there is no need to let the blood go to waste. Traditionally, blood from slaughtering larger animals was caught in a vessel and used in cooking. When I lived in Norway, I enjoyed fresh blood pudding, blood sausage, and even blood pancakes from a couple of pigs slaughtered on a neigh-

Figure 28.4. The chopping block. Photo courtesy of Mike Focazio.

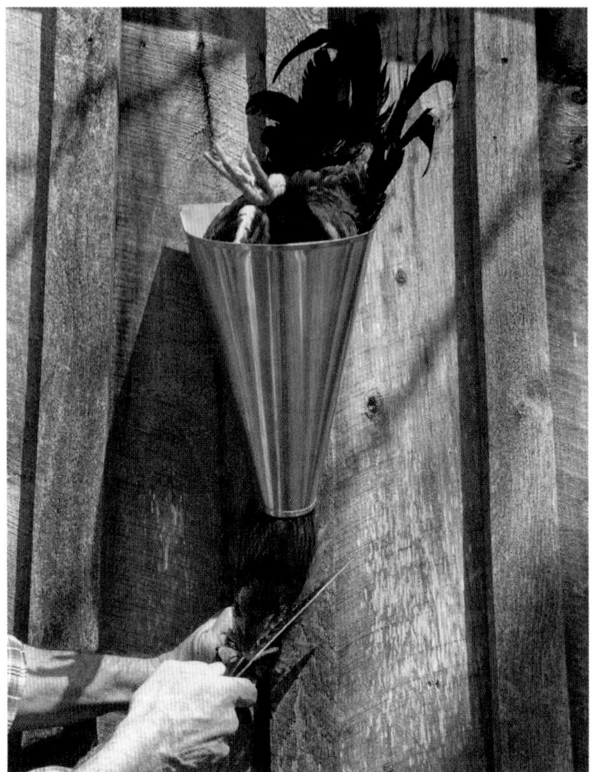

Figure 28.5. The killing cone. Photo courtesy of Mike Focazio.

boring farm. But I never met anyone who made the same culinary uses of poultry blood, and I haven't done so either. There are a couple of possibilities, however. You could catch it in a bucket of water or straw and use it as a liquid fertilizer or mulch for heavy feeders in the garden such as corn, or trees and shrubs, giving some thought to forestalling such curiosity seekers as dogs and foxes in the neighborhood. I catch the blood and pass it on to my composting worms and soldier grubs. Don't tell the contagion police, but I have even experimented with feeding the coagulated blood to my chickens, who ate this nutrient-rich food with apparently only beneficial effect.

The Chopping Block

A solid, stable round from a log makes a good chopping block (see figure 28.4). Drive a couple of large nails into the block, spaced to grip the bird's head as you pull on the feet. Under such restraint, the bird will not continue struggling. Do not rush—pull on the feet, stretching out the neck, take a breath, and steady yourself for a decisive blow of the hatchet that chops off the bird's head with one whack. Hold the flapping bird over your collection vessel, so the blood will drain without spattering.

The Killing Cone

A killing cone made of sheet metal—in different sizes, depending on the species to be slaughtered—is a useful accessory. Hang it on the side of an outbuilding or a tree. Insert the head through the hole at the bottom of the cone, pulling it to stretch the neck and draw the wings and legs more tightly into the confinement of the cone. Use your sharpest knife to make a quick, decisive cut just to the side of the "jaw." (See figure 28.5.) Note that you sever the jugular vein *only*, not the windpipe, resulting in less stress for the bird. Keep your grip on the head as the bird bleeds out thoroughly, guarding against a final spasm that might flip it out of the cone. Homemade versions of killing cones include burlap sacks or plastic buckets with holes cut in a corner of the bottom through which to pass the bird's head.

English Method

You can kill the bird by what is sometimes referred to as the English method, if it is young enough for you to break the head off the neck. I find chickens at the fryer-broiler stage, and most old hens, easy to kill with this method. I cannot break the necks of mature cocks, ducks, or geese, so for those birds I use the cone or the chopping block instead.

Holding the head of the bird firmly in your strong hand, and the feet in the other, brace the bird over one thigh, and imagine you are going to *pull it apart*. At the right point of tension—which can only be learned by experience—give a sharp twist-snap downward-outward, and the head will separate completely from the neck. Hold the bird away from your body until its spasms subside (see figure 28.6).

You will find this method difficult the first time you try it and may mistakenly conclude you are not strong enough to make it work. Trust me: It is not a matter of strength but of technique—that is, the right degree of tension and proper action in the wrist. The first time the head comes off—so easily, really, with a sort of liquid giving-way—will be an interesting surprise.

An advantage of this method is that it is less messy—the bird does bleed out, but the blood is retained inside the skin of the neck. Note, however, *it is essential that the head actually break away from the neck for this method to work*—when that happens, the jugular vein is severed as well, and the bird bleeds out properly. It is possible to kill the bird simply through trauma to the spinal cord but without breaking the jugular, resulting in a dead bird that has not properly bled out. Squeeze the skin between the head and the stump of the neck: *If you do not feel a completely flaccid, empty-balloon space at least big enough to insert three fingers, you have not properly broken off the head.*

About Using a Mask

The reflex thrashing of the bird after killing kicks loose an invisible cloud of poultry dander. If you find that your sinuses are heavily congested after slaughtering, especially during the night, use a good dust mask during this phase the next time you slaughter—and later in the plucking phase as well, if you are using a mechanical plucker. The one I use is the Respro Sportsta Mask, extremely effective, and comfortable enough to wear for all dusty chores on the homestead.

The Naked Fowl

Do not be discouraged if the work at the table is at first painfully slow—with practice the hands learn the required efficiencies, and processing becomes faster and more fluid.

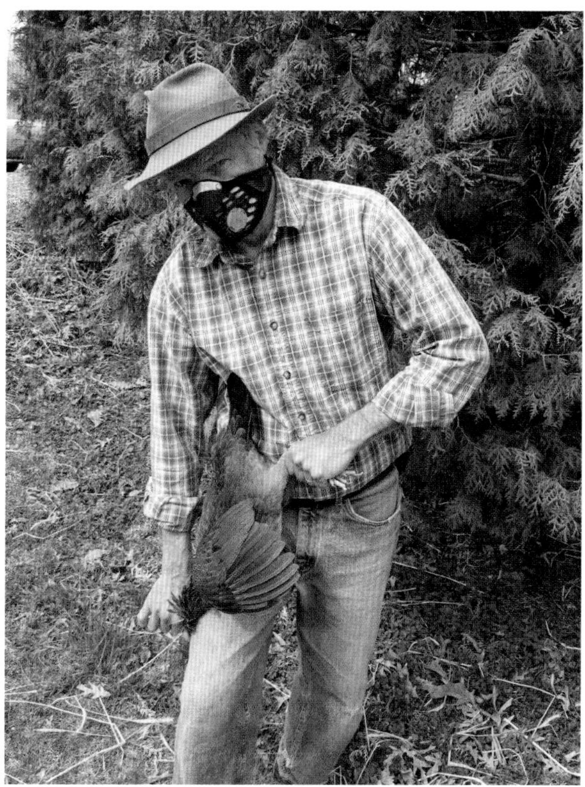

Figure 28.6. "English" method. Note use of mask to prevent inhalation of dander. Photo courtesy of Mike Focazio.

Figure 28.7. Proper scalding is the key to a clean, easy pluck.

Figure 28.8. Testing the scald.

Figure 28.9. Dunking.

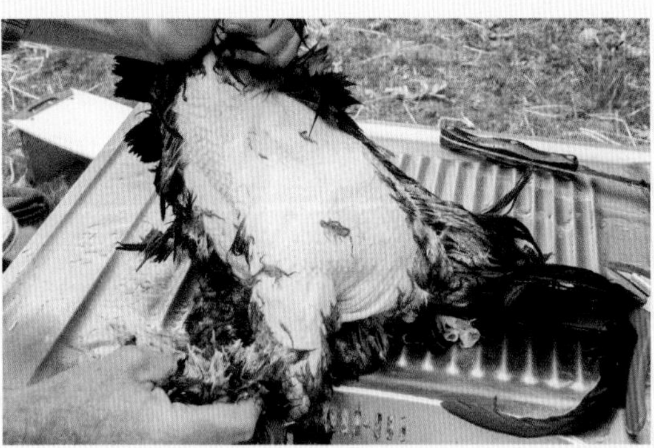

Figure 28.10. Plucking manually is easy if you have scalded properly.

Figure 28.11. A mechanical plucker speeds up the work.

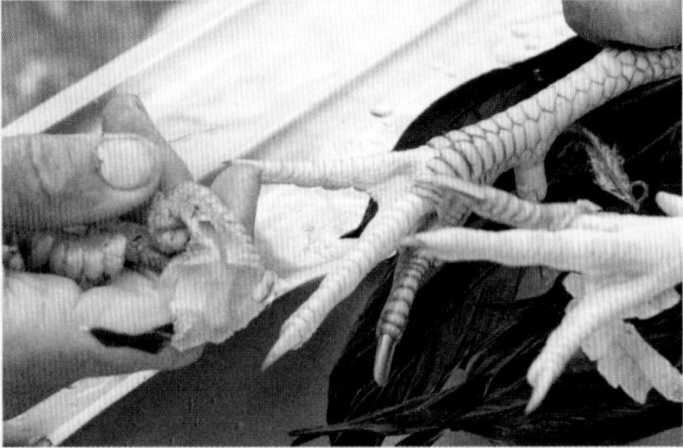

Figure 28.12. Cleaning the feet.

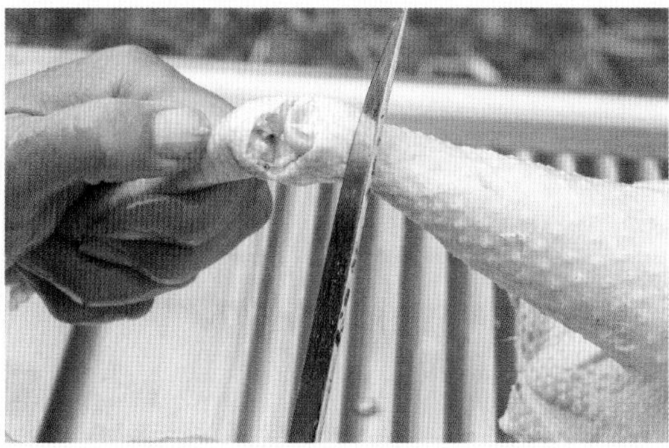

Figure 28.13. Cutting off the leg.

Figure 28.14. Starting the evisceration.

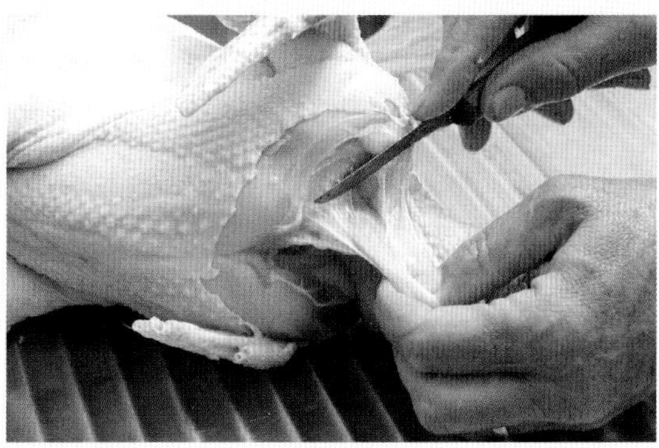

Figure 28.15. The crop—empty if the bird was properly starved.

Figure 28.16. If the bird was not starved, the full crop may spill its contents.

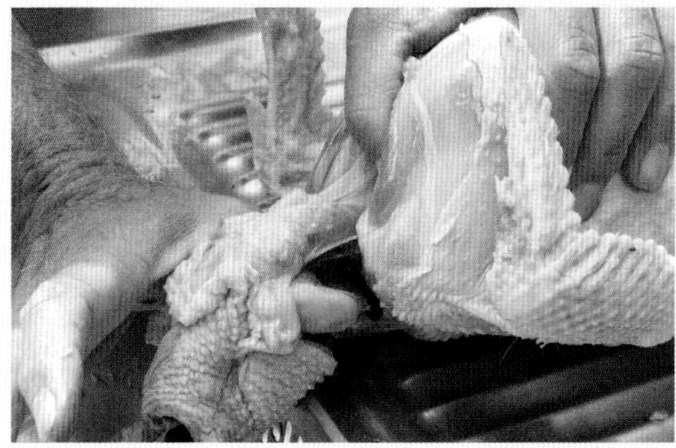

Figure 28.17. Leading with the thumb, separate crop, esophagus, windpipe, and skin from the neck.

Figure 28.18. Cut off all that tubular stuff, as close as you can to where it enters the body cavity.

Photos courtesy of Mike Focazio.

Figure 28.19. Cutting off the neck.

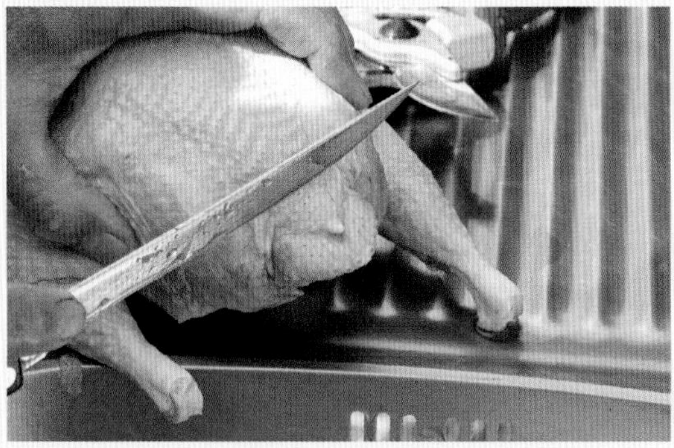

Figure 28.20. Removing the oil gland. Part 1.

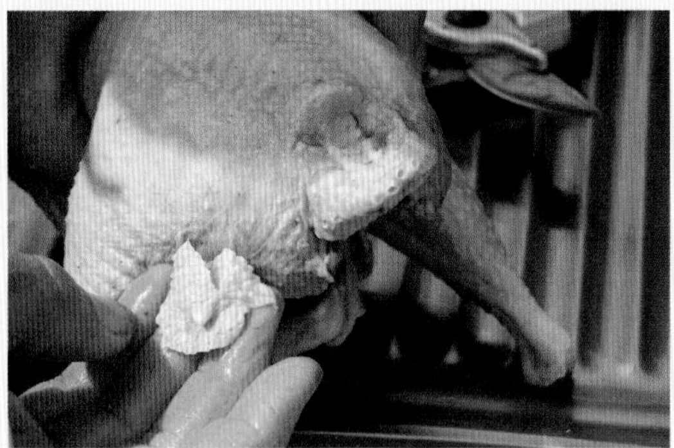

Figure 28.21. Removing the oil gland. Part 2.

Figure 28.22. Cut a small opening into the abdomen, slicing through skin and fascia only.

Figure 28.23. Insert two fingers from either direction (*left*) and give a stout pull to open the body cavity (*right*). Note that in a tough, older bird you may have to cut through the skin and muscle on either side of the body cavity to give full access to the interior.

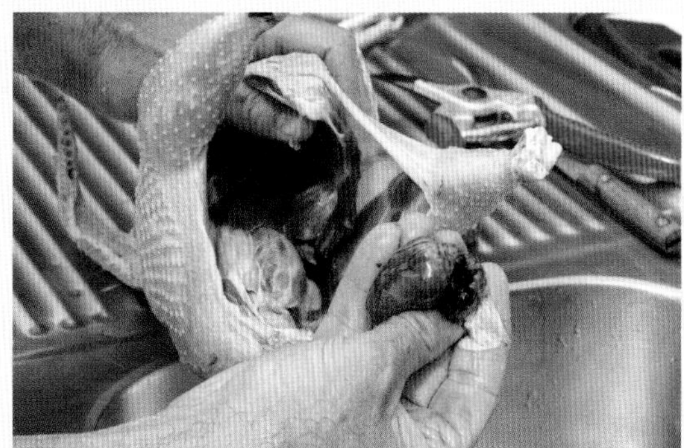
Figure 28.24. Pull out the heart.

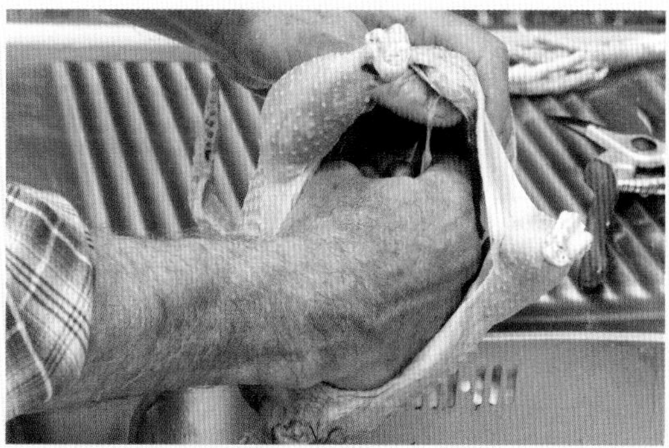
Figure 28.25. The fingernails lead the way in a scooping action along the rib cage and spine, underneath the viscera.

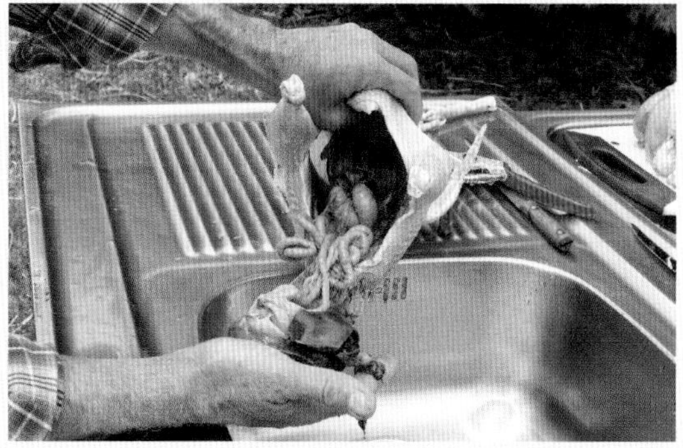
Figure 28.26. Pulling out the "package" of entrails.

Figure 28.27. Cutting the liver from the bile sack.

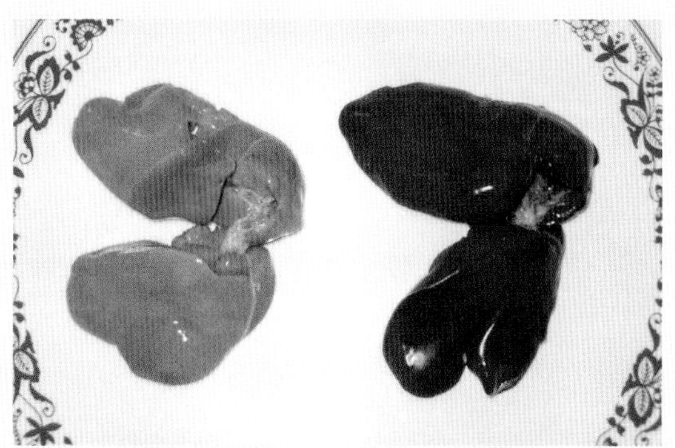
Figure 28.28. Livers of old bird (left) and young bird compared.

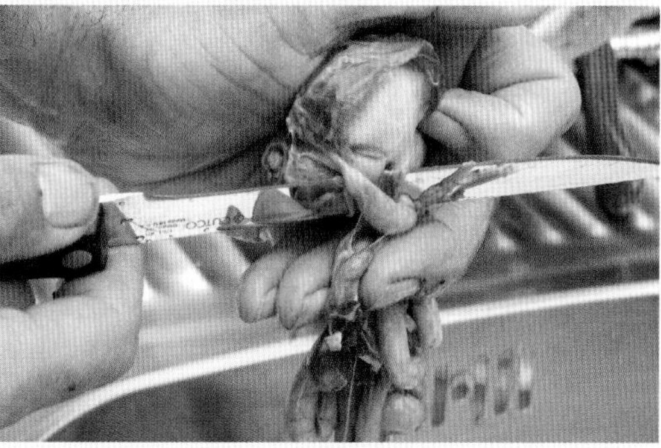
Figure 28.29. Cutting away the gizzard.

Photos courtesy of Mike Focazio.

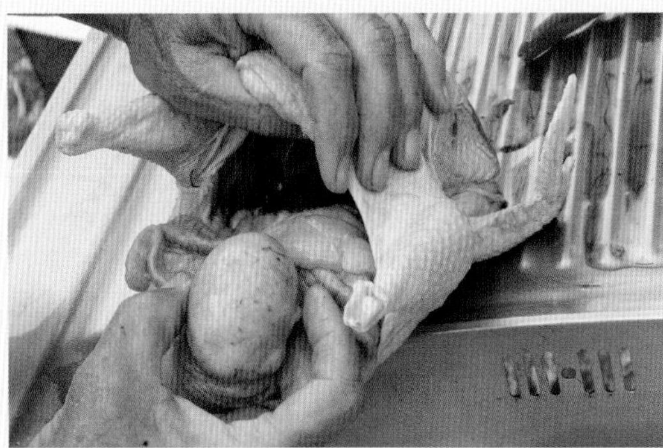

Figure 28.30. Fully developed egg in the oviduct.

Figure 28.31. It is easy to remove the fat reserve around the end of the intestine, and easy to render it into high quality cooking fat.

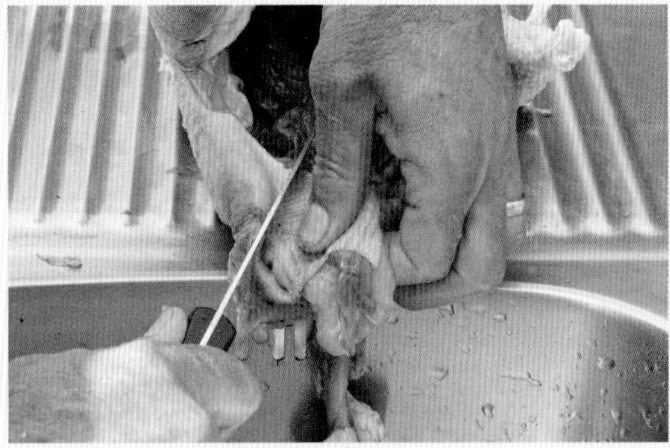

Figure 28.32. To finish removing the entrails intact, start with cuts on either side of the vent . . .

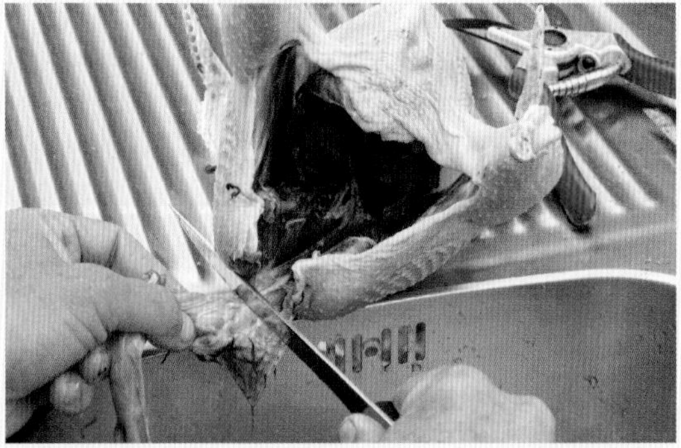

Figure 28.33. . . . and end with a slicing cut underneath both cloaca and vent.

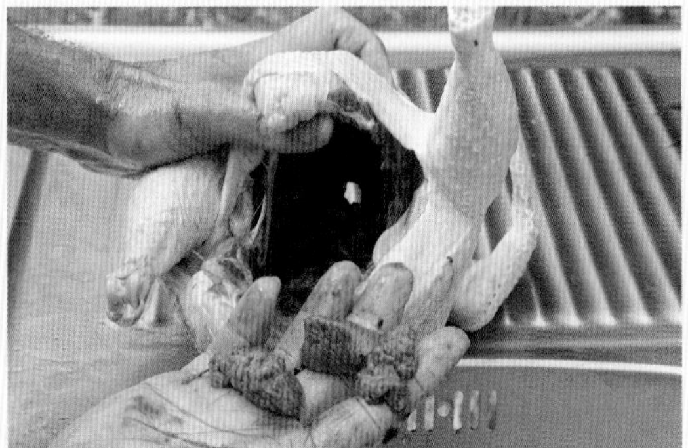

Figure 28.34. Popping out the lungs can be a bit tricky.

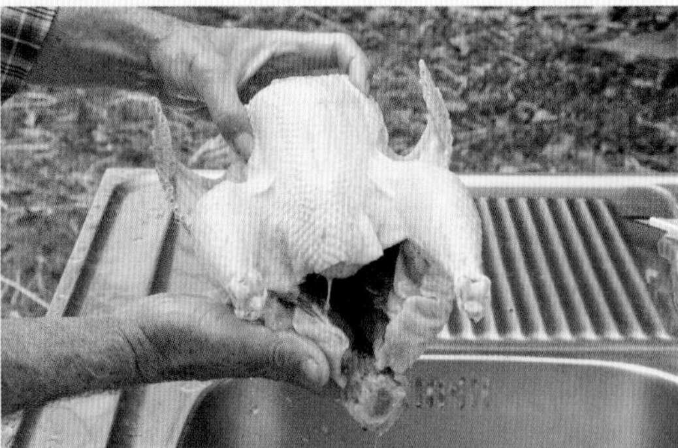

Figure 28.35. Don't forget to say "Thank you."

Photos courtesy of Mike Focazio.

Scalding

See figures 28.7, 28.8, and 28.9.

The key to a clean pluck is a good scald. Note that *scalding temperature is nowhere near the boiling point.* I set my thermostatically controlled scalder to 145°F (62.8°C), and that is a good temperature to aim for if your scalding container is over a stovetop or burner. However, it is not necessary to use a thermometer to measure the temperature. Just stick in a finger. Can you immerse the finger without getting a burn, but not for more than a second? That's proper scalding temperature.

Note that over-scalding—either through too high a temperature or scalding too long—starts to cook the skin, which then tears when you pluck. Under-scalding, on the other hand, fails to loosen the feathers sufficiently, and they are difficult to pluck. No exact formula can be given for scald time—how long to scald depends on the age of the bird, the species, the point in the plumage cycle, probably the phase of the moon. You will learn only through experience when enough is enough in the scalder.

Add a few drops of liquid soap to the scalder, to break the surface tension of the water and increase penetration to the skin—especially important with waterfowl. Put the bird into the hot water, and use an implement that enables both agitating it up and down and, later, snagging a leg. Let's call it our "poker"—mine is an old three-prong cultivator missing one tine. After agitating the bird awhile with your poker, snag a leg to pull it above the water for testing proper scald. Pinch-squeeze the scaly covering of the shank: When that covering easily breaks loose from the skin of the leg, remove the bird from the scalder and dunk in cold water, to stop the skin from overheating from the residual heat in the water under the feathers.

Plucking

See figures 28.10 and 28.11.

If you've scalded properly, it is easy to remove the feathers in handfuls. It is better to start with the largest feathers, on the wings and tail, since they are the ones that start resisting pulling out first as the follicles cool. As said, you may find a mechanical plucker a good investment.

Alternatives to Scalding

You may see references to two alternatives to the previous approach to plucking: *waxing* and *brain piercing*. You can dip the bird to be plucked—especially ducks—into melted wax, let the wax solidify, then peel it off, feathers and all. Or so it is said—I tried waxing a couple of times, and it was a disaster. But I foolishly used the sort of wax used to seal jars of jelly. If you plan to try waxing, be sure to buy proper *plucking wax*.

You can also drive the point of a stiff-bladed knife through the roof of the bird's mouth and into its brain. If done correctly, this method supposedly not only kills the bird but loosens the feather follicles, allowing easy dry plucking without the necessity of a scald. Or so it is said—again, my attempts were downright nightmarish in execution, and after all the trauma failed to yield easy dry plucking. Good luck if you try it.

There is one alternative to scalding that is sure-fire: *skinning* the bird, just as you would skin a rabbit. I prefer to keep the skin on a dressed bird, so I always scald and pluck if I'm butchering a group of birds. But if I need to dress out just a couple on the run, I skin them, in lieu of heating a large amount of water in the scalder. To skin, simply make a slice through the skin of the back large enough to hook a couple of fingers under each edge of the cut, and pull. Most of the skin pulls away easily. The skin at the base of the large feathers on the second joint of the wing takes a bit more effort; and I just cut away the pinion—the last joint—rather than trying to get its feathers off. Young growing birds are easy to skin; old stew hens and ducks require more effort but are not too difficult; but old mature cocks—forget it! They're *really* tough.

Cleaning the Feet

See figure 28.12.

Always save the feet. They are a valuable addition to the stockpot, yielding not only essential minerals but also *collagen*, beneficial for the entire digestive tract and all the body's connective tissues. Remember how you used the scaly covering of the shank to test the scald? Simply continue with the same pinching-pulling action to pull the covering off leg and toes like a glove. Be sure to pinch tightly and pull the toenails as well, and the cuticles will pop off. If the cuticle resists popping off, use your shears to snip off the entire toenail. The result is a pristine foot you will be glad to have in your stockpot.

Legs

See figure 28.13.

This is the first point to take care to avoid sharp edges of cut bone, which could puncture the wrapping if you freeze your bird. (If you're cooking it right away, of course, the point is moot, and you can cut leg or neck any way you like.) Therefore, you should not use your shears to cut through the joint at the hock to remove the leg.

Instead, lifting by the foot, pretend you are going to *break the leg sideways at the hock* with the weight of the suspended bird. The resulting tension on the joint makes it easy to slice through the skin and find the cartilage-padded interstice between this ball-and-socket joint. Once the edge of your blade has found that space, continue cutting through skin, connective tissue, tendon—anything but bone—until the leg is cut away.

Head

Unless you chopped off the head, at this point you should remove it. If you used the English method, cut through that empty-balloon segment of neck skin with a knife. If you used a killing cone, the head will still be attached—cut it off with your shears.

The head can also be reserved for the stockpot if you like. Pull off the feathers, rub the coating off the skin of comb and wattles, and pinch hard on the beak—the horny cuticle will pop off, like the toenail cuticles earlier.

Should You Singe?

The work thus far has produced a completely naked fowl, ready to eviscerate. But if you look closely, you may see fine "hairs" sticking here and there out of the surface of the skin. Some folks are repelled by these hairs (more properly called *filoplumes*) and feel compelled to singe them off to produce an absolutely pristine carcass, using the flame of a gas stove, or even a propane torch. Concern about a few (almost invisible) filoplumes is a good deal too fussy for my taste—when the bird comes out of oven or pan, nobody notices or cares. I have never taken a blowtorch to a dressed chicken.

If you do choose to singe, sweep the carcass lightly with the flame—you do not want to heat the skin—and the filoplumes burn away with the merest touch of flame. Wash thoroughly afterward, to prevent carbon residues remaining on the skin.

Evisceration

For convenience, the following descriptions assume you are right-handed. Lefties will need to switch hands and directions as appropriate.

Crop, Windpipe, and Esophagus

See figures 28.14 through 28.18.

Lay the plucked bird on its back, its long dimension perpendicular to you, its neck to the left. Start the evisceration at the skin where neck joins breast. Using your small knife, slice through the skin on the far side, and continue slicing through skin (only) to make a full circle around the neck (see figure 28.14). As you slice you expose the *crop*, a semitranslucent membranous pouch—on the bird's right (the side nearest you in this position), close beside its neck—in which the bird stores its food for pre-processing. Because you wisely starved your slaughter birds overnight, the crop is empty and this step is not messy (see figure 28.15). If the crop is full, it may rupture, spilling the fermented contents over the top of the carcass and the work surface (see figure 28.16). This is no disaster—a thorough rinse whisks them away.

Force a thumb between the neck and everything attached to it: the crop—in the process pulling it free from the top of the breast meat—the esophagus, and the windpipe. Pull all that, along with the skin, free of the neck (see figure 28.17). Separate the neck skin from the three tube-like elements, and reserve it for the stockpot as well. Then pull on the tube-y things, and cut them off as close as you can to where they enter the body cavity (see figure 28.18).

Cutting the Neck

See figure 28.19.

Do not leave a stub when you cut off the neck—the exposed edges of sheared bone could poke through your wrapping. Instead, position the blades of your shears on either side of the base of the neck, but well up between the shoulders of the wings. The resulting cut keeps the sheared edges tucked up between the shoulders, out of contact with packaging. Be sure to save the neck.

Removing the Oil Gland

See figures 28.20 and 28.21.

The oil gland, located just beneath the skin on top of the tail, secretes oil the bird uses to preen its feathers. If this gland is not removed from the carcass, it may affect the flavor of a cooked bird. To remove it, place the bird breast down, tail end to the right. To the left of the gland's nipple are two little mounds—the lobes of the gland. Just to the left of the lobes, slice down vertically until you hit bone. Then turn the edge of the blade right, toward the end of the tail, and make a scooping cut to slice under the two fatty lobes and the nipple. If you plan not to keep the tail but to cut it off when cutting away the entrails, you do not have to bother removing the oil gland.

Opening the Body Cavity

See figures 28.22 and 28.23.

Lay the bird on its back, perpendicular to you, neck end to the left. Make a shallow slicing cut into the abdomen, just to the right of the end of the keel or breastbone. Cut through skin and fascia *only*. (The fascia is the translucent membrane surrounding the inner organs.) Remember good knife technique: Avoid poking at the skin with the point of the knife—the bird's intestines are immediately below (see figure 28.22).

Sooner or later, you will have a slip of the knife and nick the intestine, spilling some of its contents into the body cavity. Do not despair—even with such a mishap, your home butchering is vastly more sanitary than that of commercial poultry. After eviscerating, simply do a more thorough rinse of the interior cavity than usual.

Make the cut into the abdomen just big enough so that you can hook two fingers into the opening from either direction. Give a stout pull, tearing a larger hole in the skin and fully opening the interior cavity. Note that, if the bird is old and tough, you may have to slice through additional skin and muscle on either side of the initial cut to make it easier to pull open the body cavity (see figure 28.23).

Please note that the pressure exerted on the intestinal tract at this point can cause what I call a "poop attack"—the forcing out of some residual fecal material from the cloaca. A poop attack is unlikely if the bird was properly starved in preparation for slaughter, but you can guard against it anytime by hanging the vent end of the bird over the drain when you pull open the carcass. If there is an expulsion, rinse it away from the vent, being careful to avoid backwash into the body cavity.

Heart

See figure 28.24.

Reach as far as you can into the body cavity to find an organ that feels like a large grape—the heart. Hook two fingers around it and pull it out. Squeeze out the remaining coagulated blood, rinse, and set aside—I use the stainless-steel bowl with lid that I mentioned—along with the neck and feet and other good things you are saving.

Liver

See figures 28.25 through 28.28.

Reach again into the body cavity, fingers and hand encircling the gastrointestinal tract. The fingernails lead the way, tight to the rib cage, finding the seam between the chest wall and the ropy tubes of the tract and other organs, the gizzard filling your grasp like a slippery apple (see figure 28.25). Grip all and pull. The tract and the organs pull free in one mass—connected only at the base of the abdomen—which you allow to hang over your drain (see figure 28.26).

The liver is the large, dark red organ beside the gizzard. The clear tissue connecting it to the other organs is easily torn away with the fingers to leave it attached at one point only—to a small, dark green sack the shape and size of a caterpillar, the bile sack. The bile it contains is essential for the bird's digestion of fats but is extremely bitter. Sacrifice a bit of the liver as you pare it away from its connection to the bile sack with your small knife (see figure 28.27). If on occasion you do have a spillage of bile, rinse with more than usual thoroughness and proceed.

Note the picture of this cut as an example of proper knife technique: The weight of the hanging gizzard is used to put a little tension on the connection between liver and bile sack, allowing the cut to be made against that tension, rather than against the steel work surface. Of course, the blade must be sharp, to enable cutting against such a tiny bit of tension.

The liver of a young, healthy bird—on the right in figure 28.28—is plump, dark red, and glistening. Save it—it is extremely nutritious. The liver of an old bird, on the left in the picture—equally healthy, equally well fed—is usually pale brown, indicating longer service as the bird's major metabolic filter. While I honor this liver for the good work it has done, I do not eat such livers.

Gizzard

See figure 28.29.

The gizzard is a large, muscular organ with a tough interior pouch, filled with bits of rock. In lieu of chewing its food, the chicken processes it inside the gizzard, using the small grinding stones and digestive enzymes. One tube goes in, one comes out. Cut off both, flush with the surface of the gizzard, and add to the other "goodies."

Testes, Eggs, and Fat Deposits

See figures 28.30 and 28.31.

If the bird you are butchering is a cockerel, the testicles—light yellowish, football-shaped, and tucked up against the spine at about its midpoint (visible in figure 28.26)—may be the size of kernels of corn. If you're eviscerating a mature cock, you may be astounded: In proportion to his body, they are enormous. Pluck out the testicles. If from a mature male, add to the goodies bowl if inclined to eat them. Toss them into the gut bucket if disinclined.

If you are butchering a hen, you will find (at about the same point as the testicles in the cockerel) a cluster of cells destined to become egg yolks. You will likely see several growing yolks of various sizes, or even a fully developed egg. The one in figure 28.30 is a finished egg ready to be laid, still inside the oviduct. Keep these eggs, and yolks the size of a pea or larger.

Especially in a hen, and even more so in hens butchered in the fall, you will find a deposit of glistening, yellow fat at the lower end of the body cavity (see figure 28.31). If you want to render it into high-quality cooking fat, it is easy to pull out.

Cutting Away the Entrails

See figures 28.32 and 28.33.

You have now drawn out the interior organs without spilling stuff-we-don't-want-on-our-meat, and they're hanging from the vent in one long, intact package. Keep them intact as you complete the evisceration. Hang all that ropy stuff to one side of the vent, and position your blade on the other side, between the vent and the sharp point of the pubic bone. Slice down until the blade hits bone. Now move the ropy stuff to the side of the vent where you just made your cut, and repeat the cut on the opposite side (see figure 28.32). Pull on the entrails and slice under the vent itself, leaving it intact along with its connection to the end of the intestinal tract, the cloaca (see figure 28.33).

If you chose not to remove the oil gland earlier, modify the above to remove the entire tail: After cutting through the skin between pubic bone and vent on both sides, grasp the tail with attached viscera in your left hand, and cut it off the end of the spine using your heavier knife.

Lungs

See figure 28.34.

Reach into the cavity one final time and remove the lungs. Again, lead with the fingernails as you follow the curve of the rib cage, finding the seam between rib and lung, and pop the spongy lung tissue free. This step takes some practice and can be a bit tricky—sometimes the lungs shred and resist coming free easily. Don't worry—leaving a little lung tissue is no problem. I remove them to make a neater carcass.

Kidneys

The bird's kidneys are lumpy masses the color of the liver, tucked either side of the backbone, near the vent end. Though removed in commercial processing, I find them quite tasty, so I leave them in place to be cooked with the bird.

Give a final rinse, inside and out, and admire the creature who has made so generous a contribution to your homestead, now ready to grace your table. Don't forget to say "Thank you!"

Cleanup

You will end your slaughter day with a pile of wet feathers and a bucket or two of entrails. I throw the feathers onto the deep litter in the poultry house. The birds eat some of them—feathers are almost pure protein—while the rest get buried in the litter, where they quickly decompose. The entrails can be composted or buried. If you compost, take care to assemble a pile that heats up rapidly and does not become an attractant to dogs and other curiosity seekers.

Sometimes I place the entrails out in the edge of our woods, as an offering to friends—Fox, Raccoon, Possum—who live in the neighborhood and who need to eat as well. It is my way of honoring our kinship and saying "Thank you" for coexisting peacefully while my flocks stay—mostly—intact. When I return to the site next day, the entrails have invariably been entirely removed—this gift is not creating a nuisance.

CHAPTER 29

Poultry in the Kitchen

I don't know if you've seen the T-shirt that reads: MY TASTES ARE SIMPLE—I LIKE THE BEST. *That* describes Ellen and me precisely, at least as regards food quality. Call us food snobs if you like, but our "simple" tastes are a major motivation for producing our eggs and chicken in our own backyard—we *will not eat* the debased equivalents sold in the supermarket.

Figure 29.1. Superfood.

The kitchen is where all the effort producing our own eggs and dressed poultry pays off. And part of the payoff is the assurance that, with the application of the plain common sense our grandmothers used in their kitchens, the foods from the flock are not a threat to our family's safety.

This chapter is an overview of handling and storage of both eggs and dressed poultry. It also offers a few of our time-tested recipes for using these gifts from the flock.

Handling and Storing Eggs

Imagine a packet of nutrients that, under the right conditions, can turn into a baby chicken—in twenty-one days. That packet would necessarily contain a powerhouse of nutrients, don't you think? It is certainly something you'd want to consume frequently. The name of that packet—one of the most perfect of all foods—is "egg."

Eggs contain all eight of the essential amino acids from which the body builds proteins, and high-quality fat that assists the absorption of not only the yolk's own fat-soluble vitamins (A, K, E, and D) but fat-soluble vitamins from other dietary sources as well. Eggs also provide B-complex vitamins and the essential minerals iron, phosphorus, potassium, and calcium. Importantly, they are a valuable source of cholesterol—a fat-soluble nutrient essential to the integrity of the walls of every

Eggs: The True Truth

CONTRIBUTED BY ELLEN USSERY

Eggs are a superfood, though they have been badly maligned over the years. Let's set the record straight.

We have been told that eating eggs will give us heart disease because they contain cholesterol, which will go directly to our arteries and clog them up just like it would clog your drain if you poured fat into the sink. This is the *diet–heart hypothesis*—which has been shown to be as untrue as it is simplistic.

Heart disease may arise from different causes depending on the person. Biochemically, we are *individuals*. The root cause of your heart disease may well differ from what caused mine. Thankfully, many medical practitioners have recognized this variability and accept it as the basis for practice. If you want to keep up with the latest understanding of heart disease, read what functional medicine doctors such as Dr. Mark Hyman have to say. If you are looking for the nitty-gritty on cholesterol itself, follow nutrition researcher and consultant Chris Masterjohn online. And do note that cholesterol is *essential* for our well-being: It is a critical component of cell membranes, the precursor of all steroid hormones, a precursor to vitamin D, and the limiting factor that brain cells need to make the connections with each other called *synapses*. Thus, cholesterol is essential to learning and memory. It is so crucial for so many metabolic functions that the liver makes more of it if the body does not have enough.

As for egg consumption and heart disease, consider this: A meta-analysis (published in *The American Journal of Medicine*, January 2021) of twenty-three prospective studies from 1966 to 2020—involving more than a million people—concluded: "Our analysis suggests that higher consumption of eggs (more than 1 egg/day) was not associated with increased risk of cardiovascular disease, but was associated with a significant reduction in risk of coronary artery disease."[1] As for the cause of the reduction, they hypothesized: "Egg consumption may reduce coronary artery disease via a mechanism of promoted carotenoid absorption, enhanced high-density lipoprotein cholesterol function and increased bioactive compounds (e.g. lutein and zeaxanthin) protecting against atherosclerosis."

Let's hope this puts the final nail in the coffin of the diet–heart hypothesis and that future studies will stop wasting time and money, and direct resources instead toward answering more consequential questions about heart disease.

You can enjoy eating your eggs without fear. Not only will they not harm you, they contribute to every aspect of vibrant health. We may never know all the reasons eggs are so beneficial, but we do know they supply many benefits:

- All eight of the essential amino acids, which provide 6–7 grams of protein per egg
- The fat-soluble vitamins A, E, and D, along with high-quality fat that is essential to absorb them
- Vitamin K_2, which directs calcium to the bones—preventing its deposition into the arteries
- The B vitamins—thiamin, riboflavin, niacin, pantothenic acid, B_6, and folate—as well as choline, which is essential to healthy brain and liver function

- The minerals calcium, magnesium, selenium, iodine, phosphorus, potassium, zinc, iron, copper, and manganese
- The antioxidant carotenoids lutein and zeaxanthin, which not only support health of the eye but help prevent oxidative stress
- The essential fatty acids docosahexaenoic acid (DHA) and arachidonic acid (AA)

Interestingly, the preponderance of these components is to be found in the yolk.

Of course, not all eggs are created the same. Pasture-raised eggs are far superior in many ways. Testing has shown that eggs produced on pasture, in comparison to eggs from industrial layer flocks, contain:

- Two-thirds more vitamin A
- Seven times more beta-carotene
- Significantly more folic acid and vitamin B_{12}
- Three to four times more vitamin D
- Two times more omega-3 fatty acids
- Three times more vitamin E[2]

Finally, pasture-raised eggs are full of flavor—boiled, scrambled, or fried. With only a little butter and salt, they will satisfy your palate and sustain you for hours. I do not think this is a coincidence. Our bodies are not mechanistic plumbing systems. They have an exquisitely balanced innate intelligence and can tell us a great deal about what we should eat. We would do well to pay attention.

cell of the body, but especially of those in the nervous system and brain—and of choline, a fatty substance that is a major component of the brain and is essential to proper functioning of the liver.

Happy, well-fed chickens, ducks, and guineas grace the table with high-quality eggs with a lot of easy culinary possibilities. Handling and storing are simple.

Is it necessary to wash eggs? If you keep your nestboxes lined with plenty of clean straw, frequently renewed, most of the freshly laid eggs will be perfectly clean. It is *not* necessary to wash such eggs. Indeed, I advise against it: The hen's just-laid egg is wet with a coating that quickly dries (the bloom)—a coating that not only eases passage for the hen but leaves an antibacterial residue on the surface of the egg to protect it from contamination. Washing it off may *decrease* keeping time for the egg. On the other hand, if an egg comes from the nest with even the smallest trace of mud or poop, I do indeed wash it. We use paper towel dipped in a half-and-half mix of water and white vinegar.

Can I use a cracked egg? An egg fresh from the nest with a small crack in the shell is okay to use right away. Do not store such an egg, and do not use at all if the membrane inside the shell has ruptured.

Is it necessary to refrigerate eggs? We never refrigerate our eggs. An egg is designed by nature to remain fresh and viable at least as long as it would take the mother bird to assemble a full clutch of eggs or considerably longer. We organize the flow of eggs so that we are always rotating out the older ones—they are never more than a week old when we either eat them or give them away. We would feel completely confident holding them up to ten days unrefrigerated if not in the full light of the sun or by a source of heat. If you need to hold them longer than that, then do refrigerate them—they will keep for weeks in the refrigerator.

Is it possible to store eggs long term? Fresh eggs are best, but if you need to store a supply for a lean period such as winter, when hens' production declines, there are several options.

1. For *freezing* whole eggs, stir the whites and the yolks together with a fork, being careful not to incorporate air into the mix. Pour into small freezer containers, or make individual cubes by freezing in an ice cube tray, breaking them out, and storing in a plastic freezer bag—do note that oiling the ice cube tray first is a *must*. You can also separate whites from yolks. Put the whites through a sieve to break up the albumen, then freeze as above. To prevent gumminess in frozen egg yolks, stir in salt (one-quarter teaspoon per eight yolks) or honey or maple syrup (1 tablespoon per eight yolks), depending on whether eventual use of the yolks will be in a savory or a sweet dish.
2. You might want to experiment with a couple of traditional means of preserving eggs. Early in the twentieth century, a strong solution of *eisenglass* or *water glass* (sodium silicate) was used to cover freshly laid, perfectly clean eggs in the shell in a jar or crock. Even at cool room temperatures, the eggs would keep several months. At temperatures just above freezing, they would keep half a year. I received a report from an experienced flockster who preserved twelve dozen eggs—same-day-fresh and not washed—in water glass. Her family ate the last of them six months later. The only noticeable difference in quality was some thinning of the whites toward the end of storage—but even so the family much preferred them to supermarket eggs.[3]
3. Another means of preserving eggs has been used in China for a long time. *Thousand-year eggs* start as fresh eggs in the shell smeared with a paste made of strong tea mixed with clay, quicklime (calcium oxide), wood ash, and salt, stored in baskets or clay jars to cure for several weeks or months. The strongly alkaline coating prevents the contents from spoiling while both yolk and white change texture, taste, and color—the whites brownish or light green, the yolks ranging from dark green or blue to almost black.
4. Hard-boiled eggs can be preserved by *pickling*. Simply boil and peel the eggs, fill a jar with them, and cover with a hot spiced vinegar. (You can also recycle the pickling solution from pickled cucumbers or beets.) Keep in the refrigerator up to six months and use in salads or sandwiches, or as a garnish to almost any dish.

Is there a good use for eggshells? We always save our eggshells. If the egg was soft-boiled and scooped out with a spoon, there will be residues of cooked albumen, which are worth feeding to the chooks—I just crush by hand and toss them out. Shells of eggs that were cracked out before cooking I usually reserve for the garden—finely pulverized and sprinkled around plants, they help deter slugs and snails. The sprinklings are especially valuable around plants that like an extra helping of calcium, such as tomatoes and figs. Those shells as well can be recycled to the hens as a calcium source. Disregard nonsense you may read about toasting them in an oven and grinding to a powder—just crush coarsely by hand and toss them out.

Butchering Day in the Kitchen

I put dressed chickens in the refrigerator as I complete them. As soon as they are well chilled, I am ready to prepare them for storage. If it has been a long day at the slaughter table, going on to the next step may be put off until the next day.

Is It Necessary to Age Dressed Poultry?

We don't consider it necessary to age carcasses of young birds before freezing or cooking. Even with older birds, that is, cull hens or mature cocks, there is no need to age if the bird is going to cook long and slow to make broth. We age only those chickens culled at around sixteen weeks to a year. These are not stew birds, but can be a bit on the tough side. They will be more tender—for braised dishes like *coq au vin*—if aged about a week in the refrigerator.

Packaging and Freezing

If promptly chilled, freshly dressed poultry can be held up to a week in the refrigerator. If you do plan to keep

The Dalai Lama's Eggplant Parmesan

PRESENTATION OF THIS RECIPE IS DEDICATED TO MY FRIEND MARYANNE CRISTELLO AND TO HIS HOLINESS THE DALAI LAMA.

I liked eggplant Parmesan the first time I tried some from an institutional kitchen—a college dining hall—which says something about how difficult it is to get this one wrong. One evening I got a hankering for it on my way home from work, so I stopped to buy the key ingredients—olive oil, eggs, eggplants, canned tomato sauce, and mozzarella and Parmesan cheeses. When I arrived home—a room in a building I shared with friends—my friend Maryanne, a second-generation Italian-American, asked how I would make the eggplant Parmesan. Based on the only version I knew, I replied, "I'll cut the eggplants half an inch thick, dredge them in seasoned flour, and fry them in olive oil. Then . . ." Maryanne shook her head, her face somewhere between pitying and appalled, and took the sack of ingredients out of my hands the way you'd take a dangerous toy from a child. She then started on the casserole.

As she worked, Maryanne continually bemoaned, "Too bad I have to do this in such a hurry—my mother would spend all day on it." As for me, I was almost in tears from hunger by the time the bubbling casserole made it to the table, about nine thirty. But the first bite told me this simple dish would henceforth be an essential part of my repertoire.

A couple of years later, when Ellen and I were in residence at Dai Bosatsu Zendo—a Zen monastery in the Catskill Mountains of New York State—His Holiness the Dalai Lama paid us a visit. Our *roshi* ("honored teacher") asked me to be head chef for the lunch we would serve and suggested that I make my "eggu-plantu," a great favorite of his. Thus it happened that I made eggplant Parmesan for the Dalai Lama—and sixty other diners.

There was much that was memorable about the Dalai Lama's visit. The most important thing for you to remember when considering this recipe, however: His Holiness had three helpings.

There are three pointers to surefire success, two of which I learned from Maryanne: Slice the eggplant *just as thinly as possible*, and fry the eggplant pieces after dipping in *plain beaten egg*. A third I've learned through experience: *Easy on the mozzarella.*

The eggplant: Your best bet is to grow your eggplants. Otherwise, buy one or two medium to large eggplants, the best you can find. I usually half-peel the eggplant—peel off alternate strips of its skin down its length. You could leave all the skin on, or peel it all off, depending on how much of its bitterness you want in the final dish.

Slice the eggplants crosswise as thinly as you can. Reread that last sentence—it is meant to be taken literally. Use a sharp kitchen knife to make slices as thin as you can manage while still getting whole slices—a too-thin slice

it that long, though, *do not keep it tightly wrapped in plastic*. Instead, set it on a plate and loosely cover with waxed paper or freezer paper.

I dry the carcass inside and out before freezing. To package, I have used loose plastic bags, double bagging if the plastic is thin; freezer paper, double wrapped; zip-seal plastic freezer bags; and shrink-wrapping. That is the order of my preference as well, from least to most desirable. Whatever method you use, *expel all the air you can.* And make sure the wrapping is thick enough is one that ends as a half-round. I might get about forty slices from a medium-large eggplant. That's what I mean by *thin*.

Some recipes call for salting the slices to pull excess water from them. Whenever I've done that, the final result was far too salty. I place the slices in single layers between paper towels, put a cookie sheet on top, and weight heavily with a stockpot full of water. The paper towel absorbs some of the water from the slices during a couple of hours.

Do *not* dredge the slices in flour—dip them in *plain egg*, beaten with a fork, *only*. Then fry them in a single layer in a heavy pan until golden brown, using good olive oil. Set the fried pieces aside as you work—this stage will take a while. And it will take a *lot* of egg—a major reason the dish is so good. For the forty slices mentioned above, I use eight eggs.

The tomato sauce: Summer is the time for tomato sauce. I start by sautéing onions and garlic in olive oil or duck or goose fat in my big, heavy-bottomed stockpot, then add fresh coarsely-cut tomatoes—maybe a gallon and a half. I cook until the sauce is as thick as I want it (rather thick for the eggplant Parmesan), adding salt and fresh herbs—oregano, parsley, thyme, tarragon, basil, any or all—during the last hour or so. Then I put the sauce through a food mill to remove the peels, seeds, and herb stems.

Of course, you can buy a prepared tomato sauce. You will use a lot of it when doing the final assembly—have on hand a couple of quarts for a good-sized casserole.

The cheeses: You will need a wedge—a couple ounces or so—of good Parmesan. Grate it finely on one of the punched sides of a kitchen grater. (If you use that pre-grated stuff in cardboard "cans," I don't want to hear about it.) Shred about half a pound of whole-milk mozzarella, using the large holes of the grater.

Assembling the casserole: Ladle sauce into the bottom of an ovenproof glass casserole dish, and cover entirely with eggplant slices in a single layer. Spread over the slices some more sauce, and sprinkle lightly with Parmesan and mozzarella. A light sprinkling is best—if you use too much mozzarella, it will form a tough "sole" in the dish as it cools. Continue with eggplant–sauce–cheese layers until the casserole dish is filled, then top with a final dollop of sauce and sprinkle of Parmesan.

Baking and serving: Bake the casserole in a 325°F (163°C) oven for forty-five minutes to an hour—until it is bubbling, not only around the edges but throughout the center as well. Set aside to cool a bit and firm up in texture a little. Serve with crusty bread, red wine or beer, and a fresh garden salad.

This is a foolproof recipe—you'll wow your guests on your first attempt. At its worst, it is *very* good. At its best, it is something like a scrumptious eggplant-tomato custard.

to prevent loss of water vapor from inside the package, through osmosis, to the drier air of the freezer. These two precautions will prevent or delay freezer burn.

When using either a loose or a zip-seal freezer bag, I use a drinking straw to suck the air from inside the package, then seal tightly with twist-ties or the zip seal. If using freezer paper, I wrap and tape as tightly as I can.

Home vacuum heat sealers, widely available where kitchen appliances are sold, come with a range of features and prices but need not be expensive. I have used one for many years—the resulting shrink-wrapped packages keep better and longer than those I produce using any other method.

The greatest limit with a home vacuum sealer is size. The manufacturers seem not to have anticipated that the homesteader might want to shrink-wrap an entire goose, a roasting chicken, or a turkey. For those use one of the other methods.

Cut or Leave Whole?

We freeze most dressed poultry whole, then cut as desired after thawing. It may be that the meat keeps better in larger pieces. What is certain is that if you cut before freezing, you create pointed ends of sheared bone that can puncture wrapping. When freezing cut pieces with exposed points of bone, I take care to tuck sharp points inside other cuts with a smooth exterior. It is occasionally necessary to fold a piece of freezer paper several thicknesses to pad an exposed point.

An exception to the leave-whole preference is the processing of duck carcasses. To be sure, we typically freeze the larger drakes, for special meals of roast duck. But I normally cut up a duck (female), using the same two knives I use for butchering: the thin, flexible one to fillet the breast off the breastbone in two portions, and the one with the heavier, longer blade to cut wings, thighs, and legs from the frame, which we reserve for the stockpot (see figure 29.2).

We love the breast fillets cooked like small steaks, pan-grilling them hot and quick in the fat that renders from the skin as grilling begins. *Duck breast is best served rare.* What I call the bits and pieces—wings, legs, and thighs—are delicious for braised dishes, for example with onion, red cabbage, and apple; or made into *confit*, preserved duck (recipe in appendix I). After a meal like that, we sigh with satisfaction and muse, "Hmmm, wonder what the rich folks are eating tonight!"

Stock Parts

Feet, heads, necks, and skin from around the necks can be saved for the stockpot. Dry thoroughly, and freeze in appropriate-sized packages. If you cut up carcasses of any sort (as for ducks, previous section), the bonier parts such as the back can also be saved for the stockpot. In addition, we always freeze the bony carcasses from roasted fowl until ready to make stock.

Giblets

If you have never liked chicken livers, this is your chance to appreciate *real* livers, not the dull yellow-brown, tired-looking livers your mother may have bought at the supermarket. Livers from your younger poultry should be dark red, plump, and glistening. Cook them as a first-day treat, and freeze extra livers in small packets for later use. For a fancy hors d'oeuvre, try Bridget's Chicken Liver Pâté (page 313).

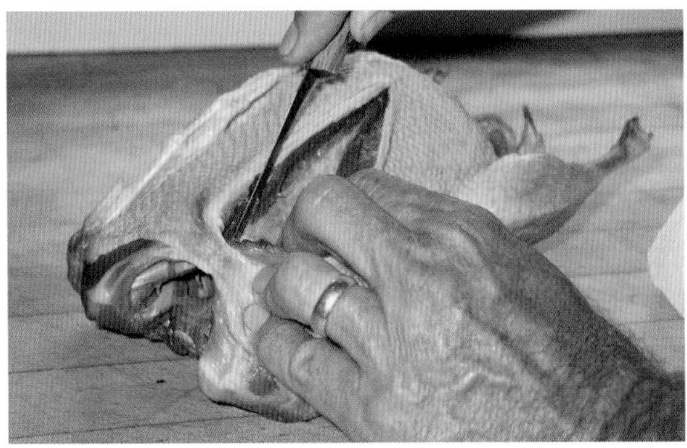

Figure 29.2. Cutting up ducks provides breast fillets for gourmet meals, bits and pieces for a variety of interesting dishes, plus additions to the stockpot.

Broth Is Beautiful

RECIPE CONTRIBUTED BY ELLEN USSERY

The best use for old hens declining in productivity is making broth. Indeed, the very best broth I ever made was from hens given to us by a friend who had kept them for years after they stopped laying.

But regardless of the precise age of the hens, chicken broth is not only a delicious base upon which to build a flavorful soup or sauce, it is itself an extremely nourishing food. Although not a complete protein, it has protein-sparing properties, meaning that it helps prevent breakdown of the body's protein when added to a mixed diet. A gelatin-rich bone broth forms hydrophilic colloids that nourish and protect the digestive tract. It contains amino acids that decrease inflammation. By providing type II collagen, hyaluronic acid, glucosamine, and chondroitin, it helps protect your joints. It contains glycine, which improves sleep quality. And modern research confirms traditional wisdom: Chicken broth does indeed help prevent and moderate colds and flu.

Now that this ancient food has reclaimed its deserved position in the food hierarchy, you can find many recipes and methods on the internet, along with some conflicting information, some of it either disproven or unproven. But two things are abundantly clear: A good gel is determinative for delivering the benefits enumerated above—and the quality of your chickens is of primary importance to production of dense gel. And lucky you —if you have the very best in your own backyard!

When slaughtering your chickens be sure to clean and save the feet, which contribute a lot of collagen, the source of that all-important gelatin—a component as well of the bones, muscles, skin, and tendons. If you don't use them in other ways, be sure to save for making broth the wings, heart, gizzard, head, and neck with its skin. If you cut your chickens into serving pieces, reserve the backs as well. In addition, the carcasses of cooked birds can be saved. I hold all these ingredients tightly wrapped in the freezer until I am ready to make broth.

I can make broth from such "spare parts" alone, but by preference I start with an old hen or rooster and surround it with saved parts and "used" bones from the freezer. *I always add some feet.* If I don't have an old bird, I use a greater portion of feet with alternatives such as carcasses and "spare parts"—to ensure extraction of plenty of gel. If you don't have enough feet, wings serve the same purpose. You will have to experiment with the ratio, but if you are using an entire old bird, addition of just its own feet is sufficient; if using say the leftover carcass from a roast chicken,

Figure 29.3. Tasty, good-for-you broth made from old stew hens and chicken feet–keep plenty on hand!

six to eight feet (or wings) should do it. Always remember—the younger the bird, the more feet or wings you should add.

For years I added some vinegar to the pot and let it rest for half an hour before cooking because "everybody" said that would increase extraction of minerals. It turns out that was more nutritional myth than nutritional wisdom. Indeed, addition of vinegar creates free glutamates, which can be problematic for some people and which nobody *needs*.

I do still cut the bones into pieces with a pruning shears reserved specifically for that purpose—using smaller pieces helps fit more bony parts into the pot. But whether cutting the bones has any nutritional benefit I cannot say.

It should be clear by now that making bone broth is a moving target. Keep an open mind and experiment to find a way to make it that fits your routine and the ingredients available to you. If you get a good gel, you can be sure you are also getting good mineral extraction, along with the abundant collagen. You can make bone broth in a stock pot on the stove, in a crock pot/slow cooker, or in a pressure cooker. You can add lots of vegetables and seasonings or none. I have had success using all these variations on equipment and methods. The most important thing is that you *always* have good broth on hand as part of your daily diet.

Most recently I have settled on Michelle Tam's recipe on her Nom Nom Paleo website using the pressure cooker feature of an Instant Pot. To whatever assortment of bones or old hen I have on hand I add an onion cut in quarters, a carrot peeled and cut into three pieces, two tablespoons Red Boat fish sauce, and three dried shiitake mushrooms—and cover ingredients with an inch or so of water. I pressure-cook on high for two hours, then let the pressure release on its own—so the whole process takes about four hours.

If you want to use the meat of the hen, cook either in a slow cooker or the stove top and remove the meat after one or two hours, then continue to cook the bones and other ingredients an additional four to twelve hours. Use of the reserved meat depends on how much flavor has been removed. Add it to soup along with rice, pasta, or potatoes and carrots plus some green vegetables for a one-dish meal. Make potted chicken by putting it in a food processor along with some seasonings and one-half to three-quarters of its weight in rendered poultry fat or softened butter; it will keep in the fridge for two weeks in a glass jar topped with a good layer of fat. If most of the flavor has been extracted, you could feed it to your dog or back to your chickens.

We often start a meal with a cup of broth—at breakfast, usually in the form of "green soup," broth blended with cooked greens. Often I just heat broth and stir in some flavorful miso, such as the red pepper and garlic miso from South River Miso Company. Otherwise I add salt and drink as is, or simmer with any of the following: a pressed or minced garlic clove and chopped parsley, trimmed shiitake mushroom stems that I have frozen, shredded spinach, or coconut cream concentrate and fresh ginger. The possibilities are endless.

The hearts and gizzards can be used as stock parts or added to soups, sauces, or stir-fries. If you have a lot of them, gizzards make a fine confit (made as described in appendix I.[4]

Processing the gizzard is complicated by its structure—a mass of lean muscle around an inner pouch with a tough lining, containing bits of stone the bird uses to grind its food. You can cut through the muscle just to the pouch lining, then peel the muscle back with your fingers to free it from the pouch, leaving a butterfly-shaped cleaned gizzard. It can be tricky peeling off the muscle without rupturing the pouch—it helps if you *chill them* before peeling. Another option is to place the gizzard on a cutting board and slice off bite-sized pieces, keeping the blade's edge away from the interior pouch and leaving it intact.

Remember Your Pets

For years we reserved necks, hearts, and gizzards for our dog Nyssa. Before dogs started eating that dry crunchy stuff, they ate raw meat, right? And bones. So we figured that Nyssa could only benefit from raw meat and bone as a substantial part of her diet. I cut the hearts and gizzards into bite-sized pieces and froze in mini-packets. I also froze the necks, tightly wrapped in pairs. Each day Nyssa got one or the other, thawed to room temperature. She loved them!

Yes, I know you have been cautioned never to feed chicken bones to dogs. That's excellent advice with reference to long bones (thigh bones, drumstick bones), cooked bones, and bones from commercial chicken generally (which are poorly mineralized, softer than home-raised chicken bones, and apt to break into long dangerous splinters). Raw chunky neck bones were never the slightest problem for Nyssa. Chewing them helped keep her teeth clean and her gums healthy, and they were unmatched for promoting good bowel function.

We used to offer our cat the same raw tidbits, but she didn't seem much interested. She was a good hunter who regularly ate her kills, so I suppose she considered our offerings a poor second best. I have read, however, of cats who relished such raw tidbits.

Rendering and Storing Fat

I hope you are not among those unfortunates still bamboozled by superstitions regarding animal fats. High-quality animal fats are in fact beneficial in the diet. Female fowl especially—whether chickens, ducks, or geese, and especially in fall—contain a heavy deposit of body cavity fat in the lower abdomen. While com-

Chicken Liver Pâté

RECIPE CONTRIBUTED BY BRIDGET CHISHOLM

I've had my feet under Bridget's table many a time, and I can assure you—this is a sophisticated hors d'oeuvre that has even folks who don't like liver coming back for more.

- 1 pound chicken livers
- 1 medium onion, thinly sliced
- 1 clove garlic, crushed
- 2 bay leaves, crushed
- ¼ teaspoon dried thyme (use 1–1½ teaspoons finely minced if fresh)
- 1 cup water
- 2 teaspoons salt, split (1 teaspoon for poaching livers, 1 teaspoon for seasoning pâté)
- 1½ cups softened unsalted butter, more or less (see below)
- Salt and fresh-ground pepper
- 2 teaspoons cognac or Scotch whiskey

Place the first seven ingredients in pot (but with only 1 teaspoon of the salt), bring to a boil, cover, and cook at a bare simmer for seven minutes. Remove from the heat, and let sit five minutes.

Strain off the liquid, then place the mixture of liver and solids in a food processor. Process, adding the butter a little at a time. You can add a little more than 1½ cups for a milder flavor, or a little less to emphasize the flavor of the liver a bit. After all the butter has been incorporated, add salt and pepper to taste and cognac. Process another two minutes until creamy and smooth.

Pour into small molds and place in the refrigerator to set. Unmold before serving with crackers or thinly sliced chewy bread. The pâté can be frozen for up to six months.

Figure 29.4. Render fat from ducks and geese (or chickens) over gentle heat.

pleting evisceration, it is easy to pull it free and store in the refrigerator until you're ready to render it. Heat a heavy cast-iron frying pan or equivalent over low heat while cutting the masses of fat into small cubes. Add the cubes to the pan and keep on lowest heat to gently melt their fat (see figure 29.4). When the contents of the pan are nothing but golden liquid fat with some crinkly remnants of the original fat tissue, pour the fat through a strainer into small jars. It will keep for months in the refrigerator or indefinitely in the freezer. Spoon the fat from the jar anytime you need a cooking fat for frying sliced potatoes or greens such as mustard or kale; for adding to chopped liver or potted meat; or even as a spread on bread or crackers, in lieu of butter. This fat is also a key ingredient for making confit (appendix I).

The shriveled tidbits remaining in the strainer are called cracklings. Save them, and use as you would croutons—that is, as garnishes for salads, soups, stir-fries, or baked potatoes, or even as a crunchy snack.

First-Day Goodies

Butchering a large contingent of fowl is hard work, and at the end of slaughter day we're not much inclined to do any complicated cooking. A couple of the more perishable goodies from butchering make great quick meals.

We prize what Ellen calls the unborn eggs from laying hens. Finished eggs we simply use as we do any other eggs; but the developing egg yolks—there may be up to half a dozen large enough to keep per hen—are both a special treat and a quick meal. We pour hot broth into mugs or bowls, add the small yolks, and wait just long enough for them to warm through. Simple, nutritious, and filling—a fine appetizer, or an entire meal if you add a quick salad.

Supper at the end of butchering day almost always centers on the fresh livers. Sauté some onions in a generous amount of butter or goose or duck fat in a heavy skillet, then sauté the fresh livers hot and quick, deglazing the pan with a little wine or sherry, maybe garnishing with bacon. Livers are rich in fat-soluble vitamins, especially A and D, antioxidants, and essential fatty acids—*and are most delicious if cooked quite rare.* The addition of steamed green beans or a green salad makes it a meal.

As said in the last chapter, the testicles of mature male fowl are as edible as those of larger animals such as lambs or calves. Those not put off by "mountain oysters" could give them a try—preparation, texture, and taste are the same, only the size is different. Older versions of *Joy of Cooking* include recipes for testicles, referred to as "fries"—we do like our euphemisms, don't we? Newer editions of *Joy* have abandoned some of the older culinary lore, so they may or may not still be a guide for this dish.

Harvey's Basic Summer Chicken

Rather than freeze all the chickens from a full butchering day, in the summer I almost always cook a couple of them right away. My approach is likely to be the same when I cull just a couple of birds in a hurry, probably skinning them for simplicity.

Since I have plenty of onions, garlic, and tomatoes in the summer, they are my starting ingredients. I sauté the alliums in poultry fat; add the cut-up chicken pieces, sometimes browning them first in the fat, sometimes not; then add coarsely sectioned tomatoes right off the vines, enough to cover the chicken. I simmer with no lid—to simultaneously cook the chicken and reduce the sauce—until the chicken is as tender as I want. How long that takes depends on the age of the birds, from less than an hour for one a couple of months old to three or so for an old stewing hen.

If the chicken pieces are cooked before the sauce is sufficiently reduced, I remove them and continue cooking down the sauce. When thick enough, I put it through a food mill to remove the peels and seeds, then return the chicken pieces.

How I finish the dish from that point depends on the mood I'm in, or what's clamoring to be used in the garden. Adding hot chili peppers and garnishing with chopped cilantro gives it a Mexican twist. Fresh oregano or thyme or tarragon allows me to pose as a hotshot French or Italian chef. If I use a *lot* of additional onion at the beginning, add several tablespoons of good paprika, and stir in cultured cream at the end, I call it Chicken Paprikash. And of course, what could be easier than working in some mixed exotic spices for a curry—especially good with the addition of fresh green beans?

CHAPTER 30

Serving Small Local Markets

Though I do not produce for market, during most of our nearly four decades at Boxwood I have produced *all* the dressed poultry and eggs we eat, year-round—and we eat a lot of both. I therefore think of myself as working right at the intersection between serious home production and small-farm production dedicated to bringing to local markets "food with a face." This chapter is addressed to the reader who is just a little more ambitious than I—one who, having mastered the inclusive, integrative approaches to poultry husbandry presented in this book to produce all the family's eggs and dressed poultry, dreams of expanding production sufficiently to earn some income—and make a difference—by offering the world's best poultry and eggs in small local markets. As you approach that level of production, I guarantee that guests or friends to whom you make gifts will say, "I've never tasted chicken like that!" or "We haven't been able to find eggs that good anywhere—can we buy from you?" Should such remarks inspire you to try your hand at marketing, I hope you will be motivated by this book's passion for healing and sustainability in a new agriculture.

And I hope your whole family will get involved. What kind of sustainability are we practicing if our children do not continue what we do on our farms? What better than a poultry marketing project to teach youngsters responsibility, independence, and business acumen?

If you are considering stepping up your poultry husbandry to production for small markets, you could not do better than to join the American Pastured Poultry Producers Association (APPPA). APPPA's members can give you better guidance—out of a deeper well of hard-won experience, intelligent observation, and constant experimentation—than any other source, bar none. Do note the *pastured* in the organization's name. APPPA producers believe passionately that the place to produce quality eggs and broilers is on good pasture—none of them would consider a confinement model. (See appendix J to learn more about the APPPA and benefits of membership.)

Relationship Marketing

Some small producers sell at the farm, others through CSAs (community-supported agriculture), and others through farmer's markets. Whatever your choice, see the contact with your customer as an opportunity to establish a *relationship*. Foster that relationship by providing as full an introduction to the quality of your foods as possible. Hand out recipes for dishes using your food products. Offer tasting samples. Garnish your sales table with pictures of happy hens on pasture and other scenes from your farm. Develop a newsletter, email list, or farm blog.

Ideally the relationship should not focus solely on the food sale of the moment—see it as a potential partnership in the transformation of farming toward more regenerative practices. Tread softly, however—nobody wants to be lectured. Without being tedious, preachy, or overbearing, see the transaction as an opportunity to *educate*—always remembering that the education opportunity is a two-way street. More customers are getting concerned, for example, about the ubiquity of soy in our modern diet and its deleterious effects. "Do you feed your chickens soy?" is a question you are bound to be asked. When you get such questions, make it a point to describe your foods, as applicable, as: pastured, GMO-free, chemical-free, antibiotic- and hormone-free, soy-free—all may be selling points with customers well informed about issues of food and health.

Customers who feel they consistently get the brush-off from the commercial food system are frustrated—a major reason they show up at your farm or sales table. Be prepared to engage their concerns seriously, and be willing to do some research to resolve their questions. Whether or not you decide to seek organic certification for your products, be prepared to explain why they are *beyond* organic—that is, even more naturally and sustainably produced than required by the National Organic Standards.

Producers in the APPPA find that interest in their products spikes every time there is a massive recall from the supermarkets or an outbreak of foodborne illness in the news. Be ready to explain to your customers why your pasture-raised products are not only tastier but *safer* and better for their health. What you tell them could be like scattering seeds—who knows when and where one will sprout?

At the same time remember Joel Salatin's advice not to waste your time with customers who only whine and complain, or who fail in their responsibilities in distribution arrangements. Customers who simply cannot see the difference between you and Walmart, or who expect you to *be* Walmart—let 'em go to Walmart!

Equipment and Logistics

If you are handy, you can save a lot of money on equipment and infrastructure costs. I know small producers who made their own range feeders, pluckers, compressed-air egg washers, ingenious mobile shelters, walk-in coolers, brooders for starting hundreds of chicks at a time, and more. (That is another reason to join APPPA—members swap information about their latest projects all the time.)

Save money on ice by making your own reusable freezer chillers: Make a saturated salt solution—as much salt as you can dissolve in hot water—and fill recycled plastic jugs of any convenient size and shape. Keep them in the freezer until needed. Because of the lower freezing temperature of the salt solution, they will be much colder than ice.

Breed, infrastructure, marketing, management—find the best overall fit for you. As one of my APPPA correspondents put it:

> *Turkeys are the most capital intensive because their equipment and brooder can only be used once a season. Cornish Cross and Pekin ducks make fuller use of brooding resources because both can be started several times successively during the season. Management level required in the brooder is the highest for starting turkeys, followed by chickens, and then ducks. Transitioning from brooder to pasture is easiest for ducks, turkeys, and then chickens. Ability to forage is best for ducks, followed by turkeys, and then chickens. Ease of processing is chickens, turkeys, and then ducks. Ease of marketing at a profitable level for us is ducks, turkeys, and then chickens. Ease of getting yourself in the door with a new customer: chickens easiest of all.*

Feeding Issues

The logistics of feeding change when the flockster starts feeding market flocks, even on a modest scale. It may no longer make sense to buy in 50-pound bags but rather to take delivery in bulk. Bulk ordering has its challenges—storage becomes a bigger part of the equation. But those getting bulk deliveries may be able to get the local feed mill to custom-make feed to their own specifications.

I'm always pushing the idea as well that flocksters—whether small-scale market producers or serious homesteaders who use a couple of tons of feed a year—band together to contract with local farmers to grow their feed grains. You buy their feed grains, they buy your chicken and eggs—how's that for keeping it local? That strategy would have a positive ripple effect in the local economy and could help keep a small farm viable. You would also know the life history of your feed grains, as opposed to buying a pig in a poke. Again, however, the big challenge would likely be how to handle storage.

Remember that feed quality can pay off, even though it costs more. One APPPA member noted that she started feeding organically raised feed grains and raised several hundred broilers more than she had the previous year—on the same amount of feed.

Regulations–and Regulators

I can see the day coming when even your home garden
Is gonna be against the law.

—BOB DYLAN

The biggest difference between producing for the family and producing for even the smallest market is being subject to applicable regulations governing food production and marketing. If we assume that the regulations are intended solely to guarantee the safety of what we put into our mouths, however, we are being naive. The regulatory process is frequently hijacked by the big producers to create impossible hurdles for the small producer. Giant, vertically integrated food processors feel threatened by the growing demand for safe, local food and want to stop it in its tracks. If they have their way, any alternative to industrial food whatever "is gonna be against the law." By all means read Joel Salatin's *Everything I Want to Do Is Illegal*—it serves up just the right mix of boiling-mad outrage and belly-laugh humor to get you in the mood for the sometimes-byzantine labyrinth of food-police regulations.

The small producer's most important contact with regulations is the regulator, often the inspector in the field. Make the regulator a friend if possible—I've had small producers report that their inspector helped them out: "Well, the regulation says you have to do *X*, but you know, you could meet the requirement by doing *Y* [minor change, maybe only a 'paper' one, to bring you technically into compliance]."

On the other hand, if your regulator turns out to be fiend rather than friend, you should know the regulations *even better* than the regulator. Again, I've had numerous reports of regulators handing out diktats off the top of their heads, with disadvantageous implications for the operation—only to be stopped cold by a level-eyed producer quoting the regulations, chapter and verse.

If you have a problem with regulatory agencies you cannot resolve on your own, call on the Farm-to-Consumer Legal Defense Fund (FTCLDF), a leading advocate for small farmers being harassed by agencies of government.[1] Intervention by one of their lawyers on your behalf may more readily sober up an agency acting arbitrarily. For example, when a health inspector disallowed use of Polyface Farm eggs by a Charlottesville, Virginia, restaurant, Joel Salatin immediately called FTCLDF. Within twenty-four hours of the Legal Defense Fund's intervention, the health department had rescinded its arbitrary ruling.

Pricing

In a market blinded by the illusion that cheap food is a blessing, there is often resistance to paying the higher price of nonsubsidized poultry products. The customers the small producer should target are those more concerned with food quality—and safety—than bargain-basement prices. And they are out there: My APPPA correspondents often report they cannot meet growing market demand for "real food."

In calculating a fair price to charge, remember to pay yourself for your labor. "Fair" must be fair *to you* as well as the customer.

Questions about price can also be opportunities for educating customers. For example, you might tactfully ask customers unhappy with your prices if they are aware that the "cheap chicken" and "cheap eggs" in the supermarket are *subsidized*—by *their taxpayer dollars*.

Complaints about price can be frustrating. But you might like to know I have had numerous reports from small-market producers about customers who worry they're not charging *enough* for their products. Such customers understand the *One dollar, one vote* rule and are willing to pay a worthy price—to ensure their supplier stays in the market.

One thing beginners should keep in mind is *never* to charge a hobby price when selling in local markets. Even if you are just dabbling with sales of a bit of surplus from home production, undercutting prices of producers who depend on sales for their livelihood is an inexcusable disservice to them. Building a sustainable agriculture that adequately rewards suppliers of quality food is serious business—do your part.

Market Offerings

Whatever poultry foods you most like to see on your own table are probably what you should consider first for offering in a market. Your own enthusiasm for the superior foods you provide your family will be infectious among your customers, and their queries about less usual fare may suggest new marketing ventures.

Eggs

Eggs are the easiest part of the typical home poultry project to expand into market sales, if only because they require so little additional in the way of infrastructure and equipment. Selling eggs may be an ideal project for children. And I don't mean just gathering the eggs—though that's a fine chore for the wee ones—but organizing and managing the entire care of the egg flock and the market sales.

You will have to check the regulations in your own state as to whether you need to be licensed or inspected, and regarding candling, grading, washing, packaging, labeling, and refrigerating eggs.[2] Talk with others serving egg markets in your area as well—in some cases actual practice, for example whether reuse of egg cartons is allowed, is more casual than is strictly set forth in the regulations. If you sell at a farmer's market, it may have requirements to be met as well.

Regarding breed choice for layer flocks, there are those in the pastured poultry community who wax dogmatic on the hybrid superlayer strains as the only practical and economical choice, because of their higher rate of lay. But many small-market producers base their market egg operations on traditional dual-purpose breeds such as Rhode Island Reds, Barred Plymouth Rocks, and Black Australorps.

Those who rely on traditional dual-purpose breeds for market eggs find they take fuller advantage of natural forages; are more easily managed; make a better-quality egg; live longer; stay healthier; and are more alert to their environments, thus more likely to escape predation. Remember as well that anyone who enters the market for eggs is after a couple of years going to be looking for a market for stewing hens—many producers note how much easier it is to sell the larger carcass of a dual-purpose hen than the smaller one of a hybrid layer.

For many it is a bonus that the dual-purpose breeds lay large brown eggs. Though egg quality is strictly a matter of diet and lifestyle, not shell color, some producers despair of trying to educate customers who stubbornly believe that brown eggs are better than white. Use of the traditional brown-egg breeds helps keep such customers coming back. (If you prefer hybrid layers, options such as Golden Comets and Red Sex Links that lay brown eggs are available.)

Some producers keep a mix of white-egg breeds, several brown-egg breeds for multiple shades of brown, even a scattering of Ameraucana hens, finding that a rainbow effect in the display carton helps spark interest. Some producers find that Ameraucana eggs especially, even at the cost of a slight decrease in overall production, may justify including to catch customers' eyes—especially those with children—and trigger questions about "those green eggs."

The Traveling Layer Flock

A mobile henhouse for the pastured layer flock—an "eggmobile"—is a great convenience. Keepers of market laying flocks make them from whatever comes to hand, from converted recreational vehicles to school buses, but the most common versions I've seen are mounted on hay wagon chassis or flatbed trailers and pulled with a pickup, ATV, or tractor.

A nearby friend, Stephen Day, assisted by his father Richard, made an "eggmobile" for a flock of one hundred layers. Stephen reports:

> *We started building the coop with no schematics and only a general idea of how it would come together. Dad and I both have prior experience putting together outbuildings, so we just brainstormed each new problem as we built. The whole idea is based on Joel Salatin's eggmobiles, and my total cash outlay was about $1,500, not including labor time.*
>
> *We started with an old 16- by 8-foot (4.8- by 2.5-m) tandem flatbed with an iron gridwork floor that we bought cheap through Craigslist. The grid floor keeps manure from piling up in the trailer and pulls plenty of air up from below and out along the corrugations of the roof panels.*
>
> *We made six sections of stud wall, two feet on center, bolted them through the floor, and braced everything with 2 × 4s diagonally. Since the trailer will never be on perfectly level ground, the roof needed no pitch—it's just six sheets of corrugated metal roofing. The walls are plywood with two chicken doors cut into them as well as a human door, two windows, two vents, and two rectangular flaps for gathering eggs from nestboxes without going inside.*
>
> *The interior has several roosts made of saplings, with room to expand to accommodate more birds. There are 25 nestboxes, which I hope will eventually serve 200 laying hens. I'm starting with 100 as an experiment to test the trailer, our predator situation, the egg market, and the chickens' impact on our pasture. We use a gravity-fed cup watering system, and the chickens are fed primarily by pasture grazing, supplemented by pellet feed. We tow the trailer several hundred yards with our pickup every week or two, to keep their pasture area fresh.*

The Cadillac of eggmobiles is the one built by Tim Koegel of Windy Ridge Farm in southwestern New York State for his market layer flock. An impressive high-tech modification is the addition of solar panels (not shown in the picture), which can be mounted on either side, depending on the orientation of the shelter in relation to the sun after a move. The solar panels supply charge to an extensive system of electric nets, as well as interior lighting for work inside during the dark early-morning and evening hours—and to provide supplemental lighting to encourage laying in winter. Deeper into winter, with reduced solar input available, Tim connects the shelter and its electric nets to grid power.

Tim's flock uses this shelter year-round. In the winter two 250-watt infrared heat lamps on a thermostatic control prevent freezing of the waterers. When he parks the mobile shelter for the winter (see figure 30.6), Tim attaches a hoophouse of the same dimension. Each is 10 by 20 feet (3 by 6 m), yielding a combined space of

Figure 30.1. Eggmobile for a modest layer flock inside an electric net fence.

Figure 30.2. Eggmobile on a hay wagon chassis.

Figure 30.3. Stephen Day's eggmobile for 100 to 200 layers on his farm in northern Virginia. Photo courtesy of Stephen Day.

Figure 30.4. Tim Koegel's year-round layer flock shelter, the Cadillac of eggmobiles. Photo courtesy of Tim Koegel.

Figure 30.5. Interior of Stephen Day's eggmobile. Note metal grill floor, which enhances ventilation and allows droppings to fall through, for less frequent cleaning. Photo courtesy of Stephen Day.

Figure 30.6. Tim's eggmobile attached to a hoophouse, which affords more space for the winter flock. A scraped yard provides even more space for exercise in the sun when weather permits. Photo courtesy of Tim Koegel.

400 square feet (37 square m) for the flock of 150 layers. Deep litter in the shelter is wood shavings; in the hoop, hay. Tim encourages scratching of the latter by feeding scratch grains there.

He pushes back the snow to provide an exercise yard, which the hens enjoy using on sunny days. The dormant sod does not furnish grazing, of course, so Tim sprouts grains for feeding.

When you have customers who depend on your eggs, managing the flock regarding when to retire older, less productive hens and bring along a new flock of layers becomes more complicated. As said earlier, you will also need to find buyers for the old stewing hens your operation generates. Some producers sell into a growing market for raw, natural pet food. The strategy for putting excess cockerels to good use (in "Value-Added Sales," page 328) would serve as well for old hens. Remember there may be some customers that especially value your old hens—enough to buy them live and save you the effort of slaughtering them (see "Other Markets," page 329).

You don't have to tell customers how different your eggs are—an attention-getting trick is a display of your pastured egg and a supermarket egg, cracked open in side-by-side saucers. Alternatively, friends of mine hard-boil some of their eggs and some of the supermarket imitations, and slice them in half for display. Differences in interior structure, yolk color, and integrity of the whites are all that you need to tell the tale.

Some producers mark their egg cartons with the laid-on date as a way of highlighting how fresh they are.

I hear mixed reports from APPPA members about getting a decent price for their eggs. Some encounter an appreciation of the obvious superior quality, and a willingness to pay the higher cost. Others find that customers, fixated on $1.29/dozen eggs in the supermarket, stubbornly refuse to pay more than an unfairly low price. Many producers find that initial obstinacy about price is followed by repeated demands for more once customers decide to give their pastured eggs a try, as in this report from the field:

> *I had a customer last year at the farmer's market who confronted me about selling pasture raised eggs for $7 doz. I told her to buy them quick because my bookkeeper was putting the price up to $8. I explained that my price has something to do with the fact that we buy organic feed, pay our crew well, and have workers' comp insurance to pay. Then I advised her that on second thought she actually shouldn't try them, since they were so good, she would never be able to eat any other eggs. She was not impressed. But later that day she came back, still indignant but willing to try half a dozen. I warned her again but to no avail. Next week she came and thrust the money into my hand and picked up a dozen. She's been a regular customer ever since.*

Broilers

If you want to supply fresh-dressed broilers to your market, the first decision is what type of bird to grow. Many pastured poultry producers grow Cornish Cross—the foundation of industrial broiler production—as their market bird. Some report that they have learned to work with the Cornish and don't see any reason to change. Others are not happy with the weaknesses of this supercharged hybrid—the leg problems, the heart problems, the laziness as foragers on pasture—but find that customers are so used to the plump, broad breast of the Cornish that they resist buying a broiler with a narrower profile.

Many of my APPPA correspondents prefer meat hybrids more suited to pasture—such as the so-called Freedom Ranger meat strains coming out of the French *Label Rouge* system of certified broiler production—and similar hybrid crosses developed in the United States. (I offered thoughts on these hybrids in the sidebar "If You Want to Grow a Meat-Bird Hybrid. . ." on page 40.) A few offer both Cornish Cross and Freedom Ranger type broilers, charging an additional 50 cents or more per pound for the latter to compensate for their longer grow-out. The occasional adventurous producer raises traditional dual-purpose breeds such as Barred Rock, Buff Orpington, or Delaware as market broilers for customers who will accept a skinny carcass with superb flavor—and who will pay a higher price commensurate with the even longer grow-out. In today's market, however, traditional-breed broilers are a hard sell.

Regarding regulations, as a broad generalization, most states follow the federal guidelines for poultry production for sale: The grower may process on the farm and sell direct to the consumer up to one thousand birds (regardless of species) per year with no inspection or regulation. The "Producer/Grower" exemption allows for production of up to 20,000 birds, with registration and periodic inspection of your processing facility but with no requirement of an on-site processing inspector. Some states have more restrictive regulations, however, so check with more experienced processors in your area, with fellow APPPA members, or with the FTCLDF about regulations you will need to follow. I can't say it enough: If you plan to serve a market, you would be foolish not to join the APPPA. Their discussion forum should prove far more effective than reading any amount of confusing bureaucratese.

Regulations can be a thorny thicket—many small producers advise keeping as low a profile as possible. If you don't have to be inspected or regulated by a given agency, avoid getting on its radar with inquiries—one of my correspondents said that's "like poking a hornet's nest with a stick to see what happens." He described calling three different state agencies to inquire about regulations for broiler production. He got three different answers—all of them wrong, as it turned out. If inquiries are necessary, make them anonymously.

Figure 30.7. Many growers of broilers like to move them onto pasture as soon as they are feathered. This "Prairie Schooner" is in effect a giant brooder for broilers too young to range. Photo courtesy of feathermanequipment.com.

Remember that if you choose to fudge existing regulations a bit in your favor, a first-offense action by the powers that be is rarely more than a cease-and-desist warning.

Major questions for setting your price for broilers have to do with whether you distribute them packaged through a CSA or farmer's market, or require customers to pick up at the farm and do their own packaging—and whether you sell them whole or cut up. Most producers sell their broilers whole, but I hear from more and more that they're responding to customer demands for cut-up chicken. If you try the latter, the increase in your own costs results from not only the added labor but also extra time and materials in packaging, weighing, and inventorying. Still, you can make money if customers are willing to pay for the convenience of getting their preferred cuts only. For example, one correspondent reports that she sells whole broilers for $4 per pound (thus getting $18 to $24 each for birds that range from 4½ to 6 pounds), while charging $13 per pound for boneless and skinless breasts.

Tooling Up

Efficient processing for any but the smallest broiler market requires automating scalding and plucking and a comfortable place to work furnished with running water. Shelter for working when the weather is not cooperative is a boon as well.

The home-scale mechanical plucker pictured in figure 28.11 speeds up the work for me. But at larger scale, a tub-style plucker is just the thing, for hands-free plucking of several birds at once. Do-it-yourself flocksters can make a mechanical tub plucker for a few hundred bucks (see figures 30.8 and 30.9) or a several-at-a-time scalder, heated either electrically or by propane gas. Commercial pluckers and scalders (see figures 30.10 and 30.11) are available in a wide range of capacity and price. Many of my APPPA correspondents report satisfaction with processing equipment from David Schafer's Featherman line (https://www.feathermanequipment.com), which targets the equipment needs of producers for local markets.

Once you are supplying a steady broiler market, you might want to install an efficient farm processing facility, such as that of my friends Matthew and Ruth Szechenyi, who serve local markets for broilers and turkeys in my area (see figures 30.12 and 30.13).

Figure 30.8. Mike Rininger's Whizbang plucker, made from a recycled food-grade plastic drum, a donated steel plate custom-cut by a metal sculptor friend, and purchased parts including electric motor.

Figure 30.9. Mike's Whizbang in action. Note that in a tub-style plucker, the stiff rubber plucker "fingers" are inset both into the rotating bottom plate and into the stationary drum sides.

Figure 30.10. Commercial stainless-steel plucker in action.

Figure 30.11. Scalder with rotating shelf that cycles up to a dozen broilers through water kept thermostatically at scalding temperature.

Figure 30.12. Matthew and Ruth Szechenyi's processing facility: a 12- by 24-foot (3.5- by 7.3-m) hoophouse adapted from a FarmTek kit for a tractor shelter on a 12- by 28-foot (3.5- by 8.5-m} concrete slab.

Figure 30.13. Interior of the Szechenyi hoop, featuring stainless-steel processing equipment and worktable, and drainage to a French drain outside for easy cleaning.

In contrast with producing market eggs, broiler production for market requires serious tooling up for more efficient processing—you either make money or give away that bird, not at the feed trough, but at the slaughter table. Market processing will likely require scalding and plucking equipment that is higher capacity and more automated than that in my backyard operation. Good poultry processing equipment is expensive if purchased new—if you're handy you might try making your own. Keep your eye out for farm sales: A friend of mine stumbled on an auction where none of the bidders was interested in an array of stainless-steel pluckers and scalders—and walked away with $20,000 worth of equipment for $400.

Turkeys

Many of my APPPA correspondents confirm that turkeys can be highly profitable. One reported that she invested $3,500 converting a mobile construction site office to a state-approved processing facility. She earned back the entire costs for her new facility that fall—as profits on sixty-five preordered Thanksgiving turkeys. She noted as well that the daily care of those sixty-five turkeys was the responsibility of her six-year-old son entirely, a big boost in his self-esteem.

If you give turkeys enough range, they will forage a large percentage of their own feed, reducing costs. Your labor costs are lower when butchering turkeys as well—processing time per bird is not much more than for a broiler, but the payoff in salable meat is so much greater.

Most growers of market turkeys stick with the Broad Breasted White, the equivalent to the Cornish Cross for commercial turkey production. But enough raise heritage breeds to merit giving them a try. Though they require more management input and produce a smaller, narrower carcass, you might turn a profit if you find customers willing to pay more for the very best.

Most small producers I know grow only the number of turkeys for which they have preorders and target their turkey grow-out to Thanksgiving and Christmas. However, a farmer friend of mine followed his customers' leads into turkey production in a greater portion of the year.

In the past he'd had to do some creative marketing of his "uglies"—carcasses with cosmetic blemishes. He salvaged skinless and boneless turkey breasts and turkey tenders from those less salable carcasses, and his customers snapped them up. He made ground meat from the legs and thighs, and found it an easy sell. The remaining legs, backs, bones, and wings he sold to his restaurant chefs for making broth. In the process of making such sales, he discovered that some of his customers would prefer to eat turkey rather than chicken year-round. One, who bought eight of his Thanksgiving turkeys to freeze, said she would prefer being able to get them fresh for more of the year. He concluded that expanding his turkey operation for out-of-season sales made sense.

Specialty Poultry

Some of the remaining options for dressed poultry get us into niche-market territory, but all are good eating whose challenges have more to do with marketing than management or table quality. You are well advised to start small with these offerings and try to line up committed orders for the dressed birds before putting in the feed and sweat to ready them for sale.

As observed in the chapter on butchering, processing *waterfowl* is little different from processing chickens—except for the plucking. One correspondent noted she can process four or five broilers in the time it takes her to do one duck. Thus the price you charge for them should be commensurately higher. Many producers who try to market ducks and geese, however, find it isn't easy to get that higher price.

Still, some find that within a narrow segment of their market, supplying *duck* can be a small but profitable sideline. The Pekin is the Cornish Cross of meat ducks, fast growing and clean plucking, and is often the breed of choice for market producers. Remember *Muscovies* when considering ducks—one of my correspondents finds that the combination of busy foraging, health, ease of raising, and the fact that the ducks are excellent mothers means Muscovies are low-input—but they generate gourmet-level prices. Another reports she gets $20 per pound for duck breasts from her Muscovy drakes.

Remember ducks as an option for producing specialty eggs as well. Two young friends of mine found eager acceptance of duck eggs in the farmer's markets they serve.

Geese require even more processing time and grunt work at the worktable than ducks. But they can be a center of highly special holiday meals not likely to be available from any other source, for which people are willing to splurge. Again, they might fit in as a profitable sideline—a friend of mine has no problem selling twenty-five as preorders for the winter holidays—made even more profitable if you raise them mostly on good pasture. The large, all-white Embdens are often the breed of choice, but based on my experience, I advise giving fast-growing Africans a try as well. There is no comparison between commercial and pastured goose as a taste sensation.

I hear from a few of my correspondents who try teasing the market with *guineas*, *pheasants*, and *quail*. Don't bet the farm on such exotics, but it could be fun experimenting, especially as an unusual project to share with children.

Finally, I recommend you consider *capons* as a meat-fowl option. In times past "caponizing" cockerels was a common practice, both because it was a profitable use for cull cockerels and because it produced a matchless roasting fowl. The term means "castration"—surgically removing the testicles of young cockerels to make for faster, plumper growth sans testosterone. A well-grown capon can dress out at 12 to 15 pounds—the size of a small turkey. Yet however large it grows, or for how long, it remains tender and succulent—and continues accumulating flavor. And it doesn't waste time and energy duking it out with the other cocks in the flock.

The purpose for caponization, then, is the same as for castration of male livestock such as bull calves. The difference is that mammalian testicles are held externally, while in avian species they are deep inside the body cavity—removing them is *major* surgery. All the same, a few producers take up this option if there is a niche demand in their market for the very finest roasting chicken. Caponizing kits containing a few simple surgical implements are available, as well as manuals describing the procedure.

The prospect of caponizing is intimidating because it is so invasive, but from my own experience I assure you that—starting with the simplest of tools and instructions—you can learn to do it successfully on your own. Here is a summary of the procedure:

- Caponize young cockerels early, within the first week to three weeks ideally.
- A heightened level of sanitation is required, but not operating-room sterility. I worked on the stainless-steel surface of my slaughter table, thoroughly cleaned and wiped with alcohol. I periodically rubbed my hands and tools with alcohol as I worked.
- Secure the bird in place on the worktable. I used looped strings with small weights on their ends.
- Make an incision between two ribs. Note that, in lieu of sewing up the incision afterward, you should first pull the elastic skin of the breast well to one side. After the procedure the skin moves back into covering position over the incision site.
- Hold the incision open with surgical retractors, and remove the testicles with forceps or a special cutting scoop.
- No anesthetic is used and, no, I cannot tell you what level of pain the cockerel experiences. I learned to avoid touching a nerve beside the testicles with the forceps, which caused him to spasm as if from a jolt of electricity. Otherwise, he reacted not at all as I made the incision and removed the testicles.
- Afterward, isolate the caponized cockerels where they remain quiet and undisturbed a few days, and monitor frequently. Their robust immune systems will do the rest.

Each flockster will of course have their own notion of whether caponizing is an unacceptably cruel procedure. My own thought is that excess cockerels must be culled in any case, and I wonder which gets the better deal—the one slaughtered at fourteen weeks, or the one who endures the undeniable stress of caponization but continues a nice life for a full year or more?

Be assured that the key challenge is not the complexity of caponizing but the necessity to continue resolutely through your first batch of cockerels *whatever* happens on the operating table. Initial failures you will have, but you will learn this skill only if you persist despite them.

Other Income Options

In addition to the offerings from your flock discussed previously, there are other ways to turn a homestead or small-farm poultry operation to profit. You will probably think of more, but here are a few to consider.

Started Birds

A couple of young friends in my area started an interesting business: Riding the current wave of interest in keeping home flocks, they furnished started birds in small, ready-to-go lots. They brooded three hundred to four hundred chicks at a time in homemade brooder sheds. Though they started chicks of some of the hybrid layers as well, they favored the traditional dual-purpose breeds. They offered a win-win to their customers, who skipped the trouble of brooding themselves and could buy just the few needed for a small flock—rather than twenty-five, the minimum for shipping day-old chicks. Also, customers had their choice of breeds from among those available in the brooder at the time—an exciting option for any children involved. My friends also custom-brooded to order—for example, fifty brown-egg layers for a customer who operates a winery tasting room in which she likes to offer "country eggs" to sippers on a day outing. Started birds remaining unsold by early maturity were no problem—the two entrepreneurs butchered them (helping at a larger operation in exchange for the opportunity to process their own birds with its equipment) and took them to farmer's markets, where they had no problem selling out. As one of them summarized: "There's always something you can do with a bird."

Value-Added Sales

Correspondents who have tried it inform me they make a lot more money on value-added products—pound cake or brownies or quiches made with plenty of eggs, or broth made with the feet and other castoffs of butchering—than they do on their prime-ingredient items (eggs, broilers, and the like). I was especially impressed with the strategy of one of my APPPA correspondents, following publication of an article I wrote on the moral quandary of contributing to the euthanizing of hundreds of thousands of cockerel chicks when we place our oh-so-convenient all-pullet chick orders. His solution, like mine, was to make straight-run orders only in the future—and to butcher the excess cockerels for making potpies to sell in his farmer's market. Now isn't *that* an interesting idea—a change of practice taken for reasons of virtue turns out to generate more income as well.

If you do try to turn a profit adding value to your poultry products, be prepared: "Did someone say *regulations*?"

Custom Processing

Efficient butchering in quantity requires equipment that is either expensive, or time consuming and challenging to make. Once you've made your own equipment investment, you might offer custom processing of others' broilers or turkeys, either as an on-farm service (you come to me) or with a mobile processing unit (I come to you).

Some states place a limit on the number of birds not raised by the farmer that can be custom-processed onsite. You didn't hear it here, but one workaround that has been used is the "sale" to the farmer of birds grown on another farm; their processing as the farmer's own; and their subsequent "sale" to their original owner, all transactions neatly documented. Such is the nature of the bureaucratic rat race.

I don't know many producers who want to do a lot of custom processing—butchering all day is grueling, and most of my correspondents don't want to make a career of it—but I do know producers who have made a mobile processing unit for their own use, and rent it out to others for a fee (see figure 30.14).

Custom Milling

As the need to generate feed increases, you might start dreaming of grinder-mixers for milling your own feeds. Once your operation is tooled up, custom milling for others in the area could become an option, in cooperation perhaps with partner farmers growing feed grains. Who knows, sensible local production and resource-sharing options may come to seem—well, sensible.

Finding Your Niche Market

Trying to compete with the big boys who are wooing customers with their cheap chicken and eggs can become frustrating. It might pay off to find niches in the market they have no interest in targeting. Put another way, your ideal market is one that *may not get served* if you do not serve it.

Restaurants

Especially at high-end restaurants, chefs tend to be fanatics about quality of ingredients—at least as much a key to winning awards as their skills in the kitchen or the number of French words on their menus. It is easy for them to procure whatever ingredients they need from industrial food distributors. But they know how mediocre most of those are—what they really want is primary ingredients as good as those that show up every day on *your* table. Those who have tried them know that pastured eggs mean more taste appeal in baked goods, which stay fresh longer, for greater net profit despite the higher cost of pastured eggs; that dressed poultry with gourmet quality does not need to be hidden under exotic sauces; and that they cannot make incomparable broth as a foundation for soups and sauces from anything other than chicken feet and stewing hens raised on pasture.

Of course, restaurants operate on thin profit margins, so some chefs are unwilling to pay the producer a worthwhile price. In such cases of price inflexibility, producers may find that, while a restaurant will not pay enough for their broilers, for example, it may prove a good outlet for small-volume, high-impact ingredients like livers and feet.

Figure 30.14. Mike and Christie Badger of APPPA rent their mobile processing unit to small-scale producers in north central Pennsylvania. Stainless-steel equipment is laid out for maximum efficiency in a small space, with the flow of processing clockwise from the killing cones on the right to the scalder and plucker and worktable, ending at the chill tank. Two onboard propane tanks provide heat for the scalder; connections to water and electricity must be provided on-site. Photo courtesy of Mike Badger.

A high-end restaurant may be a better bet for sales of specialty poultry with gourmet appeal—guinea, duck, game birds (chukar, pheasant, and both dressed quail and quail eggs)—than the general market.

Other Markets

Some of my correspondents have done well selling into cross-cultural markets for chickens and other fowl. Immigrant customers are sometimes willing to buy live and do their own slaughtering, having grown up in countries where small-scale flocks are more the norm than industrial poultry. Many of my correspondents find that customers steeped in other food cultures are eager for "real" (traditional-breed) chickens, however pointy-breasted, and for spent laying hens.

In a time of increasing sophistication about food quality and nutrition, customers are more likely to come looking for chicken feet and necks and spent stewing hens to make the fabulous chicken broth

From Waste to Resource: Vermont Compost Company

Vermont Compost Company (VCC), located in Montpelier, Vermont's capital, makes thousands of tons of high-quality compost per year from what too many of us still think of as "wastes": livestock manures and bedding, spoiled silage and hay, hardwood bark and other forest wastes, processing wastes from a big cheese producer nearby, and food residuals. The manures include those of horses, cows, donkeys, mules—and of the millions of earthworms active in the enormous compost heaps. Importantly, they include lots of chicken manure as well, *not* purchased and hauled in with big trucks from afar, but generously contributed on-site by hundreds of hard-working *composter chickens*. Oh, and by the way: *All* the flock's feed comes from working the compost heaps—and from ranging the several-acre pasture of VCC's herd of eight American Mammoth Jackstock donkeys. The chickens are a key to VCC's success.

Another component in that success is inclusion of food wastes in the composting operation. VCC contributes to the implementation of a law in Vermont banning food scraps from municipal waste streams. The operation accepts food residuals from businesses such as restaurants and food suppliers, from institutions such as schools and hospitals, and from residents who deposit their food wastes into collection bins on VCC's main composting site on the edge of Montpelier. VCC also offers a collection service to businesses that generate food residuals in Montpelier, with fees dependent on volume of waste and frequency of pickup. As for the logistics of the pickup? A total of twelve 32-gallon "totes" along the route, hauled three days a week on a trailer pulled by three of the donkeys. Now *that* sight must be a fun part of life in Montpelier, what do you think?

The food residuals, together with the assorted other wastes, are assembled into gigantic compost

Figure 30.15. Vermont Compost Company uses big-boy equipment to handle decomposables at massive scale.

Figure 30.16. But the most important "equipment" is a big flock of working chooks . . .

heaps 8 or 9 feet (2.5 to 2.7 m) high (see figure 30.15). Top to bottom, the heaps are constantly worked by the chickens, all day, every day (see figures 30.16 and 30.17).

Perhaps you assume the chooks are eating mostly food residuals—hamburger buns and the like? In fact, the compost heaps contain only 5 percent food residuals—10 percent at the most. Even this small percentage contributes a lot of energy to the intense biological activity in the heaps, and that abundant biology is what really feeds the chickens. Indeed, in a recent phone conversation with Karl Hammer, VCC's owner, he assured me that even if intake of food residuals should drop to zero, the flock would still be generously nourished by the compost heaps.

VCC has used flocks of a few hundred up to twelve hundred chickens to work the compost heaps—even in winter, when the heat of decomposition keeps the composting process going, and the chooks continue feeding on decomposer organisms and their metabolites. Karl plans to increase the flock to fourteen hundred, following the recent addition of a new layer house, and muses about increasing in the future to three thousand. Eggs produced—up to fifteen thousand dozen a year, for twenty-two years now, with zero feed bill—are sold locally for an additional income stream. Hmm, which one should we say the layer flock offers VCC's operation *for free*: Their high-octane manurial contribution to the compost? Their egg production? Or their labor working the compost?

Karl made this fascinating observation when I visited VCC: The poultry industry houses and feeds three hundred million hens in battery cages to produce the nation's eggs. Meanwhile, we create endless problems for ourselves with the huge quantities of food wastes we send to landfills. As it happens, the two about exactly match—if we released those three hundred million chickens on the nation's food scraps, using VCC's model, the chooks would clean them up with no other feed needed. In terms of the energetics involved, *America could thereby get its entire egg supply for free.*

Figure 30.17. . . . who never slack off, never break down.

Figure 30.18. And in the end, mountains of compost.

they've come to think of as a foundation of good health. Seek out local-foods groups in your area—many of them publish lists of local producers of farm foods for thoughtful and demanding consumers. Some conservation organizations have published such lists as well, having concluded that a key to land and environmental conservation is good farming, the key to which is the patronage of good farmers.

The Joys of Diversity

A final thought about serving local markets: All the small producers I talk with have no interest in becoming the next Frank Perdue. Most of them have found their broiler and turkey operations quite profitable—yet I haven't encountered a single one who wanted to specialize with either. Butchering poultry all day is *hard* work. Producers I know take in stride the pulses of intense effort required but feel it would be deadening to make that same all-out effort, day in and day out. They prefer instead to *diversify* their operations, to mix it up, keep things interesting. Intuitively they understand that for a farm, as for nature, diversity is the key to robustness and stability. Most expand to their comfort level of profitable production of broilers or turkeys, then increase no further, instead adding new components to the mix, such as lambs, dairy goats, or honey production. Many are even glad to mentor new producers entering their expanding markets.

Market-sized laying flocks often are housed in eggmobiles (as shown in the sidebar "The Traveling Layer Flock" on page 320)—recycled recreational vehicles or school buses, or rolling chicken coops mounted on a utility trailer or an old hay wagon chassis—pulled when needed to greener fields of plenty by tractor, pickup, or ATV. The eggmobile can be the key to stacking mobile flocks with other livestock species, as in the Polyface Farm model of following beef cattle with layer flocks. In winter, some producers park their eggmobiles and give the birds access from them onto a winter feeding/exercise yard such as the one I use, or into an attached hoophouse.

If you are serving a vegetable CSA, adding eggs to your weekly mix would be especially easy. Many producers making this choice offer the eggs as a separate add-on to the basic subscription. Indeed, some producers offer meat CSAs. Obviously, the greater the diversity of your food products, the more intriguing the mix of CSA foods on offer to tempt potential members.

Note the possibility of developing your own loyal community, whose members revere you as "our farm," and supplying them the full diversity of the foods you produce, changing throughout the year—while keeping in touch via a website, email list, or newsletter to let them know what is coming up and to make it easy for them to place orders.

An interesting thing about diversifying is what it does to thinking about the bottom line. As said earlier, some producers find a lot of resistance in their markets to paying a fair price for their eggs. Think what a disaster that would be for a producer who had "put all their eggs in one basket"—whose income depended solely on egg sales. One correspondent in California finds the same resistance but takes it right in stride. Referring to his large layer flock, he says: "Our profit is in the shit." That is, rotating his layers over sections of his farm gives the soil such a fertility boost—for two full years—for his vegetable CSA, it is worth it in his total operation merely to break even on egg sales.

And don't think of diversity in terms of your farm products only. Try to find ways you could put in place some of the integrating practices described in this book, but at the farm scale: stacking your poultry with other livestock species; using market-sized flocks in soil fertility and insect-control projects; cultivation of worms or soldier grubs on a scale large enough to replace a good portion of purchased feeds, or finding local sources of food residuals as a resource for feeding, using the VCC model; experimenting with winter access to the outdoors that benefits the birds, the soil, and the environment. Perhaps these ideas for becoming more independent of purchased feeds and using chickens to do real work can become part of the profit equation in your market flock.

Eating with the fullest pleasure—pleasure, that is, that does not depend on ignorance—is perhaps the profoundest enactment of our connection with the world. In this pleasure we experience and celebrate our dependence and our gratitude, for we are living from mystery, from creatures we did not make and powers we cannot comprehend.

—Wendell Berry

Acknowledgments

Thanks to Lottie and Marshall Ussery for teaching me that producing some of my food in my own backyard is a good thing; and to Harry Feldman and Fritzi Grabosky, who would have found that a peculiar idea.

Thanks to readers of the first edition of this book—its widespread acceptance and the expressions of appreciation I've received have been most gratifying. I hope readers old and new will find this revised edition useful and rewarding.

I am profoundly grateful to Chelsea Green for its commitment to regenerative agriculture and sustainable living and its dedication to old-fashioned publishing values. Thanks to the entire Chelsea Green team who helped shape and refine this book, and especially to my editor for the first edition, Makenna Goodman, for her patience and good judgment, and to Fern Marshall Bradley, for assistance in this revised edition, always with a dash of humor, good sense, and inspiration. There was a time when she believed in the project more than I did, and pulled me through a rough patch. Thanks, Fern.

Thanks to Joel Salatin for being a mentor and inspiration for yea these decades, and for all he is doing to lead the way toward a saner agriculture.

Thanks for contributions from Ellen Ussery, Don Schrider, Sam Poles, Bridget Chisholm, Stephen Day, Tim Koegel. Thanks for photo contributions from my collaborator of several years, Bonnie Long, and from Mary Perrine, Deborah Moore, Auke-Bonne van der Weide, Emmanuel Keller, Dr. L. K. Yap, and many poultry-keeping friends.

Thanks to Mike Focazio, chicken buddy extraordinaire, especially for his photos that illustrate the chapter on butchering; for help understanding complicated soil issues; and for exploring with me the ecological perspectives that underlie this book.

Federico Lucas helped with work on the homestead to give me more time at the keyboard. Gracias.

Thanks in loving memory to Marvin Grabosky for seeds he planted.

And finally, thanks to my daughter, Heather, collaborator in all things digital, and to my wife, Ellen, without whose support this revision would never have come to be.

APPENDIX A

Making Trapnests

The following instructions are for a trapnest with a falling door, as described in the "Falling Door Design" section on page 257. These instructions could also guide the assembly of ordinary nestboxes mounted on a wall. Further details on that option follow the description of the trapnest project.

During years of use of the trapnest whose construction I described in the original edition of this book, I discovered a small flaw: After being triggered, the door dropped through a vertical slot defined by four narrow wood strips. However, the strips constrained its drop front and back only—even the most minuscule deviation toward either side could cause the door to jam. Though such a jam happened only rarely, when nest-trapping is a guide to critical selection decisions in a breeding project, an "only rarely" failure is not good enough. The solution was obvious: To make the "tracking slot" taller than in the original design—and to entirely "box in" the drop slot.

I recently retrofitted my original nestboxes (first built back in 2008) using boxed-in tracking channels to create a constrained track that would ensure that the doors would always fall properly into blocking position—they would never fail. I have revised the step-by-step building instructions following to share with you this essential modification. In the sequence of pictures showing construction details, you will easily recognize the new photographs that I took during the retrofitting process—the new wood used for the revised tracking channels stands out in sharp contrast to the older material.

Materials

This materials list is sufficient for two nest units, each containing two separate nestboxes:

- One sheet of plywood, CDX (½-inch or ⅝-inch, depending on which you are confident you can edge-nail effectively)
- Small nails for edge-nailing the plywood and nailing on strips to make the tracking channels (I used 4d 1⅜-inch and 6d 1⅞-inch coated sinkers)
- Four 8-foot 1 × 4 pine boards
- A few additional small pieces of hardware cloth, any mesh (optional)
- Two 24- by 12-inch pieces of ¼-inch mesh hardware cloth
- Eight #10 2-inch self-tapping screws (i.e., not requiring a pilot hole, such as deck screws)
- Four small open screw hooks and four small screw eye hooks
- One ½-inch dowel, 36 inches long
- A few meters of strong braided string, such as mason twine

The materials listed assume starting from scratch, but of course many flocksters will have on hand enough scrap from other projects to piece together what is needed. It's okay to alter suggested dimensions to accommodate material you're working with, so long as sufficient interior space is allowed for the hens.

Assembly

Note that, based on material I had on hand, I used ¾-inch plywood for the backs and sides of my trapnests, and ½-inch for the tops. Depending on the thickness of plywood you opt to use, you may have to adjust slightly some of the measurements given in what follows.

Cut two 12- by 24-inch pieces of plywood for the backs and six 12- by 18-inch pieces for the sides and middle partitions (see figure A.1).

I cut little windows into the sides of the nests. While this step is not essential, it will increase ventilation and decrease the trapped hen's sense of isolation. I cut them in the four exterior side pieces only (not in the interior partitions), making the openings about 4 inches by 9 inches (see figure A.2).

Cut two 16¼- by 24-inch pieces of plywood for the tops. Drill two ⅝-inch holes for the trigger sticks in each of the top pieces, 6 inches in from the sides and 6 inches from the back (figure A.3). To ensure a nice sharp edge on the hole, which is important to engage the notch in the trigger stick, be sure to drill from the face of the plywood that will be to the *inside* of the nest. For good ventilation, you can add a few other ⅝-inch holes in the top as well.

Edge-nail the top onto the back and exterior sides, aligning as in figure A.4. Then nail in the interior partition in the middle of the box thus formed (see figure A.5).

Cut one of your 1 × 4s in half crosswise, then rip one of the resulting 4-foot pieces in half. From the ripped pieces, cut pieces that will stop the doors when they drop. Study figure A.6 carefully, because there's a better way to create the stops than shown there. We cut individual pieces that we inset into the spaces between the sides and interior partitions. Not only did that mean more cutting and fitting, but we had to toenail one of the inside ends. You're smarter than that, so you will cut a single ripped piece to 24 inches, and nail it underneath the front ends of the exterior sides and the interior partition. (Imagine the two interior crosspieces replaced by a single piece, and the ends of the exterior sides and interior partition sitting on top of it. Flip the unit upside down, as in figure A.5 on the right, to nail on this door-stop piece.)

Nail another of the ripped pieces, cut to 24 inches, across the front of the unit at the bottom edge, aligning as in figure A.7. This serves as the perch on which the hen stands as she checks out the interior, the perfect place to lay an egg. Remember to round off the sharp edges with a wood rasp so it is more comfortable for the hens' feet. (This task had not been completed when these pictures were taken.)

Cut nest-front pieces from (unripped) 1 × 4. The length will be: 24 inches, minus the thickness of the two sides and the interior partition, divided by 2. I used ¾-inch stock, so I cut mine to 10⅞ inches. Nail these pieces in place inside the nestbox itself to serve as retainers for nesting material (figure A.8). Precise placement is not critical—I set mine 5 inches from the front of the box. Retaining nest material deeper in the interior creates an "empty vestibule" at the front, which ensures that the hen will completely enter the nest before triggering the door. (If it triggers early and whacks her behind, she will become shy of the nest.) Note that in this case there is no alternative to toenailing the end of one of the retainer pieces where it butts to the interior partition.

If you have cut windows in the sides, you can staple hardware cloth over the windows, as shown in figure A.8, to prevent escapes of small hens through the window.

Turn nest unit upside down. Cut ¼-inch mesh hardware cloth to 24 inches wide and deep enough (about 12 inches) to span the bottom of the nesting areas only. Secure the hardware cloth to the underside edges of the nest areas, using small nails and ½-inch strips ripped from 1 × 4 stock and cut to needed lengths (figure A.9). Note that I always use ¼-inch hardware cloth for the bottoms of nestboxes, never solid bottoms. Finer, dustier material sifts out through the wire, I renew the nests with fresh straw from above, and the nests remain largely self-cleaning.

Figure A.10 shows the point in the reconstruction of my original unit where I removed the tracking strips and replaced them with tracking channels that would box the doors into a constrained track. If you are retrofitting existing nestboxes, you will have to remove the tracking strips, too. If you are building from scratch, at this point simply continue with construction of the channels: Rip from 1 × 4 stock four pieces at 22 inches by 2½ inches and eight pieces at 22 inches by ¾ inches. Using 4d 1⅜-inch nails, tack the ¾-inch strips onto the top edges of the 2½-inch strips, to yield four U-shaped channels, 1 inch wide, through which the trapnest doors will fall.

Assuming your materials are the same dimension as mine, your sides and middle partition will project beyond the top by 2½ inches. Attach assembled tracking channels onto those 2½-inch projections, taking care that they are perfectly vertical (figure A.11). Now cut two doors for the trapnest unit, at a minimum of 12 inches tall. Before cutting to width, carefully measure the width of the slot defined by your channels, and make your doors that width minus ⅜ inch (for 3⁄16-inch clearance on either side of the door in the slot). I modified the doors in figure A.11 (reduced their width to fit the new tracking channels) from doors used in my original project, which I had cut to a peak. Those doors are ¾-inch plywood, 9 inches wide, and 15 inches tall at the peaks. (Note that there is no need to shape your doors to a peak—squaring them off at the top is fine.) The 1-inch tracking slots can accommodate any thickness of door from ½ inch up to ¾.

Attach some sort of bumpers on the doorstop onto which the door will drop so the door will not bang down too forcefully or loudly and panic the hen. I used strips cut from an old bicycle inner tube (figure A.12).

Mounting the Unit and Installing Doors

The unit is now ready to install. If you attach a cleat to the wall (say, a scrap piece of 2 × 4) to support the rear of the unit while you work, you can install it by yourself (figure A.13). Otherwise, get a buddy to assist.

Attachment to the wall must be rock-solid. I used four #10 2-inch self-tapping screws for each unit (see figure A.14).

Drill a hole in the exact center of the top edge of each door for a small screw eye hook (figure A.15). Position an open screw hook overhead, screwed into any accessible structural member of the henhouse. The position of this hook is the most critical alignment in the whole setup: I used a plumb bob to ensure that the hook was precisely vertical above and in line with the centered door hook.

Cut the ½-inch dowel into two 9-inch pieces for the trigger sticks. To shape each trigger stick, start by sawing a cut 5⁄16 inch into the dowel, 3½ inches from what will be the lower end (see figure A.16). This cut is made with a saw because the "shoulder" being shaped here must be well squared, to engage the edge of the trigger hole without slipping. Drill a 3⁄16-inch hole near the other end of the trigger stick. Using a knife, whittle a notch that starts about 1½ inches toward the upper end and comes down to the inside of the shoulder of the notch.

Time to put it all together. Tie a string between the hole in the trigger stick and the hook in the door, running it through the overhead hook. (Strong braided string that will not stretch, such as mason twine, is needed.) The length of the string will depend on the height at which the overhead hook is set, of course. Hook the edge of the notch in the trigger stick into the trigger hole. The door will now be suspended from the overhead hook, hanging within its tracking slot. The bottom edge of the door should be slightly above the upper edge of the top of the nestbox, so the hen's back will not bump the door as she enters the nest, which could trigger premature release. Note that the weight of the door keeps tension on the string to hold the trigger securely notched in the trigger hole (see figure A.17).

Figure A.18 shows the lower end of the trigger stick in the "set" position from the inside. As the hen begins settling into the nest, she is certain to bump against the

(continued on page 342)

Figure A.1. Cut two 12- by 24-inch pieces for the backs and six 12- by 18-inch pieces for the sides and middle partitions. (Only four sides are shown in the picture—we later cut the additional two pieces for the partitions.)

Figure A.2. We cut little (4- by 9-inch) windows in the exterior sides.

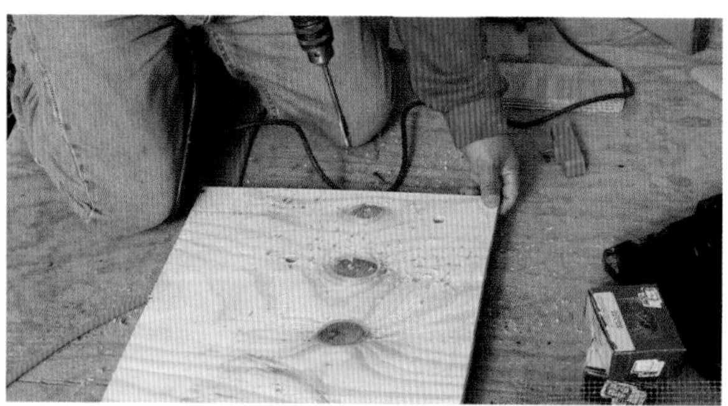

Figure A.3. Cut the top pieces and drill two ⅝-inch holes for the trigger sticks in each top piece. Position the holes 6 inches in from the sides and back, one over each nest. Start drilling from the board face that will be toward the interior of the nest. Also drill a few extra ⅝-inch holes in each top for ventilation.

Figure A.4. Nail on the top, aligning the back and exterior sides as shown.

Figure A.5. The top nailed on (*left*) and a better view of the interior partition with the unit upside down (*right*).

Figure A.6. Add the door stops (1 × 4 stock ripped in half) for the doors. *Note:* It would be simpler to nail a single piece underneath the ends of the sides and interior partition, rather than insetting two separate pieces as in this picture.

Figure A.7. Nail on another piece of the ripped stock, cut to 24 inches, for the perch from which the hen will enter the nest.

Figure A.8. Cut and nail into place 1 × 4 pieces that will retain nesting material in the interior of the nest. (Note that by this step we have stapled hardware cloth over the windows.)

Figure A.9. For a self-cleaning nest, use ¼-inch hardware cloth rather than solid bottoms. Secure with small nails and ½-inch strips. (Note that I nailed a strip onto the bottom of the interior partition as well after this picture was taken.)

Figure A.10. Assemble the tracking channels, using 22-inch strips ripped from 1 × 4 stock at 2½ inches and at ¾ inches. If you are retrofitting a set of existing nestboxes, you will have to remove the tracking strips, as shown here.

Figure A.11. Attach the tracking channels to the sides and middle partitions, taking care that their installation is perfectly vertical. Cut the doors, which will slot into the tracking channels.

Figure A.12. Attach "bumpers" (scraps of inner tubing or other cushioning material) to the door stop at the bottom of the door's track.

Figure A.13. If the wall of your coop has no structural piece (as we have here) to support the back of the unit, nail a cleat to the wall or call in an extra pair of hands.

Figure A.14. Four well-spaced #10 2-inch self-tapping screws provide rock-solid attachment to the wall.

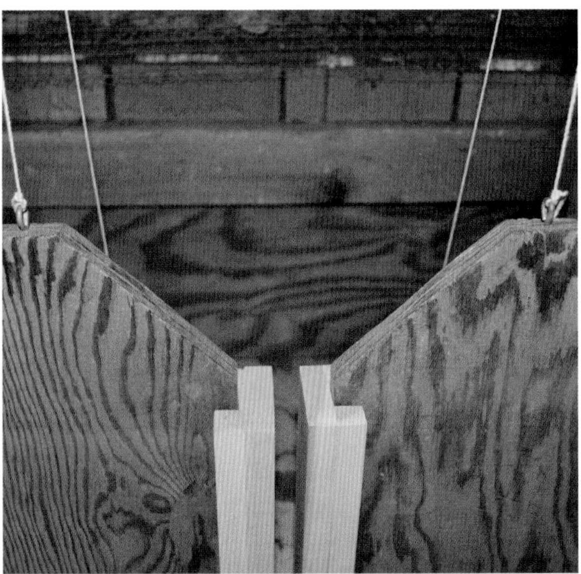

Figure A.15. Screw an eye hook into the top edge of each door at the center point. Mason's twine tied to the eye hook runs up and through an open screw hook directly above, screwed into an overhead rafter or beam. Note how doors are almost entirely enclosed within their tracking channels.

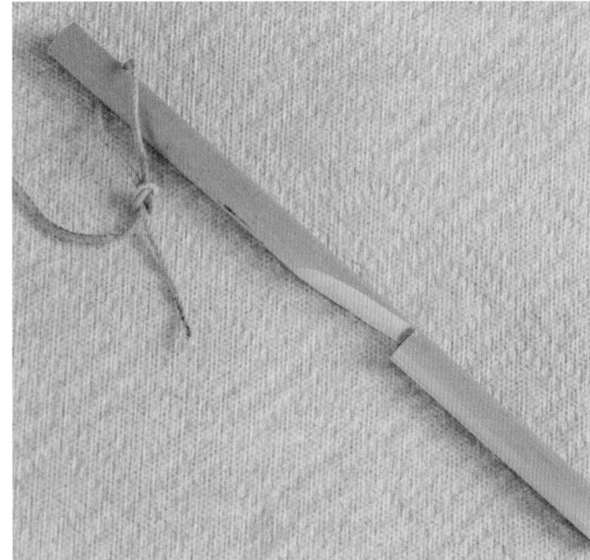

Figure A.16. Using saw and knife on the 9-inch dowel pieces, make a notch with a perpendicular edge on the lower end. The bottom edge of the notch should be 3½ inches above the lower end of the dowel. Drill a 3⁄16-inch hole near the other end of the dowel.

Figure A.17. Feed a string through the hole in the trigger stick and tie it off. Attach other end of string to the door hook, after running the string through the overhead hook as well. Weight of the door keeps trigger stick notched in place in the trigger hole . . .

Figure A.18. . . . until the hen's settling-in ritual knocks it loose, and the door drops.

Figure A.19. If the nestbox is mounted well above the floor, give the hen a step or two on which to approach the nest. This reduces the chance that she will jostle the lower edge of the door, triggering premature release.

Figure A.20. Success!

(continued from page 337)

trigger stick, knocking it loose and allowing the door to fall into blocking position.

Addition of a hop-up step (see figure A.19) reduces the chance that a more forceful landing of a hen on the entry perch will jar release of the door prematurely.

It took less than half an hour for my new nest to attract the Rhode Island Red hen in figure A.20. Had the door and trigger been in place, she would have been trapped until I could release her and record her achievement.

Troubleshooting

I say with the greatest confidence that, if this trapnest is set up properly, it *cannot* fail. Should you fail to get a friction-free drop when the door is triggered, look first to ensure that the upper hook is *precisely* positioned in a straight vertical line to the door's screw eye hook. Use a plumb bob.

Study figures A.17 and A.19 carefully, noting the *angle* of the trigger stick in its set position. The angle of the trigger stick *here* is approximately 75 degrees, and at that angle the notch locks securely onto the edge of the trigger hole when under tension from the hanging weight of the door. Obviously, the less acute that angle—the closer to 90 degrees it is—the more likely that the notch will not hold securely but will instead slip free of the edge of the trigger hole. The precise angle of the trigger stick is determined by the *height* of the overhead hook above the door. The structure in your coop may dictate a different height than I used—indeed, you might even find it necessary to run the string through an additional open screw. But whatever the routing of the string, the eventual angle of the trigger stick when supporting the hanging door must be at about the angle shown in figures A.17 and A.19—something in the range of 65 to 75 degrees.

This design assumes the nest is well filled with straw or other cushioning material. If the material breaks down over time and sifts through the wire bottom, it is possible that a small hen—standing on the thin residue at the bottom of the nest—could fail to bump the trigger stick. Keep the nest well supplied with nesting material.

Doors can be left in the closed position at night if your chickens decide they like sleeping in the trapnests. A slanted cover can be moved onto the top of the nest unit overnight to prevent the birds' roosting there.

Making Standard Wall-Mounted Nestboxes

If you want to make a set of ordinary nestboxes mounted on a wall, simply alter the instructions above in these ways:

1. A nestbox without the "vestibule" does not have to be as deep—decrease width of top and length of sides to 12 to 14 inches.
2. Cut the tops to the same depth as the sides and middle partitions. (Thus, the sides and partitions will not protrude beyond the top as they do for trapnests.)
3. Omit the additional nest-front (retainer) piece installed in the interior of the box (as in figure A.8), which provides the "empty vestibule" space I recommend for a trapnest. The nesting space will simply be the entire interior, with a 4-inch strip across the bottom in front to serve as both entry perch and retainer of nesting material.
4. The hardware cloth must cover the bottom of the entire unit.
5. Omit installation of the tracking channels, doors, and door stop.

APPENDIX B

Making a Dustbox

If it is too wet outside or the litter inside is not fine enough for dust-bathing, a backup dustbox in the henhouse ensures the chooks keep themselves free of external parasites. I have made several versions, and I caution against making one too small or too shallow—the chickens would kick out too much of the dusting materials as they bathe, requiring more frequent refilling. The dustbox I use is 24 inches by 24 inches by 16 inches, with a 2-inch lip around the top edge. Of course, you may have scrap on hand that dictates a different design from the one detailed here.

If starting from scratch, note you can make two dustboxes of this design using a single sheet of plywood. The cutting instructions to follow assume starting with one sheet of plywood to cut all the pieces needed for two dustboxes—step 1 through step 4.

For this project I chose to use ⅝-inch CDX, since I could not find a thinner plywood that was 5-ply, and I wanted to save time and effort by edge-nailing. If ⅝-inch plywood seems like overkill to you, go with whatever thickness you are confident you can edge-nail securely (or use thinner plywood attached to nailing cleats in the corners).

To keep these instructions simple, I ignore the saw's kerf—we're not making a kitchen cabinet here. I used a table saw (except for the initial cut), but you could use a handheld power saw for the entire project.

Step 1: Cut the sheet of plywood across the 48-inch dimension into four equal pieces (each 24 inches by 48 inches).

Step 2: Cut one of the 24- by 48-inch pieces in half, to yield 24- by 24-inch pieces for the two bottoms.

Step 3: Cut the remaining three 24- by 48-inch pieces 16 inches wide, across the 24-inch dimension, to yield nine pieces, each 16 inches by 24 inches. Eight of these pieces will be sides for the two dustboxes.

Step 4: Rip the remaining 16- by 24-inch piece into eight strips, 24 inches long and 2 inches wide—these will be used to make the lip around the top edge. Leave four of the resulting strips at 2 inches by 24 inches. Cut the remaining four strips to a length of 24 inches minus twice the thickness of your plywood. I used ⅝-inch plywood, so I cut mine to 22¾ inches.

Step 5: Edge-nail one of the bottom (24 × 24) pieces onto two of the side (16 × 24) pieces, aligning them as in figure B.1. I used 4d 1⅜-inch coated sinkers.

Step 6: Nail the other two side pieces to box in the open ends created by the previous step; that is, these side pieces cover the edges of both the bottom and the first two sides, as shown in figure B.2. The offset of the top edges is intentional.

Step 7: Nail the 2-inch strips cut in step 4 to form a "lip" around the inside of the top edges of the box. See figures B.3 and B.4.

Step 8: Assemble the other dustbox from the remaining pieces as described previously.

Step 9: Put 4 to 6 inches of dusting materials in the bottom of the dustbox. The ideal mix is loose, fine, and easily fluffed by the birds up under their feathers. In figure B.5 I am combining peat moss and wood ash, in a ratio of about 6:1, sifted through a quarter-inch-mesh compost riddle. You could also add dried and sifted clay soil, diatomaceous earth, or elemental sulfur (pure sulfur as a fine yellow powder). *Wear a dust mask!*

Figure B.1. Edge-nail one of the 24- by 24-inch bottom pieces onto two of the 16- by 24-inch side pieces.

Figure B.2. Assembled box. (Note offset top edges.)

Figure B.3. Nail 2- by 24-inch strips to make a "lip" around the inside of top edges, noting alignment.

Figure B.4. Close-up of alignment of "lip strips."

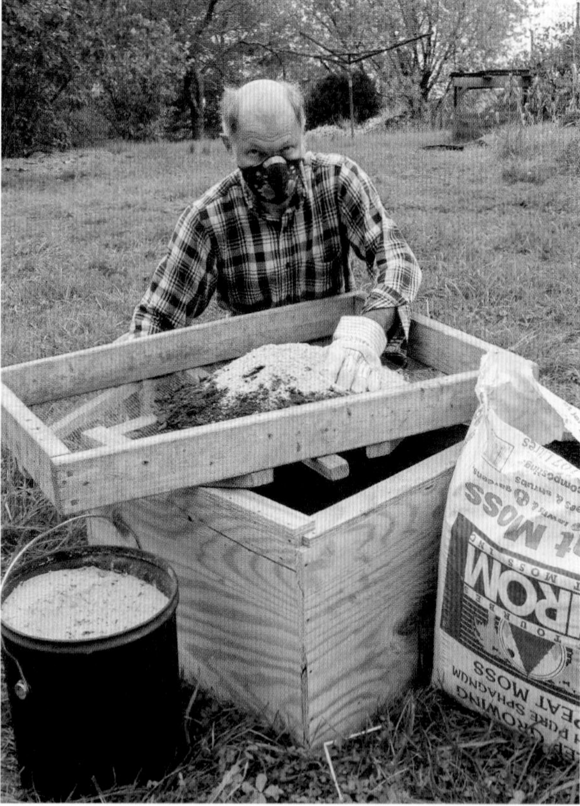

Figure B.5. Sifting for a fine dusting mix.

APPENDIX C

Making a Mobile A-Frame Shelter

Consider an A-frame for your mobile shelter—combining rigidity and reasonably light weight, with greater stability in the wind than boxier designs. Figure C.1 shows my most successful A-frame, one of my two most successful as an all-purpose mobile shelter. At 8 feet by 9 feet, it would comfortably accommodate up to fifteen layers if confined full time—I used it as a "bedroom" only for up to fifty young growing birds who had plenty of space in an electric net enclosure during the day. The roof/sides were solid, providing the flock both shelter from rain and shade when there was no tree cover, but the ends had plenty of wire-covered open framing to permit free flow of air.

This appendix shares my reflections on materials and design choices, a materials list, and the step-by-step construction of a proven shelter design. The modifications you could make are endless. If you are considering a mobile shelter similar to this one (or in the following appendix D), be sure first to review the discussions of designs and materials in chapter 12.

My previous most successful pasture shelter was the one shown in figure 12.10. The biggest design change from that old shelter was the choice of baked-enamel steel roofing instead of 24-mil woven poly to cover the shelter. It bothers me to think that the plastic we use with such convenience today could well remain somewhere on the planet in the time of our great-great-great-great-grandchildren, so I now avoid its use as much as I can. Still, woven poly fabric may be the best choice for some. It is much lighter than metal roofing—the metal roofing on my new A-frame added about 80 pounds (36 kg), while a piece of 24-mil woven poly, custom-cut to cover this same shelter, would weigh about 10 pounds (4.5 kg). Up to a point, the added weight can be an advantage—there is no question that my new, heavier shelter was more stable in the wind. Cost might also be a factor in deciding between these two coverings: I spent $112 for the metal roofing for my new shelter, including a piece of ridge vent to cap it—a piece of 24-mil white/silver woven poly sized to fit would have cost $78. A final consideration is durability. When I gave away that old A-frame (when I put this metal-roofed one into service), it was eleven years old and its poly cover was still going strong. Based on its degree of degrading from weathering at that time, I estimated a total service life of more than fifteen years. The galvanized steel roofing with baked-on enamel paint that I used in the new project is sold with a twenty-five-year warranty, so I would expect a total service life considerably more than that.

In this new version as well, for more clearance under the bottom rail I used a larger wheel diameter (10 inches rather than 8) and offset the hole for the axle bolt, as detailed in chapter 12 (see "Wheels" on page 111).

Materials

- Six 6-foot sheets of Fabral's Grandrib 3 painted steel roofing (see note 1)
- One 10-foot, 6-inch Fabral's Shelterguard AR3 Steel Ridge Cap
- Fabral's painted screws with neoprene washers (optional) or a box of neoprene washers (see note 2)
- Four 12-foot construction-grade 2 × 4s (see note 3)
- Six 10-foot 2 × 4s
- Four 8-foot 2 × 4s
- Assorted deck screws (see note 4)
- Four to eight metal corner braces (see note 5)
- Four ½-inch carriage bolts, 5 inches long
- Four each flat washers, lock washers, and hexagonal nuts for the carriage bolts
- Four wing nuts for the carriage bolts
- Four wheels with ½-inch axle bores (see note 6)
- Two pairs 1½-inch utility hinges
- One 1½-inch barrel bolt
- Four 4½-inch threaded open eye hooks
- One sheet ½-inch CDX plywood (optional) for nestboxes (see note 7)
- Small roll of ¼-inch hardware cloth (optional) for nestboxes (see note 8)
- 1 quart or more wood sealer (see note 9)
- One 10-foot roll of either 1-inch mesh poultry wire or ½-inch hardware cloth, 48 inches wide
- Small fence staples (for poultry wire) or appropriate heavier-duty fasteners (if using hardware cloth) (see note 10)
- A short length of twisted wire cable
- A scrap of old garden hose

Notes on Materials

1. I ordered the metal roofing through my local farm co-op, custom-cut to my order. There are other options for metal roofing than Fabral's Grandrib 3, of course. Check out the possibilities with your local farm supply. The width of coverage of the Grandrib 3 is 36 inches, so the choice of three pieces per side dictated a total length for the shelter of 9 feet. As for the length of the rafters, I used the Pythagorean theorem (the square of the length of the hypotenuse of a right triangle equals the sum of the squares of the lengths of the other two sides) to calculate the 6-foot length, based on a bottom rail of 8 feet 3 inches, and on a preference for cutting angles on the ends of the rafters at 45 degrees. If you are making a smaller or larger shelter, or prefer a different profile with reference to its height, use the good old Pythagorean theorem to recalculate the length of your rafters (the hypotenuse of a right triangle) and the angles at which to cut the ends of your rafters.
2. For fastening Grandrib 3 onto roof framing, Fabral offers a painted hex-head screw complete with neoprene washer. Since I had some left over from a larger roofing project, that is what I used. They are expensive, however. You could use deck screws (see note 4) for this part of the job as well, adding small neoprene washers to seal the screw holes against rain.
3. For the framing, I bought all 2 × 4s and ripped them down as needed. You could of course buy lumber already cut to your needed dimensions (2 × 2, 1 × 2, and so on) if you prefer.
4. *Do not use nails* for a pasture shelter—they will work loose as the shelter is jerked about in moving. A coarse-thread deck screw, galvanized or coated against weather, will provide much more durable framing joints and will not require the drilling of pilot holes (except at the ends of pieces to be joined). I keep a supply of self-tapping screws on hand, in a variety of lengths and shank sizes. For this project, I used, as appropriate to the join being made, all the following: #7, 1 and 1⅝ inches; #8, 1¼, 2, 2½, and 3 inches; #10, 3½ inches. (Note that, since making this A-frame more than ten years ago, I have come to prefer stainless steel, star drive screws for a structure that will be exposed full-time to the weather. The star drive head is far more slip-free in comparison to the Phillips-head screw [used in this project], and driving into the wood is much easier.)
5. I had some recycled 3½- by ¾-inch steel corner braces ("corner irons") on hand, so I used two on

each corner. If you use larger corner braces, you should need only one per corner.

6. The size wheel for your shelter is up to you. For most pasture shelters you want a solid (non-pneumatic) wheel, available from a garden or farm supply. See the discussion of wheel size in chapter 12 (page 111).
7. If you have some scrap wood on hand, use it to assemble a nestbox. If you do not, you might find it easiest to make one from plywood. If you buy a sheet of plywood, you should have at least half the sheet left over after making your nestbox.
8. I always use ¼-inch hardware cloth (welded wire mesh) for nest bottoms because it is more self-cleaning than a solid surface would be, so I had on hand the small piece required to floor the nestbox in this shelter. If you cannot find a source that will custom-cut to your length, you will probably have to buy a 5- or 10-foot roll and keep the remainder on hand for other projects.
9. The sealer I used was Cabot Waterproofing 1000, a clear silicone sealer for wood. I did not coat the interior parts of the frame that would be completely sheltered from blowing rain. I applied several liberal applications, however, to the bottom rails and end framing and the end grain of stringers and rafters. I used about 2 quarts of sealer. (In appendix D, I describe a more robust way of sealing and protecting a wood frame: *marine spar varnish*.)
10. Using poultry wire secured with fence staples is fine if a shelter will always be surrounded by electric net fencing. If not, I strongly advise using ½-inch hardware cloth to foil raccoons. You will need a supply of fender washers about 1 inch in diameter, and some weather-resistant 1¼-inch or 1½-inch screws for fastening them. Make sure the small-diameter hole in the center of the fender washer is large enough to admit the screw shank but not the screw head. More on fastening the hardware cloth on the frame in step 15 of "Construction."

Construction

Step 1: Set the rip fence of a table saw or handheld power saw to 1¾ inches, and rip all four of the 12-foot 2 × 4s in half. Cut each of the resulting pieces in half crosswise—that is, into 6-foot lengths. Reserve fourteen of these pieces for the rafters of the shelter, and set aside the other two for other uses.

Figure C.1. Ready to roll.

Figure C.2. Pythagoras insists that the alignment of the pieces for the bottom rails, shown here, is essential if the rafters as specified are to fit. The 9-foot side rail is on the left, the 8-foot front rail on the right.

Figure C.3. Attach rafters to side rails and ridge pole.

Figure C.4. Attach first pair of stringers front-to-back.

Figure C.5. Attach four diagonal braces, screwing them onto the underside of the rafters at every crossing point.

Figure C.6. Frame in openings into the interior, front and back.

Figure C.7. Set collar ties for additional rigidity, positioned to do double duty as roosts.

Figure C.8. An additional pair of stringers provide additional roosting space. Note the two upright braces in the middle that add support for the two pairs of stringers.

Figure C.9. With the end framing and top stringers to attach to, it is easy to make a simple nest box. The floor of this one is ¼-inch hardware cloth.

Figure C.10. I have added strips of scrap plywood as sides to retain nesting materials. Note the addition of a strip of pine to the top edge in front, to serve as a landing perch for hens approaching the nest.

Figure C.11. Scrap plywood panels, together with an access door (not yet in place), will keep the nest dry in blowing rains.

Step 2: Cut 12 inches off four of the 10-foot 2 × 4s, to yield four 9-foot 2 × 4s. Set aside two of these 2 × 4s to use for the side rails. With the rip fence still set at 1¾ inches, rip the other two 9-foot 2 × 4s in half. Set aside the resulting four 9-foot pieces for use as stringers.

Step 3: With the rip fence still set to 1¾ inch (4.1 cm) rip one more 10-foot 2 × 4 in half and set aside for use as collar ties and end framing.

Step 4: Reset the rip fence to 2¼ inches and rip the last 10-foot 2 × 4. Set aside the 1¼-inch-thick piece for later use. Cut the 2¼-inch piece to 9 feet, for use as the ridge pole.

Step 5: Assemble the bottom rails.

While I did say in chapter 12 that it is possible to cut down on weight by ripping the bottom rails at 2¼ inches, in the present case I was adding a lot of weight because of the choice of the metal roofing as cover. Therefore, I used full 2 × 4s for the bottom rails. No other parts of the framing needed to be full 2 × 4—various structural members were 1¾ × 1¾, 2¼ × 1½, and 1¼ × 1½ as specified in the steps following.

Lay out the bottom frame, with the two 9-foot 2 × 4s previously cut (step 2) to the *outside* and two of the 8-foot 2 × 4s to the *inside*. Study figure C.2

Figure C.12. We have attached all but one of the six metal roofing pieces.

Figure C.13. I made the nest access door from a piece of scrap ½-inch CDX plywood. A scrap of 24-mil woven poly stapled over top edge of door seals against rain. I added four strips, 1- by 1¼-inch, around the edges of the door to keep the relatively thin plywood from warping. (Along the near edge you can see one of them, and the ends of two others.)

Figure C.14. Lock ½-inch carriage bolts in the side rails, front and rear, to serve as axles for the wheels. Note that, the lower the placement of the bolt in the rail, the greater the clearance when the wheel is mounted.

Figure C.15. Lock wheel onto axle bolt with wing nut—ready to roll.

carefully and make sure that the 9-foot rails set the *length* of the bottom frame at exactly 9 feet; while the 8-foot 2 × 4s, set on the *inside* of the outer rails, make for a *width* that is 8 feet, 3 inches (8 feet plus the thickness of one nominal 2 × 4 times two). Stated another way: The 9-foot dimension will become the *sides* of the shelter; the 8 foot, 3 inch dimension, the *front and rear*. You may of course change the size of your A-frame and recalculate lengths and angles, but Pythagoras dictates that you adhere precisely to these dimensions for the bottom frame if you plan to cut and fit your rafters as directed in step 6.

We doubled up on our smaller corner braces and used #10 × 3½-inch screws for screwing into end grain and #7 × 1⅝ when screwing into cross grain. (#8 × 1⅝ would have been even better.) As mentioned in note 5, single larger corner braces would be preferable to doubling smaller ones.

When beginning assembly of the frame, it is important to keep it as square as possible. If you have a large enough completely flat surface, such as a garage floor, to work on, that is the best choice. That was not an option for us, so I simply used the most level section of lawn I have. For squaring up the corners, I laid each of the corners in turn on a sheet of plywood when joining them with the corner braces. Square your corners as well as you can (measure on the diagonal from opposite corners—the two measurements should be the same if the frame is square) before the next steps, which will lock in the structure.

Step 6: Cut 45-degree angles in opposite directions on both ends of the fourteen rafters cut in step 1. Be careful with this step: You want to end with rafters that still measure a full 6 feet on the *top* edge, with the angles coming *in* from each end. (This angling of the rafter ends shows best in figure C.4.)

Step 7: I hope you have a buddy willing to help set the first two pairs of rafters onto the ridge pole from step 4—doing that single-handedly would be more challenging than juggling while walking a tightrope. Screw the ends of all the rafter pieces solidly onto the ridge pole and the side rails, figure C.3, setting the rafters at 18 inches on center. (If you choose a different roofing material, it might require a different spacing of the rafters.) When driving screws into the ends of the rafters—or near the ends of any other pieces in the construction—I first drill a pilot hole to prevent splitting the end. (When driving deck screws into the middle of the work piece, no pilot hole is necessary.)

Step 8: Attach two of the 9-foot stringers cut in step 2 to connect the rear and the front rails (see figure C.4). I came in 32 inches from the right and left ends of the front and rear rails to set the stringers. Vary that distance depending on the size door, and access to the nestbox in the rear, that you prefer.

Step 9: Reset rip fence to ¾ inch. Rip the two remaining 8-foot 2 × 4s to make eight pieces ¾ inch by 1½ inches. (No, they're not all precisely ¾-inch thick, but we don't want to go crazy figuring the kerf here.) Set aside four of these pieces for later use. *Check a final time that the bottom frame is square.* Trim the remaining four pieces to make diagonal cross braces, from each lower corner up to the top in the middle of the structure, attaching to the *undersides* of the rafters wherever they cross, as in figure C.5. Though lightweight, these cross braces add tremendously to structural integrity.

Step 10: Using the 1½- by 1¾-inch stock cut in steps 1 and 3, cut pieces to frame for a door and access to the nestbox on the ends. Study figure C.6 for the general layout. You will have to do a bit of trial-and-error to establish the correct angles, unless you're a better carpenter than I am.

Step 11: Using the same 1½- by 1¾-inch stock, cut five additional pieces to use as collar ties, which join opposite pairs of rafters (see figure C.7). The collar ties add to structural integrity, but they can serve double duty as roosts for your birds. Exactly where you position them is up to you: The lower you set them, the longer they will be and the more useful as roosts; the higher, the easier it will be to move around inside the structure when you need to do so. A good compromise for us (based also on best use of the precut stock used in this step) was to make the collar ties 29 inches long (measured across the bottom side), which set the top 13½ inches from the ridge pole.

Step 12: To complete the framing, we added the remaining two 9-foot pieces cut in step 2 as an additional pair of stringers, 16 inches above the first pair, connecting the uprights in the framing of the end openings. These two stringers are not essential structurally, but we wanted them to provide a comfortable amount of roosting space for the birds. Remember the 1¼- by 1½ stock cut in step 4? We cut pieces from it to make two center supports for the pairs of stringers on each side. See figure C.8. (This would be a good time to use a wood rasp to round off upper edges of all pieces that will be used for roosting—specifically, the five interior collar ties and all four of the stringers.)

Step 13: I will leave it to you to design your own nestbox (see figures C.9, C.10, and C.11) and end door if you want them. For both the end door and the nestbox access door, use a pair of 1½-inch utility hinges. For latching the end door, a 1½-inch barrel bolt is sufficient. A door is less essential if electric fencing is used to protect the flock, but I always wanted the option of shutting in the birds, for a census or selection or whatever—you can see the one I made in figure C.1. For these two projects I used the ¾-inch stock generated in step 9 and the remaining 1¼-inch stock from step 4, plus—for the nest and its hinged access door—some scrap ½-inch CDX plywood. I used some ¼-inch hardware cloth on hand for the bottom of the nestbox. I used more scrap plywood (attached to the shelter at either edge of the nest access door) to protect the ends of the nestbox from blowing rain (see figure C.13).

Step 14: While all parts of the frame are still accessible, it's a good idea to apply a coat of wood sealer to all surfaces that could possibly be reached by blowing rain. Don't forget the bottoms of structural pieces: The frame at this point is rock-solid—you can flip it on its side to reach any surface that needs protection. (Again, I now prefer marine spar varnish to protect wood that will be exposed to weather.)

Step 15: Cut your preferred wire mesh to fit the mostly triangular spaces to be sealed off in the front and rear. Remember my rule of thumb: Chicken wire, stapled in place, is okay for a shelter that will remain inside electric net perimeters; otherwise use ½-inch hardware cloth. For fastening the hardware cloth: Place a fender washer over the hardware cloth and drive a screw through its hole into the wood framing below. As the head of the screw snugs up on the edges of the hole, the washer creates a fastener no raccoon can pull out.

Step 16: We attached the metal roofing using four screws per rafter (see figure C.12). As said in note 2 under the materials list, we used some leftover painted screws with neoprene washers—you could use deck screws with neoprene washers instead. The Grandrib 3 we used is designed to be laid down over horizontal roof purlins, but the addition of purlins is not needed in this simple structure. Therefore, we simply drove the screws into the rafters themselves, occasionally doing so through a ridge in the Grandrib 3—not recommended when installing on a house roof—but mostly through a flat section of its profile as recommended. We then attached a ridge cap over the peak, shown in figure C.1, to protect the interior from rain.

Step 17: Figure C.13 shows the hinged access door we added for ease of collecting eggs. I happened to have some scrap 24-mil woven poly on hand, which I used to cover the hinged top edge of the door to shed rain.

Step 18: Drill ½-inch holes into the side rails, front and rear. I drilled mine 10 inches in from each end, and 1 inch up from the bottom. Insert the ½-inch carriage bolts through the holes, from the inside of the rails. Place a flat washer, then a lock washer, then a hexagonal nut on the bolt, and tighten until the square shoulders of the carriage bolts bite firmly into the wood (see figure C.14). When ready to move the shelter, pop the wheels onto the bolts and lock them down with wing nuts (see figure C.15). Wing nuts make it possible to take off the wheels and drop the bottom rails back solidly onto the ground after the move—and to use one set of wheels on multiple shelters. If you want to install the wheels permanently, lock them on the ½-inch bolts using additional hexagonal nuts instead of wing nuts.

Step 19: Drill pilot holes, then screw in the four open hooks, two into the front rail and two into the rear, to enable pulling the shelter from either end. Make a pull using a length of wire cable and a scrap of old garden hose as padding for your hand. Twist the ends of the cable to make strong loops. Loop the ends onto the hooks, and pull to move the shelter.

Now please note: I started this step-by-step by saying that this A-frame was "one of my two most successful as an all-purpose mobile shelter." If you wondered what was the *other* "most successful," it is described in appendix D. Yes, this A-frame—tested here for years—is an excellent design. But putting on and taking off four wheels for a move each day—isn't there a better way? There is—the *lift kit* I used in my latest "mobie." As you read about it, keep in mind that if you prefer this A-frame design, you could easily mount a lift kit on *it*—and forget the tiresome mounting and unmounting of wheels on axle bolts every time you need to move it.

APPENDIX D

A Robust Mobile Shelter with a Lift Kit

There are a million ways to make a mobile chicken shelter, and each flockster looks for the best design for themselves and their flock. Last year I put into service a "mobie" rather different in design from the A-frame in appendix C, and here I share what I hope are some useful ideas to consider as you think through your design for a movable shelter.

As I've stated earlier, I recommend electric net fencing as the best solution for protecting birds, preventing their becoming a nuisance to close neighbors, and giving them plenty of room to roam. But my current situation no longer features use of big fenced enclosures, and my flock is far smaller than it used to be. Thus, I wanted to build a shelter big enough to accommodate a flock of a dozen or less without crowding; substantial enough and well defended enough to be impregnable to every predator, large or small; but easy to move every day.

Thinking Through the Design

Before you embark on this description of my new mobile shelter, I recommend you read the discussions about design and materials for mobile shelters in chapter 12 as well as the construction details of my previous "best ever" mobile shelter in appendix C.

I have always distrusted super-lightweight shelters. I wanted one far more substantial, but that required wheels for moving it, and as noted in appendix C, the on-then-off routine with four wheels got old when I needed to move the shelter every day. I find many shelters inadequately designed to foil the whole range of predators, with their astonishing array of lines of attack. And tell me, how are chickens in a shelter supposed to dust-bathe—that quintessential hygiene—when moved onto fresh grass every day?

The 8- by 10-foot mobie described here (for nine chickens, a cock and eight hens, almost nine square feet per bird) is a substantial unit—I *wanted* a substantial unit!—designed with three key features to address the shortcomings:

- I installed a heavy-duty *wheel lift kit* so that the robust shelter could be made rolling-ready literally within seconds.
- I *wired for defense*. That is, though I designed the mobie with structural barriers to exclude predators, I added charged single-strand electric wire at three

Figure D.1. I'll say it again: The ultimate mobile shelter!

levels. Never say never, but I find it hard to imagine any predators where I live—including blacksnake, raccoon, least weasel, rat, dog, fox, hawk, or even bobcat or bear—defeating its defenses.

- I installed an *onboard dustbox* for the chooks' dust baths.

One more point about choosing a mobile shelter as substantial as this one in preference to a lightweight: In my mid-Atlantic climate, keeping a micro-flock through the winter in this mobie should work fine. I did not myself keep the flock in it last winter—I moved the chooks into their "Chicken Hilton" instead. After the transfer, I set the mobie up on blocks, moved the front handle assembly inside, and installed the electric fence charger inside as well. The mobie's HotShock 5 energizer, plugged into an AC socket, powered the electric defense around the flock's winter yard (which I described in chapter 10).

If I were to use this mobie for the chooks in winter, I would stack straw or spoiled hay bales around the open sides, leaving accessible only the rear with its latched doors. Such a configuration would provide a deeply sheltered zone free of blowing rain and snow, with the wire top at the front still allowing some exposure to the sun. I would certainly put in, and regularly top off, a deep litter to keep the interior from becoming foul.

The Lift Kit

The lift kit I bought from Egg Cart'n (https://eggcartn.com) is the heart of my design, a remarkable set of powder-coated steel hardware that attaches to any solidly framed mobie design (see figure D.2). It allows a mobie up to 400 pounds to be ready to move in seconds, and the long front handle mounted above two wheels gives the mechanical advantage to easily lift and pull from the front. Be warned, however: It is expensive—about $320, plus shipping, at the time of this writing.

Construction

Please note that what follows is not intended as a detailed step-by-step regarding either materials or construction—I'm just giving an overview of the important elements of my own project. I described building the A-frame in appendix C in greater detail, but that is a more "constricted" design, with Pythagoras dictating many of the precise angles and dimensions. But design of a substantial mobie based on a lift kit could go in any direction you choose, so I leave it to your creativity. Do review design considerations in chapter 12 and construction details in appendix C as background. Good luck with your project and have fun.

Assembly

Assemble the bottom rails in a rectangle, 8 feet by 10 feet, using construction-grade 2 × 4s.

Note that the rear corner gussets are welded to lifting levers (see figure D.3), the key to the lift kit's function: to lock wheels into place with a simple push from the toe of a boot.

All four corner gussets are fitted for stout lag screws (supplied with the kit) that lock together the rails and the 36-inch corner posts, also 2 × 4 stock. Figure D.4 shows that assembly at one of the front corners.

The steel bracket, which will support the front end of the frame as it is lifted, is securely attached (both side and bottom) to the front bottom rail (see figure D.5). Note that the steel upright will subsequently be attached to a center 36-inch post. There is a hole in the bracket assembly, which will receive the pin on the end of the lift handle, for both lifting and steering the front end.

Figure D.6 shows the basic framing complete. Top rails are 2 × 4s ripped at two-thirds their width; front-to-back stringers and uprights other than corner posts are 2 × 4s ripped in half; and that all-important diagonal bracing for a rock-solid frame is the remaining one-third material made by the two-thirds rip for the top rails.

Interior Accessories

Figure D.7 shows interior accessories: My mobie is 36 inches tall, tall enough to allow for installing a set of roosts (also in that ripped-at-one-third material). It could well have been built to a 24-inch height, of course, which would have reduced weight and materials cost. It would also have reduced weight to build the dustbox and the three-nest unit from scratch, using thin plywood, but I had one extra of each (made of heavier materials) in the henhouse and chose to save time by simply screwing them into place in the mobie frame.

Not shown in any of these photographs, but essential: I later added a sloped cover to keep the chooks from roosting on top of the nestboxes and pooping the nests.

Purchased and Repurposed Components

Figure D.8 shows the components of the project: The main components I purchased were the 2 × 4s for the framing, the lift kit, and the hardware cloth. Like most working homesteads, mine has on hand materials begging to be repurposed. To save on costs, I recycled pieces of plywood plus pieces of baked-enamel steel roofing (white underside out to reflect heat away from the interior) to box in the rear of the unit against blowing rain and make the birds feel more secure at night.

Aluminum roofing, recycled from earlier projects, cut to clad the three rear access doors: the far one

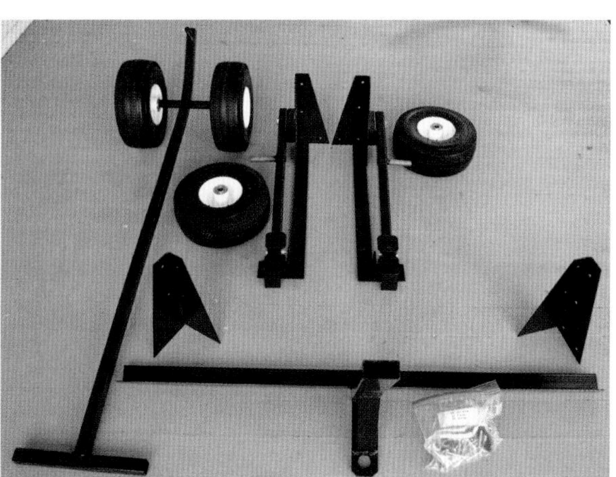

Figure D.2. The lift kit I bought from Egg Cart'n, key to this shelter's easy mobility.

Figure D.3. Rear gussets are welded to the lift lever, which will raise the mobie up onto rear wheels.

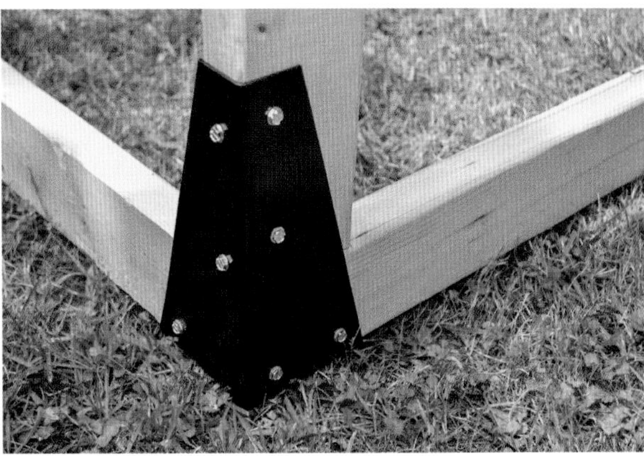

Figure D.4. Heavy steel gusset locks together the bottom rails with front corner upright.

Figure D.5. Heavy steel bracket supports both bottom rail and center upright in front. The hole in the bracket will receive the pivot pin at the end of the lift handle.

Figure D.6. Basic framing complete. Note that the handle's pivot pin is engaged with the bracket, ready for lifting and steering.

Figure D.7. All the comforts of home: dustbox, nestbox, and roosts (which add to lateral and diagonal rigidity).

Figure D.8. Almost-completed unit with three rear access doors.

Figure D.9. HotShock 5 energizer with rechargeable battery mounted inside the mobie.

Figure D.10. Ground rod and knife switch (in the open, "power off" position).

Figure D.11. Charged wires at three levels guarantee a big surprise for nosy predators.

to service the dustbox, the center for gathering eggs from the nestboxes, and the near for feeding and for replenishing the waterer. (That one also gives access to yours truly on the rare occasion I need to get a hand on the birds inside.) The doors are secured with safety hasps and clip-on carabiners (to foil clever-fingered Mister Raccoon).

The long piece of cut-to-fit metal roofing under the three doors was a gift from my buddy Mike Rininger from scrap he had on hand.

The galvanized roofing (three 8-foot pieces), surplus from someone else's project, had waited years to make its contribution. Note the overhang of the roofing at the side and rear to further block the blowing of rain into the interior.

Look closely at the near corner and notice that the lift levers are in the down position to lift the rear and lock the wheels into place: There is now just enough clearance under the bottom rails to move the mobie, but not enough to allow escapes.

Another essential component is application of multiple coats of *marine spar varnish* to protect the framing that has year-round exposure to the weather. The construction-grade 2 × 4s I used are lacking in rot resistance—especially the bottom rails, which are in direct contact with the ground most of the time. I recommend marine spar varnish, formulated for use on oceangoing vessels, as the only protective coating you should consider for ensuring long service life of a wooden mobile shelter.

Wiring for Defense

The addition of an electric deterrent against predators was easy.

- It started with a fence charger, the HotShock 5, and a rechargeable lead-acid deep cycle 12-volt battery, both mounted out of the weather inside the shelter, but high up in a corner, well above chicken level (see figure D.9). Both components are from Premier, my preferred source for electric fencing accessories. I chose the 7 amp hour battery for this small application and I bought two of them—I always have one charged and ready to swap into service when I test charge in the wires and see it has dropped down near the minimum required to repel predators.

- Good grounding is essential for maintaining a "hot" wire (see figure D.10). The ground rod here is 36 inches long, half an inch thick—driven deep enough to reach moist soil, but not so deep that it is a struggle to pull out in preparation for a move. Charge for the perimeter wires comes from inside via insulated cable to a knife switch. Pulling its handle open kills the power to the wires—I won't get a nasty surprise when feeding, watering, or gathering eggs. The bright red color and the prominent placement of the switch make it unlikely I will overlook closing the switch when I leave the shelter.
- I used plastic claw insulators attached to wood framing pieces to stand off the electric wires from the structure (see figure D.11). Wire is 14-gauge galvanized steel and runs at three levels: 3, 8, and 12 inches. I "stepped up" the lower two runs to the top insulator at the corners to route them around the steel gussets.

Figure D.12. Hanging 3-gallon waterer is easy to raise into position using two pulleys.

Waterer

A hanging waterer (attached to two pulleys for ease of raising into position) does not need to be taken out when the shelter is moved. As well, I never have to fiddle with leveling the waterer to prevent emptying of its vacuum-sealing reservoir if set on uneven ground. At first, the chickens occasionally managed to knock off the base of the waterer, releasing all the water, which could be a disaster on a hot summer day. I solved the problem by setting the waterer on a plywood platform (see figure D.12).

Rolling!

Figure D.1 shows my Rolling Fortress Gallus in action. The front end is clad with ½-inch hardware cloth (not chicken wire) to ensure the birds maximum ventilation and plenty of time in the sunshine. It is fastened with screws and fender washers, in lieu of less-secure small fence staples, to block the greedy paws of raccoons. The ½-inch mesh also excludes blacksnakes, who love eggs—if they can get to them—and chicks as well. The overhang of the roofing provides full-time shade plus extra protection from blowing rain in the sheltered interior area. I painted the roofing white, to better reflect sunlight and reduce buildup of heat inside on sunny summer days. Its corrugations vent heat from the interior as well.

The weight of my mobie is right at the limit recommended by the manufacturer of the lift kit: 400 pounds. Yet the length of the handle, with the axle of the front wheels as fulcrum, provides the mechanical advantage needed to easily lift the front. I am seventy-seven years old, and I roll it easily over flat ground. My pasture, as it happens, is on a slight but decided incline. So finding "flat ground" for me means making the daily moves on the contour of the slope, rather than either up-slope or down. The pivot pin on the end of the long handle makes it easy to steer the shelter right or left, even while backing up. (I do get assistance from my friend Federico Lucas when it is time—once a month or so—to return the unit to the top of the pasture's incline, and start again to move it gradually downslope along the contours.)

Before the daily move, I kill the power to the wires and pull up the ground rod and lay it, still attached to the ground wire, on the roofing. Then I step down on the end of the lift levers on either side to lift the bottom rail and lock the wheels into position. Within seconds the rear end is ready to roll. Pushing down on the handle over the front wheels lifts that end as I pull the mobie onto the next 8- by 10-foot plot of fresh grass. Another step down on each lift lever disengages it from its notch and returns it to the wheels-up position. Now the bottom rail is again tight to the ground around the entire perimeter.

Or is it? Yes, the bottom rail is now in blocking position for predators the size of a possum and up. But a small dip in the ground may create a bit of space under the rail, enough perhaps for incursion by a blacksnake or a least weasel. So I keep near to hand a supply of scrap wood of various lengths and thicknesses, ready to grab when needed to force into such spaces to block them off.

Now, with its physical barriers alone, the mobie is secure against any predator I've seen in my area other than a bear. But I don't want to leave even the smallest possibility a predator of any description is going to enjoy a chicken dinner at my expense. That's why I drive the ground rod into place at the new position, close the knife switch to charge the perimeter wires—and I challenge anybody to get closer to absolute impregnability than that!

Do note the importance of keeping the lowest charged perimeter wire free of grass and weeds—remember, it is only 3 inches above the surface of the ground. Contact by such greenery will not only degrade the intensity of the jolt to a potential predator but also cause more rapid loss of reserve charge in the battery. I use a power mower with the blade set low to mow perimeters all at once for several upcoming 8- by 10-foot stations. If I don't want to bother with the mower, it doesn't take long to remove grass near the bottom wire with a hand sickle.

For me, the single time in the entire year I need to get inside the mobie is when transferring the chooks to winter quarters. The *first* chicken I remove is the cock: Grabbing his ladies first would put him on outraged defense, making the move more stressful for everybody, and possibly triggering in him a permanently hostile stance.

Reducing Weight of This Design

Yes, I said previously that I could move my mobie "easily" over flat ground, though I need help pulling it back up to the top of the gradual incline on my pasture. But my ground is uneven as well as a bit sloped, and you know what? When I hit those dips and bumps, I think: *Hmm, sure wish I'd lightened up a bit on this thing!* So, if you're considering a substantial shelter using the same lift kit I use—lucky you: I have now served during a full season's use as "guinea pig" for the design as described. And can pass on some ideas for *reducing weight* you can consider when planning your project. I think a reduction from 400 to not much more than 300 pounds is possible.

- Even with a structure this size, I don't think full-width 2 × 4s are necessary for the bottom rails and the corner posts, *if* you select pieces that are completely sound—lacking in knots or other flaws. Rip your 2 × 4s for the four bottom rails and the four corner posts at two-thirds their width (just as for the top rails), and use the ripped-off pieces at one-third width for bracing, cleats, and roosts.

- The galvanized steel roofing panels I used were free but add considerable weight to the total. If you can get your hands on a few single aluminum panels, using them for your mobie's cover will add far less weight.
- My biggest weight "cost" was my choice to use a dustbox and a heavy nestbox unit, both in ⅝-inch plywood, simply because I was in a hurry and they were already assembled. Installing new-built units as part of the construction, using ¼-inch plywood and lightweight structural pieces, would have reduced weight by about 48 pounds (22 kg).

APPENDIX E

A Feed Formulation Spreadsheet

Readers who know how to use electronic spreadsheets will find it easy to design one to help formulate homemade feeds. Those who are relatively new to the game may benefit from a description of putting together a basic feed-calculator spreadsheet. The one presented here is just an illustration of how to set it up, so let's keep it simple. It would be easy to add columns for calculating percentages of fat, carbohydrate, cost of various mixes, or anything else you want to track. Indeed, you are welcome to download from my website a fully functional copy of this spreadsheet and modify it any way you like to serve your own needs: https://themodernhomestead.us/resources/downloads.

The basic design of the spreadsheet is shown in table E.1. Enter in column A ("Ingredient") all the feed ingredients on hand to use in various formulations. In the first section of the ingredients list ("Premix"), make separate listings of the fine, powdery ingredients such as fish meal and mineral supplement that you will combine into a premix for convenience. List in the lower section ("Ground/Whole portion") the bulkier components such as corn, peas, and small grains. For ingredients containing protein, I like to note percent protein—for example, "Wheat (.15)." These numbers added to entries in column A are just convenient reminders and have no calculating function. In the same way, I might add a note such as "(*Not >2.5/100!*)" to remind myself not to use more than 2½ pounds of crab meal per hundredweight of feed (to avoid excess selenium).

Note that I have highlighted two different classes of cells: one yellow, one green. The highlighting is simply to help in explaining here the function of the two types of cells. There is no need to use color highlighting in your own spreadsheet unless you prefer it.

The yellow cells are those into which I enter possible values (weights) for the different ingredients as I design a feed mix. The green cells contain mathematical formulas—that is, instructions to the spreadsheet program as to the calculations to run on the figures entered in the yellow cells. Note that the cells containing formulas do not show the formulas themselves, which are hidden in the background. Instead, each one shows "0.00," since at this point no values have been entered into the yellow cells.

Table E.2 shows the formulas I added to the spreadsheet. (*Please note* that the formulas used in this illustrative table are in the syntax required by my spreadsheet program. It is possible that your spreadsheet program requires different formula syntax. Check the Help section in your program.) Remember from table E.1 that what you will actually see as you add formulas into the green cells is "0.00." In this presentation of the spreadsheet I have printed the formulas themselves into the green cells, to show their hidden syntax.

Table E.1. Basic Feed Formulation Calculator

	A	B	C	D
1	**Ingredient**	**Pounds/100 pounds**	**% Protein**	**Pounds/25 pounds**
2	**PREMIX**			
3	Aragonite/Feeding limestone			
4	Nutri-Balancer			
5	Kelp			
6	Fish meal (.60)		0.00	
7	Crab meal (.25) (*Not >2.5/100!*)		0.00	
8	Cultured yeast (.18)		0.00	
9	**Total premix per 25 pounds**			**0.00**
10	**GROUND/WHOLE PORTION**			
11	Corn (.09)		0.00	0.00
12	Peas (.22)		0.00	0.00
13	Wheat (.15)		0.00	0.00
14	Oats/Barley (.11) (*Not >15%!*)		0.00	0.00
15	**Total**	**0.00**	**0.00**	**0.00**

Table E.2. Formulas in the Calculator

	A	B	C	D
1	**Ingredient**	**Pounds/100 pounds**	**% Protein**	**Pounds/25 pounds**
2	**PREMIX**			
3	Aragonite/Feeding limestone			
4	Nutri-Balancer			
5	Kelp			
6	Fish meal (.60)		=B6*0.6	
7	Crab meal (.25) (*Not >2.5/100!*)		=B7*0.25	
8	Cultured yeast (.18)		=B8*0.18	
9				**=SUM(B3:B8)/4**
10	**GROUND/WHOLE PORTION**			
11	Corn (.09)		=B11*0.09	=B11/4
12	Peas (.22)		=B12*0.22	=B12/4
13	Wheat (.15)		=B13*0.15	=B13/4
14	Oats/Barley (.11) (*Not >15%!*)		=B14*0.11	=B14/4
15	**Total**	**=SUM(B3:B14)**	**=SUM(C3:C14)/100**	**=SUM(D9:D14)**

I need only one formula in column B: In cell B15 I enter "=SUM(B3:B14)"—that is, this cell will calculate the total of all values entered into any or all cells B3 through B14. Note that the title of column B is "Pounds/100 pounds." I like to calculate my formulations based on 100 pounds total weight, so that the

Table E.3. A Starter Mix for Chicks on Pasture

	A	B	C	D
1	**Ingredient**	**Pounds/100 pounds**	**% Protein**	**Pounds/25 pounds**
2	**PREMIX**			
3	Aragonite/Feeding limestone	1.00		
4	Nutri-Balancer	3.00		
5	Kelp	0.50		
6	Fish meal (.60)	6.50	3.90	
7	Crab meal (.25) (*Not >2.5/100!*)	2.00	0.50	
8	Cultured yeast (.18)		0.00	
9				3.25
10	**GROUND/WHOLE PORTION**			
11	Corn (.09)	23.00	2.07	5.75
12	Peas (.22)	30.00	6.60	7.50
13	Wheat (.15)	24.00	3.60	6.00
14	Oats/Barley (.11) (*Not >15%!*)	10.00	1.10	2.50
15	**Total**	**100.00**	**0.18**	**25.00**

Table E.4. A Grower Mix for Chicks and Young Waterfowl on Pasture

	A	B	C	D
1	**Ingredient**	**Pounds/100 pounds**	**% Protein**	**Pounds/25 pounds**
2	**PREMIX**			
3	Aragonite/Feeding limestone	1.00		
4	Nutri-Balancer	3.00		
5	Kelp	0.50		
6	Fish meal (.60)	5.00	3.00	
7	Crab meal (.25) (*Not >2.5/100!*)	2.25	0.56	
8	Cultured yeast (.18)	2.25	0.41	
9				3.50
10	**GROUND/WHOLE PORTION**			
11	Corn (.09)	24.00	2.16	6.00
12	Peas (.22)	28.00	6.16	7.00
13	Wheat (.15)	24.00	3.60	6.00
14	Oats/Barley (.11) (*Not >15%!*)	10.00	1.10	2.50
15	**Total**	**100.00**	**0.17**	**25.00**

percent protein is obvious. Thus as I enter values in cells B3 through B14, I ensure that the resulting total in B15 always comes out to 100.

Column C ("% Protein") calculates the weight of protein in the total mix contributed by the weight of each ingredient entered in column B. For example, cell A6 lists fish meal, which contains 60 percent protein, and the pound weight of fish meal to be added goes into cell B6. Thus the formula to enter in cell C6 is "=B6*0.6"—that is, the weight of fish meal entered in B6 is automatically multiplied by its percent protein to return in C6 the amount of protein it contributes to 100 pounds of feed mix. Similarly, C11 will calculate the protein contributed by the amount of corn (A11, at 9 percent protein) entered in B11, so the correct formula is "=B11*0.09", or the weight of corn entered in B11 multiplied by 9 percent. (Note that I have not entered formulas in cells C3, C4, and C5 because Aragonite/Feeding limestone, Nutri-Balancer, and Kelp do not contribute any protein.)

The last cell in column C is different. Its formula "=SUM(C3:C14)/100" instructs the program to total all the above cells in the column, C3 through C14, then divide by 100, to give the total percentage of protein in 100 pounds of the proposed mix.

Designing the formulation on a hundredweight basis is convenient for showing total protein as a percentage, but when mixing the feed by hand, making it in batches of 25 pounds is more practical. Column D ("Pounds/25 pounds") will automatically calculate the amount of each ingredient to weigh out per batch. Note that, since I make the premix ahead of time so it can be handled as a single ingredient, its calculation in D9 must be a bit more complex, "=SUM(B3:B8)/4"; that is, the program will first sum the weights of all the premix ingredients (B3 through B8), then divide by four (there are four 25-pound batches per 100 pounds), to show the amount of premix to weigh out per batch.

Other formulas in column D are more straightforward: A formula such as "=B12/4" will calculate the value (weight) of peas entered in B12 divided by four, to show the amount of peas to use in a batch.

With the spreadsheet ready to perform calculations, it's time to populate it with values to design a feed mix. The values (weights) of ingredients must be in reasonable proportions, of course, as discussed in chapter 17. Table E.3 shows the weights of available ingredients I might use to make a starter feed for chicks on pasture. I entered reasonable proportions of ingredients in column B and watched as the spreadsheet automatically displayed in column C the amount of protein that would be contributed by each ingredient. I adjusted the amounts of ingredients in column B, always within reasonable parameters, until I hit a target protein content of 18 percent in cell C15. (That is low by the standards of commercial feeding, but this feed is for chicks on pasture with mother hens, who find high-protein animal foods like worms and insects for the chicks.) The amounts of premix and of each of the bulk ingredients needed per 25-pound batch are automatically displayed in the cells of column D.

Optimal nutrient profiles vary by age and by species, but it is easy to use the same spreadsheet to reformulate feeds as needed. Suppose I want to revise this recipe as the chicks grow older, and that I'm now running them on the same pasture with some growing ducklings and goslings. I can cut back a little on protein (to 17 percent), but I want to increase B-vitamin content, since waterfowl need more B complex in the diet. Dried cultured yeast is a good source, so I include some in the new formulation. The revised mix is shown in table E.4.

APPENDIX F

Layer Performance Spreadsheet

Even if you do not use electronic spreadsheets, you may still find some of the information that follows useful. But I'm only being honest when I tell you the functions of my layer performance spreadsheet will remain largely opaque. If you do have a basic competence using electronic spreadsheets, I strongly advise that, before continuing, you download a "live action" copy of the spreadsheet, in a version that fits the software you use. Links for downloads are at: https://themodernhomestead.us/resources/downloads. You will have the convenience of being able to click on a cell to discover "what's going on in the background" and how the spreadsheet works. For example, you can click on a cell that clearly contains a calculating formula, such as C24, for an immediate "Ah, now I understand: The value here—referenced as a factor for calculating scores of individual eggs laid in the five-day test by all hens hatched in 2016—is constantly updated as an average of all entries in the range (H16:H23,Q16:Q23,Z16:Z23,AI16:AI23,AR16:AR23). Of course." Once you understand the spreadsheet's functions, you can change it as much as you like to record and manipulate whatever data are useful in your own breeding project.

The Template

An advantage to using electronic spreadsheets is that you can create a *template*, a basic form that includes all the data fields, labeled rows and columns, and formulas needed to capture and organize the data you are interested in tracking and manipulating. Once you have the template in place, you can keep it unchanged while generating any number of copies you need for actual data entry. Let's begin with the template I made for recording egg laying performance of all hens in my flock (see table F.1).

Basic Structure

Column A is used to record a hen's year of hatch, which indicates her age. Column B gives the flock name of the hen if there is one, but, more importantly, at the end of each age cohort signals (with "AV/AGE") that the cell to the immediate right contains an important formula. Column C provides the hen's ID—the color and number of her wing band.

Following the end of each age cohort in column C is a cell containing a constantly updating calculation (signaled in the immediately left cell in column B by the abbreviation "AV/AGE"). The formula in this cell (e.g., the one in cell C24) automatically updates its value each time a data entry for an egg is entered for a hen in the relevant age cohort (in this case, the 2016 cohort). Specifically, the formula averages the gram weight of all eggs laid during the whole five-day test by all hens *in that age cohort*. Its continually

Table F.1. Layer Performance Spreadsheet Template

	A	B	C	D	E	F	G	H	I	J	K	L	M	N	O	P	Q	R	S	T	U	V	W	X
1							Date									Date								
2				SC/H	TE		E	GR	AV/AGE	SZ	OS	SS	SQ	SC/E		E	GR	AV/AGE	SZ	OS	SS	SQ	SC/E	
3	Copy formulas here → →						1			##				###		1			##				###	
4	2012	Blær	R102	##	0																			
5		AV/AGE	###																					
6	2013		B103	###	0																			
7	2013	Cherry	R103	###	0																			
8		AV/AGE	###																					
9	2014		G104	###	0																			
10	2014	Solana	G106	###	0																			
11	2014		B104	###	0																			
12		AV/AGE	###																					
13	2015	Avena	G108	###	0																			
14	2015		R115	###	0																			
15		AV/AGE	###																					
16	2016		B129	###	0																			
17	2016		B135	###	0																			
18	2016		B140	###	0																			
19	2016	Frijola	R135	###	0																			
20	2016		G128	###	0																			
21	2016		G131	###	0																			
22	2016	Sami	G144	###	0																			
23	2016	Stella	R158	###	0																			
24		AV/AGE	###																					
25	2017		R122	###	0																			
26	2017		R123	###	0																			
27	2017		R124	###	0																			
28	2017		G122	###	0																			
29	2017		B111	###	0																			
30	2017		B113	###	0																			
31	2017		B117	###	0																			
32	2017		B119	###	0																			
33	2017		B147	###	0																			
34	2017		B148	###	0																			
35	2017		R174	###	0																			
36	2017		R177	###	0																			
37	2017		R182	###	0																			
38	2017		R188	###	0																			
39	2017		G165	###	0																			
40	2017		G167	###	0																			
41	2017		G173	###	0																			
42	2017		G177	###	0																			
43	2017		G178	###	0																			
44	2017		G180	###	0																			
45	2017		G182	###	0																			
46	2017		B154	###	0																			
47	2017		B156	###	0																			
48	2017		B157	###	0																			
49		AV/AGE	###																					

Y	Z	AA	AB	AC	AD	AE	AF	AG	AH	AI	AJ	AK	AL	AM	AN	AO	AP	AQ	AR	AS	AT	AU	AV	AW	AX
Date									Date									Date							
E	GR	AV/AGE	SZ	OS	SS	SQ	SC/E		E	GR	AV/AGE	SZ	OS	SS	SQ	SC/E		E	GR	AV/AGE	SZ	OS	SS	SQ	SC/E
1			##				###		1			##				###		1			##				###

recalculated value is fed as a factor into the relevant "AV/AGE" cell in columns I, R, AA, AJ, or AS, with the result that size for each egg is *weighted* (in the adjacent "SZ" column) as a comparison to average egg size for the hen's age cohort, not to average size of all eggs laid in the flock.

In this template you see many cells containing hash marks ("##" or "###"), which indicate that they contain a formula. When values are entered into cells to which a formula refers, the calculated value is displayed in the formula's cell, instead of hash marks.

The basic colors (of the three groups in my clan mating system) are Red, Green, and Blue—that is, a hen's band color shows her clan. Note that I use variants of the three basic colors to highlight broody hens in column C. Specifically, I highlight those hens who are not only layers but proven broodies by using colors a bit lighter than the basic shade. Examples are R102 (Blær, a Red broody) and G144 (Sami, a Green broody). You don't see any hen labels shaded lighter blue because there were no Blue broodies in this test sequence.

Test Sequence Sheets

The template has been set up to generate the sheets needed for a particular test sequence. Typically, I would run three test periods of five consecutive days each, maybe three weeks apart, for a total of fifteen test days. (For simplicity there are only two test periods in this set, shown in tables F.2 and F.3.) Note the five spaces labeled "Date" above each set of data-entry columns in table F. 1. After a sheet is generated from the template for a test period, the five test dates will be entered in these spaces. So long as the makeup of the flock does not change, the template can be used to generate data sheets for any number of subsequent testing periods. Changes in the flock—following culling, losses to predators, passing on stock to other flocksters—require changes to the template, specifically to the list of hens in columns A through C. *Remember as well* that some cells containing formulas will need to be updated regarding *ranges of cells* to which they refer.

Column Headers in Row 2

Column D is for SC/H, the score for all eggs laid by this hen, averaged here as an overall score for this test period. This average score for this specific hen continues to update, automatically, every time an egg from her is recorded, through the entire (five-day) test period.

TE in column E is the total number of eggs laid by this hen in this period, which increments automatically each time another egg laid is entered in the columns to the right.

The rest of row 2 contains column headers for five groups of columns into which detailed information about each egg laid in the test period will be entered. (More information about the exact formulas in cells, and their functions, is given in "Data-Entry Sheets.") There are eight columns in each group, labeled as follows:

E = the value entered per egg. If an entry is made for that day, this value will always equal 1, of course. (As entries are made for additional eggs laid, they auto-accumulate into the TE value in column E, total number of eggs laid in the test period.)

GR = weight of egg in grams. I find that grams provide a clearer comparison than ounces; for example, 56 grams and 64 grams versus 2.0 ounces and 2.286 ounces. You are free to use a different weight unit if you prefer of course.

AV/AGE = average size of egg in relation to age cohort. Remember the special cells in column C (flagged on the left as "AV/AGE") which *average all eggs laid by all hens in the associated age cohort in this test period* (as C24 for example does for all eggs laid in the test period by hens hatched in 2016). A cell in a column under the AV/AGE header can *reference* one of those special averaging cells in column C—that is, can *bring in* its current value as a *factor* in further calculations of a *score* for *this specific egg*. (More on making this happen to follow.)

SZ = egg size as weighted value. This value is calculated automatically if "GR" value and "AV/AGE" cell reference are properly entered. Because it is calculated with reference both to the weight of *this* egg and to the "AV/AGE" *average*, the result is a numerical comparison (continually updating) of *this* egg to the average of *all eggs* laid by *this age cohort* in the test period. Note as well that this is a *weighted* value, weighted by factor "*.3" ("times three tenths") in the relevant formula. (The degree to which it is weighted will be a factor in the score per egg in the SC/E column to the right). If you wanted to give greater importance to egg size as a basis for selection, you could change the factor in this formula to "*.5" or "*.7" or other.

OS = overall shape. Each egg is judged for shape regarding three characteristics, with a blank entry as the default, indicating a utilitarian/acceptable shape.

1. Too-round or too-evenly-oval: value −.2, though if air cell in the egg is completely ambiguous without candling: −.5
2. Any other shape abnormalities, for example, lopsided: −.2 to −.6 depending on severity.
3. Especially appealing shapes: assign positive (+) values .2 to .6.

SS = surface shape. This is an assessment of ridges and bumps with rankings from −.2 to −.6 depending on severity (and −.1 if flaw is perceptible only on close inspection).

SQ = shell quality. Each egg is judged regarding:

1. Visible shell thinning: −.2 (if otherwise okay on candling)
2. Calcium deposits: −.8 if severe, −.1 if only a dot of calcium.
3. Fault line: −1.0 (that's "minus one"—this is a serious flaw)

SC/E = score/egg. This is the overall score for this specific egg, based on the values entered for its specific characteristics. The score is the sum of the values under E + SZ + OS + SS + SQ; that is, a numerical summary of "one egg laid" plus its weighted size plus its various shell qualities.

Note that these illustrative sheets are from a test sequence in the middle of winter, so hens are being evaluated for winter laying in addition to all the other egg traits discussed here.

Data-Entry Sheets

After data-entry sheets have been generated from the template (using the save-as function), and labeled in row 2 with the dates of the test period (TP), they are ready for data entry as eggs are laid. I show two examples of my own data in tables F.2 and F.3. This description of using the spreadsheet assumes that the user is *nest-trapping* all layers in the flock, for the entire test period, and noting on the egg in pencil the hen's ID and date collected.

Table F.2 shows the sheet for TP1 (Test Period 1). If no egg is laid by a hen in a given test day, *no entries of any sort are made.* For each egg laid, make data entries as follows.

- Note the "Copy formulas here" at the left end of row 3, pointing to sets of eight grouped cells, one at the top of each group of columns for a particular test day (cells G3 through N3, P3 through W3, etc.). Note that the left-most cell in any "copy" group of cells, for "E," conveniently has the default value "1" already entered—we are about to make an entry for *one* egg. Before making the entries for an egg, *copy* the eight cells at the top of the columns for *this day* of the test period, and paste it into the individual hen's row below, in her columns for *this* test day. Note that the group of eight cells has a common border—when pasted below, it serves to highlight a bit the entry being made.
- Assume for example that the first three hens to lay are all in the same clan, Green, and the same age cohort, hatched 2016. Before making the first entry, for G128, I copy the preformatted formulas, with border, for the eight cells at the top in the "Copy formulas here" for *this test day* and paste into her row 20.

(*continued on page 374*)

Table F.2. Layer Performance Test Period 1 Data Sheet

	A	B	C	D	E	F	G	H	I	J	K	L	M	N	O	P	Q	R	S	T	U	V	W
1							Tuesday, December 12, 2017									Wednesday, December 13, 2017							
2				SC/H	TE		E	GR	AV/AGE	SZ	OS	SS	SQ	SC/E		E	GR	AV/AGE	SZ	OS	SS	SQ	SC/E
3	Copy formulas here → →						1			##				###		1			##				###
4	2012	Blær	R102	0.73	4											1	46.4	46.05	.30	−.5	−.1		0.70
5		AV/AGE	46.05																				
6	2013		B103	1.41	2		1	48.1	49.86	.29	.1			1.39									
7	2013	Cherry	R103	1.06	3											1	49.3	49.86	.30		−.1		1.20
8		AV/AGE	49.86																				
9	2014		G104	1.04	3											1	51.3	48.30	.32		−.1		1.22
10	2014	Solana	G106	1.24	2		1	43.7	48.30	.27	.1	−.1		1.27									
11	2014		B104	0.69	1											1	47.2	48.30	.29	−.2	−.2	−.2	0.69
12		AV/AGE	48.30																				
13	2015	Avena	G108	####	0																		
14	2015		R115	1.15	2											1	49.1	49.85	.30	−.2	−.2		0.90
15		AV/AGE	49.85																				
16	2016		B129	0.50	1																		
17	2016		B135	1.38	2											1	45.3	48.94	.28	.2	−.1		1.38
18	2016		B140	1.08	1																		
19	2016	Frijola	R135	1.14	1																		
20	2016		G128	1.08	3		1	56.0	48.94	.34		−.1		1.24									
21	2016		G131	0.96	3		1	53.8	48.94	.33		−.2		1.13									
22	2016	Sami	G144	1.20	4		1	43.1	48.94	.26		−.1		1.16		1	44.0	48.94	.27	.1	−.1		1.27
23	2016	Stella	R158	####	0																		
24		AV/AGE	48.94																				
25	2017		R122	1.32	2											1	52.2	47.02	.33	.1			1.43
26	2017		R123	####	0																		
27	2017		R124	1.18	3											1	47.5	47.02	.30	.2	−.1		1.40
28	2017		G122	####	0																		
29	2017		B111	1.24	3											1	50.7	47.02	.32		−.1		1.22
30	2017		B113	0.79	1																		
31	2017		B117	####	0																		
32	2017		B119	1.38	2		1	51.2	47.02	.33	.2	−.1		1.43									
33	2017		B147	####	0																		
34	2017		B148	1.33	3											1	44.4	47.02	.28	.2	−.1		1.38
35	2017		R174	####	0																		
36	2017		R177	1.33	2																		
37	2017		R182	####	0																		
38	2017		R188	1.28	2		1	46.1	47.02	.29		−.2		1.09									
39	2017		G165	1.36	3											1	51.6	47.02	.33	.1	−.2		1.23
40	2017		G167	1.12	3		1	49.9	47.02	.32		−.4		0.92									
41	2017		G173	0.99	1		1	45.6	47.02	.29		−.3		0.99									
42	2017		G177	1.36	2																		
43	2017		G178	1.06	1																		
44	2017		G180	1.17	3											1	47.9	47.02	.31		−.3		1.01
45	2017		G182	0.80	1																		
46	2017		B154	1.35	2																		
47	2017		B156	1.29	2																		
48	2017		B157	1.35	1																		
49		AV/AGE	47.02																				

Y	Z	AA	AB	AC	AD	AE	AF	AG	AH	AI	AJ	AK	AL	AM	AN	AO	AP	AQ	AR	AS	AT	AU	AV	AW	AX
	Thursday, December 14, 2017									Friday, December 15, 2017									Saturday, December 16, 2017						
E	GR	AV/AGE	SZ	OS	SS	SQ	SC/E		E	GR	AV/AGE	SZ	OS	SS	SQ	SC/E		E	GR	AV/AGE	SZ	OS	SS	SQ	SC/E
1			##				###		1			##				###		1			##				###
1	45.5	46.05	.30	−.5	−.1		**0.70**		1	45.5	46.05	.30	−.5	−.1		**0.70**		1	46.8	46.05	.30	−.5			**0.80**
									1	52.9	49.86	.32	.2	−.1		**1.42**									
1	47.2	49.86	.28				**1.28**											1	51.8	49.86	.31	−.5	−.1		**0.71**
									1	50.9	48.30	.32		−.2		**1.12**		1	47.9	48.30	.30	−.5			**0.80**
																		1	48.8	48.30	.30	−.1			**1.20**
									1	50.6	49.85	.30	.2	−.1		**1.40**									
									1	49.0	48.94	.30	−.2	−.4	−.2	**0.50**									
1	46.2	48.94	.28	.2	−.1		**1.38**																		
1	46.4	48.94	.28	−.1	−.1		**1.08**																		
									1	39.6	48.94	.24		−.1		**1.14**									
1	53.9	48.94	.33	−.5			**0.83**											1	59.2	48.94	.36	−.2			**1.16**
									1	56.7	48.94	.35		−.2		**1.15**		1	51.5	48.94	.32	−.5	−.2		**0.62**
1	46.2	48.94	.28		−.2		**1.08**		1	43.2	48.94	.26	.2	−.2		**1.26**									
									1	49.4	47.02	.32	.1	−.2		**1.22**									
1	50.6	47.02	.32	−.2	−.2		**0.92**		1	47.0	47.02	.30		−.1		**1.20**									
									1	48.1	47.02	.31	.1	−.1		**1.31**		1	47.0	47.02	.30	−.1			**1.20**
																		1	44.7	47.02	.29	−.5			**0.79**
									1	51.3	47.02	.33	.1	−.1		**1.33**									
1	46.0	47.02	.29	.2	−.1		**1.39**											1	48.3	47.02	.31	.1	−.2		**1.21**
1	44.5	47.02	.28	.2	−.1		**1.38**		1	43.5	47.02	.28				**1.28**									
									1	42.3	47.02	.27	.2			**1.47**									
									1	51.7	47.02	.33	.2	−.1		**1.43**		1	50.8	47.02	.32	.1			**1.42**
1	44.0	47.02	.28	.1			**1.38**											1	42.9	47.02	.27		−.2		**1.07**
									1	49.1	47.02	.31	.2	−.1		**1.41**		1	47.0	47.02	.30				**1.30**
									1	40.2	47.02	.26		−.2		**1.06**									
1	46.1	47.02	.29	.2	−.2		**1.29**											1	46.3	47.02	.30	.1	−.2		**1.20**
																		1	46.6	47.02	.30	−.2	−.2	−.1	**0.80**
1	45.7	47.02	.29				**1.29**											1	48.8	47.02	.31	.2	−.1		**1.41**
1	46.8	47.02	.30	.2	−.1		**1.40**											1	44.6	47.02	.28	−.1			**1.18**
																		1	39.3	47.02	.25	.2	−.1		**1.35**

Table F.3. Layer Performance Test Period 2 Data Sheet

	A	B	C	D	E	F	G	H	I	J	K	L	M	N	O	P	Q	R	S	T	U	V	W
1							**Wednesday, January 3, 2018**									**Thursday, January 4, 2018**							
2				**SC/H**	**TE**		**E**	**GR**	**AV/AGE**	**SZ**	**OS**	**SS**	**SQ**	**SC/E**		**E**	**GR**	**AV/AGE**	**SZ**	**OS**	**SS**	**SQ**	**SC/E**
3	Copy formulas here →→						1			##				**###**		1			##				**###**
4	2012	Blær	R102	0.70	4		1	52.3	47.63	.33		−.2		**1.13**		1	47.1	47.63	.30	−.5	−.5		**0.30**
5		**AV/AGE**	47.63																				
6	2013		B103	1.28	3											1	56.1	50.85	.33	.2	−.1	−.1	**1.33**
7	2013	Cherry	R103	1.18	3		1	46.7	50.85	.28	−.2	−.1		**0.98**									
8		**AV/AGE**	50.85																				
9	2014		G104	1.00	3											1	46.6	48.55	.29		−.1		**1.19**
10	2014	Solana	G106	1.20	2											1	47.9	48.55	.30	.1	−.1		**1.30**
11	2014		B104	0.49	1																		
12		**AV/AGE**	48.55																				
13	2015	Avena	G108	1.07	2																		
14	2015		R115	1.13	2		1	52.7	48.95	.32	−.1	−.1		**1.12**									
15		**AV/AGE**	48.95																				
16	2016		B129	0.49	1																		
17	2016		B135	####	0																		
18	2016		B140	1.02	2		1	46.7	51.34	.27	−.2	−.1		**0.97**									
19	2016	Frijola	R135	1.28	5		1	44.0	51.34	.26		−.1		**1.16**		1	44.5	51.34	.26	.1	−.1		**1.26**
20	2016		G128	0.86	4		1	60.3	51.34	.35	−.5	−.1		**0.75**		1	55.4	51.34	.32	−.5	−.1		**0.72**
21	2016		G131	0.95	5		1	53.2	51.34	.31		−.2		**1.11**		1	52.8	51.34	.31	−.5	−.1		**0.71**
22	2016	Sami	G144	####	0											Went broody this date							
23	2016	Stella	R158	1.07	2		1	54.5	51.34	.32	−.1	−.2		**1.02**									
24		**AV/AGE**	51.34																				
25	2017		R122	1.21	3		1	51.2	47.04	.33		−.1		**1.23**									
26	2017		R123	1.23	2																		
27	2017		R124	1.11	3		1	48.1	47.04	.31	−.1			**1.21**									
28	2017		G122	####	0																		
29	2017		B111	1.17	3		1	48.8	47.04	.31	−.2	−.1		**1.01**		1	46.1	47.04	.29	−.1			**1.19**
30	2017		B113	1.36	3		1	46.3	47.04	.30	.2	−.1		**1.40**		1	44.3	47.04	.28	.1	−.1		**1.28**
31	2017		B117	1.48	2											1	51.7	47.04	.33	.2			**1.53**
32	2017		B119	1.13	3											1	53.1	47.04	.34		−.4		**0.94**
33	2017		B147	1.27	3											1	44.3	47.04	.28	.1	−.2		**1.18**
34	2017		B148	1.24	2											1	47.6	47.04	.30	.1	−.1		**1.30**
35	2017		R174	1.29	3		1	49.3	47.04	.31	.1			**1.41**									
36	2017		R177	0.99	2		1	45.2	47.04	.29	−.5	−.1		**0.69**									
37	2017		R182	####	0																		
38	2017		R188	1.19	3											1	47.2	47.04	.30	.2	−.2	−.1	**1.20**
39	2017		G165	1.09	3											1	49.0	47.04	.31	−.5	−.1		**0.71**
40	2017		G167	1.13	2		1	45.8	47.04	.29	.1	−.2		**1.19**									
41	2017		G173	1.21	3											1	47.7	47.04	.30	.1	−.2		**1.20**
42	2017		G177	1.25	3		1	47.2	47.04	.30		−.1		**1.20**									
43	2017		G178	1.25	2											1	45.4	47.04	.29	.1	−.2		**1.19**
44	2017		G180	1.24	2		1	48.8	47.04	.31	.2	−.1		**1.41**									
45	2017		G182	0.80	1																		
46	2017		B154	1.20	2											1	48.6	47.04	.31	.1	−.2		**1.21**
47	2017		B156	1.28	3		1	43.4	47.04	.28	.2	−.2		**1.28**									
48	2017		B157	1.15	1		1	39.1	47.04	.25	.1	−.2		**1.15**									
49		**AV/AGE**	47.04																				

Y	Z	AA	AB	AC	AD	AE	AF	AG	AH	AI	AJ	AK	AL	AM	AN	AO	AP	AQ	AR	AS	AT	AU	AV	AW	AX
Friday, January 5, 2018									Saturday, January 6, 2018									Sunday, January 7, 2018							
E	GR	AV/AGE	SZ	OS	SS	SQ	SC/E		E	GR	AV/AGE	SZ	OS	SS	SQ	SC/E		E	GR	AV/AGE	SZ	OS	SS	SQ	SC/E
1			##				###		1			##				###		1			##				###
1	45.0	47.63	.28	−.5	−.1		0.68		1	46.1	47.63	.29	−.5	−.1		0.69									
									1	54.1	50.85	.32	.1	−.1		1.32		1	50.7	50.85	.30		−.1		1.20
1	50.0	50.85	.29		−.1		1.19											1	47.5	50.85	.28	.1			1.38
1	52.4	48.55	.32	−.5	−.2		0.62											1	48.0	48.55	.30		−.1		1.20
									1	49.0	48.55	.30		−.2		1.10									
									1	47.4	48.55	.29	−.2	−.2	−.4	0.49									
1	42.5	48.95	.26		−.2		1.06											1	45.6	48.95	.28		−.2		1.08
																		1	55.0	48.95	.34		−.2		1.14
1	49.2	51.34	.29	−.2	−.4	−.2	0.49																		
																		1	46.9	51.34	.27	−.1	−.1		1.07
1	47.5	51.34	.28	.2	−.1		1.38		1	43.7	51.34	.26	.1	−.1		1.26		1	44.4	51.34	.26	.2	−.1		1.36
1	59.3	51.34	.35		−.1		1.25		1	53.7	51.34	.31	−.5	−.1		0.71									
1	58.4	51.34	.34		−.3		1.04		1	52.6	51.34	.31	−.1			1.21		1	52.3	51.34	.31	−.5	−.1		0.71
									1	56.0	51.34	.33		−.2		1.13									
1	47.5	47.04	.30		−.2		1.10											1	49.2	47.04	.31	.1	−.1		1.31
1	43.5	47.04	.28				1.28											1	45.8	47.04	.29	.1	−.2		1.19
1	47.4	47.04	.30		−.2		1.10											1	52.1	47.04	.33		−.3		1.03
1	46.2	47.04	.29				1.29																		
1	45.9	47.04	.29	.2	−.1		1.39																		
1	51.5	47.04	.33	.2	−.1		1.43																		
									1	50.3	47.04	.32	.1	−.2		1.22		1	52.1	47.04	.33	−.1			1.23
1	40.6	47.04	.26	.2	−.1		1.36											1	44.1	47.04	.28	.1	−.1		1.28
									1	44.6	47.04	.28		−.1		1.18									
1	51.3	47.04	.33	−.1			1.23											1	52.7	47.04	.34		−.1		1.24
									1	46.3	47.04	.30	.1	−.1		1.30									
1	48.1	47.04	.31	.1	−.1		1.31		1	42.1	47.04	.27	−.2			1.07									
									1	52.2	47.04	.33	.1	−.1		1.33		1	52.6	47.04	.34		−.1		1.24
									1	43.1	47.04	.27		−.2		1.07									
1	48.9	47.04	.31	.1	−.2		1.21											1	47.6	47.04	.30	.1	−.2		1.20
1	44.5	47.04	.28	.1			1.38											1	43.5	47.04	.28		−.1		1.18
									1	47.4	47.04	.30	.1	−.1		1.30									
									1	42.9	47.04	.27		−.2		1.07									
									1	46.8	47.04	.30	−.2	−.2	−.1	0.80									
									1	45.2	47.04	.29	.1	−.2		1.19									
1	41.7	47.04	.27	.1	−.2		1.17											1	44.3	47.04	.28	.2	−.1		1.38

(*continued from page 369*)

Following this, pasting the default value "1" (one egg laid by this hen) is already entered in G20. Now I enter the weight in grams of this egg, 56.0, in H20. Note how cell E20, in the "TE" column, automatically returns the total number of eggs for this hen in this test period. At this point, that value is 1—however, it will auto-increment by 1 each time I enter an additional egg in the five-day test period for hen G128. Also note that as soon as I enter the weight in H20, C24 automatically returns the average weight in grams of *all* eggs laid in *this five-day test period* by *all* hens in *this age cohort*. That value, with this first entry, shows as 56.0, but it will auto-update with each additional egg entered.

- The next step is to enter a *reference* into I20 to the averaging formula in C24 for hens in the 2016 cohort. Specifically, into I20 we simply enter "=C24", which brings the *current value* of that all-cohort average into I20 as a factor for calculation. We also see automatic changes in D20 (average of scores for all individual eggs laid by this specific hen in the five-day period); in I20 (which will always be the same as the value showing in C24, however much it changes); J20, a value comparing the size (weight) of this specific egg with the average of all eggs for *this age cohort* in *this test period*, and which *will change* as the value of C24 changes; and finally in N20, an overall score for this egg—which again will update as the average in C24 changes but *at this point* is 1.34.
- But we notice that this egg is not flawless—the shell has a bit of a ridge. Not a big one—we notice it only on close inspection, so we assign this small flaw a −.1 (minus one-tenth) in L20 (in the column for "surface shape" flaws). Oh!—now the overall score in N20 has dropped to 1.24.
- Now we again copy from "Copy formulas here" and paste into the appropriate cell ranges for hens G131 and G144. We repeat the previous steps and watch as the averaging value in C24 changes with each entry. Note that G144's egg also has an almost unnoticeable shell shape flaw, so we enter a −.1 value in L22. But the egg G131 laid shows a more obvious shape flaw. It's not terrible, but it is noticeably unsightly. In this case we enter a demerit value of −.2 in L21. If the shell were even more malformed, we could assign demerits of −.4 or even −.6. And yes, these values are subjective. But, over the long run, they accumulate to useful measures of "how short of (or close to) *perfection*" this hen's eggs are.
- Similarly, in column M for shell quality, we might enter negative values to denote flaws such as calcium deposits or a thin shell more subject to cracking. But note in column K, while we might assign demerits for a shape too close to perfectly round, or lopsided, we can assign *positive* values (bringing that egg's overall score up) in cases where the egg is unusually aesthetically pleasing, the perfect expression of the classic egg shape. (An example is the positive .2 value in cell K32 for hen B119.)
- The second day's entries continue as for the first, with the proviso that the "Copy formulas here" operation starts with the cells at the top of *the second day's columns*, not those for the first.

At the end of the five days of the first test period, set up a second sheet in readiness for TP2, which might follow three weeks later. Table F.3 is an example showing data from my flock.

Score Sheet

What do I do with all the data I collect? I compile it in a score sheet (see table F.4) that summarizes the data for an entire test sequence as a guide to selection of breeders.

This score sheet includes, for individual hens, their averaged scores and total eggs laid in *the entire test sequence*—copied over from my data spreadsheets. For simplicity, this example score sheet shows data for two test periods only, although typically I do three test periods in a test sequence.

Table F.4. Layer Performance Score Spreadsheet

	A	B	C	D	E	F	G	H	I	J	K	L	M	N
1	Age	Flock ID	Band	TEST PERIOD 1		TEST PERIOD 2		Average Score	Total Laid	Weighted Rate of Lay	Weighted by Age	Broody?	Band	Status
2				Score	#	Score	#							
3	2012	Blær	R102	0.73	4	0.70	4	.71	8	5.7	10.3	Yes	R102	Good broody!
4														
5	2013		B103	1.41	2	1.28	3	1.35	5	6.8	10.8		B103	Good older layer
6	2013	Cherry	R103	1.06	3	1.18	3	1.12	6	6.7	10.8	Yes	R103	Good broody!
7														
8	2014		G104	1.04	3	1.00	3	1.02	6	6.1	8.6		G104	Good older layer
9	2014	Solana	G106	1.24	2	1.20	2	1.22	4	4.9	6.8	Yes	G106	Good broody!
10	2014		B104	0.69	1	0.49	1	.59	2	1.2	1.7		B104	*Cull*
11														
12	2015	Avena	G108		0	1.07	2	1.07	2	2.1	2.6	Yes	G108	Good broody!
13	2015		R115	1.15	2	1.13	2	1.14	4	4.6	5.5		R115	Give to Gary
14														
15	2016		B129	0.50	1	0.49	1	.49	2	1.0	1.0		B129	*Cull*
16	2016		B135	1.38	2		0	1.38	2	2.8	2.8		B135	Give to Gary
17	2016		B140	1.08	1	1.02	2	1.05	3	3.2	3.2		B140	Give to Gary
18	2016	Frijola	R135	1.14	1	1.28	5	1.21	6	7.3	7.3	Yes	R135	Good broody!
19	2016		G128	1.08	3	0.86	4	.97	7	6.8	6.8		G128	Good 2nd year layer
20	2016		G131	0.96	3	0.95	5	.96	8	7.7	7.7		G131	Good 2nd year layer
21	2016	Sami	G144	1.20	4	Setting!			Setting!		Setting!	Yes	G144	Good broody!
22	2016	Stella	R158		0	1.07	2	1.07	2	2.1	2.1	Yes	R158	Good broody!
23														
24	2017		R122	1.32	2	1.21	3	1.27	5	6.3	6.3		R122	Good 1st year layer
25	2017		R123		0	1.23	2	1.23	2	2.5	2.5		R123	Give to Gary
26	2017		R124	1.18	3	1.11	3	1.14	6	6.9	6.9		R124	Good 1st year layer
27	2017		G122		0		0	0	0	.0	0		G122	*Cull*
28	2017		B111	1.24	3	1.17	3	1.21	6	7.2	7.2		B111	Good 1st year layer
29	2017		B113	0.79	1	1.36	3	1.07	4	4.3	4.3		B113	Give to Gary
30	2017		B117		0	1.48	2	1.48	2	3.0	3.0		B117	Give to Gary
31	2017		B119	1.38	2	1.13	3	1.25	5	6.3	6.3		B119	Good 1st year layer
32	2017		B147		0	1.27	3	1.27	3	3.8	3.8		B147	Give to Gary
33	2017		B148	1.33	3	1.24	2	1.29	5	6.4	6.4		B148	Good 1st year layer
34	2017		R174		0	1.29	3	1.29	3	3.9	3.9		R174	Give to Gary
35	2017		R177	1.33	2	0.99	2	1.16	4	4.6	4.6		R177	Good 1st year layer
36	2017		R182		0		0	0	0	.0	0		R182	*Cull*
37	2017		R188	1.28	2	1.19	3	1.24	5	.0	.0		R188	Good 1st year layer
38	2017		G165	1.36	3	1.09	3	1.23	6	7.4	7.4		G165	Good 1st year layer
39	2017		G167	1.12	3	1.13	2	1.13	5	5.6	5.6		G167	Good 1st year layer
40	2017		G173	0.99	1	1.21	3	1.10	4	4.4	4.4		G173	Give to Gary
41	2017		G177	1.36	2	1.25	3	1.31	5	6.5	6.5		G177	Good 1st year layer
42	2017		G178	1.06	1	1.25	2	1.15	3	3.5	3.5		G178	Give to Gary
43	2017		G180	1.17	3	1.24	2	1.20	5	6.0	6.0		G180	Good 1st year layer
44	2017		G182	0.80	1	0.80	1	.80	2	1.6	1.6		G182	*Cull*
45	2017		B154	1.35	2	1.20	2	1.28	4	5.1	5.1		B154	Good 1st year layer
46	2017		B156	1.29	2	1.28	3	1.28	5	6.4	6.4		B156	Good 1st year layer
47	2017		B157	1.35	1	1.15	1	1.25	2	2.5	2.5		B157	Give to Gary

Copying the Data

Modern spreadsheet programs make it easy to copy and paste entire ranges of figures. Copying from the data-entry sheets to the scoring sheet is detailed next.

- To begin, copy from sheet TP1 all the *values* for "SC/H" (average score for each hen) in column D and all the *values* for "TE" (total number of eggs laid by each hen) in column E. Paste them into columns D and E on the score sheet to record overall results for the first test period. Repeat the equivalent copy and paste between TP2 and columns F and G on the scoring sheet, to transfer results for the second test period. *Be careful when pasting these two columns into the scoring sheet:* The cells of columns D and E on the test-period sheets contain *calculating formulas*, and a straight copy-and-paste will paste their *formulas* into the cells of the scoring sheet—a result that would be meaningless for use there. Instead, do a "paste special" into the relevant cells on the scoring sheet, choosing to paste the *values* of the copied cells only, *not* their formulas.
- Enter in H3 an averaging function for the two individual scores in D3 and F3: "=AVERAGE(D3,F3)". Then copy that cell, select appropriate cells below in the same column, and paste the formula from H3 into them all—the spreadsheet will automatically change the reference cells; for example, it will paste the formula as "=AVERAGE(D17,F17)" into H17 and "=AVERAGE(D31,F31)" into H31.
- In I3 enter the formula to calculate total eggs laid during both periods, E3 plus G3: "=SUM(E3,G3)". Again, copy that cell into the cells below, for calculations adjusted to those specific rows.
- The next column, column J, is for what I call the "weighted rate of lay." The formula entered into J3 multiples the total of eggs laid (I3) by the hen's all-egg average score (H3) for the test periods: "=H3*I3". And then of course "copy downward." It is obvious from studying this column that a hen who lays fewer eggs, with few or slight flaws, may score a higher weighted rate of lay than a hen who lays more eggs, with significant flaws.
- Finally, configure a calculation that relates a hen's production to her age. Specifically, what is needed here is a factor for *weighting* laying performance in relation to age. Note the impressive performance of Blær, the only hen remaining in the 2012 cohort (that is, the cohort of my original breeders)—eight eggs in ten days, in the middle of winter. That performance is extraordinary, given that at the time of this test she was six years old—so I *weight* it in relation to younger cohorts to reflect her value as a breeder who holds production as she ages. Specifically, I factor her weighted rate of lay (J3) by a multiplier of 1.8 (=J3*1.8) into K3, for an age-weighted score of 10.3. The weighting factor for hens in the 2013 cohort is 1.6; for 2014, 1.4; for 2015, 1.2; and for hens in their first and second laying seasons (2016 and 2017 cohorts), presumably the best they will ever have, the multiplier in the equation is simply "1."

Using the Data for Selections

This score sheet highlights hens who are not only being tested as layers but who are *also* proven broodies. I don't assign a numerical value for broodiness, though I do track performance of broodies minutely, using the separate "Broody Performance Spreadsheet" described in appendix G. I never cull a good broody. Although I do test her performance as a layer, her mothering skills trump laying numbers—though of course the ideal breeder scores high for both. (See the data in the score sheet for Cherry, R103, as an example of a top-of-the-class broody who also laid well for her age.)

Because I selected hatching eggs *using trapnests*, I had the option of setting clutches of eggs *from superior layers only* or *from superior broodies only*—or a random mix of both—depending on needs and goals at a given point in the breeding program.

Speaking of broodies, if a hen went broody during a test sequence, I simply dropped her from the rest of that sequence, as denoted by the word "Setting!" in the score sheet. (Note that these entries "Setting!" for Sami,

G144, are entered here, in this test period in the middle of winter, just by way of illustration of this important point: I wouldn't *penalize* a hen for volunteering for mother duty during an egg-laying test period. I would, however, select strongly against a hen who insisted on going broody in winter.)

The information payoff from all this recordkeeping is the judgments we can make about hens based on the test data. Some examples show in column N of the score sheet. In my flock, hens who have served well as mothers ("Good broody!") will be retained even without stellar performance as layers. Tracking age cohorts separately makes possible judgments such as "Good 2nd year layer" or "Good 1st year layer." And obviously, given the scheduling of this test in the middle of winter, any hen with a high score has proved her ability to maintain good winter production.

Note that the results of the test (of a total of forty hens) allowed for selection of twenty-four hens I wanted to retain for my own breeding, and eleven I was willing to pass on to another flockster interested in raising Icelandics ("Give to Gary"). Not surprisingly, I retained the best for my own breeding program—but the ones I passed on were good birds with at least decent laying performance. There were five hens who failed the test this cycle, and I passed them on to the stockpot (*Cull*).

The sharp-eyed reader may notice that the total number of layers in this, their first laying season (2017 age cohort), is twenty-four, and that I retained thirteen in my own breeding project. Such a reader might say, "Ah, you're keeping over 50 percent—nowhere near as low as the 10 percent or 20 percent you've recommended for retaining as breeders." But note that, prior to this test of hens in this age cohort as *layers*, I had *already* culled many pullets before they reached onset of lay—for shortcomings such as undesirable rate of growth, slight crook in keel or crooked toe, and other factors. Given that I hatched almost one hundred chicks in 2017, my retention of thirteen hens as breeders is not that far above an ideal of 20 percent females retained each breeding season discussed elsewhere.

100 chicks / 2 (gender factor)
 = approximately 50 females
13 females retained / 50 total females
 = 26 percent retention rate

Note that culling continues in following seasons, if hens perform poorly *in comparison to their advancing age cohort*—as seen in N10 and N15 in table F.4. So yes, the hatchet was indeed playing a large role in my breeding project.

In Conclusion

My intention in presenting this discussion in such detail is to give you an idea of the breadth and depth of the data you can capture using well designed electronic spreadsheets, and the degree to which you can tailor results as guides to your own breeding goals. It would be easy to modify these sheets to capture and manipulate a different data set—one targeted for example at meat qualities: rate of growth, weight at slaughter, carcass conformation, width of skull, heart girth, and more.

APPENDIX G

Broody Performance Spreadsheet

Table G.1 summarizes the performance of fourteen Icelandic hens (one of whom later set a second clutch) doing "broody duty" in a typical season, separately evaluating performance on the nest and subsequently as a mother. Table G.2 (see page 381) expands on the raw scores for each hen, for a fuller sense of her performance in each phase.

Incubation

The following is an explanation of the codes for evaluation factors found in column A of table G.1.

BL:I = Baseline: Incubation. The hen gets a baseline score of 10 if she successfully completes incubation and hatches a clutch of chicks, indicating her basic competence at providing the conditions for embryos to grow and hatch. Values in the cells that follow add to or subtract from the baseline score, as a measure of performance superior to or below the norm. (Note: even if a hen was "fixed" on a nest with fake eggs, in preparation to be a foster mother for chicks coming through the mail, she would still receive evaluation scores for her performance in the "setting" phase up until it was time to leave the nest with her adoptive chicks.)

+EB = Early broodiness. My strong preference is for mother hens to go broody early in spring so that all required hatching occurs in one big "wave" at the beginning of the green season. I therefore give extra points for early settling into broodiness (and subtract for tardiness): Any time in March = +4; first week in April = +3; second week = +2; third week = +1; fourth week = 0; any time thereafter = −1.

+PB = Prior broodiness. I especially prize repeat service as a reliable broody. One point is added for each prior hatching season in which the hen served as mother.

−Flirt = "Flirting" with broodiness. The hen gets demerit points if she "flirts" with broodiness before truly settling. Perhaps she gives every indication of having entered broodiness, even remaining in the egg nest overnight rather than going to roost—but I find she has failed to remain on the nest the following night, when I go out to move her to the broody box. Drawn-out flirting is a distraction and a drag on management time. Note that "flirting" behavior may also manifest as laying of any additional eggs after being moved to the broody box, even though prior behavior indicated she had thoroughly entered full broodiness.

−Panic = Panic behavior before move. In this case, even as a hen settles into broodiness in the egg nest, she freaks out whenever I come near (e.g., to gather eggs from under her). The good broody has

Table G.1. Broody Performance Spreadsheet

	A	B	C	D	E	F	G	H	I	J	K	L	M	N	O	P
1		Sami G144	Sara R124	Cherry R103	Avena G108	Blær R102	Azula B148	Hugla R122	Jezabel G167	Dapp B111	Stella R158	Freyla R177	Maggie B154	Frijola R135	Lady Bug B156	Sami G144 2nd Brood
2	BL:I	10	10	10	10	10					10		10	10	10	10
3	+EB	4	4	4	4	4	3	2	2	1	0	0	0	−1	−1	2nd brood
4	+PB	1		4	2	5					1			1		2
5	−Flirt	−1				−1		−1	−1	−2	−1	−1			−1	
6	−Panic						−4		−1			−2				
7	−Agit							−4	−1	−2		−1	−1			
8	−RM															
9	−NN															
10	−Poop	−1	1	1	−2	1								−1		
11	+/−@C	−1	0	1		−1		Note	Note							
12	+/−@H	1	1		1	1										
13	+/−Other	−1		7		2	Left nest!		Left nest!	Left nest!	5	−4	4	1	2	2
14	**Incubation Score**	**12**	**16**	**16**	**15**	**21**	**Culled**	**Culled**	**Culled**	**Culled**	**15**	**Culled**	**13**	**10**	**10**	**14**
15																
16	BL:M	10	10	10	10	10					10		10	10	10	10
17	+/−MMj			2	1						−1		2		2	−1
18	−Nurt	−2				2										
19	−Prot															
20	−Forg															
21	−Ag:H															
22	−Ag:C															
23	+/−Other	2	2			2					−1		2	2	2	4
24	**Mothering Score**	**10**	**12**	**12**	**11**	**14**					**8**		**14**	**12**	**14**	**13**
25	**Combined Score**	**22**	**28**	**38**	**27**	**35**	**Note**	**Note**	**Note**	**Note**	**23**	**Note**	**27**	**22**	**24 See note!**	**27**

the instinct to rely on stillness and/or flaring at the intruder as her protection, and that is best for the cooperative working relationship I value.

−Agit = Excessive agitation after move. I assign demerits if the hen is unduly agitated following the move, and takes too long getting settled in the broody box.

−RM = Refusing move to broody box. Once a hen clearly enters broodiness in the egg nest, it is a serious demerit if she refuses the move—that is, refuses to settle in the broody box. I am unlikely to give such a hen a second chance (and yes, that means culling her to the stew pot). I might allow that second chance, however, if she is a first-timer who seems in other ways to be "getting the idea."

−NN = "Nervous Nellie." I distinguish this from the Agit category because the "Nervous Nellie" is the hen who continues to be excessively fearful or easily disturbed following the move to the broody box. She is nervous whenever I monitor her or service the broody box, though the behavior doesn't rise to the level of failure to hatch.

−**Poop = Failure to relieve herself off the nest.** Negative points if the hen poops the nest. No demerit if she poops on the edge of the nest, so long as the result isn't soiled eggs: −1 to −3 depending on how badly she has stained or covered the eggs with excrement. (While the score is usually a negative here—for poor nest sanitation—I might assign a positive value to a hen who keeps the nest unusually clean.)

+/−**@C = Excessive nonviable eggs at candling.** At candling, I give the broody the benefit of the doubt—she may have been given an infertile egg. But too many nonviable eggs (when other broodies show better numbers) could be her fault. Thus: Plus 1 if eggs under her are 100% viable at candling. 1 egg discarded out of 11, no fault. Minus 1 for each nonviable egg over 1.

+/−**@H = Hatching success after candling.** I'm more confident about assigning responsibility to the broody for failure to hatch eggs that were viable at candling. A 100% hatch of all eggs retained in the nest after candling (even if total number is only 10 or 9) = +1. Each egg that fails to hatch after candling = −1. Note that a weak, struggling chick not vigorous enough to go outside with the family immediately is counted as a no-hatch.

+/−**Other. Other reasons to favor or deprecate incubation performance.** If hatching is unusually prolonged—if it takes her an additional day or even longer to hatch her chicks—the broody receives negative points. On the positive side, a hen might have had to remain "fixed" on fake eggs before starting incubation, to help with a scheduling problem. For example, if she set with patience for a week and a half in "pre-incubation" and then completed a three-week hatch, that is a plus.

Mothering

BL:M = Baseline: Mothering. Ten points are assigned for basic competence raising a clutch of chicks until they're ready to be on their own. Values in following cells add or subtract based on performance notably above or below the norm.

+/−**MMj = "Mama mojo."** Some hens just have more "mother power" than others—like a magnet, such a hen draws chicks from another with less. I give a hen who loses chicks to another a −1; one who draws chicks from others, +1. Having a lot of "mama mojo" is not necessarily a good thing. But I note it because in my experience a "mojo mama" is more likely to accept a graft of "foreign" chicks. One who does so in an emergency gets a higher + score—+2, maybe more.

−**Nurt = Failure in nurturing.** Attentive nurturing of her clutch—being a diligent, instructive, go-getter mother—is baseline, negative points for being more lackadaisical. (While the score is usually a negative here—for lackadaisical mothering—I might assign a positive value to a hen with extraordinarily high mothering skills.)

−**Prot = Failure in protecting.** Defending her chicks is baseline; negative points for cowardice in the face of threats.

−**Forg = Failure in teaching foraging skills.** Skillful foraging for her clutch is baseline; negative points for lackadaisical rather than diligent performance.

−**Ag:H = Undue aggression, other hens.** Some aggression among hens is normal behavior. Becoming the terror of the barnyard is not, especially if aggression is pushed to the point of possible injury.

−**Ag:C = Undue aggression, any chick.** Any level of aggression toward other hens' *chicks* is another matter. This is a serious demerit—and it is a judgment call when viciousness toward "other" chicks requires culling this hen. (A hen who *kills* any chick should be culled immediately.)

+/−**Other.** Other reasons to favor or deprecate mothering performance

Table G.2. Notes on Broody Performance

Sami G144	**Incubation: −1 Flirt:** Serious flirting several days (but settled immediately when moved). **−1 Poop:** Pooped into one end of nest. Some staining of the eggs resulted; not serious. **−1 @C:** 9 eggs viable at candling. **+1 @H:** All 9 eggs viable at candling hatched. **−1 Other:** Clutch took more than 2 full days to hatch. **Mothering: +2 Other:** Accepted graft of 6 chicks already 5 or 6 days old. **−2 Nurt:** Evening of 3/30 (about 7:30), Sami had gone to one of the nestboxes, abandoning her (clearly chilly) chicks huddling on the litter. Placed the chicks under her. Negligence did not repeat–subsequently, she bedded down with her chicks in the same corner each night.
Sara R124	**Incubation: +1 Poop:** Excellent nest sanitation–left the nest entirely every time she relieved herself. Eggs were ***clean!*** and much easier to candle. **Zero "@C"** reflects 10 viable embryos; **+1 @H**, that all those viable eggs hatched. **Mothering: +2 Other:** Especially strong performance as devoted and protective mother from the beginning (especially for a first-timer). Overall, an impressive addition to the broody group.
Cherry R103	**Incubation: +1 Poop:** Excellent nest sanitation. **+7 Other:** Reflects outstanding performance in two crises. Day 17, remained off nest all day following a nest break (*entirely my fault*); returned her to broody box in the evening, re-settled into broodiness immediately, set her on clutch of the Buff Orp eggs that proved under several hens to be of close to zero hatchability. Morning after candling indicated only one BO embryo viable, 22 OEG chicks arrived in the mail, earlier than hatchery had promised, and after 3 days in transit. Several chicks severely stressed, looked very much "on the way out," despite tonic drink of honey/garlic/apple cider vinegar–risked a daytime graft of the chicks. Cherry "didn't even blink"–splendid response as a more than willing mother. Next morning, I couldn't even tell which of the chicks had been so close to expiring–their survival not just a matter of food and water but of *mother love*. **Mothering: +2 MMj:** Cherry welcomed 6 chicks unable to get back to their mother after dark (problem in setup on my part), kept them warm–and alive–until next morning, when they returned to their own mother. **Conclusion:** For maintaining disciplined broodiness more than 6 weeks in the broody box, despite extraordinary challenges, and for her generosity as "mother in a pinch," Cherry is SuperMama of the season!
Avena G108	**Incubation: −2 Poop:** Poor nest sanitation, with significant dirtying of the eggs. Despite that, **+1@C:** all viable at candling and **+1 @H:** hatched 11 for 11. **Mothering: +1 MMj:** Formed unusually tight bond with her chicks from day one.
Blær R102	**Incubation: −1 Flirt:** Couple of days' "flirting" including laying of two additional eggs after move to broody box; then went full-blown broody and set like a rock. **+1 Poop:** Excellent nest sanitation. **−1 @C:** Deducted because 2 were nonviable at candle. However, note that I did set 12 eggs in this clutch rather than the usual 11. **+1 @H:** Hatched all 10 of the remaining viable eggs. **+2 Other:** Accepted a move directly onto a "live" clutch (rather than initially onto fake eggs) because of time pressure for that clutch of eggs. **Mothering: +2 Nurt:** Caught outside by unexpected downpour (1 inch in 20 min) but kept chicks *fluffy dry*. Astounding! **+2 Other:** As always, high marks for consistently being the best of all my mother hens.
Azula B148	**Incubation: −4 Panic:** Every time I serviced her nest, flew out in freaky panic. (Had accepted the initial move okay.) **Abandoned nest Apr 18. Culled April 20.**
Hugla R122	**Incubation: −1 Flirt:** For two or three days; but settled immediately when moved to broody box. **−4 Agit:** Several times while servicing, found her not fixed on the nest–just standing in a corner and staring. **Note re @C:** "Fixed" on fake eggs in preparation for fostering chicks thru the mail. **Culled Apr 30.**

Table G.2 (*continued*)

Jezabel G167	**Incubation: −1 Flirt:** Off and on several days. **−1 Panic:** Freaked out when I got her to the broody box, but then seemed to settle in. **−1 Agit:** Off the nest the day after being moved but settled by nightfall. **Note re @C:** "Fixed" on fake eggs in preparation for fostering chicks thru the mail. **Abandoned nest April 29. Culled Apr 30.**
Dapp B111	This hen was questionable from the start: Flirting quite a while; agitated after move; finally **abandoned the nest. Culled Apr 27.**
Stella R158	**Incubation: −1 Flirt:** Laid an additional egg the day after being moved. **+1 Other:** Went broody from one day to the next and settled immediately after being moved. **+4 Other:** Accepted a critical *daytime* graft of chicks that came in the mail, after only 13 days in the broody box. **Mothering:** Did a merely adequate job with the grafted chicks; probably with a clutch she had hatched herself would have performed better (as indeed she did last year, her first time as a broody). But with the challenged graft of chicks thru the mail, she didn't rise to the level of performance exhibited by Maggie B154 **re MMj and Other: −1 on each.**
Freyla R177	**Incubation: −1 Flirt:** Intermittent flirting over many days before settling. **−2 Panic:** Unusual level of fly-off-the-nest panic when approached in nest. **−1 Agit:** Restless next morning after move, but settled by the end of the day. **−4 Other:** Tried to graft 5 chicks on Freyla when the shipment arrived by mail May 9. She totally freaked out when I placed the chicks with her—despite the 14 days she'd been on the nest. Returned later and she seemed to be displaying normal aggressive protectiveness—but on the next visit, she was again settled, this time in a different part of the nest, and *not* covering the chicks. I've concluded I cannot work with this hen. **Culled May 12.**
Maggie B154	**Incubation: −1 Agit:** Restless morning after move, but settled by the end of the day. **+4 Other:** Accepted a critical *daytime* graft after only 10 days in the broody box, and later accepted the 5 chicks Freyla R177 had failed to mother. **Mothering: +2 MMj:** Drew the mail-delivered chicks from Stella in a way that was crucial, given that she had to overcome a weak bond and "confusion" from the chicks' side. **+2 Other:** Truly impressive performance for a first-time mother, especially considering the (only) 10 days on the nest before receiving adoptive chicks.
Frijola R135	**Incubation: −1 Poop:** She did poop the nest, but mostly away from the clutch, some soiling of the nest but not severe. **+1 Other:** Excellent "defend the nest" spirit on the nest (before move) and complete absence of flirting. **Mothering: +2 Other:** Unusually high level of attentive nurture.
Lady Bug B156	**Incubation: −1 Flirt:** Moderate flirting before move to broody box May 13, then settled rock-solid immediately. Fixed on fake eggs (3½ weeks) in order to foster 14 of the 28 chicks arriving by mail June 7. **+2 Other:** Accepted grafted chicks immediately. **Mothering:** For 10 days Lady Bug was a marvelously attentive mother: **+2 MMj** for "mama mojo-ing" grafted chicks from both Stella and Sami; and **+2 Other** for overall superior performance. ***Sadly, night of June 17 a blacksnake tried to swallow Lady Bug and killed her in the attempt.***
Sami G144 2nd Brood	**Incubation:** Moved to broody box about a week and a half before receipt of chicks thru the mail. Set with 14 of the 28 chicks rec'd June 7. **+2 Other:** Despite the abbreviated "hold" time in the broody box, accepted chicks immediately. **Mothering: −1 MMj:** Sami lost her chicks to Lady Bug's greater mama mojo (I had not provided a bonding period before setting up her and Lady Bug in the same enclosure in the henhouse)—not a great fault, given how unsettled was the introduction of mailed-in chicks to foster mothers. **+4 Other:** Sami pined for her lost chicks. So when Lady Bug was killed by the blacksnake, I put Sami with all the chicks who had been with Lady Bug, and she began mothering them without missing a beat.

APPENDIX H

Hatching Chicks in an Incubator

Contributed by Don Schrider

My delight in working with mother hens is such that I have never used an incubator for hatching. Yet an incubator may be the best option in your situation. If that is the case, learn from a master, my friend Don Schrider, who uses two incubators to hatch many dozens of chicks a year.

Over the centuries various methods have been successfully used to incubate the eggs of fowl. The Egyptians built what amounted to ovens, attended to by crawling inside and adjusting the temperature by feel. In America, early breeders used broody hens to hatch the eggs of nonsitting breeds of chicken. Later, kerosene and electric incubators were developed. I have even known someone who tossed his eggs into his compost pile only to find chicks springing forth three weeks later. The point is, you should understand that nature has designed the egg such that it will hatch even when conditions are not exactly perfect.

Successful incubation actually starts prior to the egg being laid: Be sure the hatching eggs you use are from healthy, vigorous stock that have been fed well. Remember that the egg is the sole nutrition for the development of the chick, as well as its food for the first three days after hatch. Commercial layer mash is hardly suitable to producing eggs with sufficient nutrition to produce vigorous chicks.

Proper handling and storage of hatching eggs is important. Think of a fertile egg as being pregnant; it is growing ever so slowly at room temperature, and rough treatment and temperature extremes will damage it. Keep hatching eggs in a location with a fairly constant humidity and temperature, ideally about 60°F (16°C), where the eggs will not be disturbed—a basement works well, as long as it is not overly damp, cold, or drafty, as does a pantry or a closet so that sunlight does not heat up the eggs during the day. Monroe Babcock, a famous commercial Leghorn breeder of the past, recommended storing the eggs in an enclosed container with little or no ventilation—the idea being to prevent environmental changes and prevent development in the fertilized eggs before it's time to place them in the incubator.

I like to store my eggs in egg cartons in a portable cooler. I also place a scrap piece of 2 × 4 under one end, changing ends daily, to rotate the eggs—this prevents the germ cell from sticking to one side of the egg and as we'll see is even more important during incubation.

Select eggs to go into the incubator with care. First, heavily soiled eggs should not be used. Remember, an incubator is an ideal environment to grow things, including bacteria—let's not introduce contaminants to this environment. You can remove small manure spots by lightly sanding just the soiled area of an egg.

You can also wash the eggs in a disinfectant bath, but consider: Doing so removes the "bloom"—a protective antibiotic coating applied when the egg is laid. If you do choose to wash the eggs, do so just before incubation; use water that is warmer than the eggs so that disinfectant and germs are not drawn into the egg as its temperature changes; and be sure to use a minimal amount of disinfectant so that you do not kill the embryo (Pine-Sol or chlorine bleach, about one teaspoon to a gallon of water; commercial eggwash solution, as directed). More environmentally friendly cleaning solutions include quaternary ammonia, ethyl alcohol, hydrogen peroxide, and peracetic acid, a mix of acetic acid (vinegar) and peroxide. Second, eggs that are odd in shape, rough in texture, too big, too small, or poorly colored should not be used. Such egg characteristics are highly heritable, so select eggs for the incubator that look like what you want to see in the egg basket when your future hens start laying.

Let's talk about incubator types. There are a variety of incubator designs: from those meant to hatch three eggs to big commercial units that will hatch tens of thousands per batch. Be on the lookout for great deals on old and used incubators. To decide what kind of incubator you need, let's think about how many chicks you want to produce and how often. If you use one incubator, you will only be able to set one batch every three weeks. A Styrofoam incubator with an egg turner installed, usually holding forty-two eggs, is fairly inexpensive but has many nooks and crannies and is thus hard to disinfect between batches. Brinsea makes an all-plastic incubator, which disassembles, and the parts can be put in the dishwasher for cleaning.

GQF Manufacturing makes a 400-egg incubator with a tray at the bottom that will allow you to hatch up to 135 chicks per week while continuing incubation of other batches in trays above. When operated in this way, however, the dust produced by the hatching chicks' down gets on the subsequent batches, and thus cleaning is hard to do without cooling the other eggs too much. I like to hatch a few very large batches of chicks, so I use a GQF incubator of this capacity, which performs that all-important function of turning the eggs twice a day, as my main incubator. I solve the GQF's cleaning problems by using as well an old Brower incubator made of redwood—I use it as the hatching incubator for the last week of incubation. This system allows me to clean the GQF only once per season and confine the mess to the Brower, which I clean between batches.

Incubator operation and location are important for success. Incubators come with instructions for opera-

Figure H.1. Don Schrider's two incubators, set up in a closet to keep the temperature steady. He incubates the eggs for two weeks in his GQF incubator (*right*) and completes the hatch in the Brower incubator (*left*), confining the mess to it and cleaning it between batches. Photo courtesy of Don Schrider.

Figure H.2. Hatch day in Don's Brower incubator. Photo courtesy of Don Schrider.

tion, including adjusting the temperature and humidity for best results. Most incubators have fans installed to move the warm air around the eggs. They usually operate at a temperature of 99°F (37°C) and humidity at 55 to 65 percent. Location of the incubator is of prime importance. Choose a location with a constant temperature and humidity, out of the sun, and where it will not be disturbed: basements, guest rooms, and closets all make good choices. I once had a friend who put his incubator in his garage but had bad results—the incubator could not keep up with the changes in temperature around it. Another friend also used his garage but built an insulated closet for the incubator. The best location I have used was a basement—humidity and temperature were constant, and hatches were excellent.

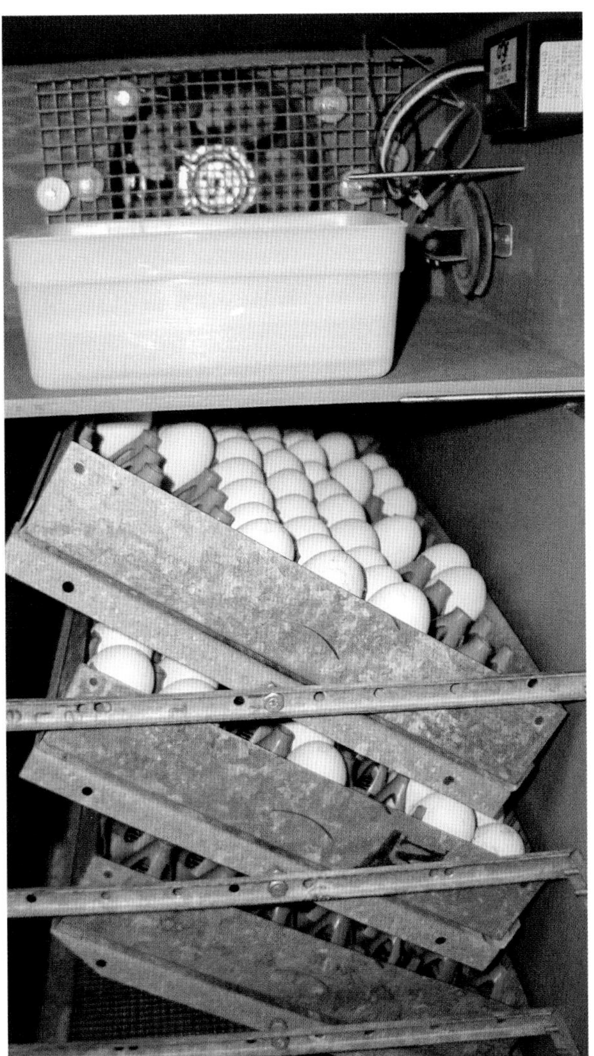

Figure H.3. Don's GQF incubator turns the eggs automatically, an essential task he would otherwise have to do by hand. Photo courtesy of Don Schrider.

Let me end by giving a number of tips I have found useful over the years:

- I use either Pine-Sol or bleach to disinfect my incubators. Add just a teaspoon of Pine-Sol to a gallon of the water used to maintain humidity, and the incubator will stay clean and fresh smelling. Bleach can also be used, at about half this rate, but it may cause galvanized parts to corrode. Use your nose! If an incubator smells bad or "off," it needs cleaning. If the disinfectant fumes burn your nose or eyes then it is way too strong for the chicks, too. (Again, more environmentally friendly cleaning solutions include quaternary ammonia, ethyl alcohol, hydrogen peroxide, and peracetic acid.)
- Save eggs for hatching for up to two weeks. You'll get best results with eggs saved for up to ten days, but 14-day-old eggs will still hatch well. After two weeks hatchability drops drastically. Eggs stored in a refrigerator can hatch, even if they do get a little too chilled, so if your family has accidentally put your hatching eggs in the refrigerator, all is not lost.
- Candle eggs at 10 to 14 days. Working at night with the lights off, I set each egg on the lens of a strong flashlight to shine its light through the egg. Remove all eggs that have stopped growing, as they will overheat the other eggs in the incubator.
- If a good egg shows a slight crack when you candle, you can still usually save it. Simply light a candle, and drip the wax over the crack. Use as little as possible so that the chick inside the egg can breathe.
- If the power goes out, wrap a blanket around the incubator to help hold in the heat. Eggs can stand a few hours with the heat off as long as they do not chill. Don't despair; they will simply hatch a little later.

- If hatching eggs have been mailed to you, let them sit at room temperature overnight for a better hatch. This allows the eggs to re-form: The egg's structures—yolk, albumen, chalazae [two spiral bands of tissue that stabilize the yolk in the center of the albumen]—have been jarred and shaken in transport and need time to "resettle." Also, the eggs gradually come up to room temperature. Giving them a "rest" this way avoids shocking the embryo at the start of incubation and yields much better results than eggs placed immediately in the incubator.
- Don't open the incubator until 24 hours after the chicks are due to start hatching. Chicks hatch with a three-day supply of food and fluid, absorbed with the last of the yolk, so they can stand the wait while their brothers and sisters hatch. But opening the door will cause a drop in humidity, with the result that chicks are unable to break the membrane holding the shell together.
- Hatching chicks is one of the wonders of having a home poultry flock. Good luck.

—DON SCHRIDER

APPENDIX I

Duck Confit, Convenience Food Extraordinaire

Recipe contributed by Ellen Ussery

Is duck confit fast food or slow food? I'd have to say it's both. It takes careful, patient preparation, but in the end you will have ready at hand the makings of a delicious and deeply satisfying meal that you can put together in no time at all.

Once you have duck confit (pronounced *con-FEE*, it means "preserved duck") in your fridge or cellar, you're ready for anything. You may find yourself tired and hungry at 6 PM and realize that you've given no thought to dinner. You can whip out some confit, heat it gently, toss a salad, add some crusty sourdough bread (or if you happen to have some leftover boiled potatoes, brown them quickly in some of the confit fat), and enjoy your dinner. After such a meal your bodily strength will be fully restored.

Any dinner guest presented with your confit will feel honored indeed, certain that such depth of flavor and silky texture could only be achieved by very hard work on the part of the cook, and that something precious is being shared.

Actually, though, making duck (or goose) confit isn't a lot of effort—you can schedule it to fit into other activities while easily producing this spectacular food.

So don't be put off by the length of these instructions. Basically there are three steps: Cure the duck pieces in salt and herbs for sixteen to eighteen hours; poach them very gently in rendered fat for three to four hours; and allow them to cool slowly to room temperature in the cooking fat.

The Ingredients

Harvey cuts up the ducks after butchering (see figure 29.2), and I freeze the wings, thighs, and legs until I have the time and inclination to make confit. When freezing the pieces, I put into each package exactly enough to fit snugly in a single layer in the cooking vessel I will use to cook them.

Rendered fat is the other main ingredient. I always make this as soon after butchering as possible (see figure 29.4). It keeps in the fridge for a long time and in the freezer at least a year. It is crucial to have enough fat to fully cover all your meat when poaching it. Sometimes your ducks will not provide enough fat. If you have geese, their fat is excellent for use in confit as well. Chicken fat is not an acceptable substitute, but good-quality lard is, especially when mixed with some duck fat. I cannot tell you how much you will need. It depends on the size of your birds, the size of your cooking vessel, and what shape container you will use to store the confit. But for a batch cooked in a 12-inch (30-cm) pan, I would have on hand 5 cups of fat at a minimum.

The only other ingredients in my confit are *salt*, *thyme*, and *garlic*. There are many recipes calling for more complicated seasonings, but we prefer to accentuate the flavor of the meat itself. I have always used coarse Celtic sea salt. Recently I ran out and used kosher salt, which is not as coarse. It seemed to penetrate the meat more completely than the coarser sea salt I was used to, so if I were to use it again, I would use slightly less. The amount again depends on the size of the container in which you will cure the meat. But for a recipe to be cooked in a 12-inch cooking pan, about a quarter cup of salt should do. I usually have fresh thyme out in the garden until late autumn, and that is what I use, but I have used dried thyme with good results. As for garlic, doesn't everybody have a guest room closet full of garlic, like we do? The more the better, I say.

The amounts that follow are what work with my 12-inch copper rondeau, pictured in figure I.2:

8–10 pieces of duck leg, thigh, and wing
¼ cup coarse Celtic sea salt
15 cloves garlic, peeled but left whole
15 sprigs fresh thyme
5 cups (or more) duck and/or goose fat
Parchment paper cut to fit atop the layer of duck pieces

Curing

Using a glass casserole dish large enough to hold all your duck pieces snugly in a single layer, scatter enough of the salt to barely cover the bottom of the pan, then half the garlic and half the thyme. Lay in the duck pieces, skin side up (see figure I.1). Sprinkle the remaining salt, garlic, and thyme on top. Cover tightly and refrigerate sixteen to eighteen hours. I usually do this around five or six in the evening, to be ready to cook the following morning.

Cooking

Rinse off all the salt from the duck pieces and dry thoroughly. Place them in your oven-proof cooking pan in one cozy layer, skin side up. A metal pan is better than glass (see figure I.2). Add about half of the garlic from the curing process. Pour in the rendered fat, which you have gently melted in a saucepan, being sure to completely cover the duck. Cut the parchment paper for a snug fit in the pan directly on top of the duck—it helps keep the duck covered in fat (see figure I.3). Place in a 200°F (93°C) preheated oven. Keep an eye on it as it heats. *What you want is for a tiny bubble or two to rise from the fat every few seconds*. You may have to raise the heat a bit, but you do *not* want the fat to reach a simmer—that is too hot. Monitor carefully until you get a steady series of bubbles. After three hours, check it by piercing with a fork. In most cases it will be tender but not falling off the bone. That is the point at which you should stop cooking. But if it still feels quite tough, then continue for up to another hour.

You can also poach the duck in a slow cooker. The advantages are that it will cook at the proper low temperature without using as much energy as an oven. Also, if you are doing this in warm weather, a slow cooker won't heat up the kitchen. In my case the disadvantage is the size of my slow cookers: Neither holds as much in a single layer as my rondeau.

Cooling and Storing

Immediately upon removing the duck from the oven, use tongs to place the pieces in the storage container. Then strain the fat into a heat-proof glass jar or measuring pitcher, working in two stages. First, stop well before any of the cloudy contents of the pan are poured out, to yield a jar of clear fat. Pour this immediately over the duck. It might not cover it completely, but it should be close (see figure I.4). Let cool gently at room temperature—a slow cooling is as important as the slow cooking for developing the flavor and texture of the meat.

Figure I.1. Place the duck pieces in a glass casserole dish with salt, garlic, and thyme.

Figure I.2. Place the cured pieces in a single snug layer in the bottom of a heavy cooking vessel.

Figure I.3. Cover with fat, then with a piece of parchment paper cut to fit.

Figure I.4. Pour in the strained duck fat.

Figure I.5. Cover completely with solidified fat. If using soon, okay to leave it in the casserole dish.

Figure I.6. You can also keep in a jar or a ceramic crock, especially for longer storage.

Now pour the rest of the fat through the strainer into your pitcher. The result will be a bottom layer of meat juices and a top layer of fat. Set the pitcher in the refrigerator for several hours until the fat solidifies. When it does, spoon it out and cover the bare spots of your duck, then store the confit container in the fridge or cellar. *Complete coverage by the fat is essential for preserving the cooked duck* (see figure I.5). The salty meat juices left in the bottom of the pitcher would sour the confit as it ages, so it's important to exclude them in this step. Use them instead for soup or sauce making.

I generally use a flat glass casserole for storing confit, since that makes it easier to remove a few pieces at a time. However, it takes more fat to fully cover the duck in a casserole dish. By using a widemouthed mason jar, you can squeeze the duck pieces more closely together, and end with a thicker layer of fat on top. It can be difficult to remove the pieces of duck from a jar without mangling them, so I set it in hot tap water till the fat softens enough to allow for removing the pieces I want. Or I might plan to use the whole jar at one time. Another option is storage in a crock (see figure I.6).

Most people store confit in the refrigerator. But traditionally in France, confit was stored all winter in a cellar with a constant cool temperature. So this is a safe choice, as long as you are certain that there are no hot-water pipes or the like that would raise the ambient temperature at any time during storage.

Though you can eat your confit as early as the next day, the flavor improves with aging in the fat. I have kept confit in the refrigerator for as long as three months. But since I don't have a usable cellar, and refrigerator space is at a premium, I find that it works best for me to eat it within a few weeks and make successive batches. An advantage of this low-heat preparation is that the duck fat has come nowhere near its 375°F (190°C) smoke point and therefore has not broken down. So it is perfectly safe to use it for a second batch of confit, and afterward for sautéing potatoes or greens and the like.

Once you have learned how to fit confit into the rhythm of your life, you won't want to be without it.

APPENDIX J

Resources

The following are some resources I have found useful regarding not only poultry husbandry but also how to support a more sustainable and regenerative agriculture.

Organizations

American Pastured Poultry Producers Association (https://apppa.org/): Consider joining the APPPA for access to a world of information on practical poultry husbandry. Founded in 1997 by Joel Salatin and other small farmers, the organization is dedicated to small farmers producing for local markets. Membership (currently $50 annually) includes a newsletter subscription and online access to archives of past newsletters and a discussion forum. Discussion is serious, respectful, and to the point—tap into a deep well of seasoned experience. Members may register to be entered into APPPA's "Find a Farmer" interactive map. Not only can the map help consumers find their way to you, but you may find other members nearby who can help steer you through the confusion and around the pitfalls of government regulations in your market.

The Livestock Conservancy (https://livestockconservancy.org): To learn more about older breeds of poultry and other livestock, join The Livestock Conservancy (formerly known as American Livestock Breeds Conservancy), which is dedicated to preserving traditional and historic breeds. TLC believes that the key to conservation of such breeds, many of which are threatened with extinction, is using them for the *economic qualities* for which they were bred—that is, for their fit in the effort to produce food for the family or local markets. TLC offers a list of rare and endangered poultry breeds and a list of conservationist breeders.

Society for the Preservation of Poultry Antiquities (www.sppa.club): The SPPA is dedicated to perpetuating and improving rare breeds of poultry. Membership in either SPPA or TLC will put you in touch with other breeders willing to supply stock. SPPA's *Breeder's Directory* is especially detailed and well organized.

Sustainable Poultry Network USA (https://spnusa.com/): SPN is a national network of breeders dedicated to "slow poultry," traditional breeds that are more resilient, forage more of their own foods, and—precisely because they grow more slowly—have superb flavor. (Yes, some SPN breeders sell traditional breeds as dressed poultry into local markets.) Many SPN members offer hatchlings for sale: chicken breeds such as Black Australorp, Wyandotte, Dorking, Buckeye, Dominique, Delaware, Plymouth Rock, and New Hampshire; Narragansett turkeys; Appleyard ducks; and more.

Books

Sustainability Issues

Read Wendell Berry! Poet, novelist, and essayist—choose the books of his that suit your particular tastes. They all offer a vision of what has gone so profoundly wrong with agriculture—with culture generally—in our time, but also of the directions we need to take for healing, of both ourselves and the Earth. Collections of essays include *The Unsettling of America* (originally published in 1977, republished in 2015 by Counterpoint Press); *Sex, Economy, Freedom & Community* (Pantheon, 1993); and *What Are People For?* (2nd edition, Counterpoint Press, 2010). *Remembering* (originally North Point Press, 1988, and now Counterpoint Press, 2008) is a wonderful novel. Do find a collection of his poems—it's *all* good.

Gene Logsdon, *Holy Shit: Managing Manure to Save Mankind* (Chelsea Green Publishing, 2010) is an informative discussion of the agricultural use of manures. I think Logsdon means both the title and subtitle literally—he forcefully, though wittily, makes the case that the contribution by livestock animals of their manures is a key to sustainable farming.

I highly recommend reading Alan Weisman's *The World Without Us* (St. Martin's Press, 2007), which imagines that *Homo sapiens* as a force in the world disappeared overnight. What would be the consequences for the further development of species and natural systems? The reader soon realizes that Weisman's thought experiment is not an idle conceit but a profound meditation on our impact in the world.

Understanding and Working with Soil

Sir Albert Howard's classics, *An Agricultural Testament* (originally published in 1940, republished in 2010 by Oxford University Press) and *The Soil and Health* (originally published in 1947, reprinted in 2006 by University Press of Kentucky, with an introduction by Wendell Berry) helped inspire the organic movement as a rejection of industrial agriculture. They seem a bit dated today but can still guide efforts to improve our soil.

Dr. Elaine Ingham has done more than perhaps any contemporary scientist to enlarge our understanding of soil life. Her *Soil Biology Primer* is an excellent brief introduction for those new to the concept of the *soil food web*. Buy a print copy ($18) from Soil and Water Conservation Society (https://www.swcs.org) or get a free pdf download at https://www.envirothonpa.org/wp-content/uploads/2014/04/7-Soil-Biology-Primer.pdf. Dr. Ingham's website is at https://www.soilfoodweb.com/.

Two discussions of the staggering complexity of soil life and fertility cycles, readily accessible to the lay reader, are *Life in the Soil* by James Nardi (University of Chicago Press, 2007) and *Teaming with Microbes* by Jeff Lowenfels and Wayne Lewis (Timber Press, 2010).

Building Soils for Better Crops: Ecological Management for Healthy Soils by Fred Magdoff and Harold van Es (4th edition, SARE, 2021) is a useful overview of soil ecology and soil care. Buy a print copy ($23) or download a free pdf version at https://www.sare.org/resources/building-soils-for-better-crops/.

If vermicomposting for bioconversion of organic "wastes" is new to you, check out *Worms Eat My Garbage, 35th Anniversary Edition* by Mary Appelhof (annotated edition, Storey, 2017). A classic in the field, with information on worm biology and setup for using earthworms to recycle kitchen "wastes." The basics are quite simple, though—you could easily find all the information you need through an online search engine. Start small, and work up to vermicomposting on any scale you like.

The best source I know of for information on vermicomposting at farm or industrial scale is Rhonda Sherman's *The Worm Farmer's Handbook* (Chelsea Green, 2018). See also George Sheffield Oliver's, "My Grandfather's Earthworm Farm," available online in its entirety at https://www.journeytoforever.org/farm_library/oliver/oliver_farm.html—and as chapter 6 of Thomas J. Barrett's *Harnessing the Earthworm* (Bookworm Publishing, 1976). Also see the "Manual of On-Farm Vermicomposting and Vermiculture," by Glenn Munroe, Organic Agriculture Centre of Canada, available in pdf form at https://iweco.org/sites/default/files/2019-03/GEF-IWCAM_Manual_Vermiculture_2012.pdf.

Poultry

Page Smith and Charles Daniel's delightful *The Chicken Book* (North Point Press, 1982) contains practical information about chickens as a backyard enterprise, to be sure, but is worth reading mostly because of its insights into the role of the chicken, and chicken lore, throughout the ages.

Judy Pangman's *Chicken Coops: 45 Building Plans for Housing Your Flock* (Storey Publishing, 2006) has blueprints and lots of ideas for coops for many flock sizes and management situations.

Fresh-Air Poultry Houses: The Classic Guide to Open-Front Chicken Coops for Healthier Poultry (Norton Creek Press, 2008) is a republication of Prince T. Woods's *Modern Fresh-Air Poultry Houses*, originally published in 1924. Woods tells of one design after another of open-front poultry housing in even northerly climes that improved results over tighter housing by increasing airflow through the coop.

Gail Damerow's *Chicken Health Handbook, 2nd Edition* (Storey Publishing, 2015) is the most complete treatment of its subject you will find, aside from highly specialized scientific works.

Dave Holderread's *Raising the Home Duck Flock* and *The Book of Geese* (Garden Way, 1978 and Hen House Publications, 1981) are excellent introductions if you are interested in raising waterfowl.

Chris Ashton's *Domestic Geese* (Crowood Press, 1999) is another useful guide to keeping geese.

Carol Deppe's *The Resilient Gardener: Food Production and Self-Reliance in Uncertain Times* (Chelsea Green Publishing, 2010) is about more than poultry, but I especially recommend her thoughts on using an all-purpose duck flock to fit into the total gardening project.

Andy Lee and Pat Foreman's *Chicken Tractor: The Gardener's Guide to Happy Hens and Healthy Soil* (Good Earth Publications, 1994) has excellent ideas for putting chickens to work in—but without trashing—the garden, using small mobile shelters the width of a garden bed.

If you are contemplating an egg or poultry marketing venture, read Joel Salatin's *Pastured Poultry Profit$* (Polyface, Inc., 1993). Read as well his *Everything I Want to Do Is Illegal* (Polyface Press, 2007). You will be amused by Joel's trademark humor, but boiling mad as you realize that many of the regulations for the sale of food products have more to do with favoring the profits of gigantic food corporations than the safety of what we put into our mouths.

J. Russell Smith's *Tree Crops: A Permanent Agriculture* (Devin-Adair Company, 1950) is not a poultry book, but I list it because it suggests foods that ranging flocks can forage from trees—acorns, beechnuts, mulberries, and more.

Other Homesteading Topics

Rolfe Cobleigh's *Handy Farm Devices and How to Make Them* (first published 1909 by Orange Judd Company, republished 1996 by The Lyons Press) has useful ideas for things you can make yourself for success on the homestead and small farm.

Dave Jacke's *Edible Forest Gardens* in two volumes, *Vision & Theory* and *Design & Practice* (Chelsea Green Publishing, 2005), is an introduction to a fascinating subject, the forest garden, which will suggest many possibilities to the creative flockster. The extensive appendices alone are worth the price of the books.

Equipment/Accessories

The **dust mask** I have used by preference for decades is the Respro Sportsta, available from many online suppliers. It is both comfortable enough to wear for long periods at a stretch and highly effective. I always wear it for the killing and plucking phase of butchering (when a good deal of poultry dander gets kicked into the air), but frequently for other dusty work as well.

If you decide to make your own feeds, you will need a **feed mill**. Unfortunately, the one I used to make many a ton of feed for more than two decades (shown in figure 17.1) is apparently no longer available. The closest equivalent I could find through online searches looks to be the Bravo Feed Grinder offered by Premier (https://www.premier1supplies.com), a supplier in which I have the highest confidence. The Bravo has a built-in motor that plugs into household current, and a collection bucket lidded against release of milling dust.

Electric net fencing is a fundamental tool that, for most of my poultry-husbandry years, was the key to management and protection of my flocks. My preferred source is, again, Premier: https://www.premier1supplies.com/. Everyone on the staff at Premier uses electric fencing to manage livestock, so products sent their way get rigorous real-world testing, and technical advice is well grounded in experience. Friends whose judgment I value also recommend Kencove: https://kencove.com/fence.

If you butcher a lot of birds, or if you plan to serve a broiler or turkey market, it might pay to tool up with a **mechanical plucker** and a **scalder** at a minimum. If you want to make your own, Herrick Kimball offers *Anyone Can Build a Whizbang Chicken Scalder* and *Anyone Can Build a Tub-Style Mechanical Chicken Plucker* at https://www.planetwhizbang.com/poultry-processing.

If you prefer to buy assembled, high-quality equipment, check out David Schafer's Featherman Equipment Company at https://www.feathermanequipment.com, which offers scalders, pluckers, and processing accessories.

The **caponizing kit** I used, from Nasco, is no longer available. At the time of this writing, Amazon offers four similar kits, ranging in price from $30 to $43. A simple instruction manual, originally written for Sears, Roebuck & Co. in 1922, is available online at www.afn.org/~poultry/capon.htm.

The **candler** I recommend is the Brinsea OvaView High Intensity egg candler, available from the manufacturer's website, https://www.brinsea.com.

The **wing band applicator** I use is the Jiffy 893, purchased (along with numbered wing bands colored to fit my three-clan mating system) from: https://www.nationalband.com/jiffy-893.

Online

I maintain a website at https://themodernhomestead.us with a couple hundred pages about gardening, soil care, all things poultry, and other homesteading topics. I hope you will find it useful from time to time. Free downloads mentioned in this book can be found at: https://themodernhomestead.us/resources/downloads.

"Pastured-Raised Poultry Nutrition," by Jeff Mattocks of Fertrell Company, originally written for Heifer International in 2002, is available as a free download in pdf format at https://attra.ncat.org/wp-content/uploads12019/05/chnutritionhpinew.pdf

See also *"Pastured Poultry Nutrition and Forages"* by Terrell Spencer, issued by National Center for Appropriate Technology in 2013, available as a downloadable pdf at https://www.sare.org/resources/pastured-poultry-nutrition-and-forages.

An excellent online introduction to the astounding diversity of breeds is http://feathersite.com, complete with thousands of photographs. It has lots of information and links regarding all things poultry. This is a site you will come back to again and again.

If my description of Icelandics as a do-it-all breed appeals to you, the best place for information and contact with breeders is: https://www.facebook.com/groups/icelandicchickens.

Journey to Forever (https://www.journeytoforever.org) has interesting information for the flockster, homesteader, and small farmer, including an online library that contains whole books free for the reading, including some of those listed above.

Aside from the mega-hatcheries, Glenn Drowns's Sand Hill Preservation Center (SHPC) offers the largest selection of traditional and historic poultry breeds on the continent—dozens upon dozens of breeds of chickens (both standard and bantam), geese, ducks, turkeys, guineas, even quail. Check out https://www.sandhillpreservation.com/poultry and browse. (SHPC also offers seeds for dozens of traditional vegetable, herb, and flower varieties—and more. Hmm, wonder what he does in his leisure time?)

Another excellent collection of free books available online is the *Soil and Health Library* at https://soilandhealth.org.

Glossary

bantam: A miniature breed of chickens. A few bantam breeds such as Sebright are naturally small. Most bantam breeds have been miniaturized by crossing in the bantam gene and selecting for it.

beak: The bony structure on the head of a bird, consisting of an upper and a lower jaw, which it uses for eating, catching and killing prey, manipulating objects, feeding young, and preening.

bile: An intensely green, bitter-tasting liquid produced in the liver, and essential to digestion of fats in the bird's food and absorption of fat-soluble vitamins.

breed: A group of domestic fowl with the same conformation, carriage, size, and (usually but not quite always) plumage pattern and colors—results of consistent selection for the same traits.

broiler: A table chicken slaughtered young, tender enough for dry-heat cooking methods such as frying, broiling, grilling, and baking.

broody, broodiness: Refer to the instinctive behaviors associated with incubating and hatching eggs and nurturing chicks. *Broody* can be used as an adjective (a broody hen) or a noun (a subflock of good broodies). A hen is said to "go broody" when entering this special state of mind, "broodiness."

cannibalism: Injurious pecking of one another among chicks or chickens, brought on by stress or inadequate protein in the diet. Well-managed flocks are unlikely to exhibit this behavior.

castings: Excreta of earthworms, an excellent soil fertility amendment.

CDX plywood: Construction grade plywood designed for exterior use (the "X" in its name).

ceca (singular, cecum): Long blind pouches attached to where the large and small intestine of a chicken join, serving various functions in the digestive process.

chook: An imported word from Australia and New Zealand, literally meaning "chicken," often used by flocksters playfully or affectionately.

cloaca: The stretchy membranous pouch at the end of a bird's digestive tract, in which the feces collect before being expelled. The reproductive tract also exits through this area, for expulsion of the egg in the case of the hen, and of semen in the case of the cock.

coccidiosis: A disease in which chickens suffer from overwhelming numbers of coccidia, up to nine species of microscopic parasitic protozoans. Chickens with healthy immune systems develop immunity through moderate exposure to coccidia.

cock: The adult male chicken.

cockerel: A male chicken up to one year of age.

coir: The granular residue from industrial extraction of fibers from coconut husks. Often used as a replacement for peat in garden applications. Usually sold as compressed briquettes or blocks, which expand greatly when soaked in water. After being "uncompressed" by hydrating, it can be dried again and then used to control excess moisture in vermibins.

comb: The fleshy appendage on top of a chicken's head, varying in shape and size by breed, larger in the cock. Radiates heat to help cool the body and probably, in the case of the male, serves as a sexual attractant.

confit: Meat, such as duck or goose, that has been cooked and preserved in its own fat.

creep feeder: Any structure that—through use of slats or mesh size—gives access for feeding to younger and therefore smaller livestock or poultry, while physically excluding adults.

crop: A stretchy membranous pouch, an enlargement of the esophagus, just to the right of the base of a chicken's neck. A storage pouch into which to put a lot of food in a hurry, and in which a start is made on digestion by softening the contents with digestive fluids.

CSA: Community Supported Agriculture, an arrangement in which consumers pay for a season's subscription to a farm's produce up front and receive in return a weekly "basket" of produce, the exact makeup of which changes as different crops are ready for harvest. CSAs in the past have typically furnished vegetable and fruit produce only, but many now offer eggs and even fresh-dressed broilers as well.

debeak, debeaking: To debeak a chick is to clip off part of the upper beak to prevent cannibalism. Such alteration should never be needed in a well-managed small-scale flock.

detritivore: An organism—such as earthworm, pill bug, cricket, or fungus—that feeds on dead organic matter in the process of decomposition.

diatomaceous earth: A powder consisting of siliceous remains of diatoms, ancient one-celled algae that formed shells from silica. Particles, sharp and cutting at the microscopic level, have a lethal effect on insects. For this reason diatomaceous earth is an effective non-toxic insecticide in the feed bin or dustbox, or as an emergency treatment for infestation by lice or mites.

drake: An adult male duck.

duck: Generically, a duck is a waterfowl, *Anas platyrhynchos*, descended from the wild mallard. If the reference is to gender, the *duck* is the adult female (as opposed to the male *drake*).

duckling: Generically, the young of ducks, up to one year of age. If it is necessary to distinguish gender at this stage, a duckling is female, a drakeling male.

duodenal loop: The upper part of the small intestine, where digestive fluids are added to the ground and mixed food passed into it from the gizzard.

dust bath, dust-bathing: Like some wild birds, chickens and other gallinaceous fowl make dusty hollows in soil and fluff dust up under their feathers to kill external parasites. As well, dust-bathing is an activity they clearly enjoy.

dustbox: A box the flockster builds with dusty materials inside, to ensure adequate access to dust-bathing. Alternatively, a bit of ground outside with a cover to keep it dry for dust-bathing even in rainy weather.

eggmobile: A movable shelter for a flock of pastured layers, complete with roosts and nestboxes.

electric net fencing: A lightweight mesh fence for confining and protecting pastured flocks within a defined perimeter, made of plastic strands interwoven with fine stainless-steel wires that carry an electric charge. Also called electric netting.

flockster: A word created by the author since no exact one existed: A person who keeps a flock of domesticated fowl, usually based on the conviction that they have a lot to offer as working partners for a more self-sufficient homestead or small farm.

Freedom Rangers: A marketing term referring not to a specific breed but to a type of ranging broiler bred for pastured production in the French *Label Rouge* certification system. In contrast to the Cornish Cross, which is the preferred broiler of the poultry industry (and of many small-farm pastured producers as well), the Freedom Ranger is colored rather than white, is more robust, takes better advantage of foraged foods, and has a longer grow-out by ten days to two weeks (with a payoff of better flavor).

gallbladder: A small pouch attached to the liver, the size and shape of a caterpillar, which stores and releases bile.

Galliformes: The avian order to which domesticated chickens belong (along with turkeys, guineas, pheasants, quail).

game: Refers to breeds of chickens the cocks of which are used in the "sport" of cockfighting, illegal in most jurisdictions. Game breeds have been used in the development of many modern utilitarian breeds.

gander: An adult male goose.

gizzard: Sometimes referred to as the "mechanical stomach." Made up of two sets of powerful muscles around a pouch containing food that has been passed into it by the proventriculus, plus grit, small bits of stone swallowed by the bird. The working of these muscles, together with the grinding action of the grit, pulverizes and mixes the food and prepares it for further digestion.

goose: Used generically, refers to a domesticated waterfowl descended from wild geese. When the context has to do with gender, *goose* means the adult female (as opposed to *gander*, the male).

gosling: Generically, the young of geese, up to one year of age. If it is necessary to distinguish gender at this stage, I suppose you could use "ganderling" for the male and "gosling" for the female, though such usage is not supported by any dictionary I know of.

grit: Small bits of stone picked up by fowl for furnishing the gizzard with a grinding agent. Crushed granite grit is readily available to purchase as a supplement, in sizes appropriate to the age and species of fowl being fed.

guinea: A galliform fowl, *Numida meleagris*, long domesticated, originally from Africa.

hackle: The rear and side neck plumage of a chicken or other fowl.

hardware cloth: Welded wire screening useful for making protective barriers in the henhouse and self-cleaning nest bottoms. Available in different mesh sizes, ½-inch and ¼-inch being the most common.

hen: An adult female chicken.

hock: On a chicken or other galliform fowl, the joint between the lower thigh (what we call the "drumstick" at the table) and the shank.

hoophouse: A shelter, either permanent or temporary, made of plastic sheeting, opaque or transparent, over an arched frame, usually of metal or plastic pipe.

humus: The final carbonaceous residue of organic matter added to soil. Enhances water retention, assists uptake of nutrients by plant roots, and in other ways improves soil quality.

keet: The young of guineas.

Marek's disease: A highly contagious disease in chickens, involving six different herpesviruses that affect especially the nerves but may affect visceral organs, muscle, and skin as well. Though the associated viruses are almost universally present where chickens are raised, well-cared-for flocks with robust immune systems are unlikely to succumb.

Muscovy: A domesticated waterfowl, sometimes called "duck," but with a different wild ancestor, *Cairina moschata*, from that of true ducks.

nestbox: An enclosure provided for laying fowl to retire for egg laying.

nest-trap: To use a trapnest to track egg production of individual hens.

ovary: In avian species the female reproductive organ that secretes hormones related to sexual function and produces ova, which start as single female sex cells and develop into yolks.

oviduct: The long, convoluted tube in a female bird in which the egg is made and prepared for expulsion.

pancreas: A glandular organ that produces digestive enzymes involved mostly in digestion of proteins.

pasting up, pasty butt: A condition in which viscous feces expelled by a chick stick to the down around the vent, occluding it. Can be caused by chilly conditions in the brooder or poor feed. Unlikely if brooder is well managed and natural feeds are given from the first day.

pinion: The third and final segment of the wing.

pipping: The breaking open of the eggshell from the inside by the chick, allowing it to hatch.

poult: The young of turkeys.

preening: Care of its feathers by a chicken or other fowl. The bird expresses oil from its preening or uropygial gland with its beak and applies the oil to its feathers to clean, maintain, and partially or fully waterproof them.

primaries: The long, stiff outer feathers of the wing, growing from the pinion, also known as the flight feathers.

proventriculus: In avian species the glandular stomach, between the end of the esophagus and the gizzard, in which hydrochloric acid and digestive enzymes are added to the food the bird has ingested and in which digestion begins.

pullet: A female chicken up to one year old.

rendering: Gently heating body fat from slaughtered ducks, geese, or chickens until it liquefies and can be saved as a cooking fat.

rooster: A euphemism for the cock (male chicken), as silly as it is prudish—female chickens *roost* as avidly as males!

saddle: The rear of the back in the cock, extending to the juncture of the back and the tail, covered with the long, pointed saddle feathers that are a distinctive part of his plumage.

set, setting: Traditionally, a hen who has "gone broody" is not said to "sit" on her clutch of eggs but to "set." She is referred to as a "setting hen" not a "sitting hen." That is the usage followed in this book. We can as well speak of "setting eggs" under a broody hen for her to incubate.

sexing: The separation of hatchery chicks by gender, making possible the shipment of all-pullet or all-cockerel orders (as opposed to straight-run orders). *Sexing* also refers to the visual inspection of the genitalia of waterfowl (ducks and geese) to determine gender.

shank: The part of the leg between the hock and the foot.

sickles: The long, curved tail feathers of the cock, a distinctive part of his plumage.

soil food web: The complex set of living organisms, visible and microscopic, in the top layers of the soil profile, which break down organic residues on or in the soil and make them available as plant nutrients.

soldier grubs: The larval stage of the black soldier fly, which can be cultivated in a bin using organic "wastes" and given as high-protein feed to poultry.

spleen: A small organ beside the liver with important functions in regulating red blood cell supply and the immune system.

spur: A horny projection from a cock's shank, above the rear toe, short and blunt in most farm breeds, long and sharp in games. Used by cocks as a weapon when fighting to assert dominance. Occasionally found in game hens.

straight run: In the natural gender ratio for chickens, which, as in humans, is approximately half and half. Hatchery chicks can be sexed (separated by gender) to furnish orders of all pullets or all cockerels. Alternatively, the chicks can be sent straight run.

trapnest: A special nestbox for tracking egg production, with a door that drops or swings into place when the hen enters, trapping her until released by the flockster. (You may see "trapnest" used as a verb, as in: "Trapnest your hens when selecting breeders." That usage is decidedly odd to my ear—I use "nest-trap" instead: "Nest-trap your hens when selecting breeders.")

uropygial gland: See *preening*.

vacuum-seal waterer: A waterer that automatically replenishes as the birds drink from it, consisting of a reservoir of water over a base with a narrow trough to prevent wading in the water. A hole in the base allows water to run out of the reservoir. When the hole is covered by the rising water level, a vacuum forms inside the reservoir, preventing additional flow until the hole is once again exposed as the birds drink. Metal and plastic versions are available, from a quart up to 7 gallons or more. You can make your own, so long as the placing of the hole in the base allows for the proper vacuum/release cycle.

vent: The anus of a fowl, through which feces—and in the case of a female, eggs—are expelled.

vermibin: Any of various designs of bins for cultivating earthworms.

vermicast: Another term for earthworm castings.

vermicomposting: The composting of organic "wastes" using earthworms, yielding castings (earthworm poop) as a soil fertility amendment, and potentially a harvest of worms to feed poultry, pigs, or farmed fish. Also called *vermiculture*.

wattle: The fleshy appendage hanging down from the upper throat of a chicken, varying in size by breed and gender, the male's being significantly larger. Like the comb, the wattle helps radiate heat to cool the body.

Notes

Chapter 1: The Abundant Ecology

1. For comparison: 4–6 percent is the percentage of organic matter content recommended for vegetable and flower gardens, with 3–5 percent being the recommended minimum. Agricultural soils considered our most productive average 3–6 percent, with highly fertile soils in the Northern Great Plains ranging from 4–7 percent. Silt loam and clay loam soils are characterized as high in organic matter at 4–5 percent and at greater than 5 percent as very high.

Chapter 2: The *Integrated* Small-Scale Flock

1. Sir Albert Howard, *The Soil and Health: A Study of Organic Agriculture* (1947; University Press of Kentucky, 2006, introduction by Wendell Berry), 11.

Chapter 3: Your Basic Bird

1. See the discussion by Page Smith and Charles Daniel on the origin of *rooster* and the symbolic role of the cock in various periods and cultures, in their delightful *The Chicken Book* (North Point Press, 1982), 51ff. H. L. Mencken discussed the euphemizing compulsion in his *The American Language: An Inquiry into the Development of English in the United States*, 2nd ed. (Alfred A. Knopf, 1921), 149, and elsewhere. Even words with the slightest hint of sexuality were subject to censorship—a bull, for example, being referred to as a *gentleman cow*!
2. See, for example, Nicholas E. Collias, "The Vocal Repertoire of the Red Junglefowl: A Spectrographic Classification and the Code of Communication," *The Condor* 89, 510–24, https://doi.org/10.2307/1368641.
3. Temple Grandin and Catherine Johnson, in "Rapist Roosters," *Animals in Translation: The Woman Who Thinks Like a Cow* (Bloomsbury, 2005), pages 70ff.

Chapter 4: Planning Your Flock

1. You will often see references to this breed as "Araucanas." Do not be deceived. The true Araucana is a small rumpless breed (lacking the protuberance at the end of the spine that supports the tail feathers of most breeds) originating in Chile (bred by the Araucana tribe there). The novelty of the Araucana's pastel-tinted eggs was its main appeal when it was introduced to the United States in the 1930s. Unfortunately, the gene for rumplessness in this breed carries a lethality factor: When the mating cock and hen share the gene, one-quarter of the resulting embryos die in the shell. Since the gene for the color of the eggshells is dominant, it was easy to retain it in various crosses between the true Araucana and other breeds. These crosses were standardized in the 1970s into the Ameraucana—larger than the Araucana and consistently having a rump, making it easier to breed (because of the absence of the lethality factor). True Araucanas are extremely rare and difficult to find, but the misnomer *Araucana* for *Ameraucana* is extremely common. I do not know of any commercial hatchery that offers true Araucanas. I carefully questioned a representative of one hatchery advertising them who insisted that their stock really was the true Araucana—even referring to the distinctive Araucana characteristics: ear tufts and rumplessness. The chicks I gullibly ordered turned out to have muffs and beards (in lieu of ear tufts)

and rumps—that is, they were in fact Ameraucanas. If you want to raise true Araucanas, breeders are listed in the *Breeders Directory* of the Society for the Preservation of Poultry Antiquities (see appendix J). In practical terms, Ameraucanas are likely to be a better fit in the productive home flock, but let's refer to them with their proper name, shall we?

2. You can find the chart at "Henderson's Handy Dandy Chicken Chart," *Sage Hen Farm*, last updated March 10, 2022, John Henderson, http://www.sagehenfarmlodi.com/chooks/chooks.html. Look for the snowflake icon in the "Egg color and productivity" column indicating breeds more likely to lay better in winter.
3. To help in your search, check www.feathersite.com for the most complete set of pictures I know of hundreds of breeds of all sorts of poultry—not only chickens but ducks, geese, turkeys, pheasants, peafowl, swans, and more. Remember the Henderson chart at www.sagehenfarmlodi.com/chooks/chooks.html for a compendium of characteristics of dozens of chicken breeds.
4. The catalog for Glenn Drowns's Sand Hill Preservation Center has a wealth of information about traditional breeds of chickens, ducks, guineas, geese, and turkeys (https://www.sandhillpreservation.com/poultry). Other extensive lists of poultry breeds, their history, characteristics, and productive uses, are at https://www.poultrypages.com/poultry-breeds.html and www.ansi.okstate.edu/breeds/poultry. The latter has many excellent color photographs and illustrations.

Chapter 6: Starting Your Flock

1. The two quotations were copied from, respectively, https://www.mcmurrayhatchery.com/vaccinate_for_marek_disease.html and https://www.idealpoultry.com/order_policies.
2. Quoted from D. C. Kennard and V. D. Chamberlin, "Built-Up Floor Litter Sanitation and Nutrition." The entire article, first published in *Ohio Farm and Home Research Bulletin* 34, no. 261, (1949), 162–66, appears within the body of an excellent discussion of deep litter at Robert Plamondon, "FAQ: Deep Litter in Chicken Coops," *Robert Plamondon's Rural life* (blog), September 27, 2016, https://www.plamondon.com/wp/deep-litter-chicken-coops/.

Chapter 8: Manure Management in the Poultry House

1. Kennard and Chamberlin, "Built-Up Floor Litter."
2. Joel Salatin discusses a "loose housing option" for a layer flock in *Pastured Poultry Profit$* (Polyface, Inc., 1993), 259–61, in which he recommends a "target minimum space allotment" of 5 square feet (.5 square m) per hen. His observations about stocking density in a flock on deep litter in relation to capping are from several of his presentations I have attended.

Chapter 10: Ranging the Flock

1. Joel Salatin has inspired countless producers of pastured broilers with his now famous 10- by 12-foot (3- by 3.5-m)mobile pens, the design and management of which are described in *Pastured Poultry Profit$*, 63.

Chapter 11: Ranging Flocks Using Electric Net Fencing

1. My preferred source for electric fencing and accessories is Premier: https://www.premier1supplies.com.

Chapter 12: Mobile Shelters

1. You should be able to find 24-mil poly sheeting at any greenhouse supply—in 100-foot (30-m) rolls. The only supplier I know of that will custom-cut to your order is Northern Greenhouse Sales (https://www.northerngreenhouse.com/).

Chapter 13: Putting Your Flock to Work

1. Wendell Berry, "Conservation and Local Economy," in *Sex, Economy, Freedom & Community* (Pantheon Books, 1993), 13–14.
2. Gene Logsdon, *Holy Shit: Managing Manure to Save Mankind* (Chelsea Green Publishing, 2010).
3. Sir Albert Howard, *An Agricultural Testament* (1940; Oxford City Press, 2010), 47. There is a

wealth of information as well from the Rodale Institute; visit https://rodaleinstitute.org.

4. A good introduction is "Using Weeder Geese," Metzer Farms (one of the major waterfowl sources in the United States), last accessed April 30, 2022, https://www.metzerfarms.com/UsingWeederGeese.cfm.
5. Logsdon, *Holy Shit*, 106, 112.

Chapter 15: Thoughts on Feeding

1. An excellent compendium about feeding chickens naturally: Jeff Mattocks "Pastured-Raised Poultry Nutrition," (Heifer International, 2002), available as a free pdf download at https://ucanr.edu/sites/placernevadasmallfarms/files/102993.pdf. See also Terrell Spencer, "Pastured Poultry Nutrition and Forages" National Center for Appropriate Technology, 2013, https://www.sare.org/wp-content/uploads/Pastured-Poultry-Nutrition-and-Forages.pdf.

Chapter 16: Purchased Feeds

1. Mattocks, "Pastured-Raised Poultry Nutrition," 9–10.

Chapter 17: Making Your Own Feeds

1. I bought the feed mill pictured in figure 17.1 from Lehman's and it served me well. At the time of this writing it is no longer offered by Lehman's—indeed, I could not find a source anywhere. The closest equivalent for serious feed milling at the small end of the scale may be the Bravo Feed Grinder offered by Premier (https://www.premier1supplies.com/p/bravo-feed-grinder), complete with built-in motor that plugs into household current and a collection bucket tightly lidded against release of milling dust.
2. I recommend the Brecknell ElectroSamson (https://www.brecknellscales.com/products/hanging-scales/electrosamson-hanging-scale.html), which can be held either by hand or an overhead hook, three models with capacities from 20 to 100 pounds, digital readout, easy to set a tare.
3. Check out New Country Organics (NCO) at https://www.newcountryorganics.com. NCO mills its own feeds at two locations (Virginia and Texas) and supplies a network of dealers throughout the United States.
4. Fertrell (https://www.fertrell.com) offers organic fertilizers and livestock supplements. Jeff Mattocks, its president, is a gold mine of information on poultry feeding and health.

Chapter 18: Feeding Your Flock from Home Resources

1. There is confusion about comfrey types and terminology. Since there are indeed some comfrey (*Symphytum*) species that can become problematic, it is important to find one developed for *garden use*, one that will not spread by seeds or runners—which is to say, one of the *Bocking clones*. Though twenty-one Bocking strains were developed, I have never seen any offered for sale other than Bocking #4 and #14. If your source offers both, study descriptions when making your choice. If only one is offered, in my experience there is little practical difference for the uses I am describing. Note that many official sources advise against feeding comfrey to livestock. Though used for centuries as fodder plant and even as human food and medicine, comfrey does contain pyrrolizidine alkaloids. Lab rats fed massive chemically pure doses of these alkaloids have developed liver disease. You will have to do your own research and draw your own conclusions. My conclusion is that there is no problem using whole comfrey leaf as fodder for livestock. See Lawrence D. Hills *Comfrey Report: The Story of the World's Fastest Protein Builder*, (Henry Doubleday Research Association, 1975). See also the discussion of the subject in an appendix to Lawrence Hills *Comfrey: Fodder, Food & Remedy* (University Books, 1976), 229–37.
2. Carol Deppe, *The Resilient Gardener: Food Production and Self-Reliance in Uncertain Times* (Chelsea Green Publishing, 2010), 188.
3. J. Russell Smith, *Tree Crops: A Permanent Agriculture* (Devin-Adair Company, 1950), 186.
4. A good overview is R. A. Leng, J. H. Stambolie, and R. Bell, "Duckweed—A Potential High-Protein Feed Resource for Domestic Animals and Fish,"

Livestock Research for Rural Development 7, no. 1 (1995): http://www.lrrd.org/lrrd7/1/3.htm.

Chapter 19: Cultivating Recomposers for Poultry Feed

1. See George Sheffield Oliver, "Lesson 6: Putting the Bluebottle Fly to Work," *Friend Earthworm: Practical Application of a Lifetime Study of Habits of the Most Important Animal in the World*, last accessed April 30, 2022, https://journeytoforever.org/farm_library/oliver/oliver2b.html.
2. Do check out the plethora of design possibilities to be found online before settling on one to try. Here's an ingenious one presented in a YouTube video: Morelia Australia, "Black Soldier Fly Larvae Container (BSFL Large System)," YouTube video, 2:37, https://tinyurl.com/7s3rhxv7. It is based on a heavy-duty storage tub and 1½-inch PVC pipe and elbow fittings. (No gluing required for any of the fittings.) My guess is the 32-gallon tub used would hold about 20 gallons of feeding medium—that is, twice the capacity in either of two premade grub bins currently in the market, both of which cost over $150. This design is typical of many homemade bin plans that use PVC pipe to "ramp" the crawl-off of the grubs.
3. Large-scale operations support breeding populations of soldier flies in enclosed spaces such as greenhouses, heated artificially as needed. However, providing for sexual activity of winged adults in such spaces is likely beyond the scale of a homestead cultivation project.
4. Mary Appelhof, *Worms Eat My Garbage* (Flower Press, 1982, revised 1997).
5. The question always comes up whether use of newspaper in vermicomposting is safe. Modern newspaper inks—even colored inks—are based on vegetable oil and have no toxins. I do not use the slick, coated-paper parts of a newspaper for bedding.
6. I have placed several orders in recent years at Uncle Jim's Worm Farm (https://unclejimswormfarm.com) and via Bentley Christie's site (https://www.redwormcomposting.com/buy-composting-worms), with equally good service and healthy worms from both. Bentley is a fanatic "worm guy," and his Red Worm Composting blog is chock-full of information about cultivating worms.
7. Bentley Christie did this calculation based on optimal conditions, a temperature of 77°F (25°C) and ample food supply ("Willa Red Worm Population Double in 3 Months?" Red Worm Composting, last accessed April 30, 2022, https://www.redwormcomposting.com/general-questions/will-a-red-worm-population-double-in-3-months): 100 adult worms producing 3 cocoons per week which after a 21-day incubation hatch 3 baby worms each, who at 42 days are sexually mature and start producing more cocoons. With those numbers driving the math, at the end of 12 weeks there could be: 2,800 adults, 5,400 juveniles, and 9,000 cocoons. Did somebody say "Be fruitful and multiply"?
8. Rhonda Sherman, *The Worm Farmer's Handbook: Mid- to Large-Scale Vermicomposting for Farms, Businesses, Municipalities, Schools, and Institutions* (Chelsea Green, 2018). An excellent online resource: Glenn Munroe, "Manual of On-Farm Vermicomposting and Vermiculture," Organic Agriculture Centre of Canada, available at https://lifeboat.com/InfoPreserver/images/a/a1/Vermiculture_farmersmanual_gm.pdf.
9. George Sheffield Oliver, "My Grandfather's Earthworm Farm," http://www.journeytoforever.org/farm_library/oliver/oliver_farm.html. It is also chapter 6 of Thomas J. Barrett's *Harnessing the Earthworm* (Bookworm Publishing, 1976).

Chapter 20: One Big Happy Family

1. My friend Don Schrider notes that in his experience it is more likely to be a toenail than a spur that injures the hen. The problem may emerge more in some breeds than in others: When he was working with Buckeyes, Don trimmed the toenails as well as the spurs on the cocks to prevent injury. In his current work with Light Brown Leghorns and Dark

Brown Leghorns, he trims spurs and toenails if he notices the backs of hens getting bald.

Chapter 21: Protecting Your Flock from Predators

1. If you ever need to foil a fence-flying fox and you get your supplies from Premier, talk to one of the old-timers there. One of them gave me hints for outwitting a fence-jumper: baiting the charged wires with bacon, installing the fencing with an outward slant, and installing extra-long posts for additional runs of netting or closely spaced lines above the level of the charged net.
2. There is one exception to my dismissal of the "high-tech gizmos" on the market for stopping predators. According to reliable reports I have received, the solar-powered Nite Guard Solar ($25 per unit) really does effectively deter nocturnal predators. See https://www.niteguard.com.
3. See a brief APPPA video, "Pastured Poultry Ecosystems Coexist with Wildlife," 4:11, August 13, 2022, https://www.apppa.org/ecosystems.

Chapter 22: Helping Your Flock Stay Healthy

1. I should add a major caveat: I have had dreadful luck trying to brood turkey poults (see "Turkeys" in chapter 24, page 229), though a few started turkeys who had "graduated" from my buddy Mike Focazio's brooder have been easy for me to grow to superstar status on the Thanksgiving table. Turkey poults are pretty fragile, and it's common to lose distressing numbers of them with little clear indication why they're dropping like flies. Then, past the brooder phase, they're tough as nails. In any case, I'm not the guy to advise you about health issues encountered when brooding turkey poults.
2. As I write, it seems that Verm-X cannot be ordered in the United States directly from its British maker. However, it is distributed (in 8-ounce bottles) from Amazon. It is certified organic, based on herbals that include cinnamon, fennel, peppermint, thyme, and garlic. There is no withdrawal period for eating eggs from treated hens.

Chapter 24: Other Domestic Fowl

1. Dave Holderread's *The Book of Geese: A Complete Guide to Raising the Home Flock* (Hen House Publications, 1981) and Chris Ashton's *Domestic Geese* (Crowood Press, 1999) both describe vent sexing waterfowl—both of ducklings and goslings, and of adult ducks and geese. Unfortunately, at the time of this writing, both books appear to be difficult to find for purchase online.

Chapter 26: Managing the Breeding Season

1. See Manoz, "The Trap Nesting Thread," Back Yard Chickens forum thread, September 5, 2011, https://www.backyardchickens.com/threads/the-trap-nesting-thread.567357 for more than a dozen designs. Be sure to follow the entire thread of the discussion (clicking on page 2 and page 3, etc.). Some of these designs are difficult to figure out and the graphics are obscure. But careful study reveals the core concepts, which with a bit of ingenuity could be adapted to fit your own preferred design.

Chapter 27: Working with Mother Hens

1. To my knowledge, the only hatcheries that offer standard (as opposed to bantam) OEG chicks are Cackle Hatchery (https://www.cacklehatchery.com/product-category/baby-chicks/standard-old-english-game) and Purely Poultry (https://www.purelypoultry.com/old-english-game-large-fowl-chickens-p-302.html). I have bought reliable stock from both.
2. See https://www.brinsea.com/p-524-ovaview-high-intensity-egg-candler.aspx. This candler is expensive at $50—but there are a lot of junky ones out there to be avoided.
3. One of the better guides to embryo development is Pyxis, "Development of a Chicken Embryo Day by Day," Back Yard Chickens forum thread, January 31, 2017, https://www.backyardchickens.com/articles/development-of-a-chicken-embryo-day-by-day.72537.

Chapter 28: Butchering Poultry

1. There are numerous YouTube videos that could be useful as you prepare for your first slaughtering day. Browse for others, but here are two that make good presentations: Tica Farm, "Detailed Chicken Processing and Butchering Step by Step," YouTube video, 12:29, September 16, 2019, https://www.youtube.com/watch?v=7l1WJTSNybk The Able Farmer, "The Complete Chicken Butchering Guide (Part 1)," YouTube video, 2:37, August 17, 2019, https://www.youtube.com/watch?v=y4IHoJhMrfE.
2. I recommend reading Stephen Budiansky's *The Covenant of the Wild: Why Animals Chose Domestication* (William Morrow, 1992; Terrapin Press, 1995), the subtitle of which encapsulates its interesting take on the origins of domestication: Certain animal species *initiated* the process even more than did the humans. Budiansky's book is not perfect—for example, he uses a single anecdote near the end of the book, about one of his ewes with a prolapsed vagina who welcomed tight confinement in a stall, as an implied rationale for the industrial CAFO model (concentrated animal feeding operation). But it imparts a more scientifically based correction to our usual human-centric concept of the origin and nature of domestication.

Chapter 29: Poultry in the Kitchen

1. Chayakrit Krittanawong et al., "Association Between Egg Consumption and Risk of Cardiovascular Outcomes: A Systematic Review and Meta-Analysis," *The American Journal of Medicine* 134, no. 1 (January 2021): 76–83.e2, https://doi.org/10.1016/j.amjmed.2020.05.046.
2. Good scientific studies comparing natural and industrial eggs are few and far between. *Mother Earth News* published a summary of their findings: Cheryl Long and Tabitha Alterman, "Meet Real Free-Range Eggs," Mother Earth News, Real Fod (blog), October 28, 2020, https://www.motherearthnews.com/real-food/free-range-eggs-zmaz07onzgoe. (See especially the concluding section, "Mounting Evidence for Free-Range Eggs.") Other studies are reported at Marta Czarnowska-Kujawska et al. "Folate Content and Yolk Color of Hen Eggs from Different Farming Systems," *Molecules* 26 (2021): 1034, https://doi.org/10.3390/molecules26041034; Julia Kühn et al., "Free-Range Farming: A Natural Alternative to Produce Vitamin D–Enriched Eggs," *Nutrition* 30, no. 4 (April 2014): 481–84, https://doi.org/10.1016/j.nut.2013.10.002.
3. Water glass is available in a 1-gallon size, enough for fifty dozen eggs, from Lehman's, https://www.lehmans.com/product/water-glass-liquid-sodium-silicate.
4. Processing the gizzard in preparation for making confit or other dishes starts with cutting the muscle tissue away from the "silver skin," its tough protective liner. To watch a video, I recommend: Fowl Fortress Farm, "UPDATED-Cleaning, Preparing, and Cooking Gizzards," YouTube video, 24:34, September 15, 2018, https://www.youtube.com/watch?v=OCUU7Fuu_XI&t=0s.

Chapter 30: Serving Small Local Markets

1. The mission of the Farm-to-Consumer Legal Defense Fund (https://www.farmtoconsumer.org/) is to defend the right of family farmers to sell their foods to consumers, as well as the right of consumers to obtain their foods directly from farmers; and to protect America's small farmers from harassment at all levels of government. Joining and donating to FTCLDF could well mean as much for sane, sustainable farming, and by extension ecological stewardship, as membership in other worthy organizations such as Sierra Club, National Audubon Society, Nature Conservancy, and Slow Food.
2. An overview of management for egg quality and egg washing at market scale Robert Plamondon, "FAQ: Egg Washing," *Robert Plamodon's Rural Life* (blog), September 5, 2016, https://www.plamondon.com/wp/chicken-faq-egg-washing.

Index

Note: Page numbers in *italics* refer to photographs and figures. Page numbers followed by *t* refer to tables.

About the Author

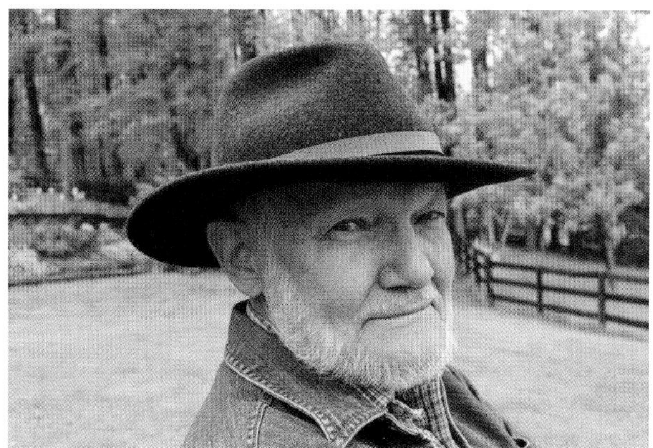

Mary Perrine

Harvey Ussery's homestead in northern Virginia has been the setting for constant experimentation toward regeneration, sustainability, and harmony with the surrounding ecology. He has shared what he has learned in numerous articles in *Backyard Poultry*, *Mother Earth News*, *Countryside & Small Stock Journal*, and *Grit!*, the newsletter of American Pastured Poultry Producers Association. Ussery has presented widely at national and local events on poultry and other homesteading topics, and maintains a highly informative website, TheModernHomestead.US. He shares the ongoing adventure with his wife, Ellen.

the politics and practice of sustainable living

CHELSEA GREEN PUBLISHING

Chelsea Green Publishing sees books as tools for effecting cultural change and seeks to empower citizens to participate in reclaiming our global commons and become its impassioned stewards. If you enjoyed reading *The Small-Scale Poultry Flock, Revised Edition*, please consider these other great books related to homesteading and agriculture.

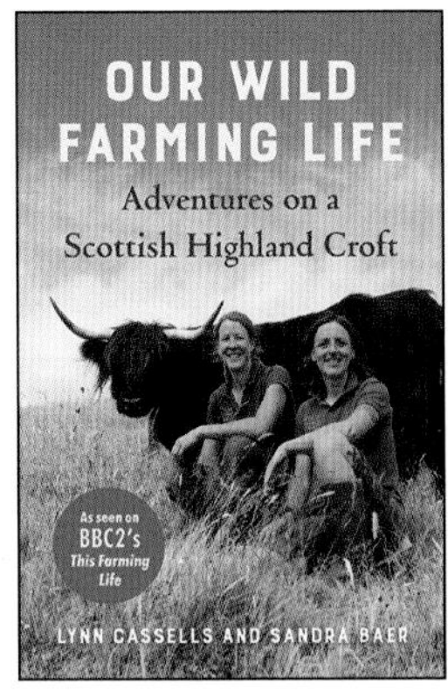

OUR WILD FARMING LIFE
Adventures on a Scottish Highland Croft
LYNN CASSELLS and SANDRA BAER
9781645020707 (UK)
Hardcover
9781645021650 (US)
Paperback

GRASS-FED BEEF FOR A
POST-PANDEMIC WORLD
How Regenerative Grazing Can
Restore Soils and Stabilize the Climate
RIDGE SHINN and LYNNE PLEDGER
9781645021247
Paperback

RAISING PASTURED RABBITS FOR MEAT
An All-Natural, Humane, and Profitable
Approach to Production on a Small Scale
NICHKI CARANGELO
9781603588324
Paperback

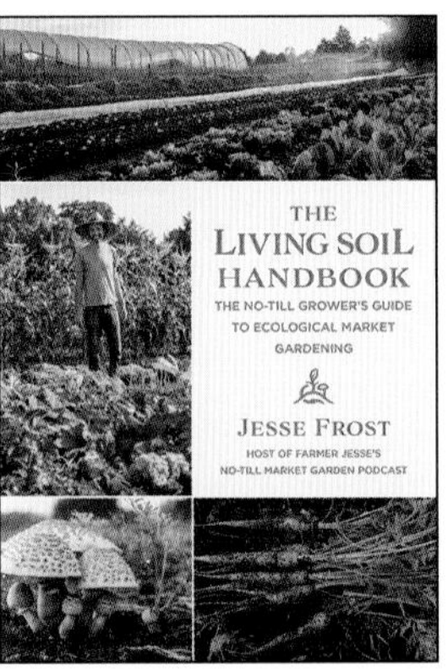

THE LIVING SOIL HANDBOOK
The No-Till Grower's Guide to
Ecological Market Gardening
JESSE FROST
9781645020264
Paperback

For more information,
visit **www.chelseagreen.com.**